BIOLOGICAL EVOLUTION

BIOLOGICAL EVOLUTION

Peter W. Price

Northern Arizona University

SAUNDERS COLLEGE PUBLISHING
Harcourt Brace College Publishers

Fort Worth   Philadelphia   San Diego   New York

Orlando   Austin   San Antonio   Toronto

Montreal   London   Sydney   Tokyo

Text typeface: Palatino

Compositor: Monotype Composition Company, Inc.

Acquisitions Editor: Julie Levin Alexander

Developmental Editor: Christine Rickoff

Managing Editor: Carol Field

Project Editor: Sarah Fitz-Hugh

Copy Editor: Mary Patton

Manager of Art and Design: Carol Bleistine

Art Director: Leslie Ernst

Art Assistant: Sue Kinney

Text Designer: Rebecca Lemna

Cover Designer: Lawrence R. Didona

Text Artwork: Stephen Price; Tad Theimer, Taina Litwak; TASA Graphic Arts, Inc.; Rolin Graphics, Inc.

Director of EDP: Tim Frelick

Production Manager: Charlene Squibb

Field Product Manager: Sue Westmoreland

Printed in the United States of America

6 7 8 9 0 1 2 3 4 0 6 2 10 9 8 7 6 5 4 3 2

ISBN 0-03-0968437

Library of Congress Catalog Card Number: 95-068932

On the Cover

Insects, bats, and birds: The cover image depicts three of the
great terrestrial radiations associated with the ability to fly.
Sources: Green lacewing (Chrysopa species) © Stephen Dalton.
Greater horseshoe bat (*Rhinoiopus ferrum equinum*) © OSF/Ste-
phen Dalton. Calliope hummingbird ♂ © Tony Stone World-
wide/Neal and Mary Mishler.

PREFACE

DINNER IN THE IGUANODON MODEL, AT THE CRYSTAL PALACE, SYDENHAM.

Dinner in a dinosaur. An engraving of a dinner given for 20 inside the incomplete construction of *Iguanodon* for the grounds of the Crystal Palace exhibition, one of several made between 1852 and 1854, which attracted immense publicity and still stand today. Richard Owen worked with the sculptor Waterhouse Hawkins to produce the life-sized reconstructions, and the two of them organized the dinner party illustrated here. (*The Mansell Collection*)

I was most fortunate to take a course on biological evolution early in my doctoral studies 30 years ago at Cornell University, from the "redoubtable" William L. Brown, as Edward O. Wilson has called him. Yes, Brown did inspire some fear, but what he mostly inspired was awe in and wonder at the subject of evolution. He made the whole world and all organisms of interest and of relevance to the study of evolution. The fascination was not so much with what was known but what was yet to be known and understood and what needed further study. Few certainties existed, but a tantalizing body of theory offered a view of the world that could sustain one's interest

indefinitely. For me it has become an ongoing quest that permeates all my teaching and research.

Consider the conquest of the air. Four great radiations of animals with four independent solutions to the complexities of flight have graced this earth: insects, birds, and bats (as illustrated on the cover of this book) and the pterosaurs. The engineering feat involved with design of a firm but flexible surface, and the articulation and musculature for flapping, coupled with an ability to fold the wings over the body, has largely defied clear understanding. Charles Darwin certainly pondered at length the question of such complex adaptations and how they could arise.

But there is no question about the evolutionary potential of flying organisms.[1] Most animals can fly: several million species of insects, 9000 species of birds, 1000 species of bats. Of the warm-blooded vertebrates, 80 percent can fly! No doubt the energetic efficiency of flight, compounded by dispersal and foraging abilities, have contributed to these spectacular radiations of life, and the grandeur seen in nature.

There is indeed grandeur in the evolutionary processes and their consequences, as noted by Darwin in the last sentence of *The Origin of Species*. Because of evolutionary theory we can understand why such extensive relationships exist among all organisms, and yet how the process of evolution can result in so many species with their own individuality. Can any human fail to wonder about the origins of life or themselves? Can we afford not to know, when our own name, *Homo sapiens*, states that we do?

All educated people should be provided the opportunity to see themselves as part of nature and as a result of the evolutionary process. It would be ideal if evolution were taught in high schools and universities as an aspect of general education, for it explains so much about ourselves and the world around us. I wrote this book because of my fascination with nature and because I feel that more people should have access to the wonders of the evolutionary process and its results. While writing I have been guided by some general principles in which I firmly believe.

ORGANIZATION

The text strives to provide a historical view of the development of ideas. History is a theme that is picked up repeatedly and developed throughout the book. The text emphasizes Darwin and his contribution at the beginning because Darwin played a central role in founding the theory of evolution as a robust scientific theory. The text then progresses from Darwin's emphasis on natural selection as an evolutionary mechanism to developments since Darwin's time, which recognized many alternative mechanisms in the evolutionary process, to a modern view of evolutionary theory.

Macroevolution

A discussion of macroevolution, the big picture of evolution, follows the discussion of Darwin and his theory. In the 1800s when Darwin lived and wrote, the big issues were macroevolutionary. How did new species originate? How did fossil records change through sedimentary sequences through time, and how were they correlated over geographic regions? What *were* those recently discovered, very large-boned organisms, lizard-like, but obviously extinct on a global scale? Recall that the "Great Exhibition" ("of the Works and Industry of All Nations") in the Crystal Palace, originally in London's Hyde Park, which opened in 1851, staged evocative and widely publicized reconstructions of dinosaurs. Gideon Algernon Mantell (1790–1852) and his wife had discovered the fossilized bones of the great extinct iguana-like lizard, *Iguanodon*. A reconstruction was displayed at the exhibition. These magnificent beasts were appreciated in the public domain before Darwin published *The Origin* in 1859. More than six million people visited the Great Exhibition, including Darwin and family in late July 1851 (Desmond and Moore 1991).[2]

Other gripping developments forced consideration of the human's place in nature. The missionary Thomas Savage discovered the gorilla in West Africa in the late 1840s. Richard Owen acquired skulls from a sea captain, and in 1855 Wombwell's traveling menagerie displayed a living specimen for the first time in England.[1] Owen took great pains to disconnect the lineages of gorilla and human, but not without stirring up considerable debate and widespread interest. The members of upper-crust society demanded that their lineage remain unsullied by such disreputable "beasts" as the gorilla. But the more objective Thomas Huxley was clear in 1858 that "to the very root and foundation of his nature man is one with the rest of the organic world."[2] Adding fuel to the debate that year was Hermann Schaaffhausen's description of a Neanderthal skull,[2] clearly a human fossil, perhaps an inevitable discovery, given the overwhelming evidence that had accumulated of stone tools—tools that could not have been made by animals other than humans. The public in England during the late 1850s, including the working class, were far more aware of macroevolutionary issues concerning humans than they are today.

Microevolution

Once the big picture was exposed, then the explanations were needed. Many theorists attempted a mechanistic explanation of how species could change through time; Darwin and Wallace were the most successful. A hereditary approach to evolution

---

[1]Dial, K.P. 1994. An inside look at how birds fly: Experimental studies of the internal and external processes controlling flight. Proc. Soc. Exptl. Test Pilots 38:301–313.

[2]Desmond, A., and J. Moore. 1991. Darwin. Penguin, London.

emerged with the work of Gregor Mendel, but did not gain full steam until the turn of the century and later.

Of course certain aspects of the study of macroevolution came much later than microevolutionary studies. However, it is my belief that the history of biological evolution should start with the big picture of evolutionary events, and then move on to the more detailed examination of the processes involved. Once we are involved with and understand the need for a more mechanistic view of evolution, the book turns to the microevolutionary aspects.

I think this is the way the modern synthesis in evolutionary biology developed as discussed in Chapter 3, and I like the way in which a macroevolutionary perspective primes us to understand the more difficult subjects involving, for example, population genetics, protein polymorphism, and genetic systems. In addition, as Mayr noted, the biological species concept and geographic speciation are cornerstones in the modern synthesis, such that I feel it preferable to discuss them early in the book. Themes through the book seem to develop naturally from macroevolution to microevolution. For example, the founder effect is treated in Chapter 5, and is picked up again in Chapter 15, starting with a simple view and moving to a more adequate mechanistic understanding. This is how science develops, as Chapter 1 explains.

The Scientist's Viewpoint

In keeping with the historical theme, I have attempted frequently to represent a scientist's opinion at the time of its writing. Some care is needed, then, to differentiate whose opinion is being represented in the text, because historical perceptions may not be appreciated today as adequate or acceptable. The discourse of reason is a central mechanism in the advance of any science, and without it many introductory texts miss the mechanism and fascination of science. This dialectic or debate keeps a discipline alive and well; it nurtures and accelerates progress. In biological evolution, new and heated debates seem to be emerging almost all the time, and capturing some of the enthusiastic argument involves stating points of view that are not necessarily currently accepted by the majority. Whether we agree or disagree with a point of view, we need to understand the argument; we must be receptive to alternative explanations.

Unanswered Questions

Science has not solved all problems, or even the majority of problems, especially in biological evolution. Therefore, I intentionally leave some issues unresolved, and I intentionally leave a subject in a state of flux. For example, in Figure 5.2 treating speciation in *Dendroica* warblers, I include Wilson's warbler as an inset that has not been influenced by glaciation in the same way as the other warblers discussed in the text. Likewise, in Chapter 4, I do not provide a certain answer on what a species is, but try to leave a message about the fascination and challenge for new research that will enlighten us on why populations stay so similar as to appear as one species.

FEATURES

Can anyone paraphrase Darwin and leave a precise understanding of what Darwin wished to convey to his readers? Can anyone today develop an argument as eloquently as D'Arcy Wentworth Thompson could? In his introduction to *On Growth and Form* the editor of the abridged edition, John Tyler Bonner,[3] remarked upon the difficulty of avoiding the use of the well-chosen words of other authors, "a temptation, as will be evident, that I have not resisted" (p. vii). Questions about what Darwin really meant in some of his passages continue to fascinate us, and my personal feeling is that this interest can be stimulated best by quoting Darwin. If we can become personally interested in *The Origin of Species* I maintain that we can become vitally involved with the theory of evolution. This argument can be extrapolated but to a lesser extent to other major figures in this field.

Language

The book strives to make the majority of subjects in biological evolution tangible and interesting, providing access to the scientific primary literature in the field. It is essential to introduce language in the field that may seem difficult at first, for scientific literacy is the centerpiece of modern views on undergraduate education in science. Consideration of technical words inevitably raises the delicate question of jargon. *Jargon* is a name given to technical terminology of specialists in an area of knowledge, and as such it is essential that a textbook expose the reader to this terminology. The derogatory use of the word *jargon* is perhaps more common, being pretentious or unnecessarily obscure terminology. I have avoided use of jargon for its own sake, but I have introduced terms that I think are useful and part of the vocabulary of scientists in this field.

---

[3]Bonner, J. T. (ed.) 1971. On Growth and Form by D'A. W. Thompson, pp. vii–xiv. Cambridge Univ. Press, Cambridge.

Figures

I have designed the figures to be understood by reading the figure and the caption. Fully explanatory captions and footnotes with the figure are an essential ingredient in providing a visual image in addition to the conceptual impression generated in the text. The visual impact of figures is probably more effective than text in stimulating interest, and in facilitating comprehension and learning. Therefore, I have invested almost as much time in developing figures as I have in developing text. I hope some of the figures will be captivating, illustrating such wonderful subjects as extinct organisms, the adaptive radiation of honeycreepers in the Hawaiian archipelago, and even pictures of species simply relevant to the development of biological evolution. The text is written with reference to figures and tables so that these components are integral and essential to the development of an argument.

References

I have used a format for references close to that in the journal *Evolution*. This is because familiarity with standard format for scientific papers is helpful in providing a guide to how papers are cited in the literature.

I have attempted to cite the large majority of sources of information I have used in the book. Citation is, first, a matter of giving rightful credit to the source; second, it provides a lead on how one assimilates knowledge; and third, it offers students an opportunity to do further reading and to study independently if they want more detail. My hope is that no one motivated to delve more deeply into a subject will meet with a dead end of information.

Geographical Interest

One of the fascinating aspects of studying evolution is its relevance to the whole Earth. Examples of concepts and ideas come from all over the globe, with a more-or-less inbuilt geography lesson. This interest I have tried to foster by giving fairly explicit locations so that they may be found in a good atlas. A summary of sites discussed in the book is provided on the world map on the inside of the front cover.

Questions for Discussion

At the end of each chapter I have suggested some questions, which could be used as food for thought, or for a discussion section that could be a part of a course. In my experience, oral debate on evolutionary issues greatly enhances a student's comprehension of and familiarity with the dialectic of biological evolution. Just the nature of the debate and the language employed are somewhat alien to many students who have experienced an education involving much straightforward descriptive science and much learning by rote. Encouraging the reader to become involved in the intellectual development of the science has been one of my themes.

ACKNOWLEDGMENTS

It is with great pleasure that I recognize all my teachers, colleagues, and students, who have provided me with a unique education through publications and personal interactions. I would be remiss to mention even one by name, for to do so would diminish the others' contributions, and who can determine the relative value of all the learning experiences in a career?

More tangible is the dedication and expertise of Louella Holter, with her almost flawless word processing skills and her goodwill whenever I asked for yet another draft of the book. I am grateful to Steve Price and Tad Theimer for their scientific knowledge and research and their creativity invested in many of the illustrations in this book. My good fortune to work with the editors at Saunders College Publishing has been enriching and a personal pleasure.

Thanks also are due the following reviewers, whose suggestions have helped make the book better: Bayard Bratstrom, California State University, Fullerton; Edmund D. Brodie, Jr., University of Texas at Arlington; William J. Etges, University of Arkansas; Michael Foote, University of Michigan; Laurie Godfrey, University of Massachusetts, Amherst; Michael LaBarbera, University of Chicago; Allan Larson, Washington University; Timothy Lewis, Wittenberg University; Gary McCracken, University of Tennessee, Knoxville; Steve Murray, California State University, Fullerton; David Polcyn, California State University, San Bernardino; Chester Wilson, State University of New York, Stony Brook.

Peter W. Price

Flagstaff, Arizona

May 1995

LEGENDS TO PART OPENERS

Part I, Introduction, page 1

Iguanodon, ("iguana-tooth"), a reconstruction of the first dinosaur to be described by Mantell in 1825. (Based on a figure by John Sibbick in Norman, D. 1985. The Illustrated Encyclopedia of Dinosaurs. Salamander Books, London.)

Part II, Macroevolution, page 57

By the Carboniferous period, mighty trees grew in swamps. *Lepidodendron* (**left**) is estimated to have grown as tall as 54m, with a basal diameter of over 1m and root-like structures that extended 12m into the swamp. *Sigillaria* (**right**) had a trunk well over 30m in length, and 1m in diameter. (Based on figures in Steward, W.N. and G.W. Rothwell. 1993. Paleobotany and the evolution of plants. 2nd ed. Cambridge University Press, New York.)

Part III, Microevolution, page 293

The life cycle of a thorny-headed worm parasite is quite complex; host sites include pig intestine (**left**) and a larval beetle (**left and right**). (From Noble, E.R., and G.A. Noble. 1976. Parasitology: The Biology of Animal Parasites. 4th ed. Lea and Febiger, Philadelphia.)

Part IV, Conclusion, page 409

Self-awareness, a rock painting from Cape Province, South Africa, depicts a woman using a weighted digging stick to dig up a plant bulb for food. In understanding our origins and evolutionary future, have we progressed enough? (Based on a figure in van der Merwe, N.J. 1992. Reconstructing prehistoric diet. In S. Jones, R. Martin and D. Pilbeam (eds.). The Cambridge Encyclopedia of Human Evolution. Cambridge University Press, New York.)

BRIEF TABLE OF CONTENTS

CONTENTS

I

INTRODUCTION

1

DEVELOPMENT OF SCIENCE

The peppered moth, *Biston betularia*, with the *typica* morph on the left and the *carbonaria* morph on the right. The caterpillar appears below.

Because so much confusion seems to exist among the general public about the nature of evolutionary biology and the questions this science can address, a clear view is worth developing. What are the limits of science? How does science develop?

1.1 WHAT IS SCIENCE?

Science can be defined as a branch of study concerned with observation and classification of facts and especially with the establishment of verifiable general laws and scientific theories. This means that science discovers what is real and factual. Independent workers test and repeatedly confirm information to verify it. Facts can then be classified and formed into general laws and theories. The law of gravity, prompted by Newton's observation that an apple, when unsupported, falls to the ground, is such a general law, based on many independent observations and many independent tests.

Not all paths of research result in verifiable information. Researchers make errors and report them in a factual way. When tested again, the published data may be found to be incorrect. Thus, science has the dual roles of discovering factual information and rejecting data or concepts that are inconsistent with current data and ideas. This process of rejection or invalidation of reported information is of central importance in science. It provides a self-correcting system that helps science to move toward greater, more accurate and detailed understanding of the natural world.

1.2 THE SCIENTIFIC METHOD

Science can also be defined as the search for objective knowledge using the **scientific method.** Scientists employ the scientific method to study accessible knowledge by doing the following:

1. Recognize a problem and formulate it into a question.
2. To answer the question, collect data through observation and, when possible, experimentation.
3. Develop tentative answers to the question. These are called **hypotheses.** A hypothesis is stated in such a way that testing it can provide results consistent with it or results that are inconsistent with it, causing its rejection because it has been proved invalid. That is, hypotheses can be accepted or proven false on the basis of objective tests. If a hypothesis is proven to be false, it is said to have been falsified, in the parlance of the philosopher of science (e.g., Lakatos 1970). Such **falsification**

does not involve "cooking the books," but rather provides a clear demonstration that an apparently logical conclusion is incorrect after further inspection or testing. Thus, the most effective hypotheses are those that explicitly or implicitly provide a clear test that includes the possibility of falsification.

4. Test the hypothesis repeatedly using different methods and avenues of inquiry. If all such testing results in data that are consistent with the hypothesis and that fail to invalidate it, the hypothesis gains in credibility and stature. A series of tests yielding results consistent with the hypothesis may lead to the development of new questions to delve more deeply into a problem. Refutation may result in a rephrasing of the question and certainly the erection of an **alternative hypothesis.**

This process may go on for years, decades, or even centuries (as we shall see in one example). The result is that science cannot always provide adequate answers, or even correct answers, but it does provide the best answers available at the time, based on the scientific method.

In practice, the scientific method has serious pitfalls that not all scientists have been able to avoid. Personal biases and even political doctrines may derail the search for accessible knowledge. For example, craniometry, or the measurement of cranial volume as an index of group characteristics, was perverted, especially by Morton (e.g., 1839, 1844, 1849), by a bias that Caucasians were an intellectually superior race or even a superior species. As with any human endeavor, science is practiced in a society, and that society influences the manner in which the scientist proceeds, even if subliminally. Gould (1981, 54) noted that Morton was a "self-styled objectivist," and yet a reanalysis of Morton's data showed his summaries to be "a patchwork of fudging and finagling in the clear interest of controlling a priori convictions." Was this by design? Gould argued the alternative: "Yet— and this is the most intriguing aspect of the case— I find no evidence of conscious fraud; indeed, had Morton been a conscious fudger, he would not have published his data so openly."

Craniometry was but one aspect of the insidious argument that humans have genetically predetermined roles in life, to be leaders, servants, laborers, or landed gentry. Such **biological determinism** has provided a false crutch to a variety of political doctrines concerning social position and racial status. One such sociological viewpoint was derived incorrectly from Darwinian evolution and asserted that the upper classes in a social hierarchy had biological attributes conferring superiority in the struggle for existence. This argument, called **social Darwinism,** provided the wealthy and powerful a prerogative, which appeared to be scientifically founded, to exploit and demean the lower classes.

Extrapolating from such examples of blatant misuse of the scientific method, we must acknowledge that scientists as a group are not as objective and free of bias as we would like to think. "Science's potential as an instrument for identifying the cultural constraints upon it cannot be fully realized until scientists give up the twin myths of objectivity and inexorable march toward truth" (Gould 1981, 23). Gould bolsters his position with a quote from Myrdal (1944):

> A handful of social and biological scientists over the last 50 years have gradually forced informed people to give up some of the more blatant of our biological errors. But there must be still other countless errors of the same sort that no living [human] can yet detect, because of the fog within which our type of Western culture envelops us. Cultural influences have set up the assumptions about the mind, the body, and the universe with which we begin; pose the questions we ask; influence the facts we seek; determine the interpretations we give these facts; and direct our reaction to these interpretations and conclusions.

If we acknowledge that human biases and social and cultural influences are impediments to the search for general and factually based scientific theory, should we beware of other problems with the scientific method? A major challenge facing scientists is the search for generality. This search for patterns in nature, especially broad patterns, is a preoccupation. For example, there is a pattern in the design of any phylum of animals or any division or class of plants. Because early systematists determined the order of taxa on the basis of only a small number of sample species, inevitably classification had to change with increased sampling of the world's fauna and flora. The accumulation of information through time results in continual reassessment and modification of perceived patterns.

Thus, a central objective of the scientific method is to make the transition from specific examples to general patterns. We try to extrapolate beyond observations to make broadly general statements about nature, then test them. The scientific method necessarily involves two very different forms of logic, or formal reasoning.

The first, **induction,** is the process of reasoning from a part to a whole, or from particulars to generalities. Since scientists who are developing new understanding never have all the information that will ultimately be available, such **inductive logic** is central in the scientific world. Frequently, however, induction is founded on insufficient sampling and is premature, in which case it may be a diversion from the mainstream of understanding.

It is inductive logic that enables a scientist to move from a series of observations to a hypothesis that provides a general but tentative answer to the questions raised—i.e., to proceed from step 2 to step 3 in the foregoing list. From a collection of observations the scientist develops a generalization.

In practice, considerable uncertainty about the real nature of an explanation may prompt the scientist to erect several alternative hypotheses, each one in need of testing. A common approach is to erect a **null hypothesis,** which assumes the *absence* of the generalization under consideration—say, statistically different attributes between groups of organisms purported to demonstrate different patterns in design. For example, at first sight a primate might seem inseparable from other mammals because of a common set of attributes such as a body covered in hair; an integument with sweat, scent, sebaceous, and mammary glands; a four-chambered heart; and nonnucleated, biconcave red blood corpuscles. Using only such a set of characteristics, an investigator would conclude that the null hypothesis could not be proven invalid. On closer scrutiny, however, new characteristics might force the alternative hypothesis that primates are distinct from other mammals because of a unique combination of especially large cerebral hemispheres; fore- and hind-limbs with five digits, usually with flat nails rather than claws; and limbs adapted for grasping, originally associated with arboreal life. Induction is a very creative process because it forces us to consider as many alternative hypotheses as possible that may account for a particular set of observations. However, our inductive abilities also may be constrained by biases or lack of knowledge, which inevitably slow our progress with the scientific method.

Once we have a general principle or hypothesis to work with, we can test it using the process of **deduction,** the second form of logic in the scientific method. From the general case we can infer something about particular cases. Such **deductive logic** enables us to refine our observations and experiments so that we subject a hypothesis to closer scrutiny. From the general hypothesis that hooked hairs or trichomes on a plant act as defenses against certain insect herbivores, it can be deduced that for any plant species with hooked trichomes there are likely some important herbivores that are susceptible to immobilization or puncture by such a coating. Specific cases can be examined and experiments conducted, as by Gilbert (1971), with herbivores found in the vicinity. In an experiment a member of the passion vine genus, *Passiflora adenopoda,* was not damaged by herbivores that typically defoliate members of the genus: the larvae of *Heliconius* butterflies. Gilbert found that *Heliconius* larvae placed on *P. adenopoda* became impaled on the trichomes and died rapidly. In this case the general hypothesis was supported and not refuted or invalidated, adding to the credibility of the argument.

This process of observation, induction resulting in generalizations, and deduction leading to detailed testing with observations and experiments can be repeated in sequence with increasingly more accurate and increasingly broader generalizations.

1.3 QUESTIONS RELEVANT TO SCIENCE

We may then ask what kinds of questions can be answered by the scientific method. One question might be "What causes sickness in humans?" This inquiry recognizes a problem (step 1 earlier). In response, researchers can collect data on the symptoms, the infected organs, and the localities and ages of people getting sick, and microscopic studies may reveal the presence of pathogenic organisms (step 2). Scientists may then perform experiments involving either treatment of patients with chemicals that are toxic to the presumed pathogen or inoculation of experimental animals with the pathogen (step 2). A hypothesis is then erected explaining the cause of the sickness (step 3). Other investigators can repeatedly test this hypothesis until the pathogen, symptoms, and disease become established as fact (step 4). Thus, the scientific method can address questions about disease very effectively, as we have seen in research on acquired immune deficiency syndrome (AIDS).

Another question might be "Did humans evolve from apes?" The problem was recognized in Victorian England and elsewhere after the publication of *The Origin of Species* in 1859 and especially another book by Charles Darwin, *The Descent of Man and Selection in Relation to Sex*, in 1871. Darwin (1871, 10) argued that the bodily structure of humans is very similar to those of other animals.

It is notorious that man is constructed on the same general type or model with other mammals. All the bones in his skeleton can be compared with corresponding bones in a monkey, bat, or seal. So it is with the muscles, nerves, blood-vessels and internal viscera.

He also argued that embryonic development of humans is very similar to that of other animals; "The embryo itself at a very early period can hardly be distinguished from that of other members of the vertebrate kingdom" (14). He illustrated human and dog embryos to show how similar they are. A third line of argument was that rudimentary structures in humans are well developed and used in other animals. For example, the nictitating membrane in birds became vestigial in humans; it is the semilunar fold in the inner corner of the eye. This thin membrane can extend across the whole delicate eyeball of a bird, providing protection for it, but in humans it is reduced to a small, apparently nonfunctional, roughly crescent-shaped pad.

By similar arguments Darwin went on to reason that the human's closest relatives are the apes.

It is therefore probable that Africa was formerly inhabited by extinct apes closely allied to the gorilla and chimpanzee; and as these two species are now man's nearest allies, it is somewhat more probable that our early progenitors lived on the African continent than elsewhere. (Darwin 1871, 199)

Darwin used facts from comparative anatomy very forcefully to argue the descent of humans. He collected evidence to argue his case. More recent studies all have confirmed Darwin's general conclusions. Chimpanzees and gorillas are indeed our closest living relatives, based on all biochemical tests on hemoglobins and other proteins, immunological tests, and repeated DNA sequences. Skulls of early humans and their supposed ancestors have been found in Africa. Darwin's conclusions have been tested and retested and still they stand as established scientific fact (see Chapter 12, "Human Evolution"). Of course, experiments addressing the descent of humans cannot be conducted and we may never know the complete phylogeny of our species. Many questions remain. Still, the scientific method enables us to probe the distant past and advance knowledge. In the absence of experimental approaches, the **comparative method** in science provides a potent alternative, as Darwin demonstrated. In later chapters (e.g., Chapter 7), we will see that in biological evolution the comparative method is indispensable for probing past relationships among organisms.

This book addresses many questions about evolution, although the power of the scientific method varies significantly with the kind of question asked. We can say with assurance which living animal is most closely related to the human animal, but we are far from a clear picture of our human ancestors (Chapter 12). We can describe alternative modes of speciation, but our knowledge of which are most common and which modes apply to certain taxa remains debatable (Chapter 5). We can document the genetic differences between organisms, but the existence of such differences is often inexplicable (Chapters 12, 14, 15, and 16).

Bringing all the available evidence to bear on a particular issue is impractical in a textbook, so only one viewpoint may be expressed, or only one small piece of a puzzle may be discussed. Not all alternative hypotheses are explored. Nevertheless, throughout this book our discussion of the scientific method and questions relevant to science should prompt inquiries on the success of the method in answering particular questions, and the extent to which each subject has reached the status of a well-tested hypothesis or a scientific theory.

1.4 QUESTIONS OUTSIDE THE REALM OF SCIENCE

Many other questions cannot be answered by the scientific method. They are therefore outside the realm of science. Was President Reagan an effective president? Does sparing the rod spoil the child? Does God live in heaven? Metaphysics, philosophy, ethics, politics, and religion are beyond the domain of science simply because the questions they raise are not testable by the scientific method. They are not subject to empirical verification by observation and experiment. We cannot establish by the scientific method that a god exists or that the earth was created by a god. We cannot establish by the scientific method that capital punishment is a proper treatment of criminals. Nor can we establish that Buddhism is a more valid religion than Christianity. Thus, many valid issues of daily concern to us fall outside the range of science.

It is important to retain a clear picture of what science can do and what it cannot do. Such clarity is necessary for making logical decisions relating to debates such as that between creationists and evolutionists—in which many are misled into thinking that the philosophy of religion can be integrated with the philosophy of science. Clearly, the two are largely separate human endeavors—both important, both valuable, but different. We must keep asking: Can this idea, this hypothesis or this theory, be tested by the scientific method? Is there a way of verifying this statement? Can we test alternative hypotheses to find them valid or invalid? Can the hypothesis be falsified?

1.5 CHALLENGES IN THE ADVANCE OF SCIENCE

This book discusses some problems and challenges in applying the scientific method. For example, the effort to discern how life could have evolved some 3.5 to 4.0 billion years ago meets with significant difficulties. It is hard to reconstruct the phylogenetic relationships of dinosaurs, which were common a mere 150 million years ago. We cannot yet be sure why many large mammals as recent as 11,000 years ago—such as mastodons, saber-toothed tigers, and giant kangaroos—went extinct. We will attempt some answers and erect some hypotheses, bearing in mind

that scientists differ in point of view and argue the validity of many hypotheses.

Some people perceive this debate as a weakness of science. But they miss the point. This is the strength of science. We continue to debate issues, and we continually collect evidence until a formerly controversial point becomes almost universally agreed upon. New evidence may then undermine that view, and a new debate is started.

Science progresses by a series of approximations toward the truth, or fact. In evolutionary biology we are probably farther removed from much real truth and ultimate understanding than in many disciplines of science, such as physics or chemistry. This fact however does not make evolutionary biology unscientific. Rather, it is a young science addressing very complex issues. And yet, after little more than 100 years, it has provided incredible insights into the fascinating natural world.

To understand how slowly facts have been established and how tortuous is the path of investigation, let us follow the development of one idea, the concept of **spontaneous generation.** This idea is an old one that has been debated through the ages. The philosophers of ancient Greece were bent on finding the underlying causes and principles of the real world. Their method of search was logical reasoning rather than observation of fact. This course is perhaps a risky one, since the reasoning can be only as good as the facts on which it is based. As Bertrand Russell said, "Every advance in knowledge robs philosophy of some problems which formerly it had."

1.6 THE DEBATE ON SPONTANEOUS GENERATION

One of the problems faced by early philosophers studying biological phenomena was to explain where fairly small organisms came from. Where did flies come from? Where did the mold on their food come from? Aristotle (384–322 B.C.) and his student Theophrastus (370–287 B.C.)—as well as everybody who thought about the problem for the next 1,700 years—agreed that flies arose out of nothing, that they were generated spontaneously. This may seem a ridiculous conclusion in our time, but more than 2,000 years ago no evidence existed to suggest otherwise. Spontaneous generation was clearly the most logical answer available because small organisms seemed to be present suddenly, with no evidence of origin.

But note the first element in the scientific method: Ask a significant question. The question about the origins of small organisms revealed perception and intellect, and it has beguiled science ever since it was first asked.

Only after the invention of the microscope permitted a closer look at small organisms' life cycles could a better explanation be advanced. For example, Marcello Malpighi, a 17th-century Italian physiologist (1628–1694), was one of the first to use a microscope. He saw that galls on plants did not arise spontaneously but were caused by the larvae of insects (Farley 1974). Malpighi's contemporary and compatriot, Francesco Redi, was the first to test experimentally the question of spontaneous generation and reported his results in 1668. He wanted to see whether flies developed spontaneously or, if not, where they came from. He performed the second element in the scientific method, a critical experiment, to answer the question on spontaneous generation. He set up vessels containing meat, some closed and sealed and others covered with a net, and he compared those with dead animals in open boxes (Figure 1-1). In the open boxes Redi noticed eggs, maggots, and flies, while the sealed vessels generated no flies. Flies deposited eggs on the net cover, but no maggots appeared on the meat. Redi concluded that no flies were produced in meat on which flies could not alight and oviposit. This finding contributed to the generalization that all living matter is the product of living matter. A significant breakthrough in understanding had occurred.

The next mystery to consider was that meat continued to putrefy with time even though it was kept covered. Science had pushed back the unknown, leading to deeper understanding, but it had also prompted new questions. What was happening to the meat? As the microscope was refined, the putrefying agents were observed to be bacteria, which seemed to be generated spontaneously. In the 1700s the argu-

FIGURE 1-2

Lazzaro Spallanzani's experiment on the spontaneous generation of bacteria.

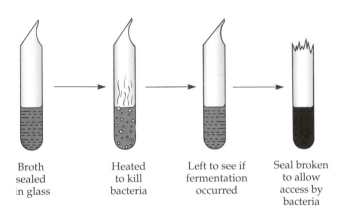

| Broth sealed in glass | Heated to kill bacteria | Left to see if fermentation occurred | Seal broken to allow access by bacteria |

ment about spontaneous generation raged again, and the explanation for the phenomenon of putrefaction came nearer to being exact. Science was now moving by a series of experiments toward a complete explanation of a biological phenomenon. Lazzaro Spallanzani (1729–1799) performed a clever experiment to see if microorganisms were generated spontaneously. Spallanzani sealed vials containing liquids that would normally putrefy rapidly and then heated the liquids. The contents remained in the vials indefinitely without fermenting, but as soon as the vials were opened the contents started to putrefy (Figure 1-2). Although this result was fairly convincing proof against spontaneous generation, the experiments were criticized by skeptics on the grounds that heating had altered not only the infusions themselves but the air contained within the vials. For example, Needham (1748) argued that heat destroyed the "vegetative force" in matter and the "elasticity of air" (Farley 1974).

It was not until the next century that Louis Pasteur proved to the majority's satisfaction that microorganisms could be derived only from other microorganisms (Pasteur 1859). He showed the hypothesis of spontaneous generation to be invalid, but not for some 200 years after Malpighi had challenged the concept of spontaneous generation. Nowadays we accept as fact that organisms arise from other organisms, yet it took civilized people thousands of years to establish the fact for all life, so that now we can state accurately that all living things are the products of other living things.

Louis Pasteur knew of Spallanzani's experiments and the criticisms that had been leveled against them. Pasteur's goal was to design an experiment in which a putrefiable liquid could be heated without the possibility that air in the vessel would be changed. The

FIGURE 1-1

Francesco Redi's experiment on the spontaneous generation of flies. The closed, sealed vessel produced no maggots or flies, nor did the vessel with a net cover. In the open box, flies deposited eggs from which maggots hatched. The maggots developed into adult flies.

| Closed and sealed | Net cover | Open box |

apparatus he used in one experiment was very simple (Figure 1-3). In swan-necked flasks he boiled liquids such as yeast water, urine, beet juice, and pepper water. By so doing he killed all living organisms in the medium and expelled germs in the air above the liquid with the steam. Pasteur showed that the nutrient solution did not putrefy even though the air was natural. All dust and microorganisms were trapped in the neck of the flask. Pasteur showed convincingly that there was no spontaneous generation but that liquids fermented because microscopic organisms were carried in the air, settled on unprotected surfaces, and caused the fermentation.

From this brief history we see that a series of increasingly refined experiments elucidated the real source of small organisms. Through well-designed experiments, researchers could make objective observations and reproduce them so that even the most skeptical were convinced that spontaneous generation does not occur. (Farley [1974] provides a more detailed account of the spontaneous generation controversy, encapsulating the historical debates that provided the energy for progress in understanding.)

The fascination of science lies in discovering accurate explanations for the natural phenomena we see around us. Many questions await answers, for we probably can never gain complete knowledge about the natural world. The satisfactory resolution of one question—as happened with spontaneous generation—usually evokes another question, and the scientific method is again pressed into action. For example, if spontaneous generation does not occur today, how could it have given rise to the first life on earth? The answers are not clear, but the question has been researched for many years and some fascinating possibilities exist (see Chapter 6, "The Origin of Life and the Emergence of Early Eukaryotes").

1.7 THE DEBATE ON INDUSTRIAL MELANISM

One more example illustrates that simple observations and questions can generate a large research effort that reveals increasing complexity. In many moth species that fly by night and rest on surfaces during the day, black or melanic forms have appeared over the last 140 years. One example is the peppered moth, *Biston betularia*—so called because of its black spots on a white background (Figure 1-4), which provide

FIGURE 1-3

Louis Pasteur's experiment on the spontaneous generation of microorganisms, which provided the final refutation of the concept.

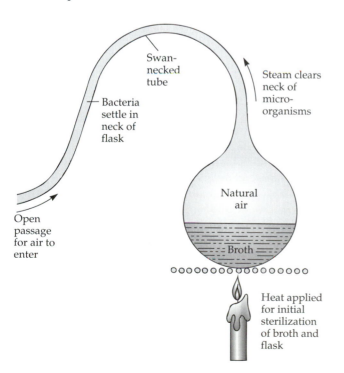

FIGURE 1-4

Two morphs of the peppered moth, *Biston betularia*. The *typica* morph appears on the left and the *carbonaria* morph on the right on a darkened tree trunk in a polluted woodland. *(From Kettlewell 1959.)*

excellent camouflage on light bark, such as that on birches, and on lichen-encrusted tree trunks and branches during the day. The first recorded melanic form of this moth was found in 1848, and in urban areas of England the proportion of melanics increased to 90 percent or higher by 1900. Such evolution of dark forms within species in polluted areas, especially areas polluted by the emission and deposition of dark particulates (soot), is generally known as **industrial melanism.**

This change in frequency from mostly light forms to mostly black forms raised a question about the mechanisms involved. Did pollutants cause a color change directly, through the action of "melanogens," as some suggested? Did the local cooling due to smoke cause selection for a moth that could absorb more heat? Were melanic moths surviving better because they were now better camouflaged against predatory birds than the light forms on the darkened tree trunks now devoid of lichens killed by pollution? These were the questions that were debated.

During this controversy, Kettlewell decided that experiments were essential to resolve the issues (Kettlewell 1956, 1959, 1973). He released known numbers of both typical and melanic moths in a polluted woodland and recaptured them over subsequent days. He noted that the melanic form, *carbonaria*, was cryptic on darkened tree trunks, whereas the peppered form, *typica*, was conspicuous. Predators such as English robins and hedge sparrows saw the moths in a similar manner. The net result was that typical moths were eaten more frequently than the *carbonaria* morph, and when moths were recaptured *carbonaria* had survived the selective predation of birds better than the *typica* morph. In similar studies in an unpolluted location, the typical form was more cryptic and survived better (Figure 1-5). Kettlewell had demonstrated that he could simulate local frequencies of morphs in his experiments and that selective predation by birds was responsible. Birds caused natural selection against the less conspicuous morph in a locality. In populations in polluted locations, natural selection resulted in evolution toward a higher frequency of melanic moths. In unpolluted areas, melanic moths were kept at low levels by the continual attrition. This process had occurred during Darwin's lifetime and provided a tangible example of evolution at work, causing Kettlewell to call it "Darwin's missing evidence" (Kettlewell 1959).

But these elegant and important experiments did not provide all the answers. If melanism was beneficial in polluted areas, why did the *typica* morph persist in these populations? Why were there two melanic morphs, the black *carbonaria* and the more mottled dark form, *insularia*? Why were melanic moths found in some parts of Britain away from in-

FIGURE 1-5

Results of Kettlewell's experiments on the effects of selective predation by birds on the *typica* and *carbonaria* morphs of the peppered moth. Note that recaptures of *carbonaria* were higher than expected (↑) on the basis of equal survival in the polluted area, and the *typica* morph survived less well than expected (↓). Kettlewell observed the reverse result in the unpolluted woodland. OBS = observed number of recaptures; EXP = expected number of recaptures. *(Based on Kettlewell 1956.)*

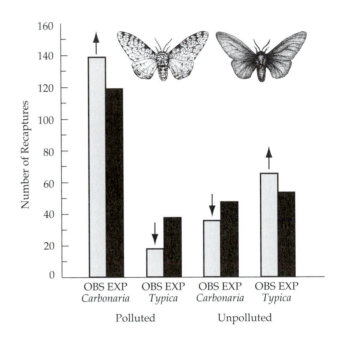

dustrial pollution (e.g., Steward 1977)? A series of studies was launched to examine the details of peppered moth morph frequencies (e.g., Bishop and Cook 1979, 1980), and we will not have reached a full understanding of the causes for evolution in the peppered moth by the end of the 20th century. Is evolution in the peppered moth driven largely by visually hunting birds or by nonvisual factors such as physiological traits (Mikkola 1984, Cook et al. 1986)? Is change in lichen cover on trees central to an explanation of morph frequency changes, or is change in the number of silver birch trees, on the bark of which both morphs are more or less cryptic, the important factor (e.g., Grant and Howlett 1988)?

A notable discovery is that the evolution of the peppered moth is continuing today. Legislation in Britain has significantly reduced atmospheric pollution, so that if a cause-and-effect relationship exists between pollution and melanism, we should expect a decline in the proportion of melanics in a population. Such decline has indeed taken place (Bishop and Cook 1975, 1980, Cook et al. 1986) (Figure 1-6).

FIGURE 1-6

The changing frequency of the peppered moth's melanic morph, *carbonaria*, near Liverpool in response to reduced pollution, indicated by sulfur dioxide (SO_2) levels, from 1959 to 1975. *(Based on Bishop and Cook 1980.)*

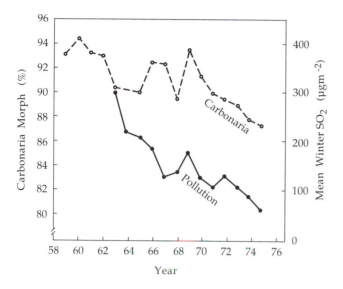

Whereas the melanic morph increased from 0 to 98 percent of the population in Manchester from 1848 to 1895, it decreased from 94 to 70 percent in nearby localities in the years 1959 to 1984 (Clarke et al. 1985). We see populations evolving continually in response to a changing environment.

1.8 MICROEVOLUTION AND MACROEVOLUTION

Industrial melanism illustrates biological evolution in its simplest possible form—a change in gene or allele frequency in a population or species through time. We will see later in this book that many other processes can be involved. In *Biston betularia* the gene conferring black color increased in frequency in polluted areas, and the gene for the typical morph declined. Such evolutionary changes can be small or large and are distinguished by the terms **microevolution** for relatively small changes below the species level in populations (as in the peppered moth) and **macroevolution** for relatively large changes sufficient to produce new species and higher taxa. Macroevolution therefore encompasses some of the most fascinating and beguiling aspects of biological evolution, including major breakthroughs in evolutionary history, adaptive radiation, and mass extinction and its implications. This book covers the grand view of evolution before it deals with microevolution because it was, after all, the unfolding "big picture" that stimulated the search for mechanisms and the finer details of microevolution.

Clearly, it is much easier to devise experiments on microevolution than on macroevolution. It is relatively easy to place two populations under different conditions to see if they evolve in response to those conditions. It is much more difficult to set up an experiment to study the evolution of a new species from a preexisting species.

So many experiments have demonstrated microevolution that there is proof that it occurs. For example, when two populations of mice were kept in identical conditions, and **artificial selection** by humans weeded out the largest mice in each generation in one population and the smallest mice in the other population, the two populations diverged rapidly in weight (Falconer 1953) (Figure 1-7). Another example concerns artificial selection for reduced intercrossing between strains of corn (Paterniani 1969) (Figure 1-8). Two pure strains were planted in mixtures, and the seeds selected for replanting were from plants that had the lowest proportion of seed fertilized by the other strain. By the fifth generation the percentage of intercrossing was reduced to less than 5 percent

FIGURE 1-7

Falconer's (1953) experiment on artificial selection for large mice and small mice, with weights at 6 weeks of age provided. The dashed line indicates the common weight at the start of the experiment (generation 0).

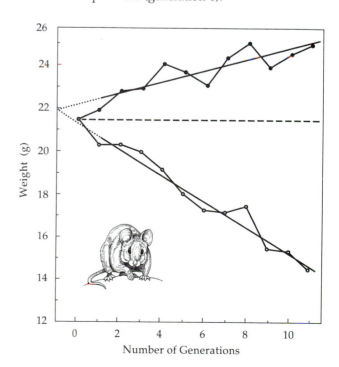

FIGURE 1-8

Paterniani's (1969) experiment on artificial selection for reduced intercrossing, or increased reproductive isolation, in strains of corn (maize), *Zea mays*. Open circles are for yellow sweet corn and closed circles are for white flint corn. The regression lines indicate the general trends of rapidly declining interfertility between the two strains.

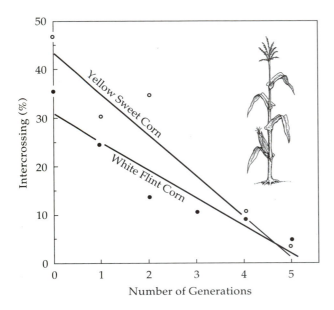

because flowering times diverged and receptivity of stigmas to pollen was reduced. In fact, when two strains of one species no longer exchange pollen, they actually behave as two new species (see Chapter 5, "The Origins of New Species"). This experiment therefore comes close to showing a macroevolutionary event.

In general, then, we will see that as we move from microevolution to macroevolution conclusions are often more contested and uncertainty becomes greater, because the scientific method becomes harder to apply in the most rigorous sense of conducting controlled experiments. Such debates make evolutionary biology a lively, fascinating subject; it is a young and vigorous science, and it concerns us all.

1.9 SCIENTIFIC LAWS AND THEORIES

In evolutionary biology, as in other sciences, the major preoccupation is to develop generalities, which are usually stated as scientific laws or scientific theories. A **scientific law** is a statement of an order or a relation of phenomena that, so far as is known, is invariable under given conditions. Newton's law of gravitation is an example. There is no requirement that the phenomenon be understood in terms of how it works; rather, it is a description of what always happens or what the relationship always is. It is a statement about the invariable behavior of certain natural phenomena. A **scientific theory** is much more than a law and may be defined as a conception, proposition, or formula relating to the nature, action, cause, or origin of a phenomenon or group of phenomena, formed by deduction and generalization from facts. That is, the scientific theory provides a mechanistic explanation for observed phenomena, based on facts. It is not a figment of imagination, an idea, or a hypothesis. A theory is based on large amounts of information assembled as a unified explanatory argument, such as the theory of gravitation, the theory of relativity, and the theory of evolution.

1.10 WHY IS THE THEORY OF EVOLUTION SO IMPORTANT TO BIOLOGY?

The theory of evolution has been called the "golden thread of biology." Dobzhansky (1973) argued that "nothing in biology makes sense except in the light of evolution." Darwin (1859, 489–490) once contemplated "an entangled bank," no doubt while walking near his residence in southern England. Such banks of soil around fields and along roads and paths support a rich array of plants and animals. And in spite of the complexity of interactions he witnessed, he recognized the simplicity of the theory that accounted for such natural richness:

> It is interesting to contemplate an entangled bank, clothed with many plants of many kinds, with birds singing on the bushes, with various insects flitting about, and with worms crawling through the damp earth, and to reflect that these elaborately constructed forms, so different from each other, and dependent on each other in so complex a manner, have all been produced by laws acting around us. . . . Thus, from the war of nature, from famine and death, the most exalted object which we are capable of conceiving, namely, the production of the higher animals, directly follows. There is a grandeur in this view of life.

There is grandeur, indeed, in a theory that unites the whole of biology and all living organisms into a coherent and comprehensive body of knowledge. No theory in biology before or since has achieved so much. The evolutionary synthesis provides a unification of biology (Mayr and Provine 1980) and a perspective from which to evaluate the growth of biological thought (Mayr 1982). The theory provides avenues for comparison across the biological sciences, including such topics as the evolution of form and function (Thompson 1942), the evolution of life cycles from bacteria to baleen whales (Bonner 1965), linkages between plant and animal populations (Harper

1977), common threads in the biology of parasites (Price 1980), the sociobiological aspects of plants and animals (Willson and Burley 1983), and the comparative biology of very small and very large organisms (Andrews 1991).

The advance of evolutionary theory provides tangible benefits to humankind beyond the cohesion of biological knowledge. The implications of the spontaneous generation debate may seem abstruse relative to our daily lives, but it was the solution to the controversy by Pasteur's experiments that established that microorganisms in the air could cause putrefaction of liquids—and wounds of the human body. Rapid application of this knowledge by Joseph Lister in 1865, in his use of carbolic acid or phenol to cleanse surgical utensils, wounds, and operating rooms, revolutionized medical practice. Countless millions of lives have been saved because some Greeks raised an insightful question. Many of our current concerns revolve around health, food, and very small organisms whose populations have enormous evolutionary potential. The practice of medicine is influenced by the evolution in bacteria of resistance to antibiotics, the practice of agriculture is influenced by the evolution in insect pests of resistance to pesticides, and the design of homes and cities attempts to maximize the separation between the worlds of humans and microorganisms by the promotion of hygienic practices.

Is any aspect of our lives uninfluenced by the facts of our evolutionary history? Our intellect, our manual dexterity, our creativity, even our spiritual perceptions, all stem from the unique evolutionary history of the human species. The pursuit of knowledge and understanding that illuminates the history of our existence on earth poses many more challenges than the exploration of the solar system or the mysteries of our physical environment. The theory of evolution provides a method and a perspective for understanding ourselves.

SUMMARY

Science is a human endeavor concerned with the observation and classification of facts, with emphasis on the establishment of verifiable general laws and scientific theories. By use of the **scientific method,** factual knowledge advances toward the truth as a series of approximations. The method clearly limits the realm of science to objective, verifiable fact and the search for facts. Other important human endeavors—in the areas of politics, ethics, and religion, for example—cannot be addressed by science, because personal opinion and belief play significant roles, and alternative perspectives are fully viable. The history of the debate on **spontaneous generation** illustrates how the process of science is self-correcting and, even if it takes 2,000 years or more, general facts can be revealed. Generalities, however, frequently stimulate new debates—such as the debate on the origin of life on earth—providing science with the energizing excitement of the unknown and of discovery.

The phenomenon of **industrial melanism** illustrates the factual nature of evolution with continuing evolutionary change in response to the changing environment. Such **microevolution** within populations and species has a robust factual basis because the experimental approach works effectively on this smaller scale of evolutionary processes. When addressing the origins of new species, genera, or families of organisms—i.e., problems in **macroevolution**—the scientific method is adequate for the task, but reality and fact are harder to achieve. When an accumulation of facts reveals invariable patterns, a **scientific law** can be generated, and the mechanistic explanation of such a law or laws produces one of the ultimate achievements of science, a **scientific theory.**

The **theory of evolution** is such a scientific theory. It provides a mechanistic explanation for the amazing unity of design seen in nature and at the same time explains the incredible diversity of nature. It unifies all aspects of the natural sciences because all become necessary elements in solving the puzzles of life on earth. All aspects of biology—from biochemistry, morphology, and physiology to behavior, ecology and sociobiology—are essential ingredients in a robust theory of evolution.

QUESTIONS FOR DISCUSSION

1. Is there any clear route by which science and religion can blend into one philosophy, or are they inevitably so distinct that no synthesis is possible?

2. Is it possible or likely that science will be able to study and incorporate what is now thought to be beyond physical nature, or metaphysical, such as so-called

extrasensory perception (ESP), ghosts, or polter-geists?

3. Can you think of fields other than evolution that pro-vide an equally coherent synthetic thread through the biological sciences?

4. Industrialization is not the only cause of selection for melanism in animal populations. Can you think of other environments that might give rise to selection of me-lanic forms?

5. Some people argue that evolution and the religious belief in Creation should be given equal time and credi-bility in the classroom. Is this a rational approach for a scientific subject or for an equally rational subject such as human thought or human belief?

6. Is it possible that at this time new life could develop spontaneously on earth or on any other planet in the solar system?

7. Do you think that it is reasonable, or good science, to argue that Pasteur falsified (i.e., invalidated) the con-cept of spontaneous generation with a few experiments in a laboratory?

8. How does the practice of science correct for the personal biases and cultural constraints of investigators, or is no such correction possible?

9. In your opinion, has evolutionary thought permeated the lives of most people, providing some understanding of, say, human design, societal evolution, human inter-actions, or life and death?

10. In what ways do we tend to use the processes of induc-tive and deductive logic in our daily lives?

REFERENCES

Andrews, J. H. 1991. Comparative Ecology of Microorgan-isms and Macroorganisms. Springer, New York.

Bishop, J. A., and L. M. Cook. 1975. Moths, melanism and clean air. Sci. Amer. 232(1):90–99.

———. 1979. A century of industrial melanism. Antenna 3:125–128.

———. 1980. Industrial melanism and the urban environ-ment. Adv. Ecol. Res. 11:373–404.

Bonner, J. T. 1965. Size and Cycle: An Essay on the Structure of Biology. Princeton University Press, Princeton, N.J.

Clarke, C. A., G. S. Mani, and G. Wynne. 1985. Evolution in reverse: Clean air and the peppered moth. Biol. J. Linn. Soc. 26:189–199.

Cook, L. M., G. S. Mani, and M. E. Varley. 1986. Postindus-trial melanism in the peppered moth. Science 231:611–613.

Darwin, C. 1859. The Origin of Species. Murray, London.

———. 1871. The Descent of Man, and Selection in Relation to Sex. Murray, London.

Dobzhansky, T. 1973. Nothing in biology makes sense ex-cept in the light of evolution. Am. Biol. Teacher, March 1973: 125–129.

Falconer, D. S. 1953. Selection for large and small size in mice. J. Genet. 51:470–501.

Farley, J. 1974. The Spontaneous Generation Controversy from Descartes to Oparin. Johns Hopkins University Press, Baltimore.

Gilbert, L. E. 1971. Butterfly-plant coevolution: Has Pas-siflora adenopoda won the selectional race with heliconi-ine butterflies? Science 172:585–586.

Gould, S. J. 1981. The Mismeasure of Man. W. W. Norton, New York.

Grant, B., and R. J. Howlett. 1988. Background selection by the peppered moth (Biston betularia Linn.): Individual differences. Biol. J. Linn. Soc. 33:217–232.

Harper, J. L. 1977. Population Biology of Plants. Aca-demic, London.

Kettlewell, H. B. D. 1956. Further selection experiments on industrial melanism in the Lepidoptera. Heredity 10:287–301.

———. 1959. Darwin's missing evidence. Sci. Amer. 200(3):48–53.

———. 1973. The Evolution of Melanism: The Study of a Recurring Necessity. Clarendon, Oxford.

Lakatos, I. 1970. Falsification and the methodology of scien-tific research programmes. pp. 91–196. In I. Lakatos and A. Musgrave (eds.). Criticism and the Growth of Knowledge. Cambridge University Press, Cam-bridge, England.

Mayr, E. 1982. The Growth of Biological Thought: Diversity, Evolution, and Inheritance. Belknap Press of Harvard University Press, Cambridge, Mass.

Mayr, E., and W. B. Provine. 1980. The Evolutionary Synthe-sis: Perspectives on the Unification of Biology. Harvard University Press, Cambridge, Mass.

Mikkola, K. 1984. On selective forces acting in the industrial melanism of Biston and Oligia moths (Lepidoptera: Geometridae and Noctuidae). Biol. J. Linn. Soc. 21:409–421.

Morton, S. G. 1839. Crania Americana, or A Comparative View of the Skulls of Various Aboriginal Nations of North and South America. Pennington, Philadelphia.

———. 1844. Observations on Egyptian ethnography, de-rived from anatomy, history, and the monuments. Trans. Amer. Phil. Soc. 9:93–159.

———. 1849. Observations on the size of the brain in vari-ous races and families of man. Proc. Acad. Nat. Sci. Philadelphia 4:221–224.

Myrdal, G. 1944. An American Dilemma: The Negro Prob-lem and Modern Democracy. Harper, New York.

Needham, J. T. 1748. A summary of some late observations upon the generation, composition, and decomposition of animal and vegetable substances. Phil Trans. Roy. Soc. 45:615–666.

Pasteur, L. 1859. Nouveaux faits pour servir a l'histoire de la levure lactique, pp. 34–36. *In* Pasteur Vallery-Radot (ed.) (1922). Oeuvres de Pasteur, Vol. 2. Masson, Paris.

Paterniani, E. 1969. Selection for reproductive isolation between two populations of maize, *Zea mays* L. Evolution 23:534–547.

Price, P.W. 1980. Evolutionary Biology of Parasites. Princeton University Press, Princeton, N.J.

Steward, R. C. 1977. Industrial and non-industrial melanism in the peppered moth, *Biston betularia* (L.). Ecol. Entomol. 2:231–243.

Thompson, D'A. W. 1942. On Growth and Form. 2d ed. Cambridge University Press, Cambridge, England.

Willson, M. F., and N. Burley. 1983. Mate Choice in Plants: Tactics, Mechanisms, and Consequences. Princeton University Press, Princeton, N.J.

2

DARWIN'S TIMES, HIS LIFE, AND HIS THEORY

The fantail pigeon, one of many breeds that aroused Darwin's interest in evolution by human selection. "Great as are the differences between the breeds of pigeon, I am convinced that the common opinion of naturalists is correct, namely, that all are descended from the rock-pigeon (Columba livia) . . ." *(Darwin 1872)*

(Illustration by Stephen Price)

As in other human endeavor, great advances in science do not erupt in a void. They have long histories and frequently occur in social milieus in which creative talents can reach beyond a background level of scientific progress to develop something remarkable. Such was the case with the theory of evolution. It is valuable for us to understand the background of Darwin's breakthrough and place the theory in perspective without detracting from its immense contribution to science.

2.1 HISTORICAL PERSPECTIVE

Charles Darwin was born at the Mount, a house above Shrewsbury in England, on February 12, 1809, in the reign of George III (r. 1760–1820). During King George's reign the population of Great Britain nearly doubled, from 7,750,000 to almost 14,500,000. The **industrial revolution** started and gained enormous momentum. With it came an **agrarian revolution** that applied scientific methods to agriculture, causing impressive increases in productivity.

King George's forces lost the American Revolution, or War of Independence, in 1781. The many European revolutions of that era climaxed in the French Revolution (1787–1799). After Napoleon Bonaparte gained control of France in 1799, the Napoleonic Wars continued until 1814, and the Battle of Waterloo resulted in Napoleon's final abdication in 1815. The French threat to British independence was gone, and the peace that would lull Victorian England was established 22 years before Victoria's ascension to the throne in 1837.

Exports from Britain increased by 75 percent from 1790 to 1801 and by another 51 percent from 1801 to 1814. In the early 19th century, Britain was a wealthy nation and was first in the cultivation of trade and industry. Napoleon called England "a nation of shopkeepers." He wished for an economic boom in his own country that could sustain his military efforts in a war dominated by the impressive buying power of the British.

Thus, from 1811, when George III lapsed into insanity and the prince regent began to oversee the country, to 1820, when the mad king died and the prince became King George IV, Britain experienced rapid change. Regency England was a time of discovery, reaction, and counterreaction.

The prince regent set an example of gluttony, drinking, and whoring. "Where royalty went, and what royalty did, the great middle class could go and do likewise, after its fashion" (White 1963, 133). This middle class was a product of the industrial revolution and included businessmen, merchants, traders, shopkeepers, officers of the armed forces, parsons,

lawyers, doctors, teachers, and scientists, many of whom also made up the intelligentsia. The relaxed morality of Regency England was clearly a reaction against the old king's overemphasis on decorum and taste.

The rise of utilitarianism demanded by an industrial age contrasted with the previous lack of connection between services rendered and society's needs. The new middle class demanded schooling for children and adults in useful subjects such as geography, history, the sciences, and modern languages as opposed to classical disciplines such as Latin, Greek, Hebrew, divinity, and moral philosophy. By 1815, 20 English towns had schools for adults and a large part of the middle class was relatively well educated. Debating societies formed and libraries were founded. Country millwrights and clockmakers turned into engineers and toolmakers, forming an elite class of civil engineers and mechanics, key professionals in an industrial society.

The educated middle class, who by and large formed the reading public, in 1801 founded the *Edinburgh Review*, which addressed itself specifically to the well educated. By 1813, the journal was selling 13,000 copies and was read by at least 40,000 people. Its authors remained anonymous so that they might express liberal ideas, and they strongly advocated Thomas Malthus's views (Eiseley 1958). Darwin's family almost certainly subscribed to the *Review*, and so Darwin probably read a significant paper published in the journal in 1831, before he set sail on the *Beagle*.

The loose morals of the Regency period, the wars from 1775 to 1815, and the French Revolution stimulated a reaction by the Anglican Church and a puritan revival. Attendance at church increased sharply, family prayers became common, and 214 new churches were built in London after passage of the Church Building Act in 1818. The Regent's government associated itself with the cause of virtue and truth, and anybody who undermined the union of church and state was vulnerable to repudiation and desertion. However, because of the stupidity of many religious tracts published between 1807 and 1820, the evangelical movement alienated intellectuals. The Reverend Sydney Smith observed, "No Christian is safe who is not dull" (White 1963, 146). Even so, the evangelical movement left its mark on Victorian England, with the themes of moral responsibility, respectability, and philanthropy pervading the lives of the middle classes.

The union of puritanism and utilitarianism made public lectures very popular with the educated public in London around 1815. Humphrey Davy, inventor of the miner's safety lamp, demonstrated the new wonders of chemistry to 300 people in one sitting.

Laughing gas, or nitrous oxide, was a tremendous success. Many people of public stature, including Josiah Wedgwood, Darwin's maternal uncle, tried a sniff of nitrous oxide in a silken bag as a recreation and palliative. In 1812, the prince regent knighted Humphrey Davy for his popularization of science. It was becoming clear that the development of science and technology would have vital national importance in industry and agriculture. In his inaugural lecture as Chair of Chemistry in the Royal Institution, Davy insisted on the utilitarian value of science. The Institution requested that he develop such subjects as the chemistry of tanning, the analysis of minerals, and the application of chemistry to agriculture.

The University of London, established in 1820, pioneered the approach of the modern university (in contrast with Oxford and Cambridge) by teaching important practical subjects for an industrial age. Pioneers of reform were also penetrating the public schools. Significantly, perhaps, one such educational reformer, Dr. Samuel Butler, was hired at Shrewsbury School in 1798. Darwin was to attend this school from 1818 to 1825 while Dr. Butler was still headmaster. One must wonder, however, how much of a pioneer Butler really was, since in his autobiography Darwin (1892) tells us that Butler rebuked him for wasting his time on such useless subjects as chemistry. Darwin had become interested in the subject through his brother, who had set up a laboratory in the family's tool shed. Darwin assisted his brother with experiments and the making of gases and as a result was nicknamed "Gas."

2.2 DARWIN'S LIFE

The major and permanent social changes that took place in England during Charles Darwin's lifetime greatly influenced his work. Science and technology were in their heyday, and Darwin's family members were aware of developments in those fields. Darwin was born into the middle class in the sense that he neither was of "blue-blooded" aristocratic descent nor labored in the mills or mines. His father, Robert Darwin, was a successful physician (Figure 2-1). Charles grew up to enjoy reading books of all descriptions at a time of scientific ferment. Excavation of canals and other structures tremendously expanded the understanding of geology, the discovery of fossils, and the development of uniformitarianism, the concept that earth's processes are uniform throughout time. Baron Cuvier (1769–1832) had fathered the study of paleontology, and James Hutton (1726–1797) had founded historical geology and "time boundless and without end" (Eiseley 1958, 65), so essential to

the uniformitarian concept of geology. William Smith (1769–1839), nicknamed "Strata Smith," found that each individual stratum contained distinct fossils, a fact that enabled the characterization of geological periods (see Chapter 7). Alexander von Humboldt (1769–1859) had made his famous voyage to Spanish America from 1799 to 1804 and after his return continued to have a powerful influence in European science—and on Darwin, who read Humboldt's *Personal Narrative* in his last years at Cambridge University (Darwin 1892). The enclosure of farms was accelerating, and selective breeding of crops and animals was rapidly improving yields. Regency England had seen a growth of liberalism, dissension from the established church, and the alienation of many intellectuals by the evangelical movement. Journals such as *The Edinburgh Review* catered specifically to the liberal intelligentsia. Debating, lectures, and libraries were all a part of middle-class life, and intellectuals were migrating from the provinces to London. Charles Darwin's grandfather, Erasmus Darwin (1731–1802), had been a member of the Lunar Society (based in Birmingham), one of the congregations of thinkers that proliferated as means of communication increased. After 1815, France and the rest of Europe became accessible for travel and intellectual exchanges (which Charles Lyell exploited), and England settled into peaceful times. The period R. J. White called "the age of seriousness" (White 1963, 151) had begun.

Shrewsbury School

All young men of the middle class were exposed to the aforementioned changes. What set Charles Darwin apart and gave him his enormous insights into the processes of evolution? It was certainly not his formal education, if his autobiography is to be believed (Darwin 1892). By the time he was eight years old he was an avid collector of shells, eggs, and minerals and was generally interested in natural history, taking long solitary walks and fishing. He entered Shrewsbury School in 1818. On leaving it in 1825 he felt that nothing could have been worse for his intellectual development. He regarded his schooling as strictly classical except for a little exposure to geography and history. At Shrewsbury School Darwin read Gilbert White's (1789) *The Natural History and Antiquities of Selborne* and watched birds and noted their habits, as White had done. He later regarded his exploits in chemistry with his brother as the best part of his education while at school, because from them he learned the meaning of experimental science. When he left Shrewsbury, he thought he was regarded as ordinary and perhaps below par in intelligence. Nevertheless, he was fond of reading, and such

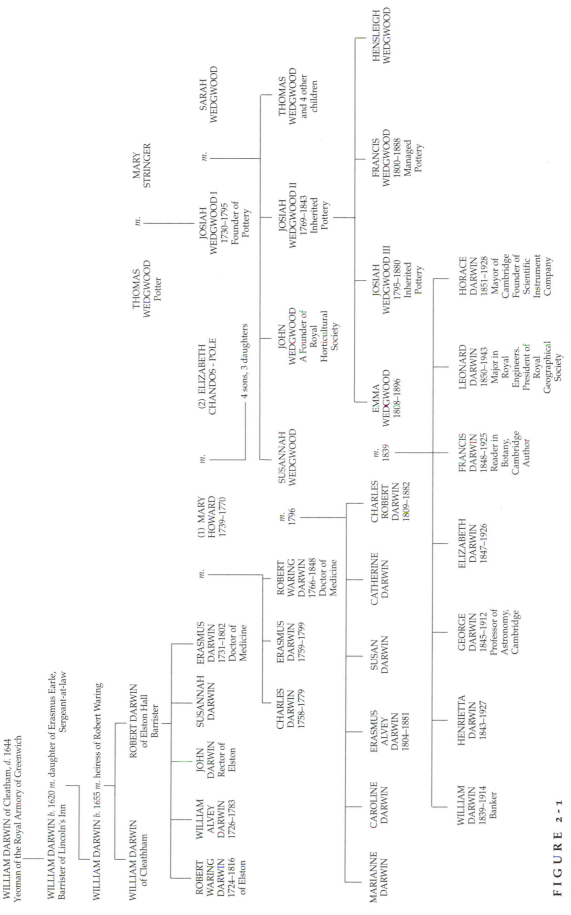

FIGURE 2-1

Part of Charles Darwin's family tree, with notes on professions and associations, up to the seven children that survived after his death. Based on Dobson 1971, Wedgwood 1909, Kelly 1975, Raverat 1952, and the *Historical and Descriptive Catalogue of the Darwin Memorial at Down House, Downe, Kent* (Anonymous 1969). Note that Josiah Wedgwood I was a contemporary and acquaintance of Erasmus Darwin, Charles's grandfather. The symbol *m.* indicates marriage, *b.* born, and *d.* died. Erasmus Darwin married twice, with spouses numbered 1 and 2.

books as *Wonders of the World* aroused his ambition to travel. He enjoyed shooting and persisted in collecting minerals. He was fascinated by insects, although his early collections were limited to already-dead insects, since his sister had impressed upon him the impropriety of killing creatures merely to collect them, and this moral responsibility extended even to the invertebrates. Darwin's attitude toward collecting was to change, to his detriment.

Edinburgh University

When Darwin was 16, his father sent him to Edinburgh University to study medicine. Soon after Charles's arrival, he discovered that his father would leave him enough property to live on comfortably, and thus ended his serious efforts to master medicine. All subjects were taught through lectures, and Darwin later realized that practical laboratory work could have helped him greatly. Only chemistry was excepted from his description of the lectures as being intolerably dull. Reading was far more enjoyable, and he seems to have used the library more than most students.

In his second year at Edinburgh, Darwin became friendly with a Dr. Grant, who on one occasion praised Lamarck and his evolutionary theory. Darwin had already read and admired his grandfather's *Zoonomia* (Darwin 1794), which expressed similar views. In his autobiography Charles remarked, "It is probable that the hearing rather early in life such views maintained and praised may have favoured my upholding them under a different form in my Origin of Species" (Darwin 1892, 13). He also attended many lectures at societies, including some by John James Audubon on the habits of North American birds, and he learned from a local taxidermist how to stuff birds.

Cambridge University

After two years at Edinburgh, when it was clear that medicine was not for Charles, his father suggested that he study for the clergy at Cambridge University. After due consideration, and finding that he could fully accept "the strict and literal truth of every word in the Bible" (Darwin 1892, 17–18), Charles agreed. He attended Cambridge from 1828 to 1831 and found it a complete waste. He spent a goodly amount of time shooting, riding, and drinking and even more time collecting beetles; by this time living specimens were fair game. As an example of his enthusiasm for beetles, Darwin told of a collecting trip in which he removed some old bark from a tree. He saw two rare beetles and captured them, one in each hand. Then he saw a third species, and so he popped one of his captives into his mouth to free his right hand. "Alas! it ejected some intensely acrid fluid, which burnt my tongue so that I was forced to spit the beetle out, which was lost, as was the third one" (Darwin 1892, 21).

During his stay at Cambridge, Darwin met many knowledgeable men—partly through Professor Henslow, whose lectures and excursions in botany he enjoyed. He did obtain a degree. He continued his avid reading and was most impressed by Humboldt's *Personal Narrative*. His enthusiasm for contributing to the natural sciences increased.

Through Henslow, Darwin met the geologist Adam Sedgwick, who later introduced the Cambrian and Devonian Periods into geology and described the distinctive fossil faunas that contributed to the characterization of each geological period (see Chapters 7 and 9). Together Darwin and Sedgwick toured North Wales, where Sedgwick was making his famous geological studies of the rock strata. One evening, Darwin mentioned to Sedgwick his discovery of a tropical shell in a gravel pit close to Shrewsbury. Sedgwick replied that it would be the greatest misfortune to geology if it were really embedded there, as it would overthrow all that was known about the superficial deposits of that area. Years later Darwin found broken arctic shells in the same gravel pits and realized that the tropical shell had no doubt been carried there by humans and discarded. Impressed by Sedgwick's uninterest in the tropical shell, he realized that "science consists of grouping facts so that general laws or conclusions may be drawn from them" (Darwin 1892, 25).

The Voyage of the *Beagle*

On his return from Wales, Darwin received a letter from Henslow saying that Captain Fitz-Roy of H.M.S. *Beagle* wished for the company of a young naturalist on his next voyage. Significantly, Darwin's uncle, Josiah Wedgwood (son of the Josiah who founded Wedgwood Pottery), persuaded Robert Darwin to let Charles go on the voyage. An interview in London with Fitz-Roy secured the position.

The *Beagle* set sail December 27, 1831, when Darwin was 22, and returned on October 2, 1836. The voyage was the most important educational experience of the naturalist's career. The voyage would have been fruitless, however, without his knowledge of and interest in natural science, gained largely by reading, discussion, and rumination during his school and college days.

Darwin took with him the first volume of Lyell's *Principles of Geology* (1830), which spelled out the uniformitarian concept of geology and identified the almost limitless time over which forces shaped the earth. Darwin was to rely on Lyell's principles as a source of ideas for many years to come (Eiseley 1958). Eiseley (1958, 353) defined **uniformitarianism** as that scientific school of thought generally associated with the names of James Hutton and Sir Charles Lyell which assumed that geological phenomena were the product of natural forces operating over enormous periods of time and with considerable, though not necessarily total, uniformity. With modifications it has become the geological point of view of the twentieth century. In the early nineteenth century it stood in considerable opposition to Catastrophism . . . and Progressionism.

(Definitions of the last two terms appear in Section 2.3.)

Wherever the *Beagle* put to shore, Darwin studied and described the area's geology, including stratification, rock types, and fossils. He avidly collected all types of animals, and here his shooting skill came in handy. He took horseback and boat trips into the hinterland and continued to study, collect, and record.

The mission of the voyage was to complete a survey of Patagonia and Tierra del Fuego, Chile and Peru, and some Pacific Islands and to make measurements around the world in order to fix the points of longitude. During the voyage Darwin sent letters and a collection of fossil bones to Professor Henslow. Henslow read some of Darwin's letters in 1835 to the Philosophical Society of Cambridge and had them printed for private distribution. Apparently these letters made a deep impression on some great minds; while on Ascension Island, Darwin heard that Sedgwick thought him worthy of a place among the great scientists of his day. Darwin was then 27 years old. This early recognition was no doubt important in establishing his credibility as a scientist, which would be tested in years to come.

We will discuss more of the scientific aspects of the *Beagle*'s voyage when we consider the ideas Darwin used to develop his theory. For now, let us concentrate on his life and his other works.

Toward the Origin of Species

On his return from the *Beagle* voyage, Darwin studied his collections in Cambridge and then moved to London, where he was active in the Geological Society. He was writing manuscripts on his voyage and became well acquainted with Charles Lyell. In July 1837, he started his first notebook on facts relating to the origin of species. He had already considered the subject for some time and continued to work on it for another 20 years. In 1838 he entertained himself by reading Thomas Malthus's 1798 "An Essay on the Principle of Population as It Affects the Future Improvement of Society." This essay had an important impact. It made Darwin realize that through the struggle for existence "favourable variations would

tend to be preserved and unfavourable ones to be destroyed. The result of this would be the formation of new species. Here, then I had at last got a theory by which to work" (Darwin 1892, 43). Darwin made these remarks in 1838, only 15 months after he had opened his first notebook on the transmutation of species and 21 years before he published *The Origin of Species* (in 1859). "Darwin gained a unique understanding from Malthus" (Desmond and Moore 1991, 265).

Charles married his cousin, Emma Wedgwood, in January 1839 and led a happy married life, with nine children born between 1839 and 1851 (Raverat 1952) (Figure 2-2). His beloved first daughter, Anne

FIGURE 2-2

Watercolor portrait of Charles Darwin, made in 1840 by George Richmond. Of the ten known portraits taken from life (Darwin 1892), this earliest work always seems the most significant, for it shows a youthful Darwin at the time he was creating his theory.

Elizabeth, died in her tenth year in 1851, causing the grieving Darwin to abandon his religious faith. His second daughter, Mary Eleanor, born in 1842, lived only briefly. His own poor health was aggravated by sociable living, so in 1842 the family moved to Down House in Downe, Kent—even now a small village well out of the London suburban sprawl. There they led a quiet life, and Darwin devoted his energies to research and to publications, which were diverse and extensive.

Darwin's major, four-volume work on barnacles, begun in 1846 and completed in 1854, set out his description of the group (Darwin 1851a,b, 1854a,b), which is still accepted today. In his autobiography Darwin evaluates this work as having been of considerable value to science and to him, by teaching him the principles of natural classification (which he went on to discuss in *The Origin of Species*). He doubted, however, that these volumes were worth eight years of his life.

Darwin had continued to collect notes on the transmutation of species, and in 1854 he concentrated on sorting those piles of information. In 1856 Lyell strongly advised him to prepare a substantial version of the theory, and Darwin started on his "big book," which would have been about four or five times longer than *The Origin* had he finished it. He was halfway through it in 1858, when Alfred Russell Wallace sent him an essay that duplicated his theory: "On the Tendency of Varieties To Depart Indefinitely from the Original Type," from Ternate in the Malay Archipelago. Lyell and Hooker persuaded Darwin to publish an extract of his own book along with Wallace's essay in the *Journal of the Linnean Society*, and both appeared in 1858. Darwin agreed reluctantly because he felt his extract was badly written, while "Mr. Wallace's essay, on the other hand, was admirably expressed and quite clear" (Darwin 1892, 44). Every reader of these two short papers has probably heartily agreed with Darwin's evaluations.

The papers actually received little attention. Later in 1858, Darwin started work on *The Origin of Species* proper, which was published in November 1859. The first printing of 1,250 copies was sold out on publication day, and the second printing of 3,000 copies, within a short time. By 1876, 16,000 copies had been sold in England alone. The book was translated into many languages, and Darwin regarded it as his most important contribution to science, as no doubt it was.

The Big Debate

Darwin received high praise for his book from many intellectuals; others ridiculed him mercilessly. Tempers were apt to run high on both sides. A curious, famous, and perhaps even important episode in this "debate" occurred seven months after publication of *The Origin*, on June 30, 1860, at the British Association meetings in Oxford. It was known that on this Saturday a Dr. Draper from New York was to present a paper titled "Intellectual Development of Europe Considered with Reference to the Views of Mr. Darwin." In addition, Bishop Wilberforce of Oxford, considered by A. D. White (1896) to be the most brilliant dignitary of the Anglican Church, was to use this occasion to roundly criticize *The Origin of Species*.

The previous day Thomas Huxley had happened to meet in the street Robert Chambers, who in 1844 had anonymously published "The Vestiges of the Natural History of Creation" and thereby raised the ire of scientist and layman alike—including Huxley, who had then written a scathing response. When the two men encountered each other in 1860, Huxley said he would not be going to the next day's British Association meeting because he was too tired. But Chambers was most animated in his criticism of Huxley as a deserter of the evolutionists' cause and persuaded Huxley to attend the meeting (Darwin 1892).

The excitement at the meeting was intense. Something like 700 to 1,000 people crammed the library of the museum. Huxley was there. Bishop Wilberforce was there. Professor Henslow was the chairman. The by-then Admiral Fitz-Roy, still an ardent fundamentalist, was there. Charles Darwin was absent.

The bishop spoke for half an hour and ridiculed Darwin and Huxley "but all in such dulcet tones, so persuasive a manner, and in such well-turned periods" (Darwin 1892, 251). However, the bishop showed himself to be quite ignorant of the details of Darwin's book and said,

> "I should like to ask Professor Huxley, who is sitting by me, and is about to tear me to pieces when I have sat down, as to his belief in being descended from an ape. Is it on his grandfather's or his grandmother's side that the ape ancestry comes in?" (Darwin 1892, 251–252)

The bishop concluded with the point that Darwin's ideas ran counter to the "revelations of God in the Scriptures" (252).

Huxley was reluctant to reply, but eventually he said, "I am here only in the interest of science and I have not heard anything which can prejudice the case of my august client" (252). He then demonstrated the bishop's poor understanding of Darwin's thesis and concluded with a reference to his descent from a monkey.

> "I asserted, and I repeat, that a man has no reason to be ashamed of having an ape for his grandfather. If there were an ancestor whom I should feel shame in recalling, it would be a *man*, a man of restless and versatile intellect, who, not content with an equivocal

success in his own sphere of activity, plunges into scientific questions with which he has no real acquaintance, only to obscure them by an aimless rhetoric, and distract the attention of his hearers from the real point at issue by eloquent digressions, and skilled appeals to religious prejudice." (253)

As A. D. White said in 1896 about Huxley's retort, "This shot reverberated through England, and indeed through other countries" (Appleman 1970, 423). There was no doubt about who had won the combat, and since that time Darwin's theory has been gaining ground. Even today, however, we see religious beliefs attempting to compete with scientific theory for a place in modern biology.

After 1860

From 1860 onward, Darwin concentrated on writing books:

1862   *On the Various Contrivances by Which British and Foreign Orchids Are Fertilized by Insects, and on the Good Effects of Intercrossing*

1868   *The Variation of Animals and Plants under Domestication*

1871   *The Descent of Man, and Selection in Relation to Sex*

1872a   *The Expression of the Emotions in Man and Animals*

1875   *The Movements and Habits of Climbing Plants*

1875   *Insectivorous Plants*

1876   *The Effects of Cross and Self Fertilization in the Vegetable Kingdom*

1877   *The Different Forms of Flowers on Plants of the Same Species*

1880   *The Power of Movement in Plants*

1881   *The Formation of Vegetable Mould, Through the Action of Worms, with Observations on Their Habits*

Charles Darwin died on April 19, 1882. After a funeral in Westminster Abbey, his body was interred in the nave of the abbey, close to the grave of Sir Isaac Newton.

Darwin's contributions to science were actually much broader than his theory. "Darwin had naturalized Creation and delivered human nature and human destiny into" the hands of a new scientific community unshackled from politics and religion (Desmond and Moore 1991, 677). He was a central force, a rallying point, in the move toward a purely secular science, separated from religious influence, that was engineered by younger colleagues including Thomas Huxley, Joseph Hooker, Francis Galton, and John Lubbock. It was perhaps ironic that a man who had argued against the Creation, and had been an agnostic since his daughter Annie's death in 1851,

should be buried in Westminster Abbey. But "the dignitaries of science and State liked the idea of worldwide homage to an English naturalist," and "the scientists' moral duty in furthering human evolution was best exercised in harmony with the old religious ideals 'upon which the social fabric depends' " (Desmond and Moore 1991, 674). Galton had realized, and had prevailed in his argument, that Darwin's burial in the abbey would confer enormous respectability on the new approach to science and the hallmark of that science, the theory of evolution. Was Darwin the "devil's chaplain" (*cf.* Desmond and Moore 1991)? No, he unyoked science from religion so that each could develop in its proper and independent way.

2.3 CONCEPTUAL BACKGROUND IN DARWIN'S TIME

The formulation of Darwin's theory occurred against a conceptual background prevalent during his early life that related to the major controlling factors of the natural world, which were, not surprisingly, in accord with the Church's teaching of a divine Creation. These factors may be summarized under the headings catastrophism, progressionism, and the fixity of species.

Catastrophism

Catastrophism was an explanation of the stratigraphic features of the earth that postulated sudden, violent, and drastic changes interspersed between long periods of stability through geologic history (Eiseley 1958). The catastrophes were imagined to be global and devastating to most or all life on earth. The catastrophes were so vast, and so foreign to current experience, that supernatural forces were seen as the common source of their power. After each cataclysm life was created again, presumably in some supernatural way. This concept united the growing knowledge of geology (due to Hutton, Smith, Cuvier, and others) with **"mosaic geology,"** or the account of the Creation in Genesis.

The concept of catastrophism is easy enough to understand in its historical context, for extensive excavations had exposed many rock strata piled upon each other, with dramatic shifts in rock type and fossil ingredients over a few centimeters in the profile (Figure 2-3). Those who subscribed to catastrophism could easily argue that each stratum of rock was the result of a catastrophe that had completely changed the substances making up the soil. Volcanic eruptions, earthquakes, tidal waves, and storms, all on a

FIGURE 2-3

A cross section of rock strata in the Paris basin, showing dramatic and sudden changes in rock types, drawn by Cuvier and Brongniart (1822). Note, for example, that in the group 5 deposits a layer with oyster fossils (*Ranc d'huitres*) is suddenly followed above by a thick layer of micaceous sand (*Sable micacé*), followed by a sandstone without shell fossils (*Grès sans coquilles*) and then marine sandstone with shells (*Grès marin superieur*).

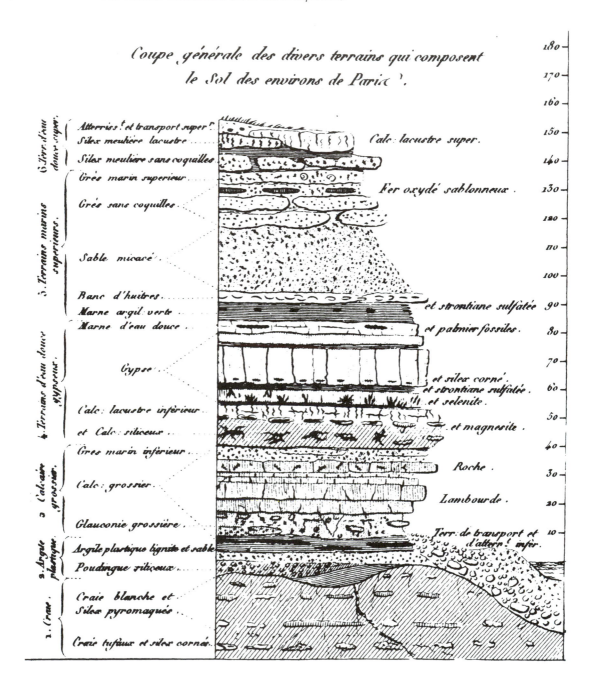

massive scale, were presumably responsible and likely had been supernaturally influenced.

Yet Cuvier, the founder of comparative anatomy, had identified a unity of nature in the anatomies of living animals. Faced with a pile of fossilized bones

dug up from excavations in the Paris basin, he said, according to Eiseley's translation (1958, 85–86),

We will take what we have learned of the comparative anatomy of the living and we will use it as a ladder to descend into the past. All our information, scanty though it may be, leads us to assume that the same unity of design of which we observe evidence in the modern world extends also across the enormous time gulfs of the past. My key, my principle, will enable us to restore the appearance of those long vanished beasts and relate them to the life of the present.

By 1837, and in *The Edinburgh Review,* Cuvier (1837, 23–24) was quoted as saying (as translated by Buckland), "At the voice of comparative anatomy every bone and fragment of a bone resumed its place." This order and close relationship between the living and the dead posed a problem for the catastrophists: although their concept accounted for disarray or pandemonium in the fossil record, it could not account for order.

Progressionism

The concept of **progressionism** neatly joined Cuvier's findings with the catastrophists' and creationists' views on the origin of species. Progressionism took the view that, after each cataclysmic episode on the earth, life was created again but in a more advanced form; thus, life had progressed from simple forms to more complex organisms across the geological past (Eiseley 1958). These progressive creations had ultimately led to the creation of humans, and the unity of nature was the manifestation of a supernatural plan.

The Fixity of Species

Linnaeus, a very influential scientist in the late 18th century, had emphasized the constancy and discreteness of species. His view, now known as the **typological species concept,** was that a whole species was represented by a single **type specimen,** which pleased both church and state and was widely subscribed to in Darwin's early life. Each species was a discrete entity, sharply different from others, and invariable. Accordingly, the creationists', catastrophists', and progressionists' views on the origin of species supposed that, once a species was present, it remained the same indefinitely. This concept has become known as the **fixity of species.**

The wide acceptance of the fixity of species posed the original problem for the uniformitarians, who sought currently acting causes for existing phenomena (uniformitarianism was defined in Section 2.2). If one believed in species, one could not believe in change or evolution, and since Linnaeus had described thousands of species, could there be much doubt about the typological species concept and the constancy of species?

Scientific Advances

Darwin formed his theory of natural selection and the origin of species against this background of established views. He did not do it alone, as Eiseley (1958) has taken great pains to show, but benefited from many influences. Eiseley lists some:

1. In the 18th century, scientists advanced theories of cosmic evolution.
2. The study of fossils forcefully suggested a world of change through many millions of years.
3. The microscope opened up an incredible unseen world.
4. In France, many great minds were interested in man, and early essays on the human population supplied Thomas Malthus with much fuel for his "Essay on the Principle of Population," with its concept of survival of the fittest.
5. The plants named by Linnaeus could be seen in gardens and greenhouses throughout England. Plants exhibited variation within a species, and gardeners and farmers practiced artificial selection (selection by humans of the best individuals for breeding) with rapid and remarkable results.

Evolutionary Ideas

In Darwin's time, evolutionary ideas were expressed more and more frequently (Eiseley 1958). The concept of evolution had found moderate acceptance in academic circles since the 18th century, but the **mechanistic explanation** of evolution was hotly debated, and no generally accepted synthesis or theory of evolution emerged until Darwin disclosed his mechanistic theory of evolution in 1859. In 1748, Benoit de Maillet published his *Telliamed,* which included ideas on the origin of man and animals and took an essentially uniformitarian viewpoint. In 1749, the Compte de Buffon started publishing his *Histoire Naturelle,* which, according to Eiseley (39), mentioned "every significant ingredient which was to be incorporated into Darwin's great synthesis of 1859," although Buffon did not pull them into a coherent scenario. In 1794, Erasmus Darwin published his *Zoonomia.* In later works he would note the similarity between organisms and conclude that all organisms are related and derived from one parent. But Erasmus Darwin believed in the inheritance of acquired characteristics, as did Lamarck.

Jean-Baptiste Lamarck

Lamarck, who had studied Buffon in detail, published his *Philosophie Zoologique* in 1809. Today we remember him more for his errors than for his impressive contributions to biology. Nevertheless, "Lamarck's great systematic study of invertebrate zoology for instance, had been the indispensable foundation for the study of invertebrate fossils and their use in the identification of strata" (Wilson 1972, 116). And Lyell took Lamarck's views seriously enough to devote an extended portion of *Principles of Geology*, volume 2, to refuting them (Burkhardt 1984). "What Lyell did do in the context of Lamarck's theory was to set out clearly and forcefully the species problem that Darwin eventually solved" (Hull 1984, xiv). Thus, a constructive critic must view Lamarck as being on a direct and important line of thought to Darwin's theory (Corsi 1978, Hull 1984).

Lamarck subscribed to the idea of **transmutation of species** but not to their extinction. As the environment changed, so the species changed and tended to diverge. In translation, Lamarck (1984, 112–113) argues as follows:

1. Every fairly considerable and permanent alteration in the environment of any race of animals works a real alteration in the needs of that race.
2. Every change in the needs of animals necessitates new activities on their part for the satisfaction of those needs, and hence new habits.
3. Every new need, necessitating new activities for its satisfaction, requires the animal, either to make more frequent use of some of its parts which it previously used less, and thus greatly to develop and enlarge them; or else to make use of entirely new parts, to which the needs have imperceptibly given birth by efforts of its inner feeling.

Lamarck incorporated these ideas into his first and second laws of nature, and in the latter he concluded that "all these [changes] are preserved by reproduction to the new individuals which arise, provided that the acquired modifications are common to both sexes, or at least to the individuals that produce the young" (113).

Robert Chambers

Passing over some significant evolutionists discussed by Eiseley, we come to the interesting case of Robert Chambers. In 1844 Chambers anonymously published *The Vestiges of the Natural History of Creation*. The book was widely attributed to Prince Albert, Queen Victoria's husband since 1840, and thus became a national sensation. In spite of his naivete and a certain amount of progressionism, Chambers was the first to unite the contributions of Hutton, with his seemingly endless time; "Strata" Smith, who noticed the distinctness of fossils in adjacent rock strata; and Baron Cuvier, who saw organization and unity through comparative anatomy and paleontology (Eiseley 1958). Chambers decided that both cosmic and organic evolution were realities. Darwin read *The*

Vestiges with great attention and carefully annotated his copy, as did Wallace his (McKinney 1972). Eiseley's (1958) analysis is that *The Vestiges* received such wide attention—the subject of evolution having suddenly become "public property"—that those who might have criticized *The Origin* sapped their energy by leveling condemnation at this easier target. Interest in *The Vestiges* hardly lagged, and in 1854 the tenth edition was released leaving the public little time to lose interest before *The Origin* was published. The public was primed to receive an authoritative presentation of an evolutionary theory.

2.4 MAJOR IMPACTS ON DARWIN'S THINKING

Darwin had formulated his **theory on the transmutation of species** six years before the publication of *The Vestiges*, and by the latter date he had a 230-page manuscript on *The Origin of Species*. In his autobiography Darwin identifies three major elements that contributed to the development of his theory. In chronological order, they are (1) Lyell's *Principles of Geology*, volume 1, published in 1830; (2) the voyage of the *Beagle* (Figure 2-4)—in particular, (a) St. Jago in the Cape Verde Islands, (b) the Pampaean formation behind Montivideo with all its large fossil bones, and (c) the animals of the Galapagos Islands; and (3) Malthus's "An Essay on the Principle of Population," published in 1798 and read by Darwin in October 1838.

Lyell's *Principles of Geology* and Uniformitarianism

Henslow recommended to Darwin that he read Lyell's *Principles*, and this Darwin did early in the voyage of the *Beagle*. In spite of Henslow's advice not to accept Lyell's views, Darwin noted in his autobiography (36): "I am proud to remember that the first place, namely St. Jago, in the Cape Verde Archipelago, in which I geologized, convinced me of the infinite superiority of Lyell's views over those advocated in any other work known to me." Lyell wrote his book as a set of principles for the work of geologists, and the subtitle of his book sets out the main principle (Rudwick 1970): *Being an attempt to explain the former changes of the earth's surface, by reference to causes now in operation*. He set out to establish uniformitarianism, a principle that is essential for a scientific approach to the natural world.

Lyell stressed the enormous time spans involved in geological changes. He attacked progressionism by advancing his steady-state theory of continual cli-

FIGURE 2-4

The voyage of the *Beagle*.

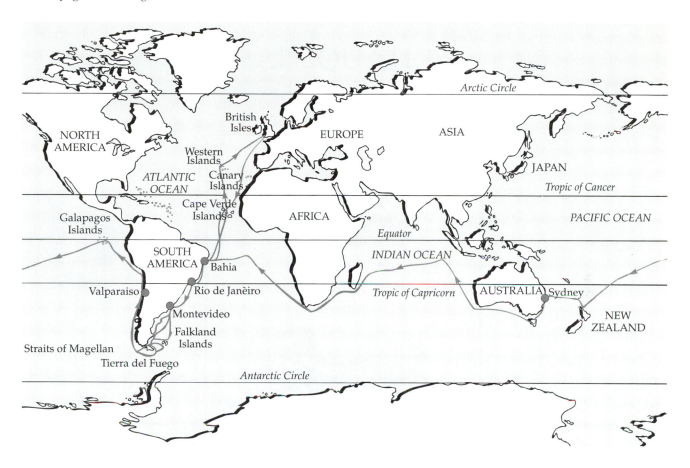

matic fluctuation about a mean and by examining the fossil evidence and the recent emergence of man, while denying any sign of a progressive system. Thus he debunked "mosaic geology," catastrophism, and progressionism. Still in the first volume he started on his own views of uniformitarianism, and the latter part of that volume and all of volume 2 describe and analyze geological processes now in operation. Lyell attempted to demonstrate that current processes were sufficient to account for all past geological phenomena and that such processes did not change in either kind or degree; rather, natural forces persisted over time. In 1842 (41–42) Lyell noted that trilobites had eyes and therefore "the ocean must then have been transparent as it is now; and must have given a passage to the rays of light, and so with the atmosphere; and this leads us to conclude that the sun existed then as now and to a great variety of other inferences." In fact, some trilobites had very large eyes (Figure 2-5). Such simple arguments that conditions on the earth have been uniform over long periods of time were very compelling.

Lyell's volume 1, however, was restricted to geological phenomena with reference to his uniformitarian concept. The frontispiece of *Principles* (1830) illustrates the remains of the Temple of Serapis in the Bay of Naples (Figure 2-6). It shows the three remaining pillars, each with a band of marble, 12 feet high and 12 feet above the pedestal, that had been deeply bored by the marine bivalve *Lithodomus*. Lyell (1830, 453–454) explained:

The perforations are so considerable in depth and size that they manifest a long continued abode of the Lithodomi in the columns; for, as the inhabitant grows older and increases in size, it bores a larger cavity, to correspond with the increasing magnitude of its shell. We must consequently infer a long continued immersion of the pillars in sea-water, at a time when the lower part was covered up and protected by strata of tuff and the rubbish of buildings, the highest part at the same time projecting above the waters, and being consequently weathered, but not materially injured.

FIGURE 2-5

Examples of trilobites. Note the conspicuous eyes on all adult specimens. Most trilobites had large eyes (for their sizes) and good vision, although some living in the deep sea were blind. **Top left:** Sketch of *Panderia megalophthalma* as it may have been situated in sediments. **Top right:** *Paedumias transitans*, known from the Cambrian Period, in the larval (left) and adult (right) stages. Adults were about 5 cm long. **Center:** *Flexicalymene meeki* in its completely rolled, flexed position. The genus is known from the Ordovician Period (Llandeilo Epoch) to the Silurian Period (Wenlock Epoch), and specimens were up to 10 cm long. **Bottom:** Three sketches of *Phacops rana*, the little froglike trilobite, showing its large eyes and stages in the flexion of its body. The genus is represented in Devonian Period deposits (Siegenian to Famennian Stages), and members were up to 6 cm long. (See Table 7-1 for a geological time scale.) *(Geological times and fossil sizes from Murray 1985. Illustration by Stephen Price.)*

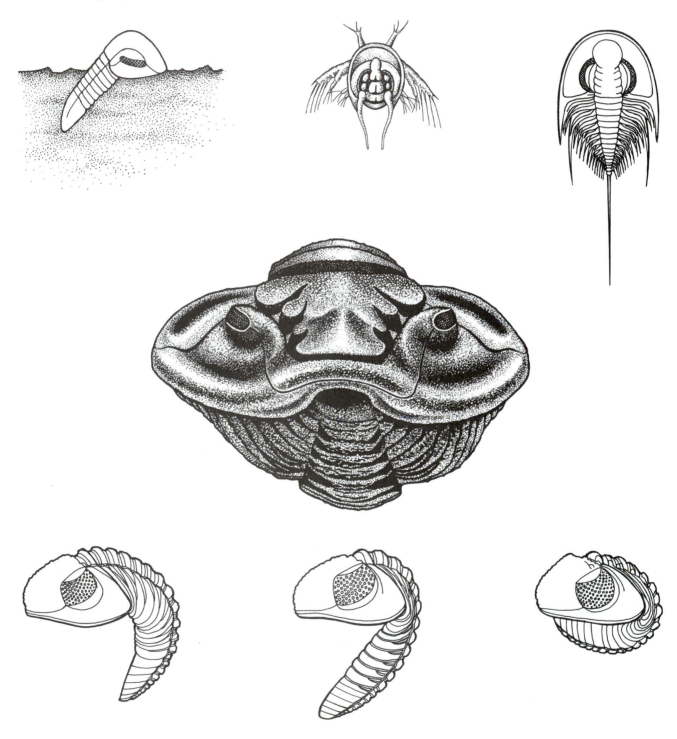

FIGURE 2-6

Remains of the Temple of Serapis in the Bay of Naples, as illustrated on the frontispiece of the first volume of Lyell's *Principles of Geology*, 1830. Note the band of bored marble 12 feet above the pedestal.

With the high-tide water level at 1 foot above, and the top of the *Lithodomus* activity 23 feet above high tide (24 feet above the temple floor), Lyell perceived that the temple built on dry land must have subsided well over 24 feet into the bay and subsequently was elevated by 23 feet to its present position, probably during the earthquake of 1538. This magnitude of subsidence and elevation, especially since it was so local in extent (Wilson 1972), impressed Lyell and clearly demonstrated that significant movements of the earth had occurred in the historical past.

Experiences on the Voyage of the *Beagle*

St. Jago and the Pampas

Darwin tells us very little about his visit to St. Jago or about his evaluation of Lyell's views as being superior to other geologists'. The ship stayed at port on the island for 23 days. Writing to his father, Darwin said, "St. Jago has afforded me an exceedingly rich harvest in several branches of Natural History," and "Geologizing in a volcanic country is most delightful" (Darwin 1892, 136), but provided few details. In *The Voyage of the Beagle*, he noted the interesting calcareous cliffs, with many shells embedded, lying on ancient volcanic rocks and covered by a layer of basalt. He noted that the hot basaltic lava had changed the chalk into a crystalline limestone. Perhaps it was there that Darwin realized that no cataclysm need have made the changes in the rock strata—all that was necessary to explain the white cliffs of St. Jago was a small volcanic eruption with lava flowing down into the sea and running over shell-filled deposits, followed by slow elevation of the mass and erosion by raindrop and sea. Uniformitarianism, or presently observable causes, readily accounted for all that Darwin saw. Lyell was right!

Riding over the Pampas behind Montevideo (Figure 2-4), Darwin saw many bones and teeth of extinct animals (Figure 2-7). He found

> some large portions of the armour of a gigantic armadillo-like animal, and part of the great head of a Mylodon. The bones of this head are so fresh, that they contain, according to the analysis of Mr. T. Reeks, seven percent of animal matter; and when placed in a spirit lamp, they burn with a small flame. The number of remains embedded in the grand estuary deposit which forms the Pampas and covers the granite rock of Banda Oriental, must be extraordinarily great. . . . We may conclude that the whole area of the Pampas is one wide sepulchre of these extinct gigantic quadrupeds. (Darwin 1860, 155–156)

These fossils, with their obvious relationships to living species, were an important point of departure for his ideas on the origin of species (Darwin 1892).

The Galapagos Islands

Even more impressive to Darwin was the fauna of the Galapagos Islands. The facts he collected from the Pampas and the Galapagos, especially the latter, were "the origin of all my views" (Darwin 1892, 150). In *The Voyage of the Beagle*, Darwin's (1860; see also 1839) first entry on animals (378–379) summarizes his feelings:

> The natural history of these islands is eminently curious, and well deserves attention. Most of the organic productions are aboriginal creations, found nowhere else; there is even a difference between the inhabitants of the different islands; yet all show a marked relationship with those of America, though separated from that continent by an open space of ocean, between 500 and 600 miles in width. The archipelago is a little world within itself, or rather a satellite attached to America, whence it has derived a few

FIGURE 2-7

An artist's reconstructions of some animals whose fossilized bones Darwin found on the Pampas and in other areas of South America. *Macrauchenia* appears on the left in the middle distance, with a glyptodont in the left foreground. On the right, a sloth, *Mylodon*, feeds on a plant, and *Toxodon* stands in the water. *(Illustration by Tad Theimer)*

stray colonists, and has received the general character of its indigenous productions. Considering the small size of these islands, we feel the more astonished at the number of their aboriginal beings, and at their confined range. Seeing every height crowned with its crater, and the boundaries of most of the lava-streams still distinct, we are led to believe that within a period, geologically recent, the unbroken ocean was here spread out. Hence, both in space and time, we seem to be brought somewhat near that great fact—that mystery of mysteries—the first appearance of new beings on this earth.

Darwin saw the large Galapagos tortoise. He saw two iguanalike lizards, unique to the islands, that may reach 3 feet long and weigh about 20 pounds. He noted that one lived on land and fed on cacti; the other was marine and fed on seaweed (it is the only lizard known to do this). In his diary of the voyage (390), Darwin remarked in his sober way, "It is very interesting thus to find a well-characterized genus, having its marine and terrestrial species, belonging to so confined a portion of the world."

Darwin found 26 species of land birds, 25 of them only on the islands, and 11 species of water birds; but only three were new species. In his diary he mentioned the "law of aquatic forms," which states that aquatic forms are "less peculiar at any given point of the earth's surface than the terrestrial forms of the same classes" (381), and noted its applicability to the birds, shells, and insects of the archipelago.

Of the 26 species of land birds, Darwin saw 13 species that formed "a most singular group of finches, related to each other in the structure of their beaks, short tails, form of body, and plumage" (380). All were peculiar to the archipelago (Figure 2-8). He remarked in his diary (381), "Seeing this gradation and diversity of structure in one small, intimately related group of birds, one might really fancy that from an original paucity of birds in this archipelago, one species had been taken and modified for different ends."

Malthus's Essay on Population

Darwin was convinced that transmutation of species did occur, that new species did arise through natural causes, that "evolution," although he did not use the term then, was a reality. He also knew that artificial selection by man could bring about this transmutation. How, then, could transmutation occur in nature? The question seemed unavoidable in 1835, and it gravely puzzled him. In 1838, he happened to read Malthus's 1798 essay on population. After observing population growth in England accelerating since about 1750, Malthus had written,

> I think I may fairly make two postulata. First, that food is necessary to the existence of man. Second, that the passion between the sexes is necessary and will remain nearly in its present state. . . . Assuming then, my postulates as granted, I say, that the power of population is infinitely greater than the power in the earth to produce subsistence for man. Population, when unchecked, increases in a geometrical ratio.

FIGURE 2-8

Illustrations of Galapagos finches used by Darwin in *The Voyage of the Beagle* (1860). "The largest beak in the genus Geospiza is shown in Fig. 1, and the smallest in Fig. 3; but instead of there being only one intermediate species, with a beak of the size shown in Fig. 2, there are no less than six species with insensibly graduated beaks. The beak of the subgroup Certhidea is shown in Fig. 4" (380–381).

Subsistence increases only in an arithmetical ratio. A slight acquaintance with numbers will show the immensity of the first power in comparison of the second. By that law of our nature which makes food necessary to the life of man, the effects of these two unequal powers must be kept equal. This implies a strong and constantly operating check on population from the difficulty of subsistence. This difficulty must fall some where and must necessarily be severely felt by a large portion of mankind.

Through the animal and vegetable kingdoms, nature has scattered the seeds of life abroad with the most profuse and liberal hand. She has been comparatively sparing in the room and the nourishment necessary to rear them. The germs of existence contained in this spot of earth, with ample food, and ample room to expand in, would fill millions of worlds in the course of a few thousand years. Necessity, that imperious all pervading law of nature, restrains them within the prescribed bounds. The race of plants, and the race of animals shrink under this great restrictive law. And the race of man cannot, by any efforts of reason, escape from it. Among plants and animals its effects are waste of seed, sickness, and premature death. Among mankind, misery and vice.

2.5 DARWIN'S THEORY OF DESCENT WITH MODIFICATION THROUGH VARIATION AND NATURAL SELECTION

Darwin was struck by the competition and waste in the natural world, as stressed by Malthus, and he later wrote in his autobiography (42–43):

And being well prepared to appreciate the struggle for existence which everywhere goes on from long-continued observation of the habits of animals and plants, it at once struck me that under these circumstances favourable variations would tend to be preserved and unfavourable ones to be destroyed. The result of this would be the formation of new species. Here, then, I had at last got a theory by which to work; but I was so anxious to avoid prejudice, that I determined not for some time to write even the briefest sketch of it.

Only in 1842 did he write his first, brief (35-page) draft.

Darwin calls his theory **"the theory of descent with modification through variation and natural selection"** in the final chapter of *The Origin of Species* (Darwin 1859, 459; Darwin 1872c, 163). Early in the book, he sets out to establish the reality of variation within populations both in domestic organisms (chapter 1) (Figure 2-9) and in nature (chapter 2).

No case is on record of a variable being ceasing to be variable under cultivation. Our oldest cultivated plants, such as wheat, still often yield new varieties; our oldest domesticated animals are still capable of rapid improvement or modification. (Darwin 1859, 8)

And of natural populations he writes:

The many slight differences which appear in the offspring from the same parents, or which it may be presumed have thus arisen, from being observed in the individuals of the same species inhabiting the same confined locality, may be called individual differences. . . . These individual differences are of the highest importance to us, for they are often inherited, as must be familiar to everyone; and they thus afford materials for natural selection to act on and accumulate, in the same manner as man accumulates in any given direction individual differences in his domesticated productions. (Darwin 1872b, 32)

Variation is a reality that provides the substrate for natural selection.

In the third chapter, devoted to the struggle for existence, Darwin writes:

Again, it may be asked, how is it that varieties, which I have called incipient species, become ultimately converted into good and distinct species . . . ? . . . All these results . . . follow inevitably from the struggle for life. Owing to this struggle for life, any variation, however slight and from whatever cause proceeding, if it be in any degree profitable to an individual of any species, in its infinitely complex relations to other organic beings and to external nature, will tend to the preservation of that individual, and will generally be inherited by the offspring. The offspring, also, will thus have a better chance of surviving, for, of the many individuals of any species which are periodically born, but a small number can survive. I have called this principle, by which each slight variation, if useful, is preserved, by the term Natural Selection, in order to mark its relation to man's power of selection. (Darwin 1859, 61)

But the expression often used by Mr. Herbert Spencer of the Survival of the Fittest is more accurate, and is sometimes equally convenient. We have seen that man by selection can certainly produce great results, and can adapt organic beings to his own uses, through the accumulation of slight but useful variations, given to him by the hand of Nature. But Natural Selection, as we shall hereafter see, is a power incessantly ready for action, and is as immeasurably superior to man's feeble efforts, as the works of Nature are to those of Art. (Darwin 1872b, 47)

In his chapter 4, "Natural Selection, or The Survival of the Fittest," Darwin expands on this theme:

Let it also be borne in mind how infinitely complex and close-fitting are the mutual relations of all organic beings to each other and to their physical conditions of life; and consequently what infinitely varied diversities of structure might be of use to each being under

FIGURE 2-9

Varieties of pigeons produced from the wild rock dove, *Columba livia* (top right), through artificial selection by humans. Top left, the jacobin; bottom left, the Indian fantail; bottom right, the English pouter, or Norwich cropper. *(Illustration by Stephen Price.)*

changing conditions of life. Can it, then, be thought improbable, seeing that variations useful to man have undoubtedly occurred, that other variations useful in some way to each being in the great and complex battle of life, should occur in the course of many successive generations. If such do occur, can we doubt (remembering that many more individuals are born than can possibly survive) that individuals having any advantage, however slight, over others, would have the best chance of surviving and of procreating their kind? On the other hand, we may feel sure that any variation in the least degree injurious would be rigidly destroyed. This preservation of fa-

vourable individual differences and variations, and the destruction of those which are injurious, I have called Natural Selection, or the Survival of the Fittest. Variations neither useful nor injurious would not be affected by natural selection. (Darwin 1872b, 61)

Darwin continues with chapters on the laws of variation; difficulties with the theory; instinct; hybridism; the geological record and its succession of organisms; geographical distribution; mutual affinities seen in morphology, embryology, and rudimentary organs; and finally recapitulation and conclusion. The theory is essentially intact by the end of the fourth chapter. Yet we may well ask: How do new species arise? Darwin explained how species change but not how new species develop. Is *The Origin of Species* a misnomer, or has Darwin's theory been insufficiently interpreted? The next chapter, which considers difficulties with the theory, will examine these questions.

Let us conclude this chapter with a simple statement that Darwin might have made around 1870 if asked to express in a few sentences the essence of his theory of descent with modification through variation and natural selection. Darwin never did write such a brief statement, and the following paragraph in italics reflects what the author of this book would like to think of as Darwin's précis, knowing full well that Darwin would never have agreed to such gross condensation.

Variation between individuals occurs in all species and is caused by the variability of the environment to which a population is exposed. Individual variation is inheritable since, as my provisional **hypothesis of pangenesis** *explains, each part of the body emits invisible gemmules that pass through the bloodstream to collect in the germ cells, carrying with them information on the exact nature of the body part from which they came. This slight variation that benefits an individual in its struggle for existence enables it to survive longer and reproduce more successfully than others. Conversely, individuals with injurious variations perish. This principle I have called* **natural selection.** *Due to the inheritance of beneficial variations, natural selection also acts on further variation in successive generations, so that initially small beneficial traits may in time become major through continued* **descent with modification.** *Since the individuals whose favorable variations diverge most markedly from the parent stock receive the most benefit, those most divergent variations are preserved and accumulated by natural selection. After much variation has accumulated, fairly* **well-marked varieties** *and then species may be noted, since* **species** *are only strongly marked and well-defined varieties. In this way, transmutation of species occurs and new species originate. The process resulting in new species, repeated all over the earth and over great spans of time, has produced the present diversity of life, the clear relationships between species alive and extinct, and the descent of all life from some primordial species.*

In the next chapter we will see that the modern theory of evolution, as it was developed by the 1940s and 1950s, differs in four essential points from the preceding paragraph:

1. The origin of variation in a species
2. The mechanisms that maintain variation
3. The importance of geographic isolation in the speciation process
4. The nature of the species

Even so, by developing the concept of natural selection, Darwin made a great contribution that stands intact. The next chapter describes the refinement of that concept.

SUMMARY

Charles Darwin was born in 1809 into a world of rapid social change stimulated by the **industrial revolution.** A middle class was forming as a result of the growing need for businessmen, merchants, master craftsmen, shopkeepers, teachers, and scientists. Demand for more practical education in schools and universities resulted in an informed public interested in science and technology. Darwin's scholasticism at Shrewsbury School, Edinburgh University, and Cambridge University was unspectacular, but his love of reading and natural history and his informal associations with professors at Cambridge provided a sound basis for his career.

The voyage of the *Beagle* from 1831 to 1836 was marked by key experiences of lasting importance to Darwin's academic stature and influence. He read Lyell's *Principles of Geology* and espoused its uniformitarian view of the earth. The animals whose fossils Darwin found in South America, so clearly related to living species, suggested change in form through time. And in the Galapagos Islands, which are of relatively young volcanic origin, Darwin saw so much evidence of rapid divergence from a common stock that he was compelled to consider the reality of **the origin of new species.**

The mechanisms promoting change in species, and the origin of new species, remained elusive for Darwin until he read Malthus's essay on population growth in 1838. That essay noted the overproduction of individuals in any population relative to the resources that could sustain them, and Darwin recognized that slight differences between individuals affected their probability of survival or death. Nature acted as a selective agent on such variation, just as humans performed artificial selection on domesticated plants and animals to improve the stock. **Natural selection** was the mechanism that caused the transmutation of species, or the formation of new species.

Darwin married Emma Wedgewood in 1839, and they moved to Down House in Kent in 1842. He spent much of his time on his classic work on barnacles, but the **transmutation of species** became a priority in 1856 and urgent in 1858, when Wallace sent Darwin a well-researched exposition on the divergence of species. *The Origin of Species* was published in November 1859. It was widely read and in 1860 was hotly debated at the British Association meetings in Oxford. Thomas Huxley's effective defense of *The Origin* was noticed around the world. Darwin continued to publish extensively and revised *The Origin* almost until his death in 1882.

As knowledge about geology and fossils increased, scientists modified the concepts of the origin of life and change in form from a biblical perception of creation to **catastrophism** and from **progressionism** to **uniformitarianism.** Catastrophism noted the dramatic change in rock strata and assumed that cataclysmic causes were responsible. But the order of

closely related fossils up a stratigraphic column and the general design in common with living species suggested a progression of creations conforming to some unified plan. Ultimately, Lyell made a convincing case for a world dominated by processes largely uniform through great expanses of time, which were therefore amenable to study with the scientific method.

Scientific advances and evolutionary ideas in the 18th and 19th centuries set the stage for a comprehensive theory. Although Linnaeus's view of the **fixity of species** posed a dilemma, the facts of change in the fossil record, change due to artificial selection, and the apparent origin of new species in the Galapagos Islands were compelling enough to spur Darwin to formulate his **theory of descent with modification through variation and natural selection.**

QUESTIONS FOR DISCUSSION

1. Do you think that the theory of evolution would have been developed without the industrial and agrarian revolutions?

2. In which ways did public opinion further the development of the theory of evolution?

3. If Darwin had died when young—say, in a shipwrecked H.M.S. *Beagle* in the Straits of Magellan—would the theory of evolution have emerged in the 19th century, in your opinion? Would Wallace or others have been able to establish the theory in a similar way?

4. To what extent do you think the concept of uniformitarianism is central to all of the sciences?

5. Some of Darwin's theory proved incorrect; subsequent findings have falsified some of it. What does this example tell us about the nature of progress in science and the role of the empirical observations on which theory is based?

6. How would you compare and contrast the relative in-

fluences of Cuvier, Lamarck, Malthus, and Lyell on Darwin's development of a theory of evolution?

7. Do you think the phrase "struggle for existence" is misleading or helpful when used in relation to the concept of natural selection?

8. Darwin wrote, "Let theory be your guide" and "I worked on true Baconian principles," meaning that he worked from empirical observations and experiments, using inductive logic. In your opinion, how did Darwin employ these two approaches, and in what sequence did he use them?

9. The term "natural selection" is fixed in our vocabulary, but is there a better name for the process—such as a term without the implication that some force in nature makes choices and favors some lineages relative to others?

10. Can you conceive of another area of biology, or another science, that could have resulted in the separation of science and religion as the theory of evolution did?

REFERENCES

Anonymous. 1969. Historical and Descriptive Catalogue of the Darwin Memorial. Livingstone, London.

Appleman, P. (ed.) 1970. Darwin, Norton, New York.

Buffon, G. L. L. (Compte de). 1749–1804. Histoire naturelle. 44 vols. L'Imprimerie Royale, Paris. Translated from the French by W. Smellie, 1781–1812. Natural History. Strahan, London.

Burkhardt, R. W. 1984. The zoological philosophy of J. B. Lamarck, pp. xv–xxxix. *In* J. B. Lamarck, Zoological Philosophy. Univ. of Chicago Press.

Chambers, R. (published anonymously). 1844. The Vestiges of the Natural History of Creation. Churchill, London. Republished 1845 by Wiley and Putnam, New York.

Corsi, P. 1978. The importance of French transformist ideas for the second volume of Lyell's principles of geology. Brit. J. Hist. Sci. 11:221–244.

Cuvier, G. 1837. Edinburgh Rev. 65(131):23–24. [Cited in anonymous review of W. Buckland, 1836, Geology and mineralogy considered with reference to natural theology, Edinburgh Rev. 65(131):1–39.]

Cuvier, G., and A. Brongniart. 1822. Description Geologique des Environs de Paris. Dufour and d'Ocagne, Paris.

Darwin, C. 1839. Journal of Researches into the Geology and Natural History of the Various Countries Visited by H.M.S. Beagle from 1832–1836, Under the Command of Captain Fitzroy. R. N. Colburn, London. [See also L. Engel (ed.), 1962, The Voyage of the Beagle, 1860 revision, Natural History Library. Doubleday, New York.]

———. 1851a. A Monograph of the Fossil Lepadidae, or Pedunculated Cirripedes of Great Britain. Palaeontographical Society, London.

———. 1851b. The Lepadidae, or pedunculated cirripedes. In A Monograph of the Sub-class Cirripedia, with Figures of All the Species. Ray Society, London.

———. 1854a. A Monograph of the Fossil Balanidae and Verrucidae of Great Britain. Palaeontographical Society, London.

———. 1854b. The Balanidae (or sessil cirripedes); the Verrucidae, &c. In A Monograph of the Sub-class Cirripedia with Figures of All the Species. Ray Society, London.

———. 1858. Part I: Extract from an unpublished work on species, by C. Darwin, Esq., consisting of a portion of a chapter entitled, "On the variation of organic beings in a state of nature; on the natural means of selection; on the comparison of domestic races and true species." Part II: Abstract of a letter from C. Darwin Esq., to Prof. Asa Gray, Boston, U.S. Dated Down, September 5th 1857. J. Proc. Linnaean Soc. 3:45–62. [Reprinted in M. Bates and P. S. Humphrey. 1956. The Darwin Reader. Scribner's, New York.]

———. 1859. On the Origin of Species by Means of Natural Selection, or The Preservation of Favoured Races in the Struggle for Life. Murray, London.

———. 1860. The Voyage of the Beagle. Republished 1962 by Doubleday, New York.

———. 1862. On the Various Contrivances by Which British and Foreign Orchids are Fertilized by Insects, and on the Good Effects of Intercrossing. Murray, London.

———. 1868. The Variation of Animals and Plants Under Domestication. 2 vols. Murray, London.

———. 1871. The Descent of Man, and Selection in Relation to Sex. 2 vols. Murray, London.

———. 1872a. The Expression of the Emotions in Man and Animals. Murray, London.

———. 1872b. On the Origin of Species by Means of Natural Selection, or The Preservation of Favoured Races in the Struggle for Life. Vol. 1. 6th ed. Murray, London.

———. 1872c. On the Origin of Species by Means of Natural Selection, or The Preservation of Favoured Races in the Struggle for Life. Vol. 2. 6th ed. Murray, London.

———. 1875a. The Movements and Habit of Climbing Plants. Murray, London.

———. 1875b. Insectivorous Plants. Murray, London.

———. 1876. The Effects of Cross and Self Fertilization in the Vegetable Kingdom. Murray, London.

———. 1877. The Different Forms of Flowers on Plants of the Same Species. Murray, London.

———. 1880. The Power of Movements in Plants. Murray, London.

———. 1881. The Formation of Vegetable Mould, Through the Action of Worms, with Observations on Their Habits. Murray, London.

Darwin, E. 1794. Zoonomia. Johnson, London.

Darwin, F. (ed.) 1892. Charles Darwin, His Life Told in an Autobiographical Chapter and in a Selected Series of His Published Letters. Appleton, New York. Republished 1958 as The Autobiography of Charles Darwin and Selected Letters. Dover, New York.

Desmond, A., and J. Moore. 1991. Darwin. Penguin, London.

Dobson, J. 1971. Charles Darwin and Down House. Livingstone, London.

Eiseley, L. 1958. Darwin's Century. Doubleday, Garden City, N.Y.

Hull, D. L. 1984. Lamarck Among the Anglos, pp. xl–lxvi. In J. B. Lamarck, Zoological Philosophy. Univ. of Chicago Press.

Kelly, A. 1975. The Story of Wedgwood. Viking, New York.

Lamarck, J. B. 1802. Recherches sur l'Organisation des Corps Vivans, et Particulièrement sur son Origine. . . . Maillard, Paris.

———. 1809. Philosophie Zoologique. Dentu, Paris. Translated from the French by H. Elliot. 1963. Zoological Philosophy. Hafner, New York. [See also Historical Natur. Classica 10.]

———. [1809] 1984. Zoological Philosophy. Univ. of Chicago Press. Translated from the French by Hugh Elliot.

Lyell, C. 1830–1833. Principles of Geology. Vols. 1–3. Murray, London. [Reprinted 1970 in Hist. Natur. Classica 83, parts 1–3.]

———. 1842. Eight Lectures on Geology. Greeley and McElrath, New York.

Maillet, B. de. [1748] 1750. Telliamed, or Discourses Between an Indian Philosopher and a French Missionary on the Diminution of the Sea, the Formation of the Earth, the Origin of Men and Animals etc. Osborne, London. Translated from the French anonymously.

Malthus, T. R. 1798. An Essay on the Principle of Population as It Affects the Future Improvement of Society. Johnson, London.

McKinney, H. L. 1972. Wallace and Natural Selection. Yale Univ. Press, New Haven, Conn.

Murray, J. W. 1985. Atlas of Invertebrate Macrofossils. Wiley, New York.

Raverat, G. 1952. Period Piece. Faber and Faber, London.

Rudwick, M. J. S. 1970. Introduction [to Lyell's Principles of Geology]. Hist. Natur. Classica 83(1):ix–xxv.

Wallace, A. R. 1858. Part III: On the tendency of varieties to depart indefinitely from the original type. J. Proc. Linnaean Soc. 3:45–62. [Reprinted in M. Bates and P. S. Humphrey. 1956. The Darwin Reader. Scribner's, New York.]

Wedgwood, J. C. 1909. A History of the Wedgwood Family. St. Catherine Press, London.

White, A. D. 1896. A History of the Warfare of Science with Theology in Christendom. New York. [Quoted in P. Appleman (ed.), 1970, Darwin, Norton, New York.]

White, G. 1789. The Natural History and Antiquities of Selborne. B. White, London.

White, R. J. 1963. Life in Regency England. Putnam's, New York.

Wilson, L. G. 1972. The years to 1841: The revolution in geology. In Charles Lyell: The Years to 1841: The Revolution in Geology. Yale Univ. Press, New Haven, Conn.

3

THE REVISED THEORY OF EVOLUTION

A fruit bat skeleton. "In certain bats in which the wing-membrane extends from the top of the shoulder to the tail and includes the hind-legs, we perhaps see traces of an apparatus originally fitted for gliding through the air rather than for flight." (Darwin 1872) Darwin mentioned the bat in relation to difficulties with his theory.

After discussing natural selection, this chapter addresses difficulties with Darwin's theory, his perspectives, and developments in the theory of evolution from his death until 1942, when Julian Huxley called contemporary evolutionary theory the "modern synthesis" (Huxley 1942). (The synthesis remained largely intact into the early 1960s.) Subsequent chapters will discuss more recent developments, including the question of whether such a synthesis was really achieved (see especially Chapters 4, 5, 16–18).

3.1 NATURAL SELECTION

Darwin on Apparent Difficulties

The mainstay of Darwin's legacy was his concept of natural selection, which provided a mechanistic explanation for evolution. Nevertheless, he recognized many problems with his theory and used four chapters, about a quarter of the book, to treat the difficulties. In chapter 6, the opening chapter on difficulties, he had this to say (Darwin 1872b, 129):

> These difficulties and objections may be classed under the following heads:—First, why, if species have descended from other species by fine gradations, do we not everywhere see innumerable transitional forms? Why is not all nature in confusion, instead of the species being, as we see them, well defined [discussed in chapter 6]?
>
> Secondly, is it possible that an animal having, for instance, the structure and habits of a bat, could have been formed by the modification of some other animal with widely different habits and structure? Can we believe that natural selection could produce, on the one hand, an organ of trifling importance, such as the tail of a giraffe, which serves as a fly-flapper, and on the other hand, an organ so wonderful as the eye [chapter 6]?
>
> Thirdly, can instincts be acquired and modified through natural selection? What shall we say to the instinct which leads the bee to make cells, and which has practically anticipated the discoveries of profound mathematicians [treated in chapter 8 on instinct]?
>
> Fourthly, how can we account for species, when crossed, being sterile and producing sterile offspring, whereas, when varieties are crossed, their fertility is unimpaired [treated in chapter 9 on hybridism]?

These were grave questions, indeed. Darwin addressed each candidly, dispelling many of the doubts an imaginative reader might develop. Let us follow his argument on a particularly difficult case, to illustrate the dialectic he employed: frank acceptance of problems along with a convincing overall treatment of a topic that would remain a puzzle for another century.

In chapter 8 on instinct, Darwin discusses the problem of sterile castes of insects such as in ants and bees. The sterile workers never reproduce but work to help the colony and the queen produce several to many of the queen's progeny. How could natural selection favor sterility, when survival of the fittest depended upon successful reproduction (Figure 3-1)?

> I . . . will confine myself to one special difficulty, which at first appeared to me insuperable, and actually fatal to the whole theory. I allude to the neuters or sterile females in insect-communities; for these neuters often differ widely in instinct and in structure from both the males and fertile females, and yet, from being sterile, they cannot propagate their kind.
>
> . . . I will here take only a single case, that of working or sterile ants. How the workers have been rendered sterile is a difficulty; but not much greater than that of any other striking modification of structure; for it can be shown that some insects and other articulate animals in a state of nature occasionally become sterile; and if such insects had been social, and it had been profitable to the community that a number should have been annually born capable of work, but incapable of procreation, I can see no special difficulty in this having been effected through natural selection. (Darwin 1872b, 218–219)

Darwin goes on to discuss the distinction of workers from reproductive males and females. Worker ants lack wings, and their instincts result in a set of behaviors totally different from those of the reproductives. Caste systems may even include both soldiers and workers (Figure 3-2) with divergent morphology and instinctive behaviors.

> But we have not as yet touched on the acme of the difficulty: namely, the fact that the neuters of several ants differ, not only from the fertile females and males, but from each other, sometimes to an almost incredible degree, and are thus divided into two or even three castes. (Darwin 1872b, 220)

It remains remarkable that, in his first exposure of the theory of natural selection, Darwin provided a convincing explanation for sterile castes of insects, as correct today as it was in 1859.

> This difficulty, though appearing insuperable, is lessened, or, as I believe, disappears, when it is remembered that selection may be applied to the family, as well as to the individual, and may thus gain the desired end. . . . With social insects, selection has been applied to the family, and not to the individual, for the sake of gaining a serviceable end. Hence we may conclude that slight modifications of structure or of instinct, correlated with the sterile condition of certain members of the community, have proved advantageous: consequently the fertile males and females have flourished, and transmitted to their fertile offspring a tendency to produce sterile members with the same modifications. This process must have been

FIGURE 3-1

Variation in morphs of the ant species *Pheidole instabilis*, as originally illustrated in Wheeler (1910), exemplifying the problem posed for natural selection by social insects with sterile castes. (*a–f*) A continuously varying range of worker types, which Wheeler recognized as (*a*) a soldier caste involved with defending the colony; (*b–e*)intermediate workers; and (*f*) the small typical worker. Note that castes *a–f* never develop wings, and their instincts for protecting the nest, foraging for food, and caring for larvae sharply distinguish them from the reproductives. (*g*) The originally winged reproductive female, or queen, has shed her wings and is ready to construct a nest and lay eggs. (*h*) The male reproductive winged morph.

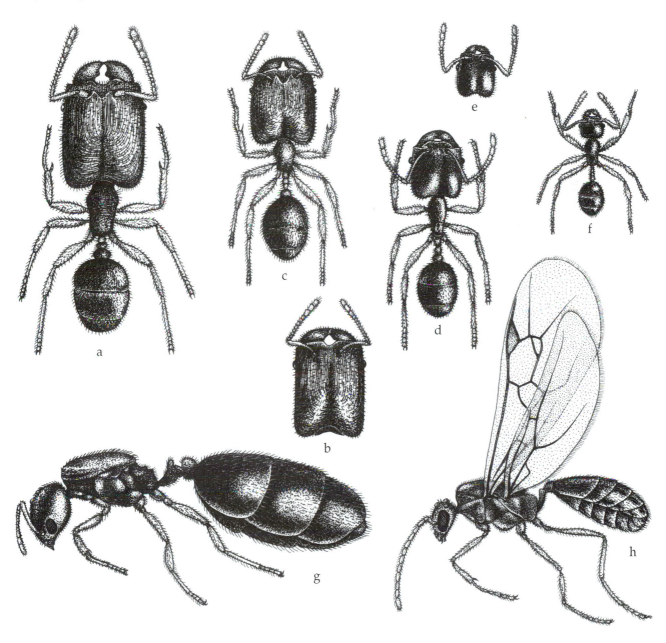

repeated many times, until that prodigious amount of difference between the fertile and sterile females of the same species has been produced, which we see in many social insects. (Darwin 1872b, 220; see also Darwin 1859, 236–242)

Selection that results in characteristics that favor genetically related individuals, as in a family, is now called **kin selection.** The concept, as Darwin explained, accounts for apparently nonadaptive **altruistic behavior,** such as giving up an opportunity to breed while helping a relative to reproduce. An individual's genes can be propagated through the agency of a close relative with identical genes, derived from a common ancestor. Therefore, **individual fitness,**

FIGURE 3-2

An original illustration from Wheeler 1910, showing sterile and reproductive castes in *Cryptocerus varians*. The soldier caste is illustrated from above (*a*) and in profile (*b*); the head is also viewed from above (*c*). Note the shield-shaped head, used to block the access of unwanted visitors at entrances to the nest. The worker (*d*) is smaller than the soldier, with a much smaller head. Note the large, flattened head of the female, used to defend her initial nest before soldiers are available for defense. (*e*) The reproductive female. (*f*) The reproductive male. Notice the general similarity between the males of *Pheidole instabilis* (Figure 3-1) and *Cryptocerus varians*, while the females, both sexuals and sterile castes, are very divergent in morphology—a feature that impressed Darwin as he tried to account for such variation under natural selection.

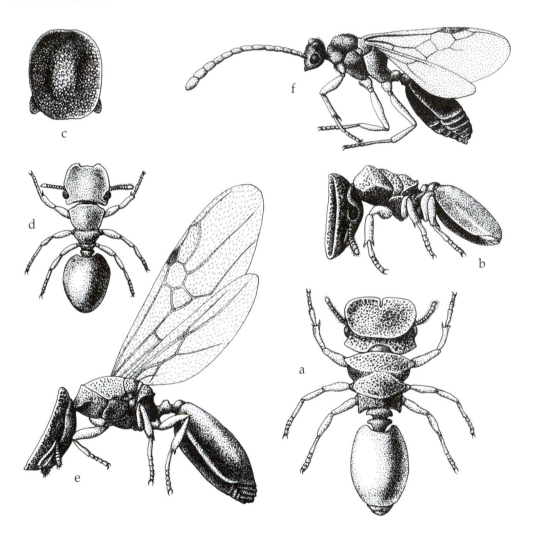

based on an individual's own reproductive success, may be increased by **inclusive fitness,** based on the individual's own fitness plus the reproductive success of relatives that carry the same genes (Hamilton 1964).

The Explanatory Power of Natural Selection

The explanatory power of natural selection as a mechanism producing all kinds of adaptation was so convincing that it seemed as if all traits and characteristics were adaptive and all could be accounted for by natural selection. Natural selection and adaptation constituted the dominant perspective in the field of evolution by the middle of the 20th century. Special cases of natural selection seemed to account for particularly difficult traits. For example, kin selection accounted for sterile castes in insects and altruistic behavior in other animals. **Sexual selection** (Darwin's term) accounted for the frequently observed morphological and behavioral differences between males and females of the same species. Male traits associated with fighting other males were attributed to **intrasexual selection.** For instance, the large mandibles, head, and body of male stag beetles have been selected for advantage in male–male jousting to acquire a female mate (Figure 3-3). In a different competitive approach, male plumage may become extremely ornate (as in the peacock) through selection by females, which mate preferentially with more vividly colored and better decorated males. Sexual selection then operates between the sexes and is called **intersexual selection** or **epigamic selection** (Huxley 1938).

The beguiling elegance and simplicity of natural selection as a force in evolution may have resulted in overenthusiasm for its use. Some cautionary notes are necessary.

Cautionary Perspectives

Scientists have offered thousands of explanations of the role of natural selection and the adaptive nature of traits, but relatively few of these explanations have been tested empirically. "Natural selection is not easy to detect" (Endler 1986, 97). Even if a study is designed explicitly to explore natural selection, it is commonly flawed as follows (Endler 1986):

1. The study does not estimate the full fitness of individuals over their complete lifetimes.
2. The study considers only one or a few traits.
3. The traits under study have unknown or inadequately known functions.

FIGURE 3-3

Males of the European stag beetle, *Lucanus cervus*, joust with large mandibles similar to antlers. Each male tries to dislodge its competitor to gain access to the female, which is smaller and lacks the mandibular trait under intrasexual selection.

Studies of natural populations that lacked the preceding flaws numbered between 120 and 130. More than 30 studies represented only two taxa, the angiosperms and the insects, although the literature on these groups is replete with apparent explanations of adaptive traits and the role of natural selection. Some groups—gymnosperms, ferns, reptiles, arachnids, and coelenterates—were represented only by studies on one or two of their species. In fact, the reports demonstrating the existence of natural selection are vastly outnumbered by the reports purporting to explain the adaptive natures of traits.

Overindulgence in the use of adaptive explanations without critical testing became so prevalent that Gould and Lewontin (1979) labeled the practice the **"adaptationist programme,"** or the **"Panglossian paradigm."** The Panglossian view—that all is for the best in this best of possible worlds—originated with the fictional character Dr. Pangloss, the overoptimistic tutor in the satirical novel *Candide,* written by Voltaire in 1759. To exemplify Dr. Pangloss's ridiculously optimistic view of life, Lewontin and Gould cited the character's remarks on his own syphilytic disease.

"It is indispensable in this best of worlds. For if Columbus, when visiting the West Indies, had not caught this disease, which poisons the source of generation, which frequently even hinders generation, and is clearly opposed to the great end of Nature, we should have neither chocolate nor cochineal."

In the Panglossian view, the disease was adaptive because with it from the New World came the benefits of chocolate and the red dye from cactus-feeding cochineal insects!

According to Gould and Lewontin, the adaptationist program proceeds from the recognition of traits that "are explained as structures optimally designed by natural selection for their functions" (585) to the proviso that any compromise of optimality is the result of trade-offs with other adaptations. And yet an organism's form and function have foundations in its phylogenetic past, and evolutionary pathways are bounded by constraints imposed by the original general plan of the organism. We must view the development of a lineage through evolutionary time as a functional system affected by many factors—of which one is natural selection—but under strong **phylogenetic constraints.** That is, each evolutionary shift occurs only in the context of an existing complex design that constrains the kind and extent of evolutionary modification. Bats evolved from a basic mammalian design that is clearly evident in their skeletons. Much of the life cycle, habits, behavior, and structure of bats reflect both a breakthrough in their way of life and the clear constraints of a mammalian heritage. "Natural selection never designs new machinery to cope with new problems" (Williams 1992).

The adaptationist program failed, in general, to view evolution in the context of a given organismal design and its phylogenetic history. It missed the opportunity to examine the fascinating interplay of phylogenetic constraints and the many factors, in addition to natural selection, that cause evolutionary change. (These factors will be the subjects of Chapters 5, 8, 12, 15, 17, and 18.) To his great credit, Darwin recognized, from the first edition of *The Origin* to the last, that natural selection was not the only mode of evolutionary change. Little could he realize, however, how great a proportion of evolutionary change occurs under forces other than natural selection.

Although Darwin's theory on the transmutation of species revolutionized natural history and to a large degree remains intact—incorporated into the modern concepts relating to natural selection and the evolution of new species—the theory contained several flaws (listed at the end of the last chapter), which can be understood only in the light of discoveries made after publication of *The Origin*. Let us treat each difficulty in turn.

3.2 THE ORIGIN OF VARIATION IN A SPECIES

The origin of variation, according to Darwin, was driven by environmental variability. Darwin opens *The Origin of Species* with his views on the causes of variation (Darwin 1872b, 5):

> When we compare the individuals of the same variety or sub-variety of our older cultivated plants and animals, one of the first points which strikes us is, that they generally differ more from each other than do the individuals of any one species or variety in a state of nature. And if we reflect on the vast diversity of the plants and animals which have been cultivated, and which have varied during all ages under the most different climates and treatment, we are driven to conclude that this great variability is due to our domestic productions having been raised under conditions of life not so uniform as, and somewhat different from, those to which the parent species had been exposed under nature.

A little later (5), he states:

> It seems clear that organic beings must be exposed during several generations to new conditions to cause any great amount of variation; and that, when the organization has once begun to vary, it generally continues varying for many generations. No case is on record of a variable organism ceasing to vary under cultivation.

Darwin had made the empirical observation that variation continues to arise in a population. The observation was central to his theory because natural selection can work only if variation occurs. In fact, this requirement was so important and fundamental that Darwin spent the first two chapters of his book establishing its credibility, and he cleverly used examples from domesticated species—those most familiar to readers—in the first chapter. Since the fact of variation was correct, even though the mechanism Darwin invoked to explain it was not, Darwin's theory was not substantially weakened, because nobody in his day could prove him wrong.

It was not until the turn of the century, with the findings of Hugo de Vries and T. H. Morgan, that the true origin of variation could be understood. De Vries published *The Mutation Theory* in German in 1901–1903 and in English in 1909–1910. De Vries noticed that new forms of an evening primrose, *Oenothera lamarckiana*, had occurred in a meadow. He realized that he might study evolution experimentally rather than use observation and inference as Darwin had done. In de Vries's rearings of *Oenothera*, new forms appeared quite suddenly among the many normal forms. De Vries coined the term "mutation" for this process whereby new species and new varieties seemed to arise suddenly in contrast with the gradual change envisaged by Darwin. Much later researchers found that *Oenothera* was a rather peculiar plant, genetically speaking, and that its so-called mutations produced no new species.

As we use the term today, a **mutation** may be defined as any novel genetic change in the gene complement or genotype, relative to the parental genotypes, beyond that achieved by genetic recombination during meiosis. Starting in 1909, T. H. Morgan and his associates observed many such mutations in *Drosophila* species, the fruit flies. Some were major changes, some minor, but none led to new species. Nevertheless, it became very evident that (1) mutations were the original sources of variation on which natural selection could act, (2) they were spontaneous genetic changes largely uninfluenced by the environment, and (3) they were more or less random in the genome and population. Morgan also saw that the small mutations were the most important in generating the variability that Darwin knew existed but could not explain correctly (e.g., Morgan 1910a, 1910b, 1912).

By 1927, H. J. Muller was producing mutations artificially by exposing *Drosophila* flies to X-rays. S. E. Luria and M. Delbrück made the colon bacterium *Escherichia coli* available for study in the laboratory by 1943, and in 1945 G. W. Beadle and E. L. Tatum detected mutants in the bread mold *Neurospora crassa*. Once scientists identified mutations, they could estimate their frequency in a population. Thus, by the early 20th century, research had revealed the adaptive nature of rare mutations. Mutation was indeed the origin of variation and provided the raw material for the evolutionary process.

3.3 THE MECHANISMS THAT MAINTAIN VARIATION

Darwin also misunderstood the maintenance of variation. He spelled out his concept of heredity in his book *The Variation of Animals and Plants Under Domestication*, published in 1868 (later than *The Origin*). This concept of heredity dated back to the Greeks, but Darwin formalized it somewhat. His provisional hypothesis, which he called **pangenesis,** was essentially preformationist and Lamarckian at the same time. In pangenesis, the "gemmules" passing from all parts of the body entered the germ cells and produced copies of the parent in the offspring. As parts became modified, additional gemmules passed down and modified the earlier gemmules, producing a copy of the parent at the instant of germ cell production. A parent's contribution to progeny was more or less pre-formed in the gemmules and included characteristics it had acquired during its lifetime. Darwin was not completely satisfied with the blending process of pangenesis. Each zygote, or offspring, was a blend of the two parents, and thus variability was lost rapidly. If each set of parents in a population produced only one offspring, variability would be reduced by half in each generation. This rapid loss was a serious problem for a theory that depended on the perpetuation of variability in all species.

Many of Darwin's contemporaries also subscribed to a belief in **blending inheritance,** because much empirical evidence seemed to support the hypothesis. Plant and animal breeders had successfully combined favorable characteristics from two breeding stocks, and children certainly possessed some characteristics of each parent. Just as a kitchen blender swirls together the several ingredients for a cake to make a homogeneous mix, it seemed that in each successive generation the process of inheritance resulted in more homogeneity and less variation on which natural selection could act.

Darwin, unfortunately, was unaware of Gregor Mendel's work between 1856 and 1865, as were the majority of scientists at that time. In 1865, Mendel reported his experiments on breeding peas to the Natural Science Society in Brünn, and he published them the following year in the society's journal (*Verhandlungen der Naturforschenden Verein in Brünn*), which was not widely circulated. What are now known as Mendel's two principles, or laws, would have saved Darwin much concern.

Mendel's Principles of Heredity

The first of Mendel's principles was the **principle of segregation of unit characters.** Mendel found, for example, that a hybrid pea from purple- and white-flowered parents was not a blend of the parents. Although the hybrid produced purple blossoms, he inferred that it then passed its purple and white properties on to different gametes. W. Johannsen in 1909 called the factors that controlled characteristics such as color **genes,** an abbreviation of the term "pangene," used by De Vries, which obviously has its roots in the concept of pangenesis (Mayr 1982). Thus, each plant in Mendel's experiment possessed two units for each characteristic, and when it formed gametes, the two units segregated, each gamete receiving only one unit for each characteristic. Such gene pairs in the parent may confer the same flower color (e.g., white) to each gamete, or they may confer different colors (e.g., white and purple). When two or more variants of a trait such as flower color occur at the same position on the chromosome in a population and segregate in a Mendelian way, they are called **alleles.** Each allele is an alternative type at the same gene location.

Mendel's second principle was the **principle of independent assortment.** He found that when he observed several pairs of genes in a plant, the progeny showed all possible combinations of genes, and he could accurately predict the genotypes and phenotypes by using the laws of probability and assuming that genes recombine at random.

Mendel's theory is a **particulate theory of inheritance.** It accounts for no loss of variability, no blending of characteristics, but only the continuing maintenance of all characteristics present in a breeding population, assuming no mortality before successful reproduction. Moreover, Mendel's laws show that the recombination of genes into new clusters generates novel genotypes and phenotypes—i.e., variability—in perpetuity. Given the numerous genes inherited by each organism and the random nature of independent assortment, every individual in a sexually reproducing population (if generated by normal cell division, gamete fusion, and zygote cleavage) is likely to be unique. Avoiding the complications of which Mendel must surely have been aware—for example, linkage of two characteristics on the same chromosome—we can see that mutation provides the initial variability, but recombination exposes endless permutations of genes to the rigors of natural selection.

Unfortunately for Darwin and the other evolutionists of the 19th century, Mendel's results

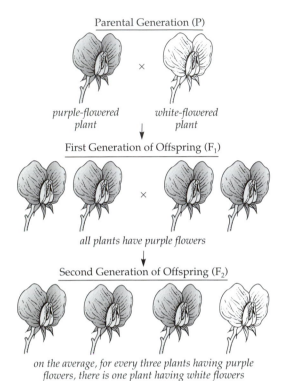

Parental Generation (P)

×

purple-flowered plant   white-flowered plant

First Generation of Offspring (F₁)

×

all plants have purple flowers

Second Generation of Offspring (F₂)

on the average, for every three plants having purple flowers, there is one plant having white flowers

Letters signify there are two alleles for the flower-color trait in each parent. Capital letter means dominance; lowercase letter means recessiveness.

During gamete formation, the alleles are separated from each other and end up in separate gametes (eggs in the case of one parent, sperm in the case of the other).

When the parent plants are crossed, their gametes fuse to form the "first filial" (F₁) generation.

The flower-color allele from one parent combines with that from the other parent in the fertilized eggs (zygotes). The Punnett square diagram shows probable combinations. In all possible combinations, the purple-flower trait will dominate this first hybrid generation.

When the first-generation plants mature and are crossed with each other, the dominant allele is paired with the recessive allele in each first-generation plant. When these pairs separate during gamete formation, the recessive trait can show up once more, as seen in the second-generation (F₂) plants. The second Punnett square shows a ratio of 1:2:1.

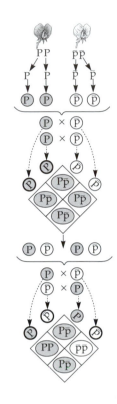

remained undiscovered by the outside world. Only in 1900 did his work become widely known, after Hugo de Vries in Holland, Carl E. Correns in Germany, and Erich Tschermak von Seysenegg in Austria confirmed the Mendelian principles.

The Emergence of Cytogenetics

Nineteenth-century cytological evidence complemented the genetic evidence. In 1848, W. Hofmeister saw that nuclei divide into small units during cell division in spiderworts (*Tradescantia* sp.), and W. Flemming in 1882 and W. Waldeyer in 1888 called these units **chromosomes.** Researchers made full descriptions of cell division, mitosis, and the reduction division of meiosis. They discovered that male and female gametes contribute equally to the hereditary material and that each carries a similar set of chromosomes. Many biologists—including W. Roux (1883), A. Weismann (1887, 1892a, 1892b), and E. B. Wilson in 1896—thus inferred that the chromosomes must be the carriers of the hereditary materials. Cytology was linked to genetics to form cytogenetics.

The cytologists discovered the physical processes by which a parental cell passes genetic material to the cells derived from it, both in normal cell division and in gamete formation. In the process called **mitosis,** the hereditary material in a diploid eukaryotic cell divides into two equal parts. Once two gametes fuse to form a zygote, the mitotic cell divisions ensure that each new cell has a replicated set of chromosomes and therefore a replicated set of genes and DNA.

One of Mendel's experiments demonstrated the principle of segregation of unit characteristics at one gene locus by crossing purple-flowered peas with white-flowered peas. On the left is the sequence of crosses as Mendel performed them, with the colors observed in the correct frequencies. On the right are the symbolized genetic structures for the parents, their gametes, the F_1 generation, and so on. Note that each parent has identical alleles for the flower-color trait—represented by either PP for a dominant purple-flowered, pure-breeding parent or $\bar{p}\bar{p}$ for a recessive white-flowered parent. Because the alleles are identical for this trait, each parent is a **homozygon**, and since genes are involved, the genetic makeup at the flower-color locus (position on the chromosome) is called the **genotype**.

In the first generation of offspring, the F_1 generation, all flowers turned out to be purple. In the second generation, or F_2 generation, 25 percent of plants had white flowers again. Thus, the white allele was not lost in the F_1 generation but persisted, even though it was unexpressed in the visual appearance, or **phenotype**, of the plant. Note that the genotypes of the F_1 generation are all the same, with the purple character (P) expressed in the phenotype and the recessive white color ($\bar{p}$) unexpressed.

The geneticist Reginald Punnett devised a convenient method for calculating the genotypes in the F_1 generation on the basis of known parental genotypes at a locus. The genes present in the gametes of each parent are used to form a matrix, and all possible combinations of gametes from each parent are recorded in the **Punnett square**. Note that with two homozygous parents (PP or $\bar{p}\bar{p}$), only one genotype ($P\bar{p}$) can be produced in the F_1 generation. Because different alleles occur at the same locus, this $P\bar{p}$ genotype is described as **heterozygous**. The heterozygotes pictured here all produce the same two gamete genotypes, P and $\bar{p}$, and the Punnett square reveals all possible combinations. There is one homozygous purple (PP), there are two heterozygous plants (genotype $P\bar{p}$) with a purple-flowered phenotype, and there is one white-flowered phenotype, which is homozygous ($\bar{p}\bar{p}$). Thus, when many F_2 plants are reared, the phenotypes appear in the ratio of 3:1 purple to white, or 75 percent purple and 25 percent white. The genotypes appear in the ratio 1:2:1, or 25 percent homozygous purple, 50 percent heterozygous purple, and 25 percent homozygous white. (Modified from Kirk et al. 1978)

DNA is replicated in the nucleus and condenses into the distinct chromosomes seen by the cytologists. Each chromosome is composed of two identical chromatids joined at the centromere. The chromosomes line up at the equator of the cell, sister chromatids are pulled apart, and each forms a new chromosome equivalent to the parental chromosome. Thus, each new cell is a duplicate of the parent cell and carries a copy of the genetic material.

Meiosis, the process of cell divisions resulting in gamete formation, is more complex. Not only are four cells produced instead of two, but accompanying the cell production are a reduction in chromosome number from diploid to haploid and an exchange of genetic material between homologous chromosomes, involving synapsis and crossing over, which results in **genetic recombination**. In the prophase, the first phase of meiosis, homologous pairs of chromosomes become aligned, and chromatids from each chromosome exchange segments, first crossing over each other, then breaking and recombining with the equivalent segment. As a result, each chromosome generally emerges from the process with a different set of genes from the parental type. Thus, meiosis does not usually preserve the parental genotype, because it alters chromosomes (see Chapter 14).

It became apparent to researchers that, although genes are preserved in a Mendelian way, considerable reshuffling of genetic material takes place between generations in sexually reproducing organisms. First, in the diploid cells of parents involved with gamete production, synapsis and crossing over cause genetic recombination during prophase 1. Then, during anaphase 1, homologous chromosomes are pulled away

FOCUS ON

♦

The Principle of Independent Assortment

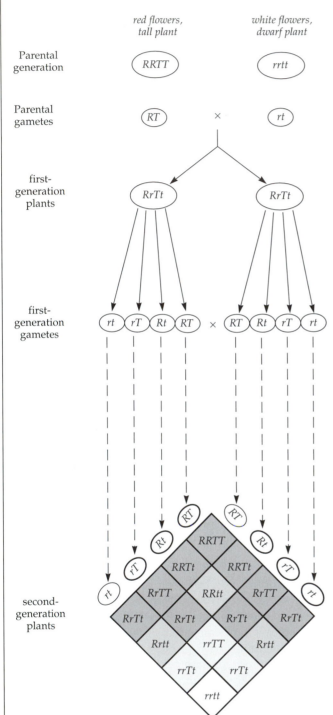

One of Mendel's experiments demonstrated the second law, or the principle of independent assortment, using two characteristics, each on a different chromosome. The crosses involved homozygous red-flowered, tall plants (RRTT) and homozygous white-flowered, dwarf plants (rrtt). Only the genotypes are illustrated.

Note that now the parental gametes are shown carrying two traits, one for flower color and one for plant height, and that the F_1 generation is composed entirely of heterozygous genotypes with the red, tall phenotype. Each of these heterozygous parents then produces four types of gametes, and so the Punnett square becomes larger and more complex, but note how the two characteristics for flower color and plant height behave independently so that all possible combinations of alleles can be found in the population. The resulting F_2 generation is therefore composed of the following phenotypes: 9 red and tall, 3 white and tall, 3 red and dwarf, and 1 white and dwarf (rrtt). Note that in the red and tall phenotypes there is only one homozygous individual (RRTT) at both loci, and the other genotypes are four RrTt, two RRTt, and two RrTT. (From Kirk et al. 1978)

from the equator, each into a newly forming cell. But the homologous chromosomes in the parents do not separate in the same way, as they were originally contributed by the grandparents in the form of one chromosome set from the male and one homologous set from the female. For each homologous pair, either may go into a new cell, whether it was derived from the male or female side of the family. This process results, again, in a probably novel combination of genes in each gametic cell.

FOCUS ON

◆

Mitosis and Meiosis

INTERPHASE   **MITOSIS**

Prophase   *Metaphase*   *Anaphase*   *Telophase*

DNA replication prior to mitosis

Four chromosomes are shown, each consisting of two sister chromatids joined at the centromere.

Chromosomes line up on equator of spindle apparatus; centromeres split.

The two sister chromatids of each chromosome are pulled apart.

Parental chromosome number is maintained in new nuclei.

MEIOSIS I

Prophase I   *Metaphase I*   *Anaphase I*   *Telophase I*   *Anaphase II*

Two pairs of homologous chromosomes are shown (dotted-to-black, white-to-striped); each consists of two sister chromatids joined at the centromere.

Homologous pairs of chromosomes synapse; crossing over can and usually does occur.

Homologous pairs of chromosomes line up on equator of spindle; centromeres do not split.

Sister chromatids of each chromosome stay together; but homologous chromosomes are separated from each other.

The unpaired chromosomes end up in separate cells (or at opposite poles if cell has not divided).

As in mitosis, the two sister chromatids of each chromosome are pulled apart.

Four haploid nuclei result.

(**Top**) A diagrammatic view of mitosis, in which a somatic-diploid cell divides into two with duplicated copies of the genetic material that was originally present in the parental cell. (**Bottom**) In meiosis, cells divide ultimately into four gametes and the chromosome number per cell is halved, from diploid to haploid. In addition, in prophase 1, homologous pairs of chromosomes synapse, crossing over occurs, and genetic recombination results. (From Kirk et al. 1978)

3.4 THE IMPORTANCE OF GEOGRAPHIC ISOLATION IN THE SPECIATION PROCESS

Darwin clearly had a most difficult time in evaluating the importance of **geographic isolation** in the speciation process. The summary paragraph on page 34 included the phrase "After much variation has accumulated." A modern biologist who subscribes to the classical model of speciation championed by Ernst Mayr (1940, 1942) would emphasize that variation can accumulate only when a large and long-term **geographical barrier** prevents or drastically limits migra-

tion between the differentiating population and the parental population.

Darwin was ambivalent in *The Origin of Species* about the importance of geographic isolation. In chapter 4 (1872b, 78–79), he wrote:

Isolation, also, is an important element in the modification of species through natural selection. . . . Moritz Wagner has lately published an interesting essay on this subject [Wagner 1868], and has shown that the service rendered by isolation in preventing crosses between newly-formed varieties is probably greater even than I supposed. But from reasons already assigned I can by no means agree with this naturalist,

47

Genetic Recombination Between Homologous Chromosomes During Meiosis

STEP 1. Synapsis

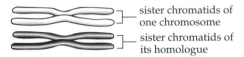

sister chromatids of
one chromosome

sister chromatids of
its homologue

STEP 2. Breakage

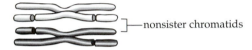

nonsister chromatids

STEP 3. Crossing Over

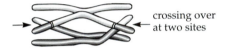

crossing over
at two sites

STEP 4. Chiasmata Formation

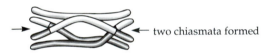

two chiasmata formed

STEP 5. Genetic Recombination Complete

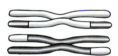

recombination
between homologous
chromosomes

Genetic recombination between homologous chromosomes during meiosis 1, prophase (Figure 3-3). The process is divided into five steps. *Step 1:* **Synapsis** involves the alignment of a pair of sister chromatids with their homologous pair. *Step 2:* **Breakage**, in which chromatids of different chromosomes break at the same location. *Step 3:* **Crossing over**, in which chromatids from different chromosomes cross over and exchange corresponding segments. *Step 4:* **Chiasmata formation**. Here, homologous chromosomes start to move apart but remain attached at points of crossing over. At each point a cross, or chiasma, is seen between chromosomes. Two chiasmata are illustrated. *Step 5:* By the time the **genetic recombination** is complete, each chiasma has broken, chromatids have separated, and the changed homologous chromosomes are reconstituted and separate. Note that the chromosomes in step 5 appear superficially (say, under a light microscope) to be identical. (Based on Kirk et al. 1978)

that migration and isolation are necessary elements for the formation of new species ... isolation will give time for a new variety to be improved at a slow rate; and this may sometimes be of much importance.

In chapter 12, Darwin (1872c, 81–82) alludes to the importance of **migration** or its lack:

The dissimilarity of the inhabitants of different regions may be attributed to modification through variation and natural selection, and probably in a subordinate degree to the definite influence of different physical conditions. The degrees of dissimilarity will depend on the migration of the more dominant forms of life from one region into another having been more or less effectually prevented, at periods more or less remote;—on the nature and number of the former immigrants;—and on the action of the inhabitants on each other in leading to the preservation of different modifications; the relation of organism to organism in the struggle for life being, as I have already often remarked, the most important of all relations. Thus the high importance of barriers comes into play by checking migration; as does time for the slow process of modification through natural selection.

In this rather obscure passage, Darwin seems to allude to (translated into modern terms) (1) the important role of **gene flow** (movement of genetic material between populations through the dispersal of individuals, gametes such as pollen, fertilized eggs, or immature progeny), in preventing isolation between populations; (2) the necessity of time for populations to accumulate differences while isolated; and (3) the factors that are important during secondary contact, after populations have diverged, such as possible hybridization and swamping of the accumulated differences, the existence of **reproductive isolation** (breeding incompatibility), and the possibility of divergence of characteristics, which would enable the coexistence of new species in different ecological niches. Chapter 5 will treat these subjects in more detail.

There is little doubt, however, that Darwin did not consider geographic isolation to be of prime importance to his theory, and if he could have been persuaded to write as brief a statement as we have, isolation would not have figured in it. Two more of his statements relegate barriers to the role of limiting dispersal rather than acting as the major factor in the speciation process:

> The great and striking influence of barriers of all kinds, is intelligible only on the view that the great majority of species have been produced on one side, and have not been able to migrate to the opposite side. (Darwin 1872c, 83)

> Hence it seems to me, as it has to many other naturalists, that the view of each species having been produced in one area alone, and having subsequently migrated from that area as far as its powers of migration and subsistence under past and present conditions, is the most probable. (84)

Darwin's Model of Speciation

Darwin never drew a diagram illustrating his concept of speciation, but a careful reading of the sixth edition of *The Origin* suggests something close to Figure 3.4, with the following important sequential ingredients for such a diagram:

1. A species spreads over a large area.
2. Environmental variation produces variations in populations, and the latter differ in each environment.

FIGURE 3-4

An attempt to illustrate Darwin's concept of the speciation process in 1872. A parental species spreads over environments A, B, C, and D. Each environment produces different variations in the populations a, b, c, and d and different pressures from natural selection. The populations diverge in characteristics and become new species. Hybrid zones between the populations in different environments are narrow because little movement between environments occurs, and hybrids are maladapted for all contiguous environments.

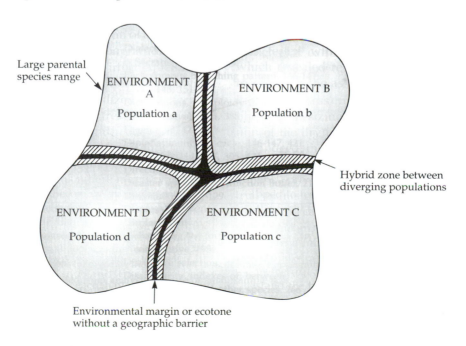

3. If little movement of individuals between environments occurs, one result is divergent, or at least independent, natural selection in each environment.

4. New species develop in each environment as a result.

5. Hybrid zones between environments result in maladapted individuals, which survive poorly in both environments.

6. Since environments are not completely discrete, continuous variation between species occurs.

7. Small populations of hybrids and competition from the parental types result in eventual extinction of intermediate forms.

Mayr's Model of Geographic Speciation

In contrast with Darwin's views expressed in 1872, **Ernst Mayr** could justifiably say in 1963 (481):

> That **geographic speciation** is the almost exclusive mode of speciation among animals, and most likely the prevailing mode even in plants, is now quite generally accepted. . . . The theory of geographic speciation is one of the key theories of evolutionary biology.

Mayr set out to describe the history of the theory and the "proofs of its correctness." He pointed out how remarkable it was that already in 1825 Leopold von Buch wrote a very lucid description of geographic speciation that "has a remarkably modern ring." Describing the organisms of the Canary Islands, von Buch wrote (as translated in Mayr 1963, 483):

> The individuals of a genus strike out over the continents, move to far distant places, form varieties (on account of the differences of the localities, of the food, and the soil), which owing to the segregation [geographical isolation] cannot interbreed with other varieties and thus be returned to the original main type. Finally these varieties become constant and turn into separate species. Later they may again reach the range of other varieties which have changed in a like manner, and the two will now no longer cross and thus they behave as "two very different species."

Mayr noted that von Buch's theory had impressed Darwin, as it fit the situation of the Galapagos finches perfectly, and Darwin had even written to Joseph Hooker in 1844, "With respect to original creation or production of new forms . . . isolation appears the chief element." By the time he published *The Origin of Species*, however, Darwin had relegated isolation to a limiting factor in dispersal rather than a creative factor in evolution, as the earlier quotations in this chapter demonstrated.

The modern theory of geographic speciation places geographic isolation in a prominent position (Mayr 1942, 155; Mayr 1963, 482):

> In sexually reproducing animals a new species develops when a population which has become geographically isolated from its parental species acquires during this period of isolation characters which promote or guarantee reproductive isolation when the external barriers break down.

Geographic speciation is also called **allopatric speciation** because during the speciation process diverging populations exist in different localities that are mutually exclusive (see Chapter 5).

We need to clarify that *geographic* isolation is a property imposed by the environment, whereas *reproductive* isolation is a property imposed by biological characteristics of the involved individuals, populations, and species. Later chapters (Chapters 4, 5, and 14) will say more on these topics. The important points to include in a modern theory of evolution are that reproductive isolation between populations must develop, in geographic isolation, through the accumulation of small differences between populations.

3.5 THE NATURE OF THE SPECIES

Darwin's Species Concept

Another major point of divergence between Darwin's theory and modern theory is the nature of the species. *The Origin of Species* clearly states **Darwin's species concept** several times in chapter 2 alone. "[Species are] only strongly-marked and well-defined varieties" (1872b, 42). "We have seen that there is no infallible criterion by which to distinguish species and well-marked varieties" (43). "Varieties cannot be distinguished from species" (44). So what is a variety, or species? "I look at the term species as one arbitrarily given for the sake of convenience to a set of individuals closely resembling each other. . . . it does not essentially differ from the term variety, which is given to less distinct and more fluctuating forms" (40). And what is the evidence of continuous variation between varieties and species?

> On the view that species are only strongly marked and permanent varieties, and that each species first existed as a variety, we can see why it is that no line of demarcation can be drawn between species, commonly supposed to have been produced by special acts of creation, and varieties which are acknowledged to have been produced by secondary laws. (Darwin 1872c, 171)

This view of the species was the obvious one for an evolutionist to embrace at a time when the majority of naturalists took the fixity of species for granted. If one could not accept fixity, one had to accept transmutation and the gradual change of one species into another in both time and space. And Darwin was not a little devious, perhaps, in linking speciation with the production of varieties that were patently "man-made" through artificial selection (e.g., Figure 2.9), without need of a mystical act of creation. This link made a potentially heretical concept an evident and tangible reality. Darwin could hardly afford to separate the evolution of new species from the production of varieties. Thus, his explanation of transmutation was an explanation of the origin of new species. *The Origin of Species* was not a misnomer, as it might seem at first glance to a modern reader who understands the discreteness of many, if not all, species.

The Biological Species Concept

Our recent understanding of the definition of species is largely due to the forceful arguments of Theodosius Dobzhansky (1937) and Ernst Mayr for the adoption of the **biological species concept,** which defines species as "groups of actually or potentially interbreeding natural populations, which are reproductively isolated from other such groups" (Mayr 1963, 19). (Mayr's background was in the systematics of birds, and he was addressing bird species when he formulated this view [e.g., Mayr 1940]. Although he emphasized the general applicability of his definition to bisexual species [Mayr 1942, 1963], his frame of reference should be borne in mind.) Any modern view of evolution must encompass these two qualities of a species: the potential for sharing in a common gene pool (all the genes in a population) and distinctness from other gene pools.

3.6 THE DEFINITION OF FITNESS

We have discussed the four points listed at the end of Chapter 2: the origin of variation, the maintenance of variation, the role of geographic isolation during speciation, and the nature of the species. Some additional, lesser issues are worth mentioning briefly. First, the use of the phrase "survival of the fittest" and the definition of "fitness" have stimulated some debate. Both Herbert Spencer's phrase "survival of the fittest" and Darwin's "struggle for existence" imply overt competition between individuals, aggressive encounters, and serious competition between species in the natural world. In *The Origin* Darwin repeatedly emphasized relationships between organisms as the important factor in natural selection. He was beginning, however, to attribute considerable importance to natural selection for adaptations related to purely abiotic factors. In the absence of clear combat or competition between organisms, the term "survival of the fittest" becomes tautological—a self-

evident redundancy—since all survivors are fit and all nonsurvivors are unfit. We thus need a definition of fitness that enables us to measure it and that does not rest on the importance of physical fitness or the attribution of survival to the presence of a certain identifiable set of characteristics, such as large muscles, brains, cunning, cold hardiness, or tolerance of thermal stress.

We can identify at least three types of fitness: that of the individual, that of the population, and that of the genotype. Encompassing the first two of these, fitness may be defined as the ability of an individual or population to leave viable progeny in relation to the ability of other individuals or populations. "The genetic fitness of a genotype . . . is measured by the contribution it makes, relative to other genotypes . . . to the gene pool of the succeeding generations" (Dobzhansky 1962, 131). Fitness is a relative measure, not an absolute measure, but it is objective. Operationally, we measure the frequencies of different alleles, or genotypes, in a population of individuals in one generation and then measure frequencies in the next and subsequent generations (see the peppered moth example in Chapter 15, Section 15.4 and Table 15.4). Thus, fitness is "the average contribution of one allele or genotype to the next generation or to succeeding generations, compared with that of the other alleles or genotypes" (Futuyma 1986, 552).

3.7 USE OF THE WORD "EVOLUTION"

A second interesting point is Darwin's use of the word **"evolution."** Gould (1974) noted that Darwin actively avoided use of the word. Even in *The Origin of Species* he used a form of the word only once, to close the book (Darwin 1872, 470).

> There is grandeur in this view of life, with its several powers, having been originally breathed by the Creator into a few forms or into one; and that, whilst this planet has gone cycling on according to the fixed law of gravity, from so simple a beginning endless forms most beautiful and most wonderful have been, and are being evolved.

Albrecht von Haller coined the word "evolution" in 1744 to describe his **preformationist** view "that embryos grew from preformed homunculi enclosed in the egg or sperm" (Gould 1974, 6). Thus, evolution meant the development of the tiny body into the embryo and finally the emergent organism. By 1859, Haller's theory was largely debunked and evolution had come to mean the occurrence of a long sequence of events, with the connotation of **progressive development** from simple to complex. Darwin rejected the notion of progress contained in this theory, however, because he believed unicellular organisms were just as well adapted to their environments as were vertebrates to theirs. The word "evolution" was already taken, in a sense, and was not available for his use. Later he purposefully assigned the title *The Descent of Man . . .* to his 1871 book, with the word "descent" representing his argument that man was a descendent of primate stock, without any implication or value judgment on whether an advance had been achieved.

According to Gould (1974), Herbert Spencer popularized the word "evolution" in his writings between 1862 and 1867, making it synonymous with development toward increasing organization and complexity. Scientists could thus use the word in place of Darwin's "descent with modification." Along with general acceptance of the word "evolution," Gould notes, has come the notion (which still exists today) that evolution involves some sort of progress and increasing complexity. This anthropocentric view cannot be supported, however, since the fossil record and the present state of many species show that many evolutionary lines have aborted, many impressive adaptive radiations were expunged before modern times, and species that have faced rapid environmental change have failed (see Chapter 9). It is safe to say that the evolutionary process results in change, but perceptions of progress are difficult to justify.

Nowadays, the term **"biological evolution"** simply means genetically based, heritable change in one or more characteristics in a population or species through time. It lacks all other dictionary connotations of the word "evolution," such as (1) change from lower, simpler to a higher more complex and improved condition; (2) gradual and peaceful social, cultural, political, or economic advance; and (3) progressive development. For example, a lineage that leads to parasitism commonly evolves from a complex design for a free-living existence to a simple design and loss of many characteristics that are no longer needed because the host provides many essentials. And, given all the problems with human design (see Chapter 12), one can hardly claim that our species is at the leading edge of a progressive evolution. Can we argue that humans are better designed than amoebae? Has the human lineage progressed, digressed, or regressed relative to other apes and the basic primate design (cf. Chapter 12)?

3.8 DARWIN'S CRITICS

To what extent did Darwin's contemporary critics contribute to the sixth edition of *The Origin*? Hull (1973) discusses the question at length. He quotes Darwin as remarking that, among the many adverse critics of *The Origin*, " 'Fleeming Jenkin has given me

much trouble, but has been of more real use to me than any other essay or review' " (Hull 1973, 302). Darwin was referring to Jenkin's (1867) critique in the *North British Review*. Fleeming Jenkin was a professor of engineering at Glasgow University, and he debated such premises as the ability of natural selection to choose and breed special varieties as humans do, the age of the earth, difficulties with classification of organic beings, and so on. A reading of his essay would not seem as profitable to a modern biologist as it would to a historian of science, since, as Hull (1973) concluded, Jenkin functioned as a coagulant and stimulated Darwin to meld various lines of investigation and view them in their proper interrelationships. The same has been said of Malthus. Some scientists would therefore argue that Malthus and Jenkin led to major conceptual breakthroughs; others, that they were of only slight help to Darwin by prompting him to organize his thoughts. Darwin himself would seem the most authoritative judge of his influences; he ascribed major status to Malthus (see also Desmond and Moore 1991) and useful status to Jenkin. Let us not deal further with Jenkin, however, since he had little impact on the modern theory of evolution.

In general, Darwin's reactions to other adverse critics ranged from tolerance to contempt. Louis Agassiz was a target of the latter: in 1860, Darwin wrote to his publisher, John Murray, that "Agassiz has denounced [*The Origin*] in a newspaper, but yet in such terms that it is in fact a fine advertisement!" (Darwin 1887, vol. 2, 64–65). In the same year, he wrote to Asa Gray: "I am surprised that Agassiz did not succeed in writing something better" (Darwin 1887, vol. 2, 124). And to Benjamin Walsh in 1864 he remarked: "I am so much accustomed to be utterly misinterpreted that it hardly excites my attention" (Darwin 1903, vol. 1, 258). For more of Darwin's views on his critics, the interested reader may consult such books as DeBeer 1964, Ghiselin 1969, Vorzimmer 1970, and Hull 1973.

3.9 THE MODERN THEORY

The modern theory of evolution is a blend of findings, already outlined in this chapter, from many disciplines within biology. A relatively brief, simple statement of modern evolutionary theory brings together those findings and can readily be compared with the statement Darwin might have made, which appears on page 34. The following statement should be dated in the 1950s or early 1960s, before enzyme polymorphism was studied (starting in 1966 with Lewontin and Hubby's work; see Chapters 15 and 17) and before molecular clocks were used to measure evolutionary time (Chapter 17).

*"New variants of **genes** are continually being added to populations through **mutation**. In addition, as explained by Mendel's two laws, recombination of genes increases the variation initially produced by mutation, since **segregation** and **independent assortment** of genes result in almost endless novelty in gene combinations and thus an almost infinite variety of possible genotypes in a population. Because of the **particulate nature of inheritance,** no loss of variation occurs from parent to progeny. Thus, each population has an enormous store of variation on which natural selection can act. The **fittest** individuals, defined as those which leave the most viable progeny that mature and reproduce, contribute the most genes to the next generation, and thus their genes become most common in the gene pool of the population. New mutations and new combinations of genes continually produce some individuals that are better adapted than others; **natural selection** favors these and selects out less fit types. Gradually, the population evolves through this differential perpetuation of genes.*

*"If a population should become divided into two by a **geographic barrier,** evolution of each new population continues independently. Differences between the two, including differences in reproductive processes, gradually accumulate such that **reproductive isolating mechanisms** become more and more effective over prolonged periods of time. Ultimately, on **secondary contact,** the two populations are reproductively isolated and two new species have evolved by the process of **geographic speciation.** Therefore, **species** can be defined as groups of actually or potentially interbreeding populations that are reproductively isolated from other such groups.*

"Speciation events lead to multiplication and diversification of species into higher taxa that retain clear phylogenetic links with parental stock, so that all species can be traced to the origin of life itself."

The **modern synthesis** was achieved not only by correcting some of the deficiencies of Darwin's theory but by resolving issues that developed subsequently (Mayr 1980, Gould 1982). Probably the most serious debate revolved around the relative roles of variation, natural selection, and (after 1900) mutation in the evolutionary process. What was the real creative force in the evolutionary process? Did variation play the central role, with selection simply weeding out the hopeless, or was variation more or less random, and only in the presence of natural selection could some directional evolutionary change occur? Did macromutation really drive the evolutionary process? Was it a sufficient cause? The rediscovery of Mendelian genetics added to the ferment of the 1890s (Mayr 1980). Its illuminating qualities were dimmed by controversy.

Fortunately, the controversy sent researchers in divergent directions that ultimately were essential to the synthesis. Although T. H. Morgan denied the importance of natural selection and emphasized the role of variation (Weinstein 1980), he became prominent in the research of mechanisms that cause variation and gene change. He established **experimental genetics** in North America. In *Drosophila* species, Morgan and his students found evidence of spontaneous variation, including **gene mutations,** and such **rearrangements of genes** as inversions, deletions, and translocations. One of the spontaneous strains enabled Muller (1927) to prove that radiation could produce mutations in *Drosophila*. All this experimental work reinforced Morgan's views.

With systematists and naturalists hardly conversing with experimental geneticists in the first three decades of the 1900s, it is hard to see in retrospect how synthesis ever could have been achieved—except for the efforts of one man, Theodosius Dobzhansky.

Dobzhansky was a practical person, a naturalist and a systematist (with a deep-seated appreciation of insects), possessing the same fundamental attributes that had made Darwin a great synthesizer. Dobzhansky worked with Morgan starting in 1927 and thus was in a unique position to unite the genetics of variation with the systematists' and naturalists' view of the discreteness of species. This he did in 1937 by writing *Genetics and the Origin of Species*.

The book undoubtedly laid the foundation for the **modern synthesis of evolution** and provided an impetus for the other synthesists (Gould 1982). Dobzhansky's novel perception was that an understanding of the discreteness of species and discontinuity among species must be central to the theory of evolution and that the involved mechanisms could be studied via experimental genetics. He argued forcefully that mutation was the basis of variation, but only with natural selection did evolution achieve any creative power. Dobzhansky acknowledged, however, that natural selection, **genetic drift** (sampling error in small populations), and dispersal between populations could all play roles in the evolutionary process, as Wright (1932) had also argued. Dobzhansky coined the term "isolating mechanisms." He divided such mechanisms into two categories, geographical and physiological. The physiological mechanisms he classified as ecological, seasonal or temporal, sexual or psychological, mechanical, gametic mortality, and inviability of hybrids. It is now recognized that the theoretical population geneticists Fisher, Haldane, and Wright, so influential in later developments, were not heavily involved with the synthesis of the late 1930s and 1940s (Lewontin 1980, Gould 1982).

Building on Dobzhansky's example, Ernst Mayr took a second step in the synthesis by unifying systematics and evolutionary theory in *Systematics and the Origin of Species* (1942). (We discussed his important contributions, the biological species concept and the geographic speciation model, earlier in this chapter.) Then came the integration of paleontology with genetics and evolution by George Gaylord Simpson in his *Tempo and Mode in Evolution* (1944), which enormously advanced the concepts of macroevolution and reinforced Dobzhansky's original view that experimental genetics was sufficient to account for micro- and macroevolution.

Next came Bernhard Rensch in 1947 with *Neuere Probleme der Abstammungslehre* (New Problems on the Science of Evolution). Its second edition was translated, on Dobzhansky's advice, as *Evolution Above the Species Level* (1959). Rensch discussed comparative morphology, systematics, and paleontology in an evolutionary context. When G. Ledyard Stebbins (1950) brought botany into the fold of evolutionary theory, the modern synthesis was largely complete.

The modern synthesis was a **synthesis of biology,** not of minds. Scientists still had areas of disagreement, as a reading of *Genetics, Paleontology and Evolution* by Jepsen et al. (1949) reveals. But the 1947 conference from which the book was derived "constitutes the most convincing documentation that a synthesis had occurred during the preceding decade. . . . Evolutionary biology was no longer split into two noncommunicating camps" (Mayr 1980, 42–43).

SUMMARY

Darwin's theory of natural selection was strengthened by his candid analysis of problems associated with application of the theory, especially problems concerning intermediate forms, ranges in complexity of structure, instinct, and hybrids. The evolution of sterile castes of social insects, such as ants, was a grave problem for Darwin, but he argued that selection could act on a family and not merely on individuals, such that all members contributed to the reproductive success of the colony. The **altruistic behavior** of sterile workers and soldiers serving the queen ant could be understood by those ants' direct relatedness to her, under the influence of **kin selection.** Thus, he showed that the genes of nonbreeding castes are passed on to the next generation through the queen.

Darwin extended the theory of natural selection to include selection between the genders of a species, or **sexual selection.** Competition between members of the same sex (usually males), involving traits for potency in fighting or threatening, is called **intrasexual selection** and is one form of sexual selection. Another form, **intersexual selection,** involves traits that increase the allure between the sexes or otherwise improve the probability that fertilization will succeed.

We must be cautious about oversimplifying the application of the theory to explain almost every trait of organisms. Natural selection is not easy to detect, and relatively few studies have quantified the force of selection in a rigorous manner. A **panglossian** view on natural selection, i.e., the **adaptationist program,** must be viewed in a realistic context with the recognition that many forces in addition to natural selection commonly influence the evolutionary process. In particular, evolution occurs in lineages with a long history of evolutionary change and **phylogenetic constraints** impose limits on the nature of adaptation. The interesting linkage among a diverging lineage, natural selection, and other evolutionary processes offers a strongly comparative approach that goes well beyond trait-by-trait examination of adaptations.

Darwin noted that all species exhibit variation of characteristics and that variation never stops. He erroneously assumed that a varying environment produced variation. Now it is understood that **mutations** are the original source of variation in a population or species. Darwin thought that variation passed from generation to generation by a process called **pangenesis,** involving gemmules passing from parts of the body into the gametes of both parents, with a blending of hereditary material at the time of fertilization. However, **Mendel's principles of heredity** described a particulate form of inheritance in which genes for a characteristic segregate during gamete formation, and genes for different characteristics are assorted independently of one another during gamete formation. The recombination of genes creates novel genotypes upon which natural selection can act. To the cytologists of the 1800s it appeared that hereditary particles must be carried on the chromosomes, and the field of cytogenetics was born.

Darwin was ambivalent about the role of **geographic isolation,** which prevents migration between diverging populations during speciation. He underestimated its importance relative to Mayr's model of **geographic speciation,** in which a geographic barrier separates two populations that evolve independently until their evolved biological characteristics—**reproductive isolating mechanisms**—prevent successful reproduction between the populations, and they function as new species. Darwin's concept of the origin of new species depended on variation in populations, produced by environmental variation. Natural selection would then act in different ways in different populations, and ultimately new species would develop in the different environments. This scenario blended well with Darwin's view of the species as just a rather well-formed variety that graded gently into another species. In contrast, the model of geographic speciation, involving reproductive isolating mechanisms, illustrated the discreteness of species and established the **biological species concept,** wherein members of the same species can breed with each other but not with members of other species.

The concept of **fitness** has become more objective since Darwin's time. Fitness is defined as the performance of individuals in one lineage relative to those in other lineages within a population, in terms of contributing progeny to subsequent generations. The word "evolution," unavailable to Darwin, now refers to biological change through the transmutation of species by natural selection. In response to criticism of *The Origin*, Darwin recognized the virtues of a few opinions, such as those of Jenkin, and the flaws in many.

The modern theory of evolution, revised to include developments into the early 1960s, incorporates the breadth of biology and represents a true synthesis of natural history, genetics, systematics, paleontology, comparative morphology, behavior, zoology, botany, and many other fields.

QUESTIONS FOR DISCUSSION

1. From your reading of *The Origin of Species,* can you suggest an alternative to Darwin's concept of the speciation process, or a more accurate scenario than the one provided in this chapter?

2. Why was Darwin's concept of pangenesis so wrong, and yet it did not seriously hamper the development of the theory of evolution?

3. What do you consider to be the key to understanding why Darwin's concept of the species was so vague?

4. Do you think that the word "evolution" still has a connotation that distorts its objective use in the literature?

5. If we accept that the modern synthesis unites all aspects of biology, should the argument be made that all courses in biology should develop a strong evolutionary theme?

6. Do you think that scientists studying evolution have used the scientific method in an ideal way? Which examples come to mind?

7. Is there evidence that some scientists have been too dogmatic in their views on evolution or that they have oversimplified issues? Which examples come to mind?

8. Can you think of examples in which a Panglossian approach to evolution has misdirected the field?

9. In your reading of *The Origin of Species,* do you think that Darwin presented the difficulties with his theory fairly and objectively?

10. If the scientific world outside Brünn (Brno) had never discovered Mendel's research on heredity, would the development of evolutionary theory have taken a different course, in your opinion?

REFERENCES

Beadle, G. W., and E. L. Tatum. 1945. Neurospora: II. Methods of producing and detecting mutations concerned with nutritional requirements. Amer. J. Botany 32:678–686.

Buch, L. von. 1825. Physicalische Beschreibung der Canarischen Inseln, pp. 132–133. Königliche Akademie den Wissenschaften, Berlin.

Correns, C. 1900. G. Mendels Regeln über das Verhalten der Nachkommenschaft der Rassenbastarde. Berichte der Deutschen Botanischen Gesellschaft 18:158–168. Translated by L. K. Piternick. 1950. G. Mendel's law concerning the behavior of progeny of varietal hybrids. Genetics (Suppl.) 35 (5, part 2): 33–41.

Darwin, C. 1859. On the Origin of Species by Means of Natural Selection, or The Preservation of Favoured Races in the Struggle for Life. Murray, London.

——. 1868. The Variation of Animals and Plants Under Domestication. 2 vols. Murray, London.

——. 1871. The Descent of Man, and Selection in Relation to Sex. 2 vols. Murray, London.

——. 1872. On the Origin of Species by Means of Natural Selection, or The Preservation of Favoured Races in the Struggle for Life. 6th ed. Murray, London.

Darwin, F. 1887. The Life and Letters of Charles Darwin. Murray, London.

——. 1903. More Letters of Charles Darwin. Murray, London.

DeBeer, G. 1964. Charles Darwin: A Scientific Biography. Doubleday, Garden City, N.Y.

Desmond, A., and J. Moore. 1991. Darwin. Penguin, London.

De Vries, H. 1900. Sur la loi de disjonction des hybrides. Comptes Rendus de l'Academie des Sciences (Paris) 130:845–847. Translated by A. Hannah. 1950. Concerning the law of segregation of hybrids. Genetics (Suppl.) 35 (5, part 2): 30–32.

——. 1901–1903. Die Mutations Theorie: Versuche und Beobachtungen über die Entstehung der Arten im Pflanzenreich. 2 vols. Leipzig. Translated by J. B. Farmer and A. D. Darbishire. 1909–1910. The mutation theory. Open Court, Chicago.

Dobzhansky, T. 1937. Genetics and the Origin of Species. Columbia Univ. Press, New York.

———. 1962. Mankind Evolving. Yale University Press, New Haven, Conn.

Endler, J. A. 1986. Natural Selection in the Wild. Princeton University Press, Princeton, N.J.

Flemming, W. 1882. Zellsubstanz, Kern und Zelltheilung. F. C. W. Vogel, Leipzig.

Futuyma, D. J. 1986. Evolutionary Biology. 2d ed. Sinauer, Sunderland, Mass.

Ghiselin, M. 1969. The triumph of the Darwinian Method. Univ. of California Press, Berkeley.

Gould, S. J. 1974. Darwin's dilemma. Natur. Hist. 83(6):16–22.

———. 1982. Introduction. pp. xvii–xii. In T. Dobzhansky, Genetics and the Origin of Species. Columbia Univ. Press, New York. [Reprint of 1937 edition.]

Gould, S. J., and R. C. Lewontin. 1979. The spandrels of San Marco and the Panglossian paradigm: A critique of the adaptationist programme. Proc. R. Soc. Lond. B. 205:581–598.

Hamilton, W. D. 1964. The genetical evolution of social behaviour, [part] I. J. Theoret. Biol. 7:1–16.

Hofmeister, W. 1848. Über die Entwicklungsgeschichte des Pollens. Bot. Zeitung 6:425–434, 649–658.

Hull, D. L. 1973. Darwin and His Critics. Harvard University Press, Cambridge, Mass.

Huxley, J. S. 1938. The present standing of the theory of sexual selection. pp. 11–42. In G. R. deBeer (ed.), Evolution: Essays on Aspects of Evolutionary Biology. Clarendon, Oxford.

———. 1942. Evolution, the Modern Synthesis. Allen and Unwin, London.

Jenkin, F. 1867. The origin of species. North Brit. Rev. 46:277–318. [Reprinted in Hull 1973.]

Jepsen, G., E. Mayr, and G. G. Simpson (eds.). 1949. Genetics, Paleontology, and Evolution. Princeton Univ. Press, Princeton, N.J.

Johannsen, W. 1909. Elemente der Exakten Erblichkeitslehre. Gustav Fischer, Jena.

Kirk, D., R. Taggart, and C. Starr. 1978. Biology: The Unity and Diversity of Life. Wadsworth, Belmont, CA.

Lewontin, R. C. 1980. Theoretical population genetics in the evolutionary synthesis. pp. 58–68. In E. Mayr and W. B. Provine, The Evolutionary Synthesis. Harvard Univ. Press, Cambridge, Mass.

Luria, S. E., and M. Delbrück. 1943. Mutations of bacteria from virus sensitivity to virus resistance. Genetics 28:491–511.

Mayr, E. 1940. Speciation phenomena in birds. Amer. Natur. 74:249–278.

———. 1942. Systematics and the Origin of Species. Columbia Univ. Press, New York.

———. 1963. Animal Species and Evolution. Belknap Press of Harvard Univ. Press, Cambridge, Mass.

———. 1980. Prologue: Some thoughts on the history of the evolutionary synthesis. pp. 1–48. In E. Mayr and W. B. Provine (eds.), The Evolutionary Synthesis. Harvard University Press, Cambridge, Mass.

———. 1982. The Growth of Biological Thought: Diversity, Evolution, and Inheritance. Belknap Press of Harvard Univ. Press, Cambridge, Mass.

Mendel, G. 1866. Versuche über pflanzen Hybriden. Verh. Naturforsch. Verein, Brünn 4:3–47. [Reprinted (in German) in J. Hered. 42(1):3–47.] English translation in E. W. Sinnott, L. C. Dunn, and T. Dobzhansky. 1958. Principles of Genetics. McGraw-Hill, New York. Also in J. A. Peters (ed.). 1959. Classic Papers in Genetics. Prentice-Hall, Englewood Cliffs, N.J.

Morgan, T. H. 1910a. Sex limited inheritance in Drosophila. Science 32:120–122.

———. 1910b. Chance or purpose in the origin and evolution of adaptation. Science 31:201–210.

———. 1912. Further experiments with mutations in eye-color of Drosophila: The loss of the orange factor. J. Acad. Nat. Sci. Philadelphia 15:321–346.

Muller, H. J. 1927. Artificial transmutation of the gene. Science 66:84–87.

Rensch, B. 1947. Neuere Probleme der Abstammungslehre. Enke, Stuttgart.

———. 1959. Evolution Above the Species Level. Translated from the German by Dr. R. Altevogt. Columbia Univ. Press, New York.

Roux, W. 1883. Über die Bedeutung der Kerntheilungsfiguren. Engelmann, Leipzig.

Simpson, G. G. 1944. Tempo and Mode in Evolution. Columbia Univ. Press, New York.

Spencer, H. 1862–1896. A System of Synthetic Philosophy. 10 vols. Williams and Norgate, London.

Stebbins, G. L. 1950. Variation and evolution in plants. Columbia Univ. Press, New York.

Tschermak-Seysenegg, E. von. 1900. Historischer Rückblick auf die Wiederentdeckung der Gregor Mendelschen Arbeit. Verhandlungen der Zoologische-Botanischer Gesellschaft in Wien 92:25–35. Reprinted 1951 as The rediscovery of Gregor Mendel's work. J. Hered. 42:163–171.

Vorzimmer, P. J. 1970. Charles Darwin: The Years of Controversy. Temple Univ. Press, Philadelphia.

Wagner, M. 1868. Die Darwinische Theorie und das Migrationsgesetz der Organismen. Duncker und Humblot, Leipzig.

Waldeyer, W. 1888. Über Karyokinese und ihre Beziehung zu den Befruchtungsvorgängen. Arch. mikrosk. Anat. 32:1–122. English translation 1889. Karyokinesis and its relation to the process of fertilization. Quart. J. Microsc. Sci. 30:159–281.

Weinstein, A. 1980. Morgan and the theory of natural selection. pp. 432–445. In E. Mayr and W. B. Provine (eds.), The Evolutionary Synthesis. Harvard Univ. Press, Cambridge, Mass.

Weismann, A. 1887. Über die Zahl der Richtungskörper und über ihre Bedeutung für die Vererbung. Fischer, Jena.

———. 1892a. Das Keimplasma: Eine Theorie der Vererbung. Fischer, Jena.

———. 1892b. Aufsätze über Vererbung und Verwandte Biologische Fragen. Fischer, Jena.

Wheeler, W. M. 1910. Ants: Their Structure, Development and Behavior. Columbia Univ. Press, New York.

Williams, G. C. 1992. Natural Selection: Domains, Levels, and Challenges. Oxford Univ. Press, New York.

Wilson, E. B. 1896. The cell in development and inheritance. Macmillan, New York.

Wright, S. 1932. The roles of mutation, inbreeding, crossbreeding, and selection in evolution. Proc. 6th Int. Congr. Genet. 1:356–366.

II

MACROEVOLUTION

4

SPECIES CONCEPTS

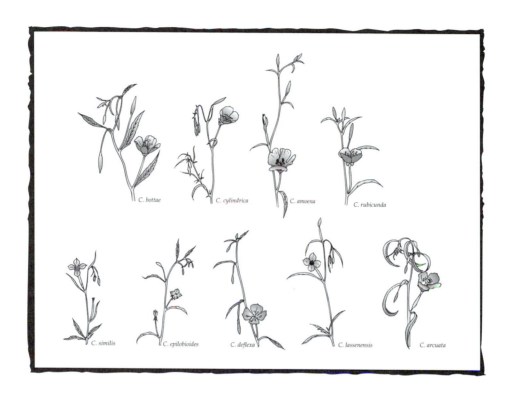

C. bottae *C. cylindrica* *C. amoena* *C. rubicunda*

C. similis *C. epilobioides* *C. deflexa* *C. lassenensis* *C. arcuata*

Nine species in the genus *Clarkia*, members of an extensive group of species in the genus *Clarkia* evolving by interspecies hybridization and by other mechanisms. *(From Lewis and Lewis 1955)*

Conceptual impressions of what a species is have changed radically through the ages, and they differ substantially depending upon the taxon of organisms involved. Because our species concept is important in framing our views on how new species arise, a historical perspective on species concepts is valuable (see also Slobodchikoff 1976, Otte and Endler 1989). This chapter therefore emphasizes an understanding of the biological nature of species and the complexity and difficulty of finding common ground for a general species concept.

Because so many evolutionary processes can result in measurable differences between populations, it is unlikely that human categorizations of species will ever be adequate. Subsequent chapters (especially Chapters 5, 14, and 15) will discuss many of these evolutionary processes. Before we get to them, it is important to attain an understanding of the ongoing debate about species concepts. First, we can detect progress toward a more unifying species concept. Second, the subject of species offers insights into the use of critical **dialectics** in science. In the Socratic sense used in science, dialectic means "discussion and reasoning by dialogue as a method of intellectual investigation" (Gove 1986, 623). It is difficult to conceive of experiments that could help with the debate on species concepts, and a strongly comparative approach to the species issue only reinforces the notion that many different kinds of entities exist in nature. Thus, rational debate becomes an important scientific undertaking, as this chapter illustrates.

4.1 THE GREEK SPECIES CONCEPT

The Greeks had an idea of how unclear the concept of species was (Zirkle 1959). Both Aristotle and Theophrastus conceived of species as unstable and highly changeable. The Latin *specere,* from which our word "species" is derived, means "to look at." So the word originally referred to the outward appearance of an organism. But it was clear that the outward appearances of organisms changed considerably over time. Children differed from their parents. Theophrastus explained how a plant specimen planted in a soil and a climate different from its parents', changed its species. This belief was akin to Darwin's view that environmental variation induced variation in a species. Theophrastus also noted that one could not use seed to propagate valuable fruit trees because the progeny would be too variable; propagation through cuttings, however, conserved the "species" that was desirable.

Aristotle believed in spontaneous generation of species, and he regarded all kinds of crosses between species as feasible and highly creative. Rampant hybridization produced new species. A camel could hybridize with a panther to produce a giraffe. An Arabian camel crossed with a wild boar produced the two-humped Bactrian camel. Oppian argued that a camel crossed with a sparrow produced an ostrich. How such crosses were consummated was not made clear. Suffice it to say that, to the Greeks, species were very unstable entities. We may call their view the **Greek species concept.**

4.2 THE TYPOLOGICAL SPECIES CONCEPT

The Greek view persisted through the Middle Ages, and only in the 18th century did it begin to change. Careful study revealed that members of a species retained a common set of characteristics, even though variation was also apparent. In the 1750s, the Compte de Buffon contributed to the idea that species were discrete, and theologians argued for the **fixity of species** (Zirkle 1959), the view that each species remained as created by God.

By the middle of the 18th century, the discreteness and stability of species were generally recognized, setting the scene for Linnaeus to collect one representative member of each species, preserve it in a museum as the **type specimen** for that species, provide a **Latin binomial** as the name for the species, and briefly describe its appearance in Latin. Thus, he developed the **typological species concept** in which all members of a species were of one basic type, represented by the type specimen, and this type did not vary significantly from place to place or through time. The naming of plants according to Linnaeus's system dates back to 1753, when he published his *Species Plantarum,* and the naming of animals dates to the tenth edition of *Systema Naturae,* published in 1758. The Linnaean system for the naming and classification of organisms remains intact today.

By 1800 the fixity of species was the prevalent view, and it formed the background against which Darwin would formulate his views on the species. The typological species concept failed to identify sibling species, but it produced distinct names for the morphs of a polymorphic organism and for species with heterogonic life cycles, or alternation of generations, such as the cynipid gall wasps.

4.3 THE BIOLOGICAL SPECIES CONCEPT

One of the major achievements of the modern synthesis was the development of the **biological species concept,** discussed in Chapter 3. That the concept cannot be applied to fossils is a limitation. Nor can organisms that reproduce using binary fission or parthenogenesis be treated under this concept. Despite its various shortcomings, to many the biological species concept is the most operational single definition available.

The Nondimensional Species Concept

One tenet of the biological species concept is that coexisting demes (populations of closely related organisms) do not interbreed if they belong to separate species. This view considers only species that are sympatric in distribution and synchronous in time. It does not consider demes that live in other places or are active at other times, and it has been called the **nondimensional species concept** (Mayr 1963). This concept avoids many problems inherent in observing dead specimens in museums. Recognition of the distinctness of species in a given area is based on any species characteristics—morphological, behavioral, ecological, or physiological. The nondimensional approach proved to be essential in the identification of

many sibling species that a typological approach had failed to discern. For example, in 1773, DeGeer named the ground cricket *Gryllus fasciatus* (Orthoptera: Gryllidae) (Figure 4-1). Audinet-Serville (1839) then transferred the species from *Gryllus* into the genus *Nemobius* (Vickery and Johnstone 1970). Much later, Fulton (1931, 1933, 1937) recognized that members of this species at the same locality differed in habit preferences and mating songs, and he differentiated three subspecies. On the basis of characteristics from morphology, ecology, and male song, Alexander and Thomas (1959) raised the status of the three subspecies to species: *Nemobius fasciatus, N. allardi,* and *N. tinnulus*. No hybrids were found in nature, although two of these species will produce fertile hybrids in captivity. Thus, the use of characteristics not available to the museum taxonomist revealed three species instead of one in the same general locality.

Thus, the nondimensional species concept emphasizes the distinctness among species in a given locality, species can be identified objectively, since by

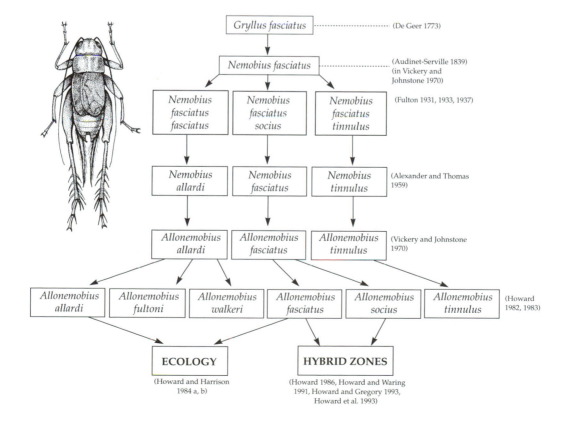

FIGURE 4-1

The development of the biological species concept for the *Allonemobius fasciatus* complex of crickets (Orthoptera: Gryllidae) over 218 years. References in parentheses cite the authorities responsible for name changes based on improved understanding of the biological species. *(Based on Howard and Furth 1986)*

Gryllus fasciatus ----------------- (De Geer 1773)

Nemobius fasciatus ----------------- (Audinet-Serville 1839)
(in Vickery and Johnstone 1970)

Nemobius fasciatus fasciatus | Nemobius fasciatus socius | Nemobius fasciatus tinnulus   (Fulton 1931, 1933, 1937)

Nemobius allardi | Nemobius fasciatus | Nemobius tinnulus   (Alexander and Thomas 1959)

Allonemobius allardi | Allonemobius fasciatus | Allonemobius tinnulus   (Vickery and Johnstone 1970)

Allonemobius allardi | Allonemobius fultoni | Allonemobius walkeri | Allonemobius fasciatus | Allonemobius socius | Allonemobius tinnulus   (Howard 1982, 1983)

ECOLOGY
(Howard and Harrison 1984 a, b)

HYBRID ZONES
(Howard 1986, Howard and Waring 1991, Howard and Gregory 1993, Howard et al. 1993)

definition one species will not interbreed with another to produce progeny as viable as the two parental species. However, the concept does not enable us to identify species that are not sympatric, or coincident in time.

The Multidimensional Species Concept

The **multidimensional species concept** considers allopatric and allochronic populations, which are populations living in separate localities or active at different times (Mayr 1963). For example, members of the genus *Nemobius* were recognized in both Europe and North America with obviously allopatric distributions, since the Atlantic Ocean separated the two groups. Close examination of the morphology of the hind tibiae (segments of the legs) revealed the presence of a glandular tibial spine in the male crickets from North America (Vickery and Johnstone 1970). Such spines were absent from European *Nemobius* and justified the separation of the crickets into distinct genera. The North American crickets were renamed *Allonemobius* in recognition of their allopatric distribution relative to *Nemobius,* which was now restricted to the Old World. Scientists examined other characteristics of the two cricket taxa, including the external male genitalia, the ovipositors of the females, and the cytology involving chromosome numbers. Such characteristics reinforced the validity of generic status for the New World species. Hence, the three species of *Nemobius* recognized by Alexander and Thomas (1959) were transferred to the genus *Allonemobius* (Figure 4-1).

The geographic approach opened another phase in the understanding of these crickets—this time in North America. Howard (1982, 1983) carried out electrophoretic studies and discovered two cryptic species within *A. fasciatus* and three within *A. allardi,* for a total of six species instead of De Geer's one. Further research based on electrophoresis, songs, and morphometric analysis recognized Howard's new species as *A. fultoni* and *A. walkeri* and reinstated what had been Fulton's subspecies as the species *A. socius* (Howard and Furth 1986). Thus, a multidimensional

examination of a genus led to a much more accurate description of part of its species complex (Figure 4-1).

Union of the nondimensional and multidimensional species concepts generated the biological species concept. The biological species concept incorporates both the distinctness between species ensured by reproductive isolation, which the nondimensional concept emphasized, and the potentially common gene pool that the multidimensional concept may recognize (Table 4-1). The view of species is local and geographical. Understanding of the distinctness of species in the *Allonemobius fasciatus* species complex continues to improve by way of ecological studies (Howard and Harrison 1984a,b) and examination of hybrid zones (Howard 1986, Howard and Waring 1991), including analysis of calling songs (Benedix and Howard 1991).

Problems with Application

The biological species concept is not without its critics, however. Sokal and Crovello (1970) concluded that the concept is not operational, it is not heuristic, and it is not of any practical value. They considered the phenetic species—based exclusively on morphological characteristics, as usually described and used by taxonomists—the most desirable species concept. Their argument is long, and so we will touch on only the major points here. Sokal and Crovello construct a rather complex flowchart, which is required for the adequate determination of a biological species, then argue that each of the seven major steps is taken largely by considering phenetic similarities and differences. During this process, many decisions are arbitrary. For example, how much phenetic homogeneity should be necessary for the assignment of two local populations to the same biological species? Because the true test of the biological species concept is so demanding, we are invariably forced to take shortcuts through the use of phenetic characteristics—the **phenetic bottlenecks** of Sokal and Crovello (1970). For example, the proper test of whether populations are actually or potentially interbreeding would take an enormous amount of time and, they

TABLE 4-1	The conditions examined by the species concepts			
Species Concept	Sympatric	Synchronous	Allopatric	Allochronic
Typological	—	—	—	—
Nondimensional	X	X	—	—
Multidimensional	—	—	X	X
Biological	X	X	X	X

claim, would be impractical. In general, then, the biological species concept is inapplicable to the real problem of classifying large numbers of populations into species, according to Sokal and Crovello. Phylogenetic analysis may be best served by an alternative species concept, discussed in Section 4.11, that emphasizes the centrality of the discovery of key diagnostic characteristics for separating species and lineages (Cracraft 1989).

Application of the biological species concept to plants also creates problems. Plant taxonomists can provide many examples of taxonomic groups that do not fit neatly into a biological species concept. According to Grant (1971), these ambiguous groups fall into two major types. First, interbreeding groups of biparental plants may have intermediate levels of reproductive isolation between the level typical of good, or proper, biological species and that of races. Some large and variable groups may be assigned to a taxonomic section or a subgenus but not be awarded specific status. Second, some form of uniparental reproduction commonly replaces biparental reproduction such that each parental stock is isolated from others and can evolve and deviate independently (Mishler and Budd 1990; see also Chapter 14 for examples in plants and animals). Such deviations lead to the evolution, in extreme cases, of semispecies and microspecies, which Sections 4.5 and 4.7 will define and discuss.

4.4 THE EVOLUTIONARY SPECIES CONCEPT

Grant (1971) argues that the biological species concept is too limiting for application to plants. He favors the **evolutionary species concept** as advanced by Simpson (1951, 1961). The evolutionary species is a population system with the following characteristics (Grant 1971).

1. Each system is made up of lineages, ancestor-descendant sequences of populations existing in space and time.
2. Each lineage evolves separately from other lineages, meaning from other species.
3. The lineage has its distinctive evolutionary role. It fits into its own particular ecological niche in the biotic community.
4. The lineage has evolutionary tendencies but is susceptible to change in evolutionary role during the course of its history.
5. A species is a group of populations or lineages under the influences of the same normalizing selective pressures.

This definition applies well to both biparental and uniparental species. The concern is not so much whether or not species hybridize but whether, when they do, they retain or lose their distinct ecology and evolution.

4.5 SEMISPECIES

Several kinds of situations complicate efforts to define the distinctness of species. **Semispecies** are groups of populations, intermediate between good races and good species, that are connected by reduced interbreeding and gene flow. They can arise during the gradual divergence of a parental stock into two good species (Figure 4-2). The sycamores in California and Arizona, *Platanus racemosa* and *P. wrightii,* appear to be good typological species. However, because they hybridize fairly effectively, they are better regarded as semispecies if one is emphasizing the biological condition more than the practicality of naming taxa.

Circular Overlap

Many groups of animals and plants illustrate **circular overlap,** "the phenomenon in which a chain of contiguous and intergrading populations curves back until the terminal links overlap with each other and behave like good species" (Mayr 1963, 664). In circular overlap, semispecies are interfertile when adjacent, but

F I G U R E 4 - 2

The names given to stages in the continuum of genotypic divergence and reproductive isolation during divergence of populations from a common ancestor. Shading indicates the potential for extensive gene flow between diverging populations. *(Based on Grant 1963)*

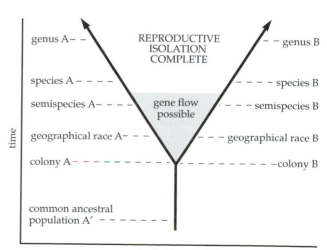

fertility is lost gradually with distance until the most distant semispecies exist in zones of overlap—that is, are sympatric—and are reproductively isolated (Figure 4-3). Such patterns can be seen in semispecies in the *Gilia capitata* group, the genus that includes skyrocket, columbines in the *Aquilegia chrysantha* group, and the pines in the *Pinus ponderosa* group (Grant 1971).

A classic case of circular overlap, however, concerns animals—the gulls in the *Larus argentatus* group around the Arctic Circle (Mayr 1963) (Figure 4-4). The European gulls, *L. argentatus* and *L. fuscus*, appear to be the terminals of an intergrading series of subspecies around the Arctic Ocean. Where *L. argentatus* and *L. fuscus* overlap along the coasts of Europe, they are found together but hybridize only rarely. They may be found in the same nesting colonies.

The original divergence from a common stock probably occurred in Pleistocene refuges in geographic isolation. A pink-footed group, *L. vegae* and its relatives, arose in eastern Asia on the Pacific Ocean, where it gave rise to the *argentatus* group. A yellow-footed group, *cachinnans*, evolved in the area of the Aral and Caspian seas and gave rise to the Atlantic *fuscus* group. *Larus cachinnans* and *L. vegae* interbreed and give rise to *L. mongolicus*, causing gene flow from Asia to North America. *Larus cachinnans* and *L. argentatus* interbreed, producing the taxon *L. omissus* in the northern Baltic Sea. The loop of interbreeding species is broken when *L. argentatus* and *L. fuscus* meet at the eastern extremity of the *L. argentatus* range and the western part of the *L. fuscus* range.

Darwin (1859) emphasized this kind of situation, in which intermediates between races and species exist and therefore "species" can be only an arbitrary designation.

Hybrid Swarms

Semispecies can also arise by hybridization between two good biological species. When reproductive isolating barriers are even 99 percent effective, hybrids can be produced and survive, especially in habitats slightly different from that of the parental type. **Hybrid swarms**—complex mixtures of parental forms, F_1 hybrids, backcross types, and segregation products—can then develop. Hybrid swarms result from crosses between the columbines *Aquilegia formosa* and *A. pubescens* in the Sierra Nevada; *Geum rivale* and *G. urbanum*, the water avens and wood avens, respectively, in England and Europe (Grant 1971); and the poplars *Populus augustifolia* and *P. fremontii* in Utah (Whitham 1989, Keim et al. 1989) (Figure 4-5).

4.6 THE SYNGAMEON

Where considerable movement of genetic material occurs between species that can be viewed as discrete because of phenetic differences, the most inclusive interbreeding group is larger than the species. For this group the term **"syngameon"** was coined by Lotsy (1925), who worked on the rather freely hybridizing birches (*Betula*) in Europe. The syngameon is "the sum total of species or semispecies linked by frequent or occasional hybridization in nature; hence, a hybridizing group of species" (Grant 1971, 54). Irises in the series Californicae studied by Lenz (1959) show an extensive array of hybrid linkages between species (Figure 4-6) and illustrate the complexity of genetic exchange in syngameons. *Euphrasia* and other hemiparasites represent examples from Europe (e.g., Yeo 1978).

4.7 MICROSPECIES

Botanists have used another term, **"microspecies,"** for any uniform population that is slightly different

---

FIGURE 4-3

An idealized pattern of circular overlap, showing a series of semispecies in which the terminals are reproductively isolated in sympatry, marked by a shaded zone of overlap. Thus, the semispecies designated A and F behave as authentic biological species in the zone of overlap even though gene flow occurs between semispecies A and B, B and C, and so on. The geographic barrier may be a large area of inhospitable land such as the Great Basin Desert in western North America, or a mass of water such as the Arctic Ocean.

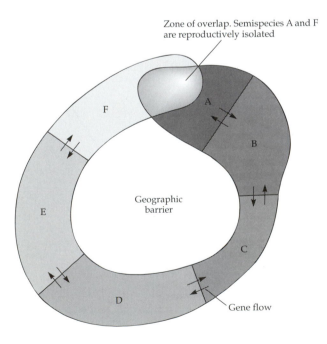

Zone of overlap. Semispecies A and F are reproductively isolated

Geographic barrier

Gene flow

Circular overlap in gulls of the *Larus argentatus* group, based on a synthesis of evidence by Mayr (1963). This figure is a simplified view from above the North Pole. Arrows indicate gene flow between species and within *L. argentatus*, but the species *Larus fuscus* and *L. argentatus*, which overlap along the coasts of Europe, are reproductively isolated.

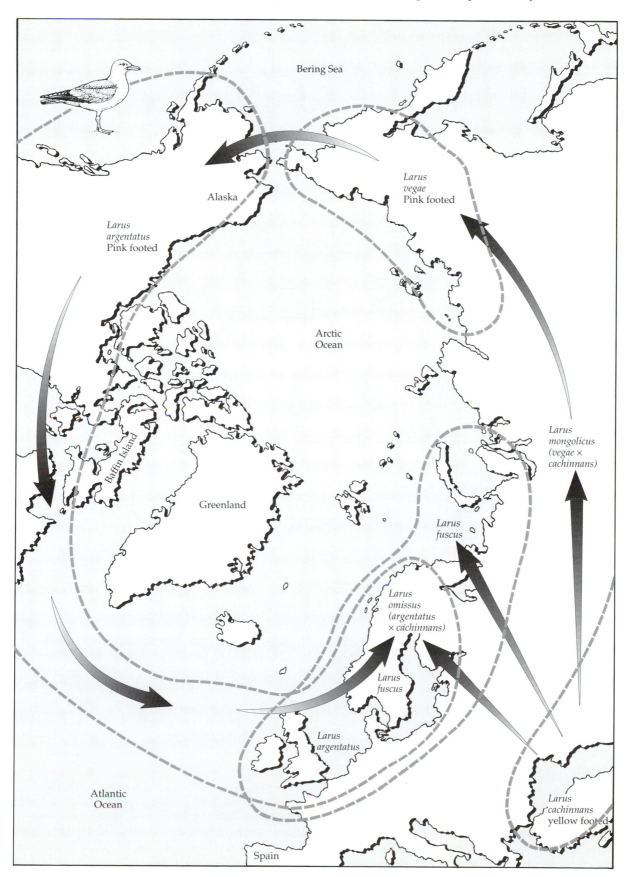

FIGURE 4-5

The Weber River drainage in northern Utah, showing distributions of Fremont cottonwood, *Populus fremontii*; narrowleaf cottonwood, *P. angustifolia*; and the hybrid swarm between them consisting of F₁ hybrids and backcrosses with the *P. angustifolia* parental type (hatched area). Representative leaf shapes in allopatric and sympatric zones are shown. *(Based on Whitham 1989 and Keim et al. 1989)*

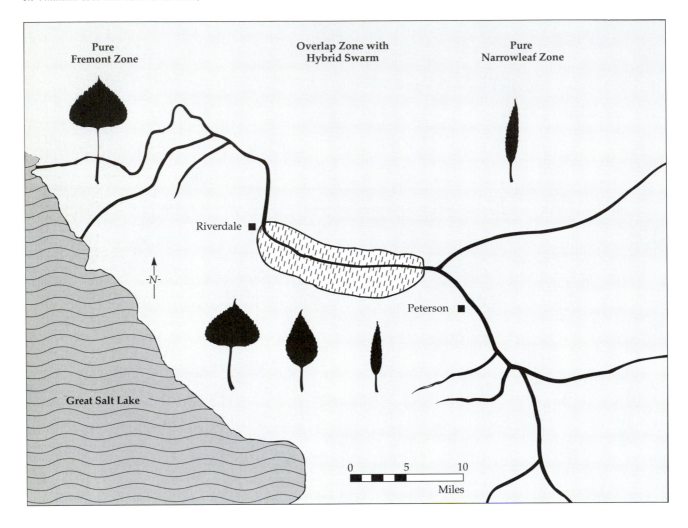

phenetically from related uniform populations. Because of its broadness, the term has been applied in a nonrigorous way to many different kinds of differentiation, such as that between local races, sibling species, and pure lines of cloning species. The only rigorous application of the term "microspecies" is to uniparental plant species, in which each lineage is reproductively isolated from others and a population may be the product of a single, morphologically unique lineage. If the morphological differences can be demonstrated to be genetically controlled, then the microspecies designation is valid.

Microspecies arise from several modes of reproduction. Clonal microspecies may result from vegetative reproduction, as in *Opuntia* cactus species in-

cluding prickly pears and chollas. Agamospermous microspecies are products of the development of viable embryos from unfertilized seeds, as in some *Rubus* species, including the brambles (Edees and Newton 1988). Autogamous microspecies result from self-fertilizing species, such as those in the grass genus *Erophila* (Grant 1971).

4.8 SPECIES OF PLANTS AND ANIMALS

In general, discrete biological species are less common in plants than in animals, and a rather specialized terminology has arisen in plant taxonomy to cope with the wide range of genetic systems and

FIGURE 4-6

The syngameon of Pacific Coast irises in the series Californicae, studied by Lenz (1959). The strengths of lines indicate the relative amounts of natural hybridization between species and semispecies. Location of these entities on the figure is for convenience and does not simulate relative geographic locations. Abbreviations are for species and semispecies names. *(From Grant 1971)*

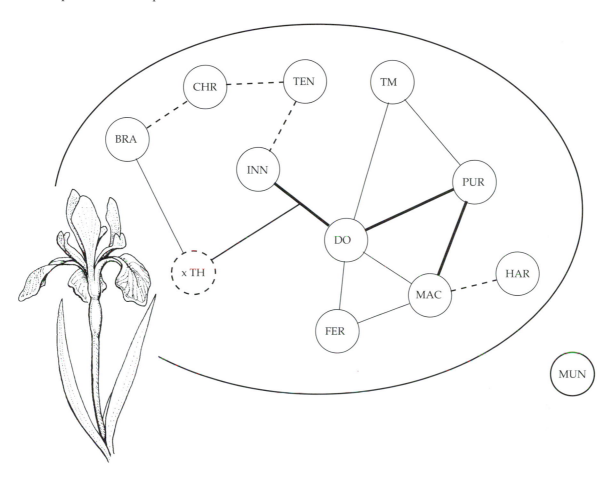

levels of reproductive isolation found in plants. The reason for this difference between plants and animals—especially when we consider the larger members of each kingdom such as angiosperms and mammals—appears to be that plants are simpler than animals and have less complex regulation of developmental processes. Certainly, animals such as mammals have much more complex behavioral repertoires involving courtship and mating, which are important in reproductive isolation. In plants, genes and dosage effects are less critically balanced than in animals. As a result, interspecific hybridization is not such a catastrophic and destabilizing phenomenon in plants as it is in animals. From the relatively open system of plant growth it is simple to develop vegetative and agamospermous forms of reproduction from which populations and species can arise. Even a single

unique individual produced by mutation or hybridization may give rise to a new species of plants. Vegetative reproduction preserves the viability of a new lineage until a new method of sexual reproduction evolves (see Chapter 5).

The genetic plasticity of plants also makes macroevolution in plants significantly different from that in animals, because not only do phylogenetic lines diverge, as in animals, but they can also combine. The lines develop into phylogenetic webs, or **reticulate phylogenies** (Figure 4-7). The complex of species in the genus *Clarkia*, studied by Lewis and Lewis (1955 and subsequently), is an example. *Clarkia* includes some showy annuals grown in gardens. The genus was named for Captain William Clark of the Lewis and Clark Expedition to the western United States (1804–1806).

FIGURE 4-7

An example of a reticulate phylogeny in *Clarkia*, based on research by Lewis and Lewis (1955). Notice that the basic chromosome number is $n = 7$, and the phylogeny has developed with events of aneuploidy, hybridization, and amphiploidy. Specific names and normal haploid chromosome numbers are given, and some examples of diploids, aneuploids, amphiploids, and hybridization are indicated. Sections of the genus are in capital letters.

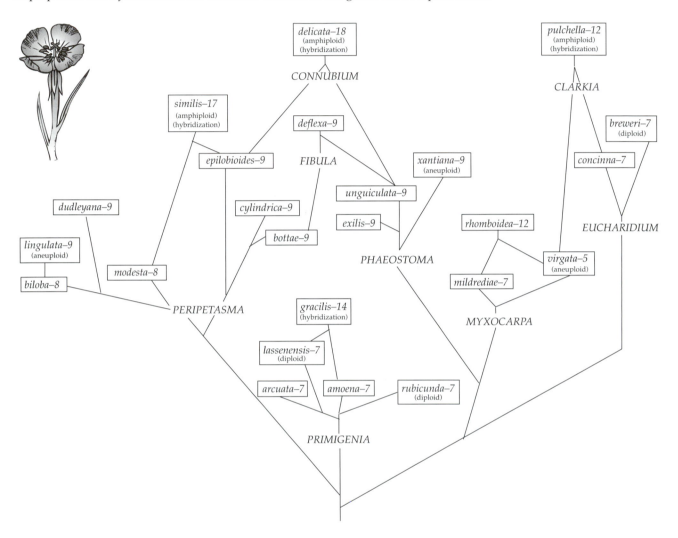

We will learn more about the nature of species as we discuss modes of speciation in the next chapter, genetic systems in Chapter 14, and change in gene frequencies in Chapter 15. Probably no species definition is adequate for all the diverse entities on this earth. Levin (1979) went so far as to say that "plant species lack reality" (381). "Whereas the processes of plant evolution are universal, the products are highly idiosyncratic owing to inherent differences in their genetic systems, sociology, and selection pressures" (384). "We create and amend species interpretations until we have a mentally satisfying organization, and this procedure works quite well for most assemblages of populations" (384).

A decade later, a symposium on clonal organisms reached little consensus on what a species really is, reinforcing Levin's attitude (Budd and Mishler 1990). The smaller an organism, the greater the problems of defining an operational species. Binary fission in bacteria results in almost endless variation among lineages (Cowan 1962), and in protozoa, morphologically similar populations classified into a single species may have many reproductively incompatible strains (Sonneborn 1957). Thus, the biological species concept may apply to many organisms, the evolutionary species concept is useful in other cases, and the typological species concept will remain as a matter of convenience and a practical concept when very little biological information on species is available.

4.9 THE RECOGNITION SPECIES CONCEPT

Despite the many complexities in nature, the typological species concept has worked well for the practicalities of sorting and naming species since Linnaeus's time. What keeps populations looking so similar that we can group them into species, whether reproduction is sexual or asexual? The obverse of the biological species concept (which emphasizes the processes involved with reproductive isolation between developing new species) is the concept that species possess many characteristics involved with fertilization and genetic compatibility. These characteristics, which keep a species a biological entity, deserve more attention in efforts to define what a species really is. In fact, Paterson (1985) made a clear distinction between the two ways of approaching the species debate: What isolating mechanisms result in new species, and what reproductive mechanisms maintain the identities and integrity of species? He renamed the biological species concept the **isolation species concept** to contrast it clearly with his own **recognition species concept** (Paterson 1985). The latter concept emphasizes the complexities of sexual biparental eukaryotic reproduction, involving all the fertilization mechanisms that result in **specific-mate recognition systems.** Courtship, timing, habitat selection, coloration, the endocrine system, copulatory organs, gamete compatibility, and other factors all enter into the ambit of fertilization mechanisms, with many adaptations ensuring that mates can recognize each other and that gamete transfer is effective and productive in terms of progeny. Thus, Paterson (1985, 25) defines a species as "that most inclusive population of individual biparental organisms which share a common fertilization system."

4.10 THE COHESION SPECIES CONCEPT

The cohesion of individuals and populations into the recognizable units we call species is in need of more study (Templeton 1989, Budd and Mishler 1990). In sexual populations, is the movement of individuals between populations part of the blending process that keeps the genetic makeup of the population similar? Is such **gene flow** important? Does natural selection work on each population to keep its characteristics from shifting off the norm? Is such **stabilizing selection** important? Or are there strong **developmental constraints** that prevent divergence in populations from the normal form? (Mishler and Budd 1990).

The **cohesion species concept** formulated by Templeton (1989) emphasizes the mechanisms that maintain the morphological stability of a set of populations. He defines a species as "the most inclusive population of individuals having the potential for phenotypic cohesion through intrinsic cohesion mechanisms" (Templeton 1989, 12). **Cohesion mechanisms** include gene flow, stabilizing selection, developmental constraints, and reproductive isolation. The cohesion species concept has much in common with the evolutionary species concept, but it emphasizes the mechanistic basis of cohesion as the biological species concept does. It also helps us to see the processes involved with breaking down the cohesion observed in so many species so that semispecies, microspecies, sibling species, and morphologically distinct species evolve.

The dialectical nature of science, the importance of alternative views, and the stepwise construction of concepts in biological evolution are of interest here. Templeton (1989), while forwarding his own species concept, recognized the value of its precursors, especially the evolutionary species concept, the isolation species concept, and the recognition species concept. Yet he was dissatisfied with those concepts, as his captions "Too little sex" (8) and "Too much sex" (10) imply. Many species do not have enough sex: they are parthenogenetic, self-fertilizing, cloning, or otherwise do not meet the criterion of biparental sexual reproduction for application of the isolation and recognition species concepts (see Chapter 14 for many examples). Many other species have too much sex: they are promiscuous beyond the bounds of species identity, forming genetically open systems that result in some of the species problems already discussed: hybrid swarms and the syngameon. The cohesion species concept overcomes these problems because, "as with the evolutionary species concept, the cohesion species concept defines species in terms of genetic and phenotypic cohesion" (Templeton 1989, 12). It emphasizes the mechanisms that produce and maintain cohesion. Cloning, asexual, and uniparental species maintain cohesion most effectively. But hybridizing species usually retain distinctive species characteristics as well, and hybrid zones are commonly narrow. The evolutionary species concept emphasizes the presence of evident cohesive units, or species, over evolutionary time, while the cohesion species concept emphasizes the mechanisms that make such cohesion possible.

4.11 THE PHYLOGENETIC SPECIES CONCEPT

A key part of tracing lineages through evolutionary time is the identification of diagnostic characteristics that clearly distinguish one lineage from another and of the nodes at which lineages diverge. The biological species concept does not provide a basis for readily following lineages in a phylogenetic analysis. In systematic biology, we need a species concept that conforms most closely to the processes employed by systematists, as discussed in Chapter 13.

Cracraft (1983, 1989) proposed the **phylogenetic species concept** to meet the needs of evolutionary analysis. He defined a phylogenetic species as "an irreducible (basal) cluster of organisms, diagnosably distinct from other such clusters, and within which there is a parental pattern of ancestry and descent" (Cracraft 1989, 34–35). Avoidance of emphasis on the evolution of reproductive isolation certainly makes the phylogenetic species concept available for more general application to all kinds of organisms: bacteria, protists, fungi, plants, and animals. Systematists and

T A B L E 4 - 2   A summary of species concepts

Species Concept	Emphases	Advantages	Limitations	Proponents
1. Greek	Instability and changeability of species	Realistic description of the apparent variation in nature	Uses term "species" in an outdated way	Aristotle and Theophrastus
2. Typological	Fixity of species	Stability in classification	Fails to document variation	Linnaeus
3. Darwinian	Species as well-developed varieties	Conformance with evolutionary concepts of change	Does not account for discreteness of species	Darwin
4. Biological (= isolation)	Common gene pool within species and reproductive isolation between species	Focus on mechanisms in speciation; testability	Applies well only to outcrossing bisexual eukaryotic species	Mayr
(a) Nondimensional	Distinctness of species overlapping in time and space	Strong focus on characteristics of living species	Does not apply to allopatric and allochronic populations	Mayr
(b) Multidimensional	Distinctness of species in different locations or reproducing at different times	Recognition of potentially common gene pool	Test results are ambiguous when proximity of individuals is forced artificially	Mayr
5. Evolutionary	Common lineage and ecological niche, normalizing selective pressure	Applicability to species with all types of reproduction	Is not mechanistic but descriptive	Simpson and Grant
6. Recognition	Species kept distinct by common fertilization system	Mechanistic basis for the unity of the species	Applies only to bisexual eukaryotes	Paterson
7. Cohesion	Mechanisms of cohesion within a species	Applicability to all types of species and modes of reproduction	Complex program of research is needed to understand cohesion mechanisms	Templeton
8. Phylogenetic	Diagnostic characteristics separating species and lineages	General applicability to all kinds of organisms; pragmatic	Mechanisms resulting in diagnosable differences are deemphasized	Cracraft

other evolutionary biologists can also employ the phylogenetic concept operationally. The biologist who emphasizes variation of diagnostic characteristics can use the concept to discern the history of divergence from a common ancestor.

The diagnostic characteristics used in phylogenetic analysis vary enormously, of course, and any traits can be scrutinized. Chapter 13, "Biological Classification," will discuss this topic further.

Table 4-2 provides an overview of major features of each species concept. Notice that every concept was perceived to have advantages when it was proposed, but limitations of the concept have emerged more recently.

SUMMARY

In ancient Greece, a species was considered to be a very unstable entity that could change appearance when placed in different environments and hybridize rampantly with all kinds of other organisms. The Greek view was replaced by the **typological species concept,** based on the **fixity of species,** such that one specimen was adequate to represent a species, describing each specimen briefly and assigning it a Latin binomial according to the system Linnaeus developed.

The **biological species concept** provides a more mechanistic view of the species, because it recognizes the coherent nature of populations maintained by interbreeding and the discreteness of species resulting from reproductive isolation. This concept addresses reproductive isolation in populations and species that live in the same locality (the **nondimensional view**) as well as in populations and species that live in different localities (the **multidimensional view**). The historical development of understanding of the cricket *Gryllus fasciatus,* from its naming in 1773 to the present recognition of a complex of six species in the genus *Allonemobius,* is an example of how the concept of a species may change.

Problems arise in the application of the biological species concept because the demand for detailed biological information becomes impractical when large numbers of species must be sorted and named. Problems also arise when hybridization is very common and when reproduction in a species is commonly asexual. The **evolutionary species concept** attempts to circumvent these problems by emphasizing the importance of the lineage, its evolutionary and ecological role, and the stability of form maintained by natural selection. The evolutionary species concept covers **semispecies,** which are intermediate between races and species and include cases of **circular overlap, hybrid swarms,** and **microspecies** resulting from asexual reproduction in lineages. Recognition of the **syngameon,** a group of many hybridizing species, requires acknowledgment of the openness of many species to gene flow within a genus. Plants appear to be more flexible genetically than animals, permitting more hybridization and the blending of once separate lineages to form **reticulate phylogenies.**

Increasing emphasis on the mechanisms that keep individuals and populations in a cohesive unit, so that species can be recognized as discrete, will promote the understanding of the species. The **recognition species concept** emphasizes the mechanisms resulting in fertilization and zygote formation, called **specific-mate recognition systems,** and views the species as the most inclusive population that shares such a fertilization system. The **cohesion species concept** emphasizes the mechanistic processes that maintain the morphological stability of a set of populations, including gene flow, stabilizing selection, and developmental constraints. A blend of species concepts promises a better understanding of the speciation process. The **phylogenetic species concept** emphasizes the need to track lineages in an evolutionary analysis, using diagnostic characteristics. The identification of such characteristics enables the development of a historical record of divergence from ancestors to descendants.

QUESTIONS FOR DISCUSSION

1. In your opinion, is the definition of a species simple for most species—and we nonetheless spend much time trying to use a definition to embrace *all* species—or is the challenge of providing a comprehensive definition of species a general and difficult one?

2. Which definitions of species do you think are aimed at identifying the core biology of a species, and which definitions seem to be aimed more at getting around the difficulty of a definition?

3. Do you think there is progress yet to be made in understanding the nature of species, and which direction would lead to increasing knowledge?

4. If you were a systematist working on a large group of species, which species concept would you employ?

5. How would you design studies over the geographic range of a species to explore the factors that keep members recognizable and functional as members of the same species over that range?

6. In your opinion, is the concept of the species important enough to spend a chapter on? Explain the pros and cons of such emphasis.

7. In a cloning species represented by many reproductively isolated lineages, which kinds of cohesion mechanisms are likely to function, and how would you investigate the relative importance of each?

8. How does the study of complex systems of species, such as the syngameon, reticulate phylogenies, hybrid swarms, and microspecies, advance scientific knowledge in a general way?

9. If Darwin could have read this chapter on species concepts (say, in 1858), do you think he would have significantly changed his concept of the species in *The Origin of Species*?

10. If species status is so dynamic, as the *Allonemobius* crickets in this chapter illustrate, is there any prospect for a stable classification of most groups in the near future?

REFERENCES

Alexander, R. D., and E. S. Thomas. 1959. Systematic and behavioral studies on the crickets of the *Nemobius fasciatus* group (Orthoptera: Gryllidae: Nemobiinae). Ann. Entomol. Soc. Amer. 52:591–605.

Benedix, J. H., and D. J. Howard. 1991. Calling song displacement in a zone of overlap and hybridization. Evolution 45:1751–1759.

Budd, A. F., and B. D. Mishler. 1990. Species and evolution in clonal organisms—summary and discussion. Syst. Bot. 15:166–171.

Cowan, S. T. 1962. The microbial species—a macromyth? Symp. Soc. Gen. Microbiol. 12:433–455.

Cracraft, J. 1983. Species concepts and speciation analysis. Curr. Ornithol. 1:159–187.

———. 1989. Speciation and its ontology: The empirical consequences of alternative species concepts for understanding patterns and processes of differentiation, pp. 28–59. *In* D. Otte and J. A. Endler (eds.), Speciation and Its Consequences. Sinauer, Sunderland, Mass.

Darwin, C. 1859. On the Origin of Species by Natural Selection, or the Preservation of Favoured Races in the Struggle for Life. Murray, London.

De Geer, C. 1773. Memoires pour servir a l'histoire des insectes. Vol. 3. Hosselberg, Stockholm.

Edees, E. S., and A. Newton. 1988. Brambles of the British Isles. Ray Society, London.

Fulton, B. B. 1931. A study of the genus *Nemobius* (Orthoptera: Gryllidae). Ann. Entomol. Soc. Amer. 24:205–237.

———. 1933. Inheritance of song in hybrids of two subspecies of *Nemobius fasciatus* (Orthoptera). Ann. Entomol. Soc. Amer. 26:368–376.

———. 1937. Experimental crossing of subspecies in *Nemobius* (Orthoptera: Gryllidae). Ann. Entomol. Soc. Amer. 30:201–207.

Gove, P. B. (ed.) 1986. Webster's Third New International Dictionary of the English Language. Merriam-Webster, Springfield, Mass.

Grant, V. 1963. The Origin of Adaptations. Columbia Univ. Press, New York.

———. 1971. Plant Speciation. Columbia Univ. Press, New York.

Howard, D. J. 1982. Speciation and coexistence in a group of closely related ground crickets. Ph.D. dissertation. Yale Univ., New Haven, Conn.

———. 1983. Electrophoretic survey of eastern North American *Allonemobius* (Orthoptera: Gryllidae): Evolutionary relationships and the discovery of three new species. Ann. Entomol. Soc. Amer. 76:1014–1021.

———. 1986. A zone of overlap and hybridization between two ground cricket species. Evolution 40:34–43.

Howard, D. J., and D. G. Furth. 1986. Review of the *Allonemobius fasciatus* (Orthoptera: Gryllidae) complex with the description of two new species separated by electrophoresis, songs and morphometrics. Ann. Entomol. Soc. Amer. 79:472–481.

Howard, D. J., and P. G. Gregory. 1993. Post-insemination signaling systems and reinforcement. Phil. Trans. R. Soc. Lond. B 340:231–236.

Howard, D. J., and R. G. Harrison. 1984a. Habitat segregation in ground crickets: Experimental studies of adult survival, reproductive success, and oviposition preference. Ecology 65:61–68.

———. 1984b. Habitat segregation in ground crickets: The role of interspecific competition and habitat selection. Ecology 65:69–76.

Howard, D. J., and G. L. Waring. 1991. Topographic diversity, zone width, and the strength of reproductive isolation in a zone of overlap and hybridization. Evolution 45:1120–1135.

Howard, D. J., G. L. Waring, C. A. Tibbets, and P. G. Gregory. 1993. Survival of hybrids in a mosaic hybrid zone. Evolution 47:789–800.

Keim, P., K. N. Paige, T. G. Whitham, and K. G. Lark. 1989. Genetic analysis of an interspecific hybrid swarm of *Populus*: Occurrence of unidirectional introgression. Genetics 123:557–565.

Lenz, L. W. 1959. Hybridization and speciation in Pacific coast irises. Aliso 4:237–309.

Levin, D. A. 1979. The nature of plant species. Science 204:381–384.

Lewis, H., and M. Lewis. 1955. The genus *Clarkia*. Univ. Calif. Publ. Bot. 20:241–392.

Linnaeus, C. 1753. Species Plantarum. L. Salvii, Holminae.

———. 1758. Systema Naturae: Regnum Animale. 10th ed. L. Salvii, Holminae.

Lotsy, J. P. 1925. Species or Linneon. Genetica 7:487–506.

Mayr, E. 1963. Animal Species and Evolution. Belknap Press of Harvard Univ. Press, Cambridge, Mass.

Mishler, B. D., and A. F. Budd. 1990. Species and evolution in clonal organisms—Introduction. Syst. Bot. 15:79–85.

Otte, D., and J. A. Endler (eds.). 1989. Speciation and Its Consequences. Sinauer, Sunderland, Mass.

Paterson, H. E. H. 1985. The recognition concept of species. *In* E. S. Vrba (ed.), Species and Speciation. Transvaal Museum Monograph 4:21–29.

Simpson, G. G. 1951. The species concept. Evolution 5:285–298.

———. 1961. Principles of Animal Taxonomy. Columbia Univ. Press, New York.

Slobodchikoff, C. (ed.) 1976. Concepts of Species. Dowden, Hutchinson and Ross, Stroudsburg, Penn.

Sokal, R. R., and T. J. Crovello. 1970. The biological species concept: A critical evaluation. Amer. Natur. 104:127–153.

Sonneborn, T. M. 1957. Breeding systems, reproductive methods, and species problems in protozoa, pp. 155–324. *In* E. Mayr (ed.), The Species Problem. Amer. Assoc. Adv. Sci. Pub. 50.

Templeton, A. R. 1989. The meaning of species and speciation: A genetic perspective, pp. 3–27. *In* D. Otte and J. A. Endler (eds.), Speciation and Its Consequences. Sinauer, Sunderland, Mass.

Vickery, V. R., and D. E. Johnstone. 1970. Generic status of some Nemobiinae (Orthoptera: Gryllidae) in northern North America. Ann. Entomol. Soc. Amer. 63:1740–1749.

Whitham, T. G. 1989. Plant hybrid zones as sinks for pests. Science 244:1490–1493.

Yeo, P. F. 1978. A taxonomic revision of *Euphrasia* in Europe. Bot. J. Linn. Soc. 77:223–334.

Zirkle, C. 1959. Species before Darwin. Proc. Amer. Philos. Soc. 103:638–644.

5

THE ORIGIN OF NEW SPECIES

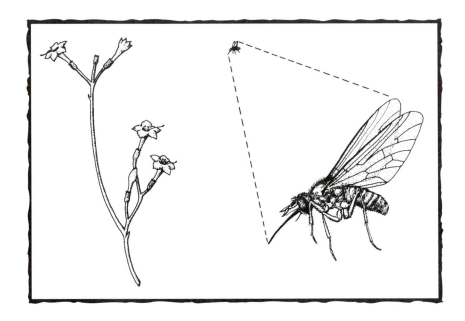

Gillia tenuiflora, a member of the phlox family, and *Oligodranes* sp. in the family of bee flies, Bombyliidae—the most numerous and effective pollinator in a population near Creston, California. *(Based on illustrations in Grant and Grant 1995)*

5.1 Geographic Speciation, or Allopatric Speciation

5.2 Sympatric Speciation

5.3 Parapatric Speciation

5.4 Speciation Mechanisms and the Kinds of Species Involved

5.5 Primary Speciation in Plants

5.6 Secondary Speciation in Plants

5.7 Agamic Systems

Summary

Darwin's (1872) **model of speciation** must be teased out of his extensive discussion on the subject, but a careful reading reveals a rather simple view of the mechanism involved with the evolution of new species. First a species spreads out over a large area (see Figure 3-4). Then environmental variation within this area produces different variations in the populations of different environments. Restricted movement of individuals between environments results in independent natural selection and finally a new species in each environment.

Hybrid zones give rise to maladapted individuals that survive poorly in both parental environments. Because environments are not completely discrete, variation between species is continuous. The hybrid populations are small and must compete with parental populations; they are likely to eventually go extinct.

It will be interesting to keep Darwin's model in mind and compare it with other speciation models that developed subsequently.

5.1 GEOGRAPHIC SPECIATION, OR ALLOPATRIC SPECIATION

After Darwin's time, a major building block of the modern synthesis was Mayr's model of **geographic speciation,** or **allopatric speciation.** The biological species concept serves as an essential component of geographic speciation, for it explains how two or more populations that once shared a common gene pool become reproductively isolated. Mayr's view that the speciation process is almost exclusively geographic has been largely accepted. As we shall see, alternative mechanisms are also important, and views on the relative contributions of those mechanisms to the diversity of life on earth are shifting. Even textbooks on evolution debate the validity of modes of speciation other than geographic speciation. Clearly, no consensus has been reached (cf. Cockburn 1991, Futuyma 1986, Minkoff 1983, Ridley 1993, Strickberger 1990, and this chapter).

We may call Mayr's views on geographic speciation the classic model of speciation, as they were largely accepted in the 1940s and are still intact. The classic model invokes a geographic barrier as an essential part of the scenario (Figure 5-1). A geographic barrier divides a population, gene flow between the two parts stops, each new population evolves independently of the other, and differences accumulate until the members of one population cannot breed with members of the other even when the ranges again overlap. Thus, two new species have evolved.

Allmon (1992) emphasizes the need for a more detailed mechanistic understanding of each phase of allopatric speciation. He notes that the process can be divided into three stages: the formation of an isolated population, the persistence of that population, and the differentiation of the population from the parental stock. The heuristic value of this subdivision lies in the increased focus on each phase. For example, in isolate formation, what are the relative roles of **extrinsic mechanisms,** such as glaciation effects, and **intrinsic mechanisms,** especially dispersal ability or vagility? Does isolate persistence depend on population size and stability, ecological amplitude or specialization, environmental rigor, niche space, or adaptation? What are the key processes in isolate differentiation? Keeping Allmon's arguments in mind will help us assess the thoroughness with which scientists have described each process of speciation discussed in this chapter.

Speciation in Warblers

Mengel (1964) developed an example of geographic speciation involving northern wood warblers (in North America) that divided into two large populations. In the black-throated green warbler group are one eastern species, the black-throated green warbler, *Dendroica virens,* and three western species with extensive distributions (Figure 5-2): Townsend's, hermit, and black-throated gray. Mengel argued that the parental species with a southeastern distribution adapted to coniferous forest during the Pleistocene Nebraskan glaciation, which compressed vegetation toward the south. During the interglacial period, the warblers expanded north and west with the conifers. The next glaciation, the Kansan, reached its southern extremity in central North America and constituted a geographic barrier separating the eastern parental species from a western population. Speciation occurred. The process was repeated in the next interglacial period, the Yarmouth, and the Illinoian glaciation, cutting off a second western species. A third warbler species in the West originated in a similar manner after the Sangamon interglacial period and the Wisconsin glaciation. This scenario also fits the distribution of species in the Nashville warbler group, reinforcing the view that large-scale geographical barriers formed by glaciation caused similar patterns of warbler distribution and (avoiding complexities not to be discussed here) explaining why three species of the black-throated green warbler group occur in the West and only one occurs in the East.

Speciation in Darwin's Finches

Alternatively, only one inseminated female or a very few individuals may colonize a new locality—perhaps an oceanic island 500 miles off the mainland—

FIGURE 5-1

Geographic, or allopatric, speciation by division of populations resulting from a geographic barrier.

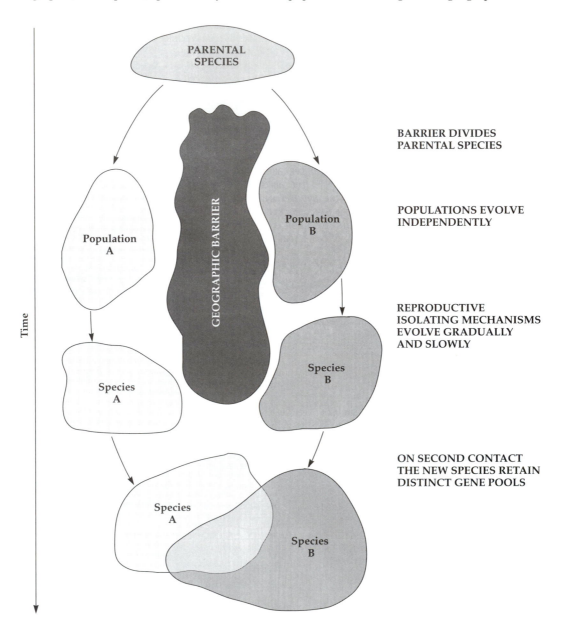

and found a new population. The **founder effect,** discussed in Section 5.4, can be defined as the founding of a new population by one or a few individuals, which carry only a small proportion of the genetic variation of the parental population. With the founder effect, sampling error becomes an important influence in evolutionary change, as Section 5.5 and Chapter 15 discuss. The new population evolves independently of the parent population and eventually becomes reproductively isolated from it.

Because the ocean is so effective as a geographic barrier, such colonization events are extremely rare.

We find an example in the original colonization of the Galapagos Archipelago by a South American finch, the formation of a new species, and subsequent speciation as new islands were colonized. The species now called Darwin's finches (Figure 5-3) are the result.

Geographic Barriers

Thus, geographic barriers may take very different forms. For large mammals and birds, they may be large expanses of ocean, high mountain ranges, or

FIGURE 5-2

Allopatric speciation of *Dendroica* warblers, based on Mengel's (1964) argument that major glaciations produced geographic barriers, resulting in a new species of warbler in western North America after each event. Note that another glaciation event is likely to compress the distribution of *D. virens* to the south and cut off another species in the west. The additional bird illustrated is Wilson's warbler (*Wilsonia pusilla*), which has a trans-American range, posing the question of how this species escaped the geographic barrier of an ice cap and consequent speciation.

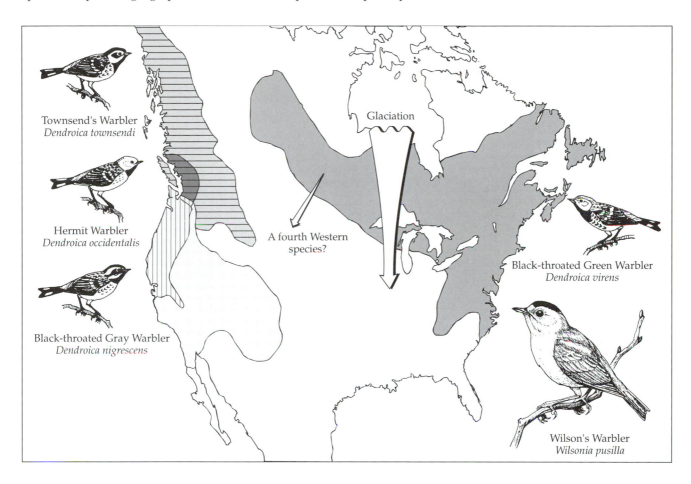

glaciers (e.g., Mengel 1964) (Figure 5-3). For terrestrial animals in general, they may be deep river valleys, wide rivers, or deserts. For fishes, they may be waterfalls, narrows, or salt water. And for smaller organisms, these and many more kinds of barriers are likely to play roles. The geographic barriers resulting in geographic isolation are clearly environmental properties that constrain gene flow between populations.

Reproductive Isolating Mechanisms, or Reproductive Barriers

Reproductive isolating mechanisms are the properties of individuals, populations, and species that constrain gene flow between populations and species.

These properties evolve during geographic isolation and may take many different forms. Mayr (1963, 91) defined reproductive isolating mechanisms as "biological properties of individuals that prevent the interbreeding of populations that are actually or potentially sympatric." His classification scheme illustrates the great diversity of the mechanisms that are potentially involved in reproductive isolation (Table 5-1). Note that in hybrid sterility, so long as the F_1 hybrid has a lower fitness than either of the parental types, the two populations remain isolated. Also, there are considerable differences between premating and postmating barriers. In the case of the former, no loss of time, energy, or gametes occurs in seasonal or habitat isolation or in ethological isolation, and some loss of time and energy may occur in mechanical isolation. Postmating mechanisms, however, waste

FIGURE 5-3

Allopatric speciation by the founder effect, based on Grant (1986) relating to speciation of Darwin's finches in the Galapagos. The bird illustrated is the large ground finch, *Geospiza magnirostris*.

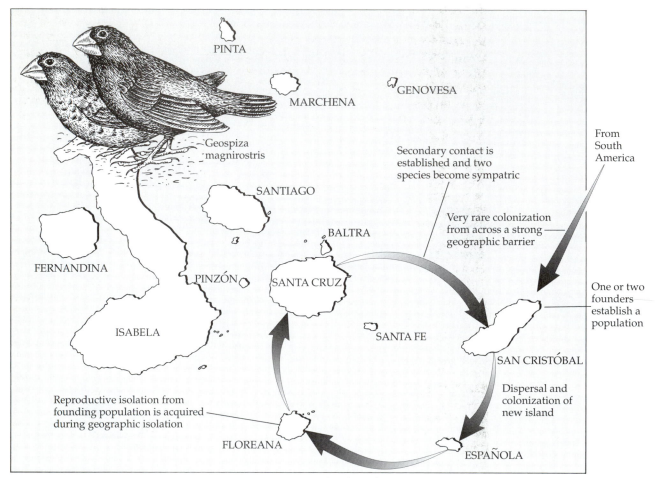

Another colonization event

time, energy, and gametes, and so these mechanisms are particularly inefficient in terms of energetics and reproductive effort. Although natural selection may be expected to increase efficiency in energy and gamete allocation and effect a transition from postmating to premating mechanisms as populations and species diverge, researchers have had difficulty finding good evidence for reinforcement of premating isolation (Butlin 1989).

One can develop a legitimate argument that isolating mechanisms are actually a whole set of characteristics that usually facilitate reproduction within a species, as Chapter 4 discusses (Paterson 1985, Templeton 1989). The emphasis should therefore be on the positive mechanisms involved with mate recognition, acceptance, and interfertility rather than on the resultant reproductive isolation between species. That is

why Paterson (1985) emphasized his **recognition species concept.** Attention to positive mechanisms may suggest a term such as **reproductive barrier** rather than reproductive isolating mechanism. However, a geographic obstacle can also act as a reproductive barrier, and so our preference is to keep the two terms distinct: a geographic barrier is a property of the environment, as described earlier, and a reproductive isolating mechanism is a property of individuals, as just explained.

Secondary Contact

After geographic isolation and the divergence of populations, secondary contact may be established. The results of such contact vary depending on the duration of isolation, the effectiveness of the geographic

TABLE 5-1   Reproductive isolating mechanisms proposed by Mayr (1963)

Type	Mechanism and Example
A. *Premating mechanisms*	Transfer of gametes is prevented.
1. Seasonal and habitat isolation	Potential mates do not meet; e.g., small-mouthed salamander, *Ambystoma texanum*, breeds in ponds, but *A. barbouri* breeds in streams (Kraus and Petranka 1989).
2. Ethological isolation	Potential mates meet but do not mate. Some species of *Allonemobius* crickets (see Chapter 4), leopard frogs (Hillis 1988). In pollination systems, a pollinating insect may carry pollen but does not deposit it on the stigma.
3. Mechanical isolation	Copulation is attempted, but no transfer of sperm takes place. Many insect genitalia require precise lock-and-key link for transfer of sperm (cf. Eberhard 1985).
B. *Postmating mechanisms*	
1. Gametic mortality	Sperm transfer takes place but egg is not fertilized. Pseudogamy in parthenogenetic *Poeciliopsis* related, for example, to the Sonoran topminnow (e.g., Vrijenhoek 1984), plants, and nematodes.
2. Zygotic mortality	The egg is fertilized but the zygote dies. Some leopard frogs in the *Rana pipiens* complex (Hillis 1988).
3. Hybrid inviability	Zygote produces an F_1 hybrid of reduced viability, or hybrid recombinations are less viable—e.g., frogs in the genus *Pseudophryne* (Woodruff 1979).
4. Hybrid sterility	F_1 hybrid is fully viable but partially or completely sterile, or it produces a deficient F_2. Crosses between horses and donkeys produce mules, which are sterile. Some hybrids in the *Rana pipiens* complex (Hillis 1988).

barrier, and ultimately the degree to which reproductive isolating mechanisms have evolved.

Let us call one scenario *case 1*. After very effective and long-term isolation, on secondary contact populations have diverged into discrete ecological niches and reproductive isolation is complete. New species become sympatric without any mutually induced evolutionary change (Figure 5-4).

Case 2 may occur after less prolonged geographic isolation or because of slower evolutionary rates, such that ecological divergence may be less complete, although reproductive isolation may be complete. During secondary contact, competition between the two new species is likely to result in mutually induced divergence in ecologically relevant traits (Figure 5-4). This process, which Darwin (1859) called **character divergence** and Brown and Wilson (1956) called **character displacement,** need not involve morphological characteristics. Rather, it can involve only characteristics of behavior, habitat use, and food choice. (Since the 1950s, the term "character displacement" has broadened to include character divergence, convergence, and parallel characteristic shifts in zones of sympatry [cf. Taper and Case 1992].)

In *case 3*, after shorter periods of geographic isolation that result in incomplete reproductive isolation, hybridization may occur between the two incipient species during secondary contact. One possibility in case 3, which Figure 5-4 illustrates, is that hybrids have equal or greater fitness than either of the parental populations. The result is complete **introgression,** or the incorporation of genes of one species into the gene pool of another. Any differences between the populations that evolved during geographic isolation would be swamped, and something akin to the old parental species would be reinstated (Figure 5-4). Another possibility is that hybrids may have low fitness, and thus introgression does not occur. Theoretically, natural selection might then improve the reproductive isolating mechanisms so that hybridization would diminish, although the reality of this process, called the **Wallace effect,** has been questioned (e.g., Butlin 1989).

Thus, the model of geographic speciation accounts successfully for many characteristics observed when species coexist, from lack of interbreeding and the presence of distinct ecological niches all the way to considerable introgression and strongly overlapping ecological niches.

FIGURE 5-4

Three possible cases during geographic isolation. *Case 1:* After prolonged geographic isolation, reproductive isolating mechanisms (RIMs) are complete and ecological divergence (ED) is complete. *Case 2:* RIMs are complete but ED is incomplete. *Case 3:* RIMs are incomplete. SC indicates time of secondary contact between diverging populations. Time passes from top to bottom in each case, and the parental species occupies niche A. Case 1 in the left-hand population illustrates ecological divergence into niche B.

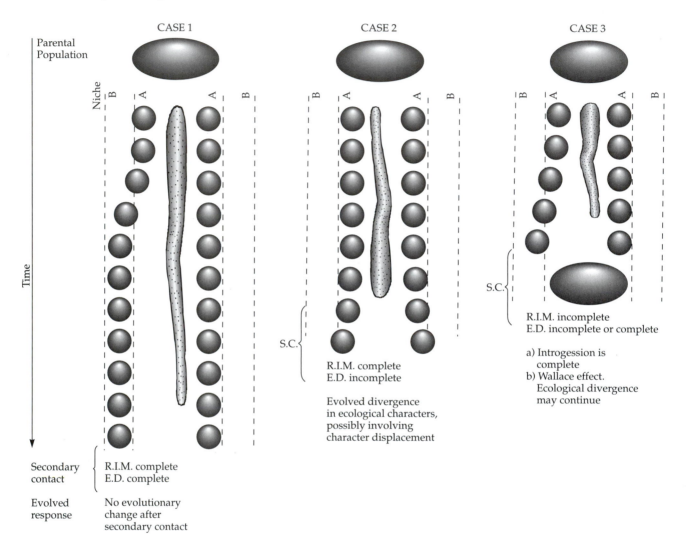

5.2 SYMPATRIC SPECIATION

Mayr (1963) pointed out that the concept of speciation without geographic isolation was much older than his geographic speciation model, as Darwin's eventual speciation model suggests. Mayr gave a brief historical account of other concepts and the negative criticism they had received. One concept seemed more viable than most others. It concerns the speciation process in rather specialized groups of insects and the evolution of new species within one locality. This **sympatric speciation model** can be defined as "the establishment of new populations of a species in different ecological niches within the normal cruising range of the individuals of the parental population" (Mayr 1963, 449) accompanied by rapid development of reproductive isolation between members of the populations in different niches.

Two points relating to specialized, or monophagous, insect groups are worth noting. First, plant-feeding monophagous insects frequently belong to large genera. Thus, their apparently rapid speciation is unlikely to have occurred during prolonged geographic isolation. For example, Ross (1962) reported that the *Erythroneura* group of leafhoppers (Cicadellidae) includes approximately 500 species, and speciation has involved about 150 shifts from utilization of one host plant species to another. Such extensive speciation is likely where many new niches exist, and with high food specificity, ecological barriers to introgression are likely to be very effective. Second, it seems unlikely that a geographically isolated monophagous species would repeatedly split into two species, each with a different host food plant. In many cases we should expect retention of the ancestral food plant by both species, but this is not common.

An Allopatric Alternative

In defense of the allopatric speciation model, Mayr (1963) pointed out that peripheral populations are likely to be exceedingly important in the speciation process. We can call this view the **marginal-population allopatric speciation model** (Figure 5-5). In a marginal population, a subsidiary host species may become the primary host because it is more abundant than the major host or is favored in other ways. (This scenario assumes that the so-called specialized insect species is not strictly monophagous but utilizes more than one host plant species, which is commonly the case.) Such a shift in hosts results in rapid adaptation to the new conditions and rapid genetic change in the small isolated population. At the same time, reproductive isolation from the parental population may develop. On secondary contact, the new species

on its new host plant remains sympatric with the parent species, and the history and mechanism of the speciation process are obliterated. Thus, the marginal-population allopatric model is a special case of the geographic speciation model.

Mayr (1963) also noted problems with applying the geographic speciation model to animal parasites. For example, a sympatric model for speciation of the human louse into the body louse and the head louse need not be complex (cf. Price 1980). Mayr, however, argues that the current species probably diverged allopatrically, the body louse evolving on the bodies of heavily clothed people such as Eskimos and the head louse on people clothed only in breechclouts and grass skirts or less.

Any discussion on speciation processes will fail to prove either the allopatric or the sympatric model beyond a doubt unless the actual process of speciation can be studied in ecological time, over a few decades or years. This kind of study, of course, is difficult to carry out, and it has seemed that speciation is such a gradual and rare event—relative to the 130 or so years since Darwin—that the issues will remain tantalizing but unresolved. Yet increasing evidence forces us to consider sympatric speciation as a reality. Because this kind of speciation can occur rapidly, we may indeed observe the process in ecological time.

FIGURE 5-5

Mayr's (1963) marginal-population allopatric speciation model.

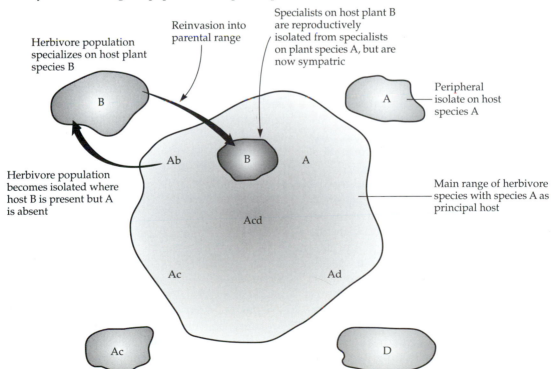

Others in addition to Mayr (e.g., Futuyma and Mayer 1980) have criticized the evidence for sympatric speciation and other modes of nonallopatric speciation. We discuss these modes here because they offer explanations of how speciation can result in very closely related sibling species, an event that would not be expected to result from allopatric speciation (cf. Bush and Howard 1986, Tauber and Tauber 1989, Cockburn 1991).

Speciation in *Rhagoletis* Fruit Flies

In the light of tantalizing debates and the skepticism of entomologists and botanists about the universality of the geographic speciation model, Guy Bush in the 1960s undertook a study of host-specific true fruit flies in the genus *Rhagoletis* and family Tephritidae. He knew that several species in this genus had recently established new host races on introduced commercial fruits. The apple maggot, *R. pomonella*, had moved from its native host, hawthorn (*Crataegus* sp.), to introduced apples in the Hudson River Valley, New York, in 1864. Then it had shifted from apples to cherries in Door County, Wisconsin, as recently as 1960 (Bush 1974, 1975a). Another host race of *R. pomonella* occurs on wild and cultivated plums, but less is known of its biology. In western North America, the western cherry fruit fly, *R. indifferens*, in Oregon, Washington, and British Columbia, shifted from the native bitter cherry, *Prunus emarginata*, to the domesticated cherries, *P. avium* and *P. cerasus*, a little before 1913, only 89 years after cherries were introduced into the region. Where bitter cherry and domestic cherries grow close together in California, *R. indifferens* occasionally attacks domestic cherries as if establishing a new host race. However, the California Department of Agriculture quickly expunges these populations.

It became evident that much information—behavioral, ecological, and genetic—was needed to establish the likelihood of sympatric speciation. Bush and his coworkers have provided such information. Figure 5-6 summarizes much of the behavior of *Rhagoletis* fruit flies in relation to host selection and mating behaviors. Note that a female fly finally accepts a fruit based on a chemical cue, which must be received by her chemoreceptor. A mutation in the fly could therefore affect the amino acid composition of a receptor protein and radically change the fly's response to a specific host-plant chemical. Also, courtship and mating occur on the host fruit, and fruits are essential to fly reproduction. Thus, host shifts often result in fly populations that are allochronic, or chronologically out of phase. Once a host shift has occurred, allochronic isolation can become a factor in limiting gene flow.

For example, the two emergence patterns of *R. pomonella* in Door County, Wisconsin, are so different that the hawthorn and cherry races of flies show practically no overlap (Figure 5-7). Emergence time is under genetic control (Smith 1988). T. K. Wood and his associates have explored in detail the processes involved in allochronic isolation in the herbivorous membracid treehopper genus *Enchenopa* (Wood and Keese 1990, Wood et al. 1990, and references therein).

The shift of *R. indifferens* from native cherry to the introduced *Prunus avium* in California involves both temporal and spatial factors. On Mount Shasta, for example, the two tree species overlap in a zone from 3,500 to 5,000 feet above sea level, but their fruiting times are so different that there is only a 2-week period in July and August in which *R. indifferens* populations overlap (Figure 5-8).

At least two genes must be involved in a host shift by an insect: one for host recognition and selection and another for larval survival once the egg is deposited. Huettel and Bush (1971) have indeed shown, in a related species of tephritid, that a single major locus controls host selection. In addition, Hatchett and Gallun (1970) found that a gene-for-gene relationship exists between host resistance and parasite survival in the hessian fly, which attacks wheat. Genes for plant resistance to the fly are matched by genes for virulence in the fly population, as described in Section 19.2. Apparently, such relationships between plant pathogens and their hosts are common (Day 1974). Thus, a shift from one host to another of similar chemistry may involve only two genes, and, because of allochronic isolation and mating on the fruit, a shift causes ecological isolation, strongly reduced gene flow between host races, and essentially instantaneous and sympatric speciation. Isolation of races is followed by changes in other loci to adapt the new race to new conditions. Thus, a factor such as larval survival may become polygenically controlled once evolution of the new race has progressed.

Bush (1975a, 199) stated,

> It is now quite clear that host races of phytophagous parasitic insects have evolved sympatrically. Furthermore, these host races are undoubtedly the progenitors of the many reproductively isolated sibling species so frequently found coexisting sympatrically on different host plants.

Even Mayr in 1970 accepted the real possibility of incipient sympatric speciation in *Rhagoletis*.

Bush (1975a) observed that sympatric speciation probably has been common in monophagous and stenophagous parasitic insects, but certain qualities of species predispose them to host shifting. (1) Mating

FIGURE 5-6

The behaviors of fruit flies, *Rhagoletis pomonella*, involved with host choice and acceptance. Male fruit flies are symbolized by ♂ and female flies by ♀.

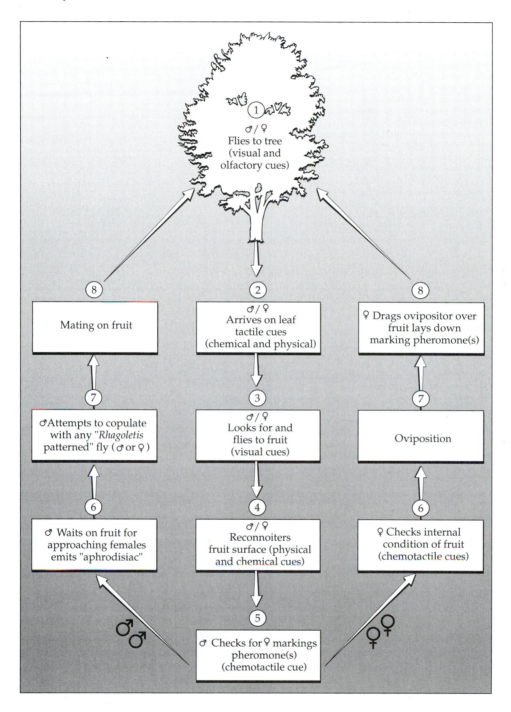

occurs on or near the host plant. (2) The adult female selects the host; the larvae have no choice. (3) The parasitic species is specialized, attacking groups of closely related host plant species. (4) Host selection and larval survival are under genetic control; other factors reinforce reproductive isolation and accelerate the speciation process. (5) As Smith (1988) demonstrated, univoltinism (a single generation per year) with genetic control of emergence time increases the probability of allochronic isolation. (6) If genetic

FIGURE 5-7

Availability of host plant fruits (top) and approximate emergence times of apple maggot adults (below) from different sympatric hosts in Door County, Wisconsin. *(Based on Bush 1975a)*

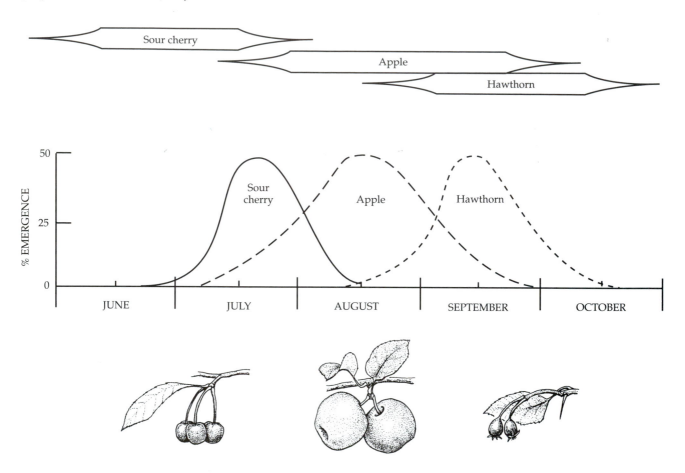

The Genetic Model

control of host-plant induction or conditioning is present, where an adult insect is induced or conditioned to return to the same host from which it emerged, fidelity to the specific ecological niche increases.

The Genetic Model

Bush (1974, 1975a) based his proposal for a generalized genetic model of sympatric speciation on the shift of *R. pomonella* from apples to cherries in Door County, Wisconsin. For a host shift to occur, mutations must produce new host selection (H) and survival (S) alleles in a race—e.g., the apple race of flies—that is homozygous at those loci: $H_1H_1S_1S_1$. H_2 and S_2 can then symbolize the mutations, which enable the bearer to select and survive on a new host—such as cherry. The new alleles might remain in the population for a variety of reasons, but certain combinations would be lethal in the older race. For instance, all H_1H_1 and H_1H_2 would probably oviposit on apple,

but any S_2S_2 larvae would die on apple. H_2H_2 females would oviposit on cherries, but their S_1S_1 progeny would die (Figure 5-9). Heterozygous H_1H_2 individuals could oviposit on both plant species. However, Huettel and Bush (1971) found that individuals were conditioned by larval food and tended to oviposit on the species from which they had emerged. This factor together with allochronic isolation would inhibit random mating, and H_1H_2 individuals would be most likely to remain on apple, where coadapted induction genes would operate effectively. Only few H_2H_2 flies that emerged early in the season would mate and oviposit on cherry (cf. Figure 5-7).

Shifts in emergence time of the new race would increase the separation of races in time. An occurrence of spread to other areas might cause divergence from a purely sympatric distribution of the new species. Investigators have documented genetically distinct host races of *R. pomonella* on sympatric hawthorn and apple (Feder et al. 1988, McPheron et al. 1988, Smith

FIGURE 5-8

Temporal and spatial availability of the fruits of *Prunus emerginata* and *P. avium*. Only a very small window, represented by a darkly shaded block, is available for a shift from the original host to the introduced *P. avium* based on synchronic and sympatric distributions. Fruiting times range from May (M) to October (O). *(After Bush 1975a)*

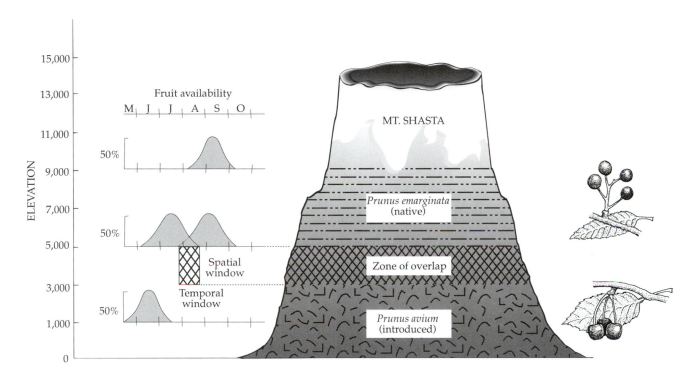

1988). The most likely mechanisms maintaining significant genetic differences are genetic control of allochronic emergence times and at least partial premating reproductive isolation because of host-plant fidelity. Diehl and Bush (1989) provide the theoretical basis for this divergence: habitat preferences of individuals and mating within habitats (see also Bush 1992, 1994).

No doubt, endless natural "experiments" in host race and species formation have taken place in wild populations, and the successes must constitute a very small fraction of the total. However, when one considers the vast number of highly specialized parasitic organisms on this earth, many of which are liable to sympatric divergence of populations (Price 1980), we must wonder about the extent to which sympatric speciation has been more common than allopatric speciation, especially when we include plant speciation, discussed later in this chapter (see also Bush and Howard 1986). As Barton et al. (1988, 14) stated,

The new acceptability of the view that speciation does not always demand a period of geographical isolation

may help to explain the origins of the tens of millions of species estimated to be alive today. Perhaps the patron saint of evolution is, after all, sympatric.

5.3 PARAPATRIC SPECIATION

White (1968, 1978) advanced a third mode of speciation that he called **stasipatric speciation;** Bush (1975b) called it **parapatric speciation.** This model defines the mechanism that accounts for a pattern of closely related species contiguously distributed in space with narrow zones of overlap (Figure 5-10). Basically, a certain set of ecological factors may limit the range of a species. A new chromosomal rearrangement or gene combination may enable a few members to colonize the unexploited habitat. Heterozygotes between the parental and progeny types are well adapted for neither the original environment nor the colonists' conditions and are rapidly selected out of the population. Premating isolating mechanisms evolve rapidly, and the new species occupies a range contiguous with that of the parental species. The new species is likely

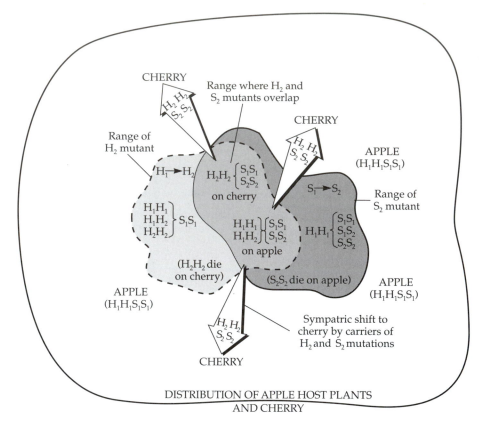

FIGURE 5-9

Bush's (1975a) genetic model of sympatric speciation, based on a shift by apple maggots in Door County, Wisconsin, from apple hosts to cherry hosts. H is the host selection gene (H₁ for apple, H₂ for cherry), and S is the survival gene (S₁ for apple, S₂ for cherry). *(After Bush 1975a)*

H = Host Selection Gene
S = Survival Gene
- - Distribution at H_2 allele
— Distribution at S_2 allele

FIGURE 5-10

A general view of parapatric speciation.

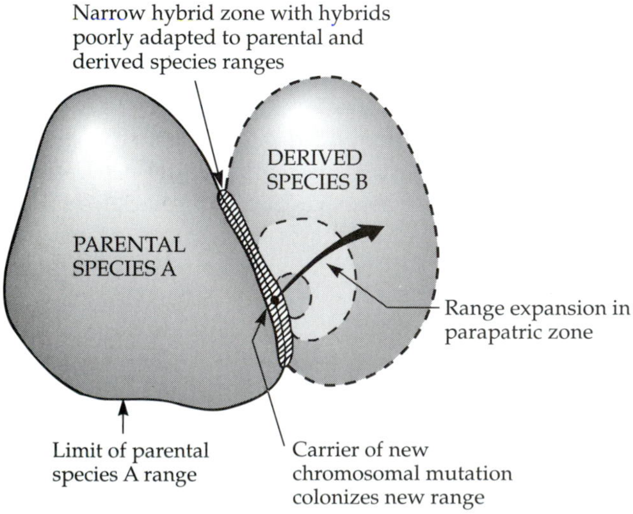

Narrow hybrid zone with hybrids poorly adapted to parental and derived species ranges

DERIVED SPECIES B

PARENTAL SPECIES A

Range expansion in parapatric zone

Limit of parental species A range

Carrier of new chromosomal mutation colonizes new range

to be largely homozygous, with little genic differences from the parental species but with chromosomal rearrangements that alter major regulatory pathways in development.

Speciation in Morabine Grasshoppers

One example of this stasipatric or parapatric model concerns the Australian morabine grasshoppers in the genus *Vandiemenella*. (The name "morabine" comes from another genus, *Morabo*.) These are all wingless grasshoppers with low vagility; consequently, they move little over a lifetime. They feed mainly on shrubs in the family Asteraceae. All the species have parapatric distributions. No two species, or even races within species, are sympatric over much of their range (Figure 5-11). Zones of hybridization are only 200 to 300 meters wide in several cases.

About 240 species of morabine grasshoppers occur in Australia. Because most of the species recognized by Key (1974) and White (1968, 1974, 1978) have not been formally described and assigned specific names, they are designated P for provisional species.

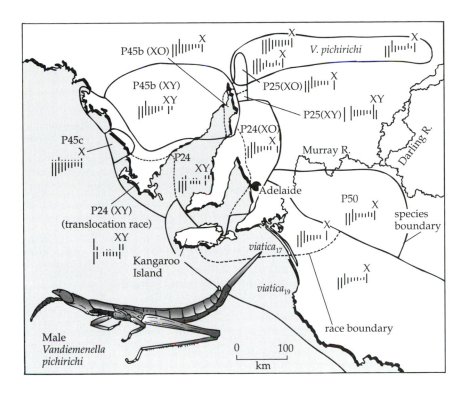

FIGURE 5-11

The distribution of species and races of *Vandiemenella* in southern Australia. A male *V. pichirichi* is illustrated in the left bottom corner. The haploid karyotype for each provisional species listed in Table 5-2 is shown, with the sex chromosome marked with an X or Y or both. *(Based on White 1978)*

In southern Australia they are numbered P24, P25, P45b, and so on (Table 5-2). The distributions of species and races are almost mutually exclusive (Figure 5-11). Each race has a unique **karyotype,** meaning a unique set of chromosomal morphologies. Figure 5-11 shows the haploid karyotype, its distribution of metacentric, acrocentric, and telocentric chromosomes demonstrating considerable variation in chro-

mosome structure. A **metacentric chromosome** has the centromere in its middle. An **acrocentric chromosome** has one arm off the centromere. A **telocentric chromosome** has two arms extending asymmetrically off the centromere. For example, the karyotype of *Vandiemenella viatica*, race 19, has 9 autosomes in the haploid state and 18 in the diploid state plus one sex chromosome, the X, adding up to 19 chromosomes. No Y chromosomes exist in this race, and so the race is designed an XO type of race (cf. Table 5-2). The nine autosomes comprise one telocentric and eight acrocentric chromosomes.

On Kangaroo Island, the distributions of *Vandiemenella* species seem very strange until one realizes that, during the Pleistocene Epoch, solid land filled what are now the channels and bays between the mainland and the island. This pattern suggests that the parapatric distributions have been remarkably stable over the past 10,000 years.

Given the lack of geographic barriers between species—indeed, the lack of any distinct factor limiting distribution—and the very local distribution of species with almost exclusive distributions of species within genera, neither the allopatric nor the sympatric model accounts adequately for the distributions of *Vandiemenella*. The parapatric model accounts for the pattern, but more research is needed on what defines the limits of species ranges.

TABLE 5-2	Species and races of some *Vandiemenella* wingless grasshoppers
Species	Races
V. viatica	19-chromosome, 17-chromosome
P24[1]	XO, XY, translocation
P25	XO, XY
P45b	XO, XY
P45c	No races
P50	No races
V. pichirichi = P26/142	3 karyotypes

See Figure 5-11.

[1] P signifies an unnamed provisional species.

5.4 SPECIATION MECHANISMS AND THE KINDS OF SPECIES INVOLVED

The allopatric, sympatric, and parapatric mechanisms of speciation are so different mechanistically and in the species distribution patterns they are likely to cause that we should expect rather different kinds of organisms to engage in each. Bush (1975b) reviewed the general attributes of the species that are likely to be generated by each mode of speciation.

Allopatric Speciation

Allopatric speciation should be subdivided into the classic model, involving **speciation by subdivision of large populations,** and **speciation by the founder effect,** in which a very small subsample of a population founds a new population in a geographically isolated area. In the classic model, an extrinsic barrier between two or more relatively large populations interrupts gene flow. In large populations evolution is likely to be slow, as many genetic changes must accumulate during geographic isolation (see Chapter 15). Thus, speciation is a long-term, gradual process. Relatively large animal species with high vagility are probably restricted to this mode of speciation. For example, the vertebrate carnivores, many birds, certain reptiles and amphibians, and most marine, lacustrine and riverine fish probably speciate slowly in geographic isolation.

Chromosomal evolution, which is a good indicator of rapid speciation (see Chapter 14), is minimal in animals with extensive ranges, such as the cats and dogs. Cats (*Felis*) in the Northern Hemisphere all have a chromosome number $2n = 38$. All dogs (*Canis*) have $2n = 78$, whereas animals such as foxes (*Vulpes* and other genera) with smaller home ranges show considerable chromosome evolution ($2n = 38, 40, 64,$ and 78 in four species). As movement becomes more local, the chance that a chromosome rearrangement will become fixed increases, and the second type of allopatric speciation becomes more likely.

The Founder Effect

Speciation by the **founder effect** may be more rapid and possibly more frequent than speciation by subdivision. A novel environment disrupts coadapted gene complexes in the parental population, and rapid evolution is likely to build to a new adaptive peak, resulting in a speciation event (Carson and Templeton 1984). Mutations in an almost empty and perhaps novel environment are more likely to be adaptive and to generate novel evolutionary steps than those in the old environment. Advantageous changes in the gene pool become fixed much more rapidly in a small

population than in the original, larger one (see Chapter 15). Major adaptive chromosome rearrangements may readily become fixed in the homozygous condition.

Organisms with moderately low vagility are susceptible to speciation by the founder effect, although it can also be effective in some very vagile animals, such as birds colonizing oceanic islands (see Chapter 8). Even subsocial or social mammals, such as primates and ungulates that typically form small, cohesive bands, can exist in a sufficiently fractionated population structure so that founder effects become important. Cave-dwelling animals with homing behavior, such as bats, are liable to found new colonies distant from parental groups. Carson (1973, 1975) suggested that speciation of some Hawaiian Drosophilidae is most likely to be by the founder effect, and many other insects must surely be equally likely candidates.

The opportunities for chromosomal evolution during this mode of speciation are great. Such rearrangements as fusions, fissions, whole-arm translocations, and pericentric inversions may cause rapid reorganization of regulatory mechanisms and thus drastic changes in the developmental process without a genic change (see also Chapters 12 and 14). Such radical changes may permit the exploitation of a set of ecological conditions or resources very different from conditions in the parental environment. Major shifts in the adaptive mode may occur during speciation by the founder effect.

Parapatric Speciation

Parapatric speciation differs from speciation by the founder effect in several important ways (Bush 1975b). No spatial isolation occurs. The involved organisms are exceptionally sedentary. Selection activates reproductive isolating mechanisms at the same time as individuals with new, highly adaptive gene combinations move into and exploit a new habitat. The organisms are sedentary enough that 100 to 1,000 meters may be the limit of movement of an individual or its progeny. Plants, terrestrial snails, pocket gophers and other fossorial rodents, morabine grasshoppers, small mammals such as *Peromyscus* species and shrews, certain lizards, mole crickets, stick insects, and many other flightless insects are likely to speciate parapatrically.

Sympatric Speciation

Sympatric speciation differs from parapatric speciation in three ways. (1) Premating reproductive isolation develops before the shift to a new niche because the new variant gene for habitat and host selection arises within the parental population. (2) Chromosomal rearrangements do not seem to be involved. (3) Speciation is likely to occur at the center of the species range instead of the periphery and thus can lead to the evolution of many sympatric sibling species.

Sympatric speciation is common among parasites of plants and animals, which constitute a very species-rich and diverse group of taxa.

Additional Speciation Mechanisms

Other common and important modes of speciation include interspecific hybridization and polyploidy, which occur frequently in plants. Association with symbionts commonly adds to the complexity of intraspecific interactions and the development of reproductive isolation, as is the case with symbiotic sterility factors in *Drosophila* (e.g. Ehrman 1983) and other pathogens that induce hybrid inferiority between host populations (Thompson 1989, 1994). Developmental pathways may change because of mutations or environment so that the timing or rates of developmental events differ from those of the same events in the parental population. The general term for developmental change that results in new morphology, life history, or behavior is **heterochrony,** meaning literally "different time" (McKinney and McNamara 1991). Clearly, the timing of events such as reproduction of morphological changes may contribute to reproductive isolation, and speciation and higher levels of macroevolution may result, as McKinney and McNamara (1991) discuss. (See also Chapter 18, "Rates of Evolution," especially Sections 8.5 through 8.7.)

Plants are so different from animals in many ways that we should examine the extent to which the modes of speciation found in these two kingdoms are similar or disparate. Grant (1971) distinguished between primary and secondary forms of speciation in plants. By **primary speciation** he meant the origination of new species from one common ancestral species, as we have discussed in the three modes of speciation—

FIGURE 5-12

Primary speciation and one example of secondary speciation recognized by Grant (1971).

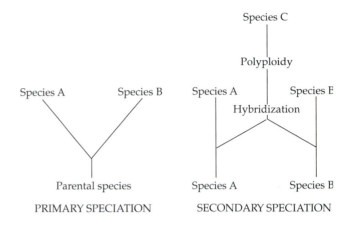

PRIMARY SPECIATION   SECONDARY SPECIATION

allopatric, sympatric, and parapatric (Figure 5-12). In **secondary speciation,** good biological species produce new species rather rapidly or even instantaneously—for example, by hybridization or polyploidy or by a shift from an outcrossing to an amictic system of reproduction. Researchers have studied such secondary speciation extensively in plants because it has been so common. Their accumulated knowledge about plant speciation can enrich our appreciation of the speciation process as a whole.

5.5 PRIMARY SPECIATION IN PLANTS

Quantum Speciation

Grant classified primary speciation into geographic, quantum, and sympatric modes (Figure 5-13). Geographic speciation is the classic model of speciation. Grant (1971, 114) defined **quantum speciation** "as the budding off of a new and very different daughter species from a semi-isolated peripheral population of the ancestral species in a cross-fertilizing organism."

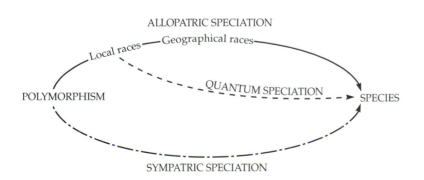

FIGURE 5-13

Modes of primary speciation as classified by Grant (1971).

In contrast with geographic speciation, quantum speciation is rapid and results in major shifts in phenotype and genotype. Two mechanisms seem to play roles in quantum speciation.

In the first mechanism, a shift from an outcrossing mode of reproduction to an inbreeding mode seems to be the major factor. A large central population of individuals is adapted for outcrossing by having genes that can effectively interact in the genome with new genes brought in during gene flow throughout the population. In this central population, inflow of new genes tends to swamp idiosyncratic variation. Once a few individuals become isolated in a peripheral population, however, inbreeding becomes common among them, with drastic genetic and phenotypic effects. Most of the change to inbreeding has strong negative effects on fitness and may cause extinction of the local population. However, in many peripheral populations, radical changes in genotype produce a few individuals with highly adaptive traits. This outcome occurs especially in peripheral loca-

tions where the parental species has presumably been an unsuccessful colonizer. Thus, the mechanism of quantum speciation fits within the umbrella of allopatric speciation by the founder effect.

Verne and Alva Grant's (Grant and Grant 1960) work on the genus *Gilia* in California (Figure 5-14) provides an example of the pattern expected from this first mode of quantum speciation, in which the shift to inbreeding is important. *Gilia tenuiflora* is a wide-ranging, predominantly outcrossing species. Two other species, *G. austrooccidentalis* and *G. jacens*, have smaller, peripheral ranges and are autogamous, or self-fertilizing. Apparently, chromosomal changes have not been important in speciation. The fact that the peripheral species live in more arid habitats than does *G. tenniflora*, the apparently parental species, suggests that a physiological shift in the adaptive mode of the species during speciation enabled a range extension beyond the parental species.

A second mechanism that Grant regards as quantum speciation fits into the parapatric model. A com-

FIGURE 5-14

Distributions of the parental species *Gilia tenuiflora* and two derived species in California. Note the two derived species in more arid locations than the parental species. *(Based on Grant 1971)*

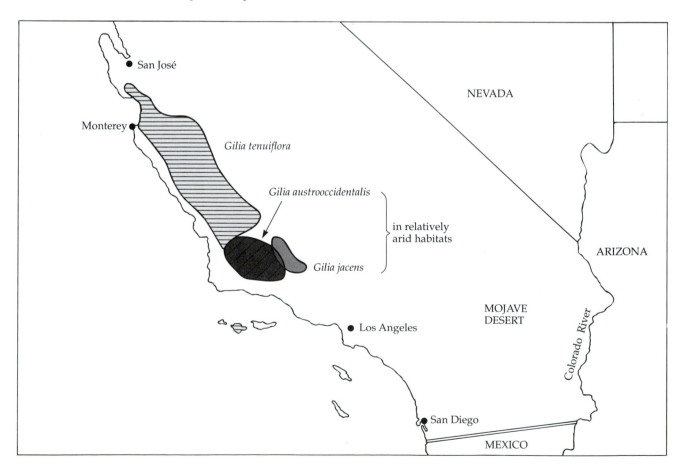

parison between *Clarkia biloba* and *C. lingulata* suggests *C. biloba* as the parental species and *C. lingulata* as the derived species (Table 5-3; see also Figure 4-7 on left). A chromosomal sterility barrier, resulting apparently from an original reciprocal translocation involving two chromosomes, and a subsequent translocation in the *C. lingulata* line reproductively isolate the species. Thus, *C. lingulata* is an aneuploid of *C. biloba*, meaning that its chromosome number is not a multiple of the parental haploid number. Because *C. biloba* is self-compatible, speciation occurred through chromosomal repatterning rather than through a shift from outcrossing to inbreeding. Lewis (1966) called this form of speciation **speciation by saltation.**

In addition to the factors already mentioned as contributing to the two mechanisms of quantum evolution, populations are likely to be very unstable. At marginal sites in severe conditions, mortality is high and population size fluctuates dramatically through bottlenecks of low numbers and rapid increases during favorable years. Thus, as Carson (1975) envisioned, populations are likely to pass through flush–crash cycles, with a founder effect after each crash stimulating rapid **genetic drift** and evolutionary change (Figure 5-15). (Lewis [1962] emphasized this evolutionary pattern in the genus *Clarkia*, arguing that **catastrophic selection** was an important ingredient in the evolutionary divergence of species.) Genetic drift is the alteration of gene frequencies through sampling error (see Chapter 15 for more detail). When populations become very small, as after a population crash, the few remaining individuals usually contain only a small proportion of the genes that were present in the largest population. Just as two crayons sampled at random from a 20-color crayon box cannot represent the color diversity in the box, so a few individuals cannot usually contain the genetic diversity of a large population. Repeated episodes of the founder effect with genetic drift result in change in gene frequencies—or evolutionary change—even in the absence of natural selection.

The Wallace Effect

Another potential factor in rapid divergence of populations during primary speciation is the **Wallace effect.** In his book *Darwinism: An Exposition of the Theory of Natural Selection,* Wallace (1889) stated that crossing between populations in the same area or adjacent areas that resulted in inferior-quality individuals would select directly for reproductive isolating mechanisms. As Grant (1971, 136–137) put it, "Those individuals in two sympatric populations which produce inviable or sterile hybrids will contribute fewer offspring to future generations than sister individuals in the same parental populations which do not hybridize." Therefore, natural selection reinforces reproductive isolating mechanisms, and speciation by reinforcement occurs (Butlin 1989). Grant (1971) argued that the Wallace effect is most likely to be important in annual plants, since reduced seed production is much more catastrophic in them than in a long-lived species with many years for seed production and many chances for crossing with compatible genotypes.

A nice example of the Wallace effect comes from Grant's (1966) studies on *Gilia.* The leafy-stemmed

---

T A B L E 5 - 3 Characteristics of *Clarkia biloba* and *C. lingulata* discerned by Lewis and Roberts (1956)

Characteristic	*Clarkia biloba*	*Clarkia lingulata*
Relationship	Presumed parental species	Derived species
Geographical range	Large area of northern Sierra Nevada and adjacent California	Very small area in one river canyon on southern periphery of *C. biloba*
Chromosome number	$n = 8$	$n = 9$
Pollination	Self-compatible	Self-compatible
Habitat	Moist, closer to ancestral habitats for *Clarkia*	Drier
Morphology	Petals heart-shaped	Petals tongue-shaped

See Figure 4-7.

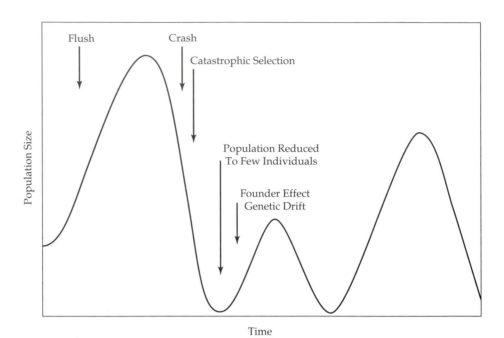

FIGURE 5-15
The flush–crash cycle with catastrophic selection and founder effects occurring at frequent intervals.

*Gilia*s can be divided into two major groups, and all are annuals. The distributions of the five foothill and valley species in California overlap extensively, with individuals of two or more species growing and flowering together in many localities. Very strong incompatibility barriers separate these species. The degree of separation can be measured conversely by the number of hybrid seeds per flower produced as a result of cross-pollination between species. In Grant's study, seed output in this group ranged from 0 to 1.2 seeds per flower and averaged 0.2 seeds per flower. Four maritime species are completely allopatric in distribution, occurring on coastal strands in North and South America. The maritime species cross-fertilize easily, yielding 7.7 to 24.8 hybrid seeds per flower, with an average of 18.1 seeds per flower.

Grant noted that in other characteristics the inland and maritime species had similar amounts of divergence between species, suggesting that the evolution of fertility barriers has accelerated greatly where species are sympatric. Butlin (1989) discusses problems with interpretation of evidence for the Wallace effect.

Sympatric Speciation

Grant (1971) found no evidence for primary sympatric speciation. However, one case suggested that detailed studies might reveal more frequent instances of the process. Gottlieb (1973) argued for the sympatric origin of a species derived from the parental stock of an annual plant, *Stephanomeria exigua*. The parental species was an obligate outcrosser with eight chromo-

somes ($n = 8$). Reproductive isolation of the new species developed sympatrically when an individual carried a new chromosomal arrangement such that hybrids had reduced pollen viability and seed set and natural selection favored the evolution of a self-pollinating breeding system.

5.6 SECONDARY SPECIATION IN PLANTS

Hybrid Speciation and Amphiploidy

The major forms of secondary speciation in plants involve a sympatric origin of new species. **Hybrid speciation,** the origination of new species directly from a natural hybrid (Grant 1971), is a secondary mode of speciation because no divergence from a common parental stock occurs. One of the problems with hybrid speciation is that two genomes must be integrated. Commonly, this occurs because segregation stops in the F_1 generation, although a great diversity of mechanisms is involved, resulting in many different kinds of genetic systems. Some methods of stabilization in hybrid reproduction include vegetative reproduction, agamospermy (the production of seeds without fertilization of the ovule), permanent odd polyploidy, and amphiploidy, or allopolyploidy. An **amphiploid,** or **allopolyploid,** is "a polyploid that originated by the doubling of the chromosomes of a zygote with two unlike chromosome sets, usually owing to hybridization of two species" (Mayr 1963, 663). **Polyploidy** is the "condition in which the number of chromosome sets in the nucleus is a multiple (greater than 2) of the haploid numbers" (Mayr 1963, 671).

The most common barrier to hybridization in viable plants is that, during meiosis, chromosomes cannot pair up because they are not homologous (Figure 5-16). Thus, hybrid species originate where this problem is circumvented by various means. In vegetative reproduction and agamospermy, meiosis is completely absent, and mitosis proceeds normally to produce vegetative individuals or to mitotically produce viable seeds. These kinds of uniparental reproduction are likely to give rise to microspecies. The development of amphiploidy results in good biological species that maintain effective sexual reproduction (Figure 5-16). Amphiploidy produces a second version of each chromosome, and so homologous chromosomes are available for normal pairing, and sexual reproduction is reestablished.

Clearly, hybrid speciation can result in very different kinds of species. Some examples follow. Common hemp-nettle, *Galeopsis tetrahit*, is a naturally occurring biological species ($2n = 32$), and Müntzing (1932, 1938) reproduced this species by crossing two diploid species of hemp-nettle ($2n = 16$), *Galeopsis pubescens* and *G. speciosa* (Figure 5-17). Despite considerable sterility between the two diploid species, Müntzing obtained F_1's, one of which was a triploid individual. He backcrossed it to the parent *G. pubescens*, which yielded a tetraploid *G. tetrahit*-like plant.

Karpechenko (1927) produced a totally artificial species by crossing the radish, *Raphanus sativus*, with cabbage, *Brassica oleracea*, to form a tetraploid, *Raphanobrassica*. Initially a diploid hybrid was formed from the parental genera. Then, this largely sterile hybrid apparently gave rise to *Raphanobrassica* after the union of two diploid gametes. Natural hybridization may produce a **hybrid complex** in which "the morphological discontinuities between the originally divergent ancestral forms" are obscured, giving rise to a species group in which discrete species are hard to identify (Grant 1971, 296).

Autopolyploidy

Another form of polyploidy in addition to amphiploidy is **autopolyploidy**, in which the same chromosome set is doubled, tripled, and so on, giving rise to new species. For example, native species of *Dahlia* have $2n = 32$, whereas the cultivated *Dahlia variabilis*, grown for show in flower gardens, has $2n = 64$ and is a tetraploid. Further multiples of chromosome number can occur, as in *Triticum*, the genus to which common wheat belongs, where $2n = 14$, 28, and 42, the last being a hexaploid ($n = 7$, $6 \times n = 42$). In *Chrysanthemum*, which includes Shasta daisies and marguerites, $2\times$, $4\times$, $6\times$, $8\times$, and $10\times$ levels of ploidy are represented. Odd numbers of chromosome sets can maintain themselves if asexual reproduction or some degenerate form of sexual reproduction occurs and meiosis is avoided. The *Rosa canina* group of dog roses in Europe has diploids ($2n = 14$), tetraploids ($2n = 28$), pentaploids ($2n = 35$), and hexaploids ($2n = 42$). The *Crepis occidentalis* group among the plants commonly called hawk's-beards has $2\times$, $3\times$, $4\times$, $5\times$, $7\times$, and $8\times$ levels of ploidy.

Many other forms of polyploidy can occur, since the presence of at least a duplicate copy effectively buffers change in chromosome number. Hence, chromosomes can be lost or added by ones and twos as well as in whole haploid sets. As a result, an almost infinite variety of karyotypes is possible in plants, and **polyploid complexes** are relatively common.

Frequency of Polyploid Plant Species

Polyploidy is so frequent in plants that it is a major form of speciation. Grant (1963) constructed a frequency distribution of chromosome numbers in flowering plants and developed the argument that many plant taxa have polyploidization events in their phylogenetic histories, resulting in chromosome numbers greater than $n = 13$. He found a modal distribution around $n = 7$, 8, and 9 and suggested that the original basic number in the angiosperms lay in this range. Then, chromosome numbers of $n = 14$ and above would all be polyploids.

FIGURE 5-16

An example of a hybridization event that results in cells with nonhomologous chromosomes, and the recovery of sexual reproduction by polyploidy. The identical pairs of letters for the parental species indicate homologous chromosomes.

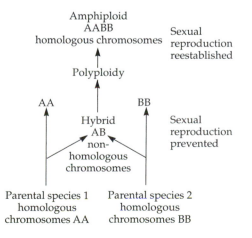

FIGURE 5-17

The possible events in the evolution of the amphiploid species *Galeopsis tetrahit*, studied by Müntzing (1932). *Galeopsis speciosa* and *G. tetrahit* are illustrated.

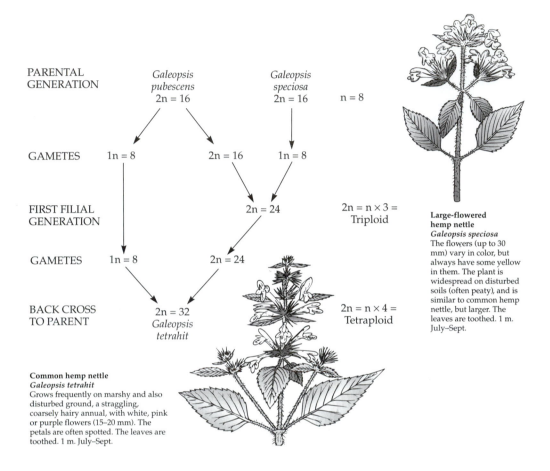

Using Grant's estimate, Goldblatt (1980) suggested that many families of monocotyledons comprise entirely polyploid species—for example, the Bromeliaceae (1,700 species), Agavaceae (300 species), and Lemnaceae (30 species). And some very large families have more than 50 percent of species that are polyploids: Cyperaceae, 73% (4,000 species), Poaceae, 60% (10,000 species), and Orchidaceae, 94% (18,000 species). In all, we regard 55 percent of monocotyledonous species as polyploids on the basis of chromosome numbers greater than $n = 13$. Lewis (1980) called this an estimate of **paleopolyploids,** meaning that polyploidization has occurred in the phylogenetic background of a species. An estimate of intrageneric polyploid species illuminates more recent speciation events, because the polyploids have arisen since the genus evolved. For monocotyledons, the paleopolyploid estimate is therefore 55 percent of species, and the intrageneric polyploid estimate is 36 percent of species (Goldblatt 1980). Both estimates

illustrate the impressive impact of polyploidy in plant speciation. In the dicotyledons, similar proportions of species are polyploid (Lewis 1980).

Polyploidy is certainly less common in animals than in plants; it occurs in only some 30 species of reptiles and amphibians (Bogart 1980) and is rare in insects (Lokki and Saura 1980). Schultz (1980) regards polyploidy in fishes, however, as a mechanism that has provided the extra genetic material for major adaptive breakthroughs. Perhaps all the salmonids and catostomids are tetraploids, for example, But it is not yet clear how important polyploidy is in animals in terms of speciation and macroevolution.

5.7 AGAMIC SYSTEMS

In addition to hybridization and polyploidy, loss of gamete fusion has occurred many times. Such **agamic systems** of reproduction result in immediate reproductive isolation from the parental stock, independent divergence of lineages, and the emergence of new species. Sympatric speciation is likely to occur. Because the nomenclature of agamic systems in plants has developed independently from that in animals, some definitions are in order (see Chapter 14). **Apomictic reproduction,** or **apomixis,** is asexual reproduction in plants in general, which may involve development of a new individual from an unfertilized egg or from a somatic cell of the parent. **Agamospermy** is seed formation without fertilization by pollen. **Vegetative reproduction** is the production of new individuals from somatic cells, not cells related to gametes. A **clone** consists of all individuals derived by uniparental reproduction from a single parental individual, the **genet** being the genotype of all individuals and the **ramet** being one individual in the clone (Harper 1977). **Uniparental reproduction** is reproduction involving one parent by vegetative means or by a degenerate form of sexual reproduction such as agamospermy. **Pseudogamy** is the necessity of pollination for seed formation in an apomictic species. (These genetic systems are discussed more in Chapter 14.)

Agamospermy is very common in plants; perhaps the best-known cases are in the family Asteraceae, in the genera *Hieracium,* the hawk weeds, and *Taraxacum,* the dandelions. It is common to find apomictic and automictic (self-pollinating) or amphimictic (outcrossing) forms of the same morphospecies. *Taraxacum officinale,* the common dandelion, has populations of apomicts and automicts in the same general locality. Most agamospermous species are the results of hybridization, and so speciation by hybridization goes hand in hand with agamospermous species. Once each lineage is reproductively isolated, the opportunities for development of many microspecies are extensive, since clones in the same locality can diverge through somatic mutations.

Overall, speciation mechanisms are very diverse in detail, although in general they may be distributed among the allopatric, sympatric, and parapatric forms of primary speciation and the hybridization, polyploidy, and agamic systems of secondary speciation (Table 5-4). Each form of speciation results in a discrete genetic system. Chapter 14 will explore the consequences of each of these genetic systems.

One of the great accomplishments of the theory of evolution is that it describes mechanisms to account for the great diversity of species on this planet. Otte and Endler (1989) and references therein discuss recent developments in the fascinating field of speciation. The question of how life originated is more difficult, but the next chapter will attempt to answer it.

TABLE 5-4   Summary of major modes of speciation in animals and plants

Characteristics	Allopatric		Parapatric	Sympatric	
	Subdivision	Founder Effect		Host Shifts	Chromosomal
Population size	Large	Very small	Very small	Very small	Very small
Major processes	Small genic differences accumulate	Genetic drift, shift in adaptive peak	Chromosal mutation	Gene mutations in host selection and survival genes	Chromosomal mutation hybrids, polyploids
Speciation rate	Slow	Faster	Very rapid	Very rapid	Very rapid
Speciation location	Allopatry	Allopatry	Edge of parental species	Center of parental range	Anywhere in parental range
Reproductive barrier	Any	Any	Premating	Premating	Postmating
Examples	*Dendroica* warblers (Mengel 1964)	Darwin's finches (Grant 1986)	Morabine grasshoppers (White 1968, 1978)	*Rhagoletis* fruit flies (Bush 1975a)	*Galeopsis* (Müntzing 1932, 1938)
		Gilia tenuiflora and relatives (Grant and Grant 1960)	*Clarkia biloba* (Lewis and Roberts 1956)		*Stephanomeria* (Gottlieb 1973), hybrid species (Grant 1971)

SUMMARY

Darwin's model of speciation lacked as an ingredient geographic barriers that isolate populations and cause them to evolve independently. The model of **geographic,** or **allopatric, speciation** emphasizes the importance of geographic isolation during which properties of populations evolve, ultimately leading to reproductive isolation and the evolution of new species. An illustration of this mechanism involved *Dendroica* warblers and glaciation acting as a geographic barrier. When one or a few individuals found a new population in isolation, as happened in the case of Darwin's finches, allopatric speciation by the **founder effect** may occur.

Geographic isolation results from environmental characteristics, such as mountain ranges, that prevent movement of individuals between populations. **Reproductive isolation** results from properties of species that constrain gene flow between populations, preventing the production of fully fertile progeny, even when the populations occur in the same locality. After geographic isolation, secondary contact between populations may indicate that the new species are fully developed in terms of both **reproductive isolating mechanisms** and **ecological divergence** into different ecological niches. If reproductive isolation is incomplete, however, secondary contact may result in **introgression** and the blending of the diverging populations.

Sympatric speciation may occur in one locality, especially if parasitic organisms, such as *Rhagoletis* fruit flies, shift to a new host. **Allopatric speciation in marginal populations** may account for the patterns of overlap in the distributions of many closely related species, although it would not lead us to expect the common occurrence of sibling species. Host shifting between closely related plant species may require only small differences involving the genes for host selection and survival of larvae in the host. The timing of fruiting and the partial overlap of host species in space may result in narrow windows in space and time for host shifts to be made, even though they occur within the cruising range of a parental population.

In **parapatric speciation,** a chromosomal mutation enables carriers to invade new territory beyond the parental range, and adjacent distributions of parental and derived species develop. Morabine grasshoppers in Australia demonstrate patterns of distribution and karyology that are best accounted for by this speciation model.

Each mode of speciation may apply to a particular set of species characteristics. For example, allopatric speciation involving extensive geographic barriers probably applies to highly mobile organisms such as birds and large mammals; sympatric speciation is likely to be common among parasitic species; and highly sedentary organisms have the potential for speciating parapatrically.

Among plants, we distinguish between **primary speciation,** involving divergence from a common stock, and **secondary speciation,** by **hybridization** or **polyploidy** or both, involving coalescing of genomes or multiplication of genomes. In primary speciation, **quantum speciation**—rapid divergence from a parental stock in a peripheral population—may include elements of the founder effect and parapatric speciation. The **Wallace effect** may promote the rapid evolution of reproductive isolation.

Secondary speciation in plants results from **hybrid speciation** usually coupled with **amphiploidy,** a form of polyploidy that stabilizes sexual reproduction by providing homologous sets of chromosomes. **Autopolyploids** are also reproductively isolated from parental diploid stock and can give rise to new species. Polyploidy has provided the basis for probably more than half the angiosperm species when the estimate includes **paleopolyploids.** Polyploidy, although a mechanism of secondary speciation, is a sympatric mode. **Agamic systems** of reproduction automatically cause reproductive isolation among lineages and the potential for evolutionary divergence until stocks can be recognized as separate species.

QUESTIONS FOR DISCUSSION

1. In your estimation, do methods exist for objectively defining or discovering the mode or modes of speciation involved with the development of a set of related species?

2. This chapter attempted a balanced view on alternative modes of speciation. Has this been a truly balanced view, or do you think that preconceived ideas and biases molded the development of this chapter?

3. Do you think that modes of speciation in plants and animals are basically the same, and should they be treated as such?

4. An enigmatic aspect of morabine grasshopper distributions is the lack of any clear physical feature at the edge of a range that may be involved in limiting dispersal. What research would you plan in order to understand the reasons for parapatric distributions in these species?

5. Do you agree that organisms that habitually live in or on other species, such as parasites, small herbivores, and mutualists, are likely to come under strong influences from the host population during the speciation process?

6. Many evolutionary biologists have expressed concern about the sympatric model of speciation and its reality in nature, because it is relatively difficult to understand how reproductive isolation could develop without some kind of physical isolation between divergent populations. What kinds of research would you propose to clarify this debate?

7. Is it possible that new modes of speciation will be discovered when we better understand the nature of the fungal and bacterial species?

8. Do you agree that the kinds of species a taxonomist studies are likely to influence views on the prevalence of certain kinds of speciation? For example, would taxonomists of birds, cloning plants, parasitic worms, and soil-dwelling fungi all agree that allopatric speciation is the usual or inevitable mode of speciation?

9. What research would you recommend be carried out on a particular group of specialized insect herbivores to determine whether its species emerged according to the marginal-population allopatric speciation model or according to the sympatric speciation model?

10. How do you think species size plays a role in speciation, given that there are many more small species than large species?

REFERENCES

Allmon, W. D. 1992. A causal analysis of stages in allopatric speciation. Oxford Surveys in Evol. Biol. 8:219–257.

Barton, N. H., J. S. Jones, and J. Mallet. 1988. No barriers to speciation. Nature 336:13–14.

Bogart, J. P. 1980. Evolutionary significance of polyploidy in amphibians and reptiles, pp. 341–378. In W. H. Lewis (ed.). Polyploidy: Biological Relevance. Plenum, New York.

Brown, W. L., and E. O. Wilson. 1956. Character displacement. Systematic Zool. 5:49–64.

Bush, G. L. 1974. The mechanism of sympatric host race formation in the true fruit flies (Tephritidae), pp. 3–23. In M. J. D. White (ed.). Genetic Mechanisms of Speciation in Insects. Australia and New Zealand Book Co., Sydney.

———. 1975a. Sympatric speciation in phytophagous parasitic insects, pp. 187–206. In P. W. Price (ed.). Evolutionary Strategies of Parasitic Insects and Mites. Plenum, New York.

———. 1975b. Modes of animal speciation. Ann. Rev. Ecol. Syst. 6:339–364.

———. 1992. Host race formation and sympatric speciation in Rhagoletis fruit flies (Diptera: Tephritidae). Psyche 99:335–358.

———. 1994. Sympatric speciation in animals: New wine in old bottles. Trends in Ecol. and Evol. 9:285–288.

Bush, G. L., and D. J. Howard. 1986. Allopatric and non-allopatric speciation: Assumptions and evidence, pp. 411–438. In S. Karlin and E. Nevo (eds.). Evolutionary Processes and Theory. Academic, New York.

Butlin, R. 1989. Reinforcement of premating isolation, pp. 158–179. In D. Otte and J. A. Endler (eds.). Speciation and Its Consequences. Sinauer, Sunderland, Mass.

Carson, H. L. 1973. Reorganization of the gene pool during speciation, pp. 274–280. In N. E. Moreton (ed.). Genetic Structure of Populations. Population Genetics Monograph 3. Univ. of Hawaii Press, Honolulu.

———. 1975. The genetics of speciation at the diploid level. Amer. Natur. 109:83–92.

Carson, H. L., and A. R. Templeton. 1984. Genetic revolutions in relation to speciation phenomena: The founding of new populations. Ann. Rev. Ecol. Syst. 15:97–131.

Cockburn, A. 1991. An Introduction to Evolutionary Ecology. Blackwell Scientific Publications, Oxford.

Darwin, C. 1859. The Origin of Species. 1st ed. Murray, London.

———. 1872. The Origin of Species. 6th ed. Murray, London.

Day, P. R. 1974. Genetics of Host–Parasite Interaction. W. H. Freeman, San Francisco.

Diehl, S. R., and G. L. Bush. 1989. The role of habitat preference in adaptation and speciation, pp. 345–365. In D. Otte and J. Endler (eds.). Speciation and Its Consequences. Sinauer, Sunderland, Mass.

Eberhard, W. G. 1985. Sexual Selection and Animal Genitalia. Harvard Univ. Press, Cambridge, Mass.

Ehrman, L. 1983. Endosymbiosis, pp. 128–136. In D. J. Futuyma and M. Slatkin (eds.). Coevolution. Sinauer, Sunderland, Mass.

Feder, J. L., C. A. Chilcote, and G. L. Bush. 1988. Genetic differentiation between sympatric host races of the apple maggot fly Rhagoletis pomonella. Nature 336:61–64.

Futuyma, D. J. 1986. Evolutionary Biology. 2d ed. Sinauer, Sunderland, Mass.

Futuyma, D. J., and G. C. Mayer. 1980. Non-allopatric speciation in animals. Syst. Zool. 29:254–271.

Goldblatt, P. 1980. Polyploidy in angiosperms: Monocotyledons, pp. 219–239. In W. H. Lewis (ed.). Polyploidy: Biological Relevance. Plenum, New York.

Gottlieb, L. D. 1973. Genetic differentiation, sympatric speciation, and the origin of a diploid species of Stephanomeria. Amer. J. Bot. 60:545–553.

Grant, P. R. 1986. Ecology and Evolution of Darwin's Finches. Princeton Univ. Press, Princeton, N.J.

Grant, V. 1963. The Origin of Adaptations. Columbia Univ. Press, New York.

———. 1966. The selective origin of incompatibility barriers in the plant genus Gilia. Amer. Natur. 100:99–118.

———. 1971. Plant Speciation. Columbia Univ. Press, New York.

Grant, V., and A. Grant. 1960. Genetic and taxonomic species in *Gilia*, [Part] XI: Fertility relationships of the diploid cobwebby Gilias. Aliso 4:435–481.

Grant, V., and K. A. Grant. 1965. Flower Pollination in the Phlox Family. Columbia Univ. Press, New York.

Harper, J. L. 1977. Population Biology of Plants. Academic, New York.

Hatchett, J. H., and R. L. Gallun. 1970. Genetics of the ability of the Hessian fly, *Mayetiola destructor*, to survive on wheat having different genes for resistance. Ann. Entomol. Soc. Amer. 63:1400–1407.

Hillis, D. M. 1988. Systematics of the *Rana pipiens* complex: Puzzle and paradigm. Ann. Rev. Ecol. Syst. 19:39–63.

Huettel, M. D., and G. L. Bush. 1971. The genetics of host selection and its bearing on sympatric speciation in *Procecidochares* (Diptera: Tephritidae). Entomol. Exp. Appl. 15:465–480.

Karpechenko, G. D. 1927. Polyploid hybrids of *Raphanus sativus* L. × *Brassica oleracea* L. Bull. Appl. Bot. Genet. Plant Breeding, Leningrad 17:305–310.

Key, K. H. L. 1974. Speciation in the Australian morabine grasshoppers: Taxonomy and ecology, pp. 43–46. *In* M. J. D. White (ed.). Genetic Mechanisms of Speciation in Insects. D. Reidel, Dordrecht, The Netherlands.

Kraus, F., and J. W. Petranka. 1989. A new sibling species of *Ambystoma* from the Ohio River drainage. Copeia 1989:94–110.

Lewis, W. H. 1962. Catastrophic selection as a factor in speciation. Evolution 16:257–271.

———. 1966. Speciation in flowering plants. Science 152:167–172.

———. 1980. Polyploidy in angiosperms: Dicotyledons, pp. 241–268. *In* W. H. Lewis (ed.). Polyploidy: Biological Relevance. Plenum, New York.

Lewis, W. H. (ed.) 1980. Polyploidy: Biological Relevance. Plenum, New York.

Lewis, W. H., and M. R. Roberts. 1956. The origin of *Clarkia lingulata*. Evolution 10:126–138.

Lokki, J., and A. Saura. 1980. Polyploidy in insect evolution, pp. 277–312. *In* W. H. Lewis (ed.). Polyploidy: Biological Relevance. Plenum, New York.

Mayr, E. 1963. Animal Species and Evolution. Belknap Press of Harvard Univ. Press, Cambridge, Mass.

———. 1970. Population, species and evolution. Belknap Press of Harvard Univ. Press, Cambridge, Mass.

McKinney, M. L., and K. J. McNamara. 1991. Heterochrony: The Evolution of Ontogeny. Plenum, New York.

McPheron, B. A., D. C. Smith, and S. H. Berlocher. 1988. Genetic differences between host races of *Rhagoletis pomonella*. Nature 336:64–66.

Mengel, R. M. 1964. The probable history of species formation in some northern wood warblers (Parulidae). Living Bird 3:9–43.

Minkoff, E. C. 1983. Evolutionary Biology. Addison–Wesley, Reading, Mass.

Müntzing, A. 1932. Cytogenetic investigations on synthetic *Galeopsis tetrahit*. Hereditas 16:105–154.

———. 1938. Sterility and chromosome pairing in intraspecific *Galeopsis* hybrids. Hereditas 24:117–188.

Otte, D., and J. A. Endler (eds.). 1989. Speciation and Its Consequences. Sinauer, Sunderland, Mass.

Paterson, H. E. H. 1985. The recognition concept of species, pp. 21–29. *In* E. S. Vrba (ed.). Species and Speciation. Transvaal Museum Monograph No. 4. Pretoria.

Price, P. W. 1980. Evolutionary Biology of Parasites. Princeton Univ. Press, Princeton, N.J.

Ridley, M. 1993. Evolution. Blackwell Scientific Publications, Oxford.

Ross, H. H. 1962. A Synthesis of Evolutionary Theory. Prentice–Hall, Englewood Cliffs, N.J.

Schultz, R. J. 1980. Role of polyploidy in the evolution of fishes, pp. 313–340. *In* W. H. Lewis (ed.). Polyploidy: Biological Relevance. Plenum, New York.

Smith, D. C. 1988. Heritable divergence of *Rhagoletis pomonella* host races by seasonal asynchrony. Nature 336:66–67.

Strickberger, M. W. 1990. Evolution. Jones and Bartlett, Boston.

Taper, M. L., and T. J. Case. 1992. Coevolution among competitors. Oxford Surveys in Evol. Biol. 8:63–109.

Tauber, C. A., and M. J. Tauber. 1989. Sympatric speciation in insects: Perception and perspective, pp. 307–344. *In* D. Otte and J. A. Endler (eds.). Speciation and Its Consequences. Sinauer, Sunderland, Mass.

Templeton, A. R. 1989. The meaning of species and speciation: A genetic perspective, pp. 3–27. *In* D. Otte and J. A. Endler (eds.). Speciation and Its Consequences. Sinauer, Sunderland, Mass.

Thompson, J. N. 1989. Concepts of coevolution. Trends in Ecol. and Evol. 4:179–183.

———. 1994. The Coevolutionary Process. Univ. Chicago Press.

Vrijenhoek, R. C. 1984. The evolution of clonal diversity in *Poeciliopsis*, pp. 399–429. *In* B. J. Turner (ed.). Evolutionary Genetics of Fishes. Plenum, New York.

Wallace, A. R. 1889. Darwinism: An Exposition of the Theory of Natural Selection. Macmillan, London.

White, M. J. D. 1968. Models of speciation. Science 159:1065–1070.

———. 1974. Speciation in the Australian morabine grasshoppers—the cytogenetic evidence, pp. 57–68. *In* M. J. D. White (ed.). Genetic Mechanisms of Speciation in Insects. D. Reidel, Dordrecht, The Netherlands.

———. 1978. Modes of Speciation. W. H. Freeman, San Francisco.

Wood, T. K., and M. C. Keese. 1990. Host-plant-induced assortative mating in *Enchenopa* treehoppers. Evolution 44:619–628.

Wood, T. K., K. L. Olmstead, and S. I. Guttman. 1990. Insect phenology mediated by host-plant water relations. Evolution 44:629–636.

Woodruff, D. S. 1979. Postmating reproductive isolation in *Pseudophryne* and the evolutionary significance of hybrid zones. Science 203:561–563.

6

THE ORIGIN OF LIFE AND THE EMERGENCE OF EARLY EUKARYOTES

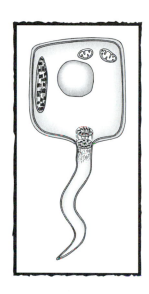

Are plant cells quadrigenomic?

As Chapter 1 noted, science advances toward factual knowledge by a series of approximations. The more difficult the question addressed, the slower the progress. One of the most difficult questions biologists ask is how life arose from the nonliving mineral earth. Added to the puzzle of how such a complex unit as the simplest living cell could have originated, the time spans involved are enormous, and the possible environments in which original life could have emerged are multitudinous.

We know what we need to understand and explain. But factual knowledge about the actual steps in the origin of life on earth is almost nonexistent. Mostly, what we have so far are hypothetical scenarios, which may or may not prove to be valid in the remotely possible event that a factually based theory of the evolution of the first living organisms emerges. It took a very long time even to get to the point where we could study the origin of life with the scientific method. First the concept of spontaneous generation (Chapter 1) had to be thoroughly debunked. Then the prevalent way of thinking had to be remolded dramatically, from views on the sudden appearance of life to belief in a gradual increase in complexity of molecular aggregates. Science had to divorce itself from religious doctrine, which it did consciously only in the latter part of the 19th century (Chapter 2, Section 2.2). Finally, a rational sequence of events replaced a train of thought dominated by the notion of spontaneity in the origin of life.

This shift to a rational approach did not occur until the 1920s, when the Russian biochemist Alexander I. Oparin and the British biologist J. B. S. Haldane broke with tradition. They independently made the novel proposition that there had occurred, over the millennia, a gradual prebiotic evolution of molecules and aggregations of molecules, which eventually resulted in self-replicating complex molecular aggregates—life itself (Oparin 1924, Haldane 1929). The novelty of their arguments changed the course of science in the field of life's origins and most research and discussion since have been focused by the perspective provided by Oparin and Haldane. The scenario we develop in this chapter is based largely on the **Oparin–Haldane hypothesis** and the developments it stimulated.

The Oparin–Haldane hypothesis, with all its ramifications, remains with us today as a very tentative perspective on the course of early prebiotic and biotic evolution. Cairns-Smith (1985) treats the origin of life as one big mystery and offers seven clues that could result in a solution. He adds to the sleuthing atmosphere by frequently quoting Sherlock Holmes. Cairns-Smith's closing quotation sets the correct tone for approaching this chapter: " 'The interplay of ideas and the oblique uses of knowledge are often of ex-

traordinary interest' " [from *The Valley of Fear* by Sir Arthur Conan Doyle]. Had Sherlock Holmes undertaken to solve this original mystery, the evolution of life, no doubt it would have challenged his reputation as a sleuth.

6.1 EVOLUTION OF THE COSMOS

The problems we encounter as we try to unravel an understanding of the evolution of life also apply to understanding the origin and development of the cosmos. This chapter briefly treats cosmic evolution only to set the scene for the origin of life, and the descriptive style should not be taken to imply fact. Rather, this account is one way of cutting through much uncertainty and debate to get to a simple view, a working hypothesis.

At the beginning of time, all forces were one: electromagnetism, the weak nuclear force, the strong nuclear force, and gravity. An explosion of pure energy known as the big bang occurred about 15 billion years ago and was the beginning of the universe. Since that time, matter has been expanding into space like a balloon being inflated from its center.

We can still see radiation that originated less than 1 million years after the big bang, when the universe became transparent to radiation (Weinberg 1977). Photons that have traveled for 15 billion years still hit the earth and register as 1 percent of the "snow" particles on a television screen. Tangible links exist among the big bang, the evolution of life, and the presence of humans on this globe.

The history of the universe is recorded in space because light travels so slowly relative to the size of the universe. Light reaching the earth from stars may be 40 million years old and can be detected by a light telescope, such as the 200-inch reflector telescope at Mt. Palomar. Quasars are even brighter than stars, and their light is detectable on earth when it is 10 billion years old (Weinberg 1985). And radiotelescopes may be able to detect 15-billion-year-old energy from the big bang in the radio spectrum.

The stages in the development of the universe may be classed as follows. *Stage 1* was the big bang, an explosion of pure energy. *Stage 2*, the cooling of the universe to a red heat, resulted in a red fog about 10 million years after the big bang. *Stage 3* was the beginning of form. About 12 billion years ago, expansion and cooling reached a stage in which hydrogen and helium gas began to condense into stars and galaxies. Darkness prevailed in space. In *stage 4*, space was again brightly lit by the intense light of early galaxies, called quasars. In *stage 5*, the first stars condensed from the early galaxies, and the first heavy elements formed from hydrogen and helium.

6.2 FORMATION OF THE EARTH

The Milky Way is the galaxy that includes our sun and solar system. The gravitational pull of denser areas in the protogalaxy of the Milky Way, which had collected gas and dust particles from space, formed the sun and the planets. As the force of gravity pulled particles together, the protoplanet earth gradually became smaller and denser and cooled until it was a molten globe of hot rock—perhaps about 4.5 billion years ago—and the earth was born.

The earth had a gravitational pull with the potential for keeping gases within its field. For example, a body must attain a speed of more than 7 miles, or 11.25 kilometers, per second to escape the earth's gravity. Currently the speed of gases in the earth's atmosphere averages about 1.25 miles, or 2 kilometers, per second for hydrogen and 0.25 mile, or 0.4 kilometer, per second for oxygen and nitrogen. Hence, we would expect the earth to retain these gases in its atmosphere. This raises a big question: If hydrogen and helium were the most abundant atoms in the protogalaxies and protoplanets, how is it that we do not have any in our present atmosphere? The answer is that the earth was very hot, and its heat energy increased the speeds of gases. At 14,500°F, for example, all the gases would have had sufficient momentum to escape the earth's gravitational field, and thus the earth's original atmosphere must have escaped into space. That atmosphere was probably composed of ammonia, hydrogen sulfide, hydrogen fluoride, hydrogen chloride, and hydrogen bromide.

The next question is: Where did our present atmosphere come from? As molten rocks cooled very rapidly, they gave off gases—water vapor, carbon dioxide, and nitrogen, in order of abundance—which dissolved in the melt. Surface temperatures were only a few hundred degrees, and under those conditions gravity could retain the gases. The result was the development of an atmosphere very different from the original. Its composition of gases was very similar to that given off by volcanoes today—perhaps 68 percent water vapor, 13 percent carbon dioxide, 8 percent nitrogen, and smaller amounts of ammonia and methane.

The oldest rocks to be found on earth are about 3.5 to 3.8 billion years old, indicating that the crust of the earth must have been in a solid state at least that long, and probably longer. As the earth cooled down, the water vapor gradually condensed into huge clouds, and for the first time rain fell on the dry earth. The massive condensation of water caused torrential rains and catastrophic erosion of the rocks, and all the water gradually collected in depressions on the earth's surface, to become our oceans and seas. The bodies of water became very rich as chemicals dissolved in the water and were concentrated by evaporation. The atmosphere lost so much water that its constitution changed to about 74 percent carbon dioxide, 15 percent water vapor, and 10 percent nitrogen. This was still very different from the present atmosphere, and still 2 to 3 billion years ago (Table 6-1). Much of the carbon dioxide in the atmosphere dissolved in rain water and would ultimately form limestone rocks made up of calcium and magnesium

TABLE 6-1   Estimated changes in the atmosphere of the earth

Gas	Atmospheres[1]			
			Cooling Earth	
	Original, 4.5 Billion Years Ago	Volcanoes Today	2 to 3 Billion Years Ago	Present
Ammonia	X	X	X	−
Methane	X	X	X	−
Hydrogen sulfide, fluoride, chloride, and bromide	X	X	−	−
Water vapor	−	68%	15%	X
Carbon dioxide	−	13%	74%	0.03%
Nitrogen	−	8%	10%	78%
Oxygen	−	−	−	21%
Argon	−	−	−	1%

[1] X indicates the presence of a gas; − indicates the absence or small to trace quantities of a gas.

carbonates, such as those that make up much of the great walls of the Grand Canyon today.

The present atmosphere is 78 percent nitrogen, 21 percent oxygen, 1 percent argon, and 0.03 percent carbon dioxide, with many additional gases, such as ozone and formaldehyde, in the upper atmosphere. We still need to know where all the carbon dioxide went and where all the oxygen came from. To understand the dramatic change in atmosphere between 2 to 3 billion years ago and the present, we need to understand the origin of life, as life itself has profoundly influenced the earth's atmosphere.

6.3 PREBIOTIC EVOLUTION

Clearly, life must have evolved in an atmosphere free of oxygen; this is an important point to remember. The oldest fossils ever found are in the oldest well-preserved sedimentary rocks on earth, which are about 3.5 billion years old (Woese 1984). They are fossils of stromatolites that formed from cyanobacteria and other photosynthesizing bacteria. Older, 3.8-billion-year-old rocks from the Isua Formation in Greenland contain evidence of life in the form of the increased isotope ratios of carbon (C^{12} to C^{13}) that are characteristic of living systems. Life is no doubt older than the oldest rocks on earth, since there must have been bacteria before photosynthetic capability evolved. Perhaps bacteria evolved 4 billion years ago on a hot, inhospitable earth, and perhaps their relatives still persist in hot springs (see Brock 1985, Haymon and Macdonald 1985).

Chemical Evolution

The spontaneous evolution of an organism from inorganic matter required the simultaneous presence of many chemicals in the correct proportions and the correct spatial arrangements. How could this have happened, and what were the basic ingredients? First,

the water and salts that would have been necessary were—not surprisingly—those found in seawater today. If life evolved in or by a sea, these basic ingredients would have been present. Even today, marine animals have body fluids very similar to salt water, and body fluids in humans and cockroaches are not so different (Table 6-2). Perhaps we preserve within our own bodies a simulation of the primordial soup in which life began and thus have another link with the early earth (see also Banin and Navrot 1975).

The other requirements for living organisms are organic chemicals, principally composed of carbon, oxygen, nitrogen, and hydrogen, all present in the inorganic compounds of carbon dioxide (CO_2), water (H_2O), and nitrogen (N_2). Because organic chemicals are now made principally by living organisms, we must try to understand how they could have arisen in a nonbiotic (nonliving) environment.

These organic chemicals can be grouped into four essential classes: the carbohydrates, fats, proteins, and nucleic acids, including DNA (deoxyribonucleic acid). Even simple sugars have complex structures; glucose, $C_6H_{12}O_6$, is an example. Protein molecules are much larger and even more complex, consisting of hundreds of amino acid residues. Such chemicals must have arisen spontaneously, by chance, to have coalesced into a living cell. One can begin to envisage the vanishingly small chance that all conditions were just right for the origin of life. And yet so much time was available—perhaps 500 million years—and there was such a vast number of possible situations, all over the surface of the earth, in which this chance could occur, that the perfect conditions for life were almost bound to occur sooner or later.

The First Organic Molecules

The stepwise course that led to primitive organisms can only be imagined, but it may have occurred as follows. Organic chemicals are made largely by living organisms, but there are exceptions. For instance,

T A B L E 6 - 2   Concentrations of ions of salts in seawater and in body fluids (millimoles per liter)

	Sodium Na$^+$	Potassium K$^+$	Calcium Ca^{++}	Magnesium Mg^{++}	Chloride Cl$^-$
Seawater	459	10	10	53	538
Sea urchin	444	10	10	50	522
Cockroach	161	8	4	6	144
Human	145	5	3	1	103

when volcanoes erupt, they release metal carbides that react with water to produce organic chemicals such as acetylene (HC≡CH). Oparin (1938) emphasized that, by the action of superheated steam in the earth's atmosphere, water molecules would have attached to unsaturated hydrocarbons such as acetylene to form acetaldehyde:

$$HC≡CH + H_2O → CH_3COH$$

Oxidation products of hydrocarbons would have included alcohols, aldehydes, ketones, and organic acids. In the presence of ammonia, other compounds such as ammonium salts, amides, and amines would have been produced. Electrical discharges in the form of lightning, working on gases in the primitive atmosphere such as ammonia, methane, hydrogen, and water vapor, would have produced organic chemicals, including the amino acids from which proteins are built.

In 1953 Stanley Miller, having worked with the Nobel laureate Harold Urey in Chicago, published the results of a pioneering experiment on the spontaneous generation of organic molecules. He simulated the primitive earth by circulating the four gases—ammonia, methane, hydrogen, and water vapor—past electrical discharges in a flask. After he allowed them to circulate continuously for seven days, he analyzed the chemicals in the brown broth that had formed and found a very rich variety of organic compounds, including some of the most important amino acids in proteins plus urea, acetic acid, and lactic acid. Such organic chemicals would have accumulated gradually and perhaps naturally in small pools of water and been concentrated by evaporation. Thus, molecules would have had increasing chances to combine in more and more complex ways.

The reducing atmosphere would have facilitated the accumulation of chemicals in a progressively richer soup (Table 6-1). No free oxygen was available to cause the breakdown of complex molecules, and no organisms were present to digest complex organic molecules for energy. Consequently, organic chemicals were probably stable over very long periods of time, provided they were protected from ultraviolet light in these "subvital territories" that Oparin (1976) recognized as so important.

Because we now have oxygen in our atmosphere, there is no chance for the accumulation of this primeval broth of chemicals, because the presence of oxygen so commonly results in the breakdown of organic compounds. As a result, there is little chance for the spontaneous generation of life. Louis Pasteur's conclusion about spontaneous generation holds today, but it does not hold for an oxygen-free atmosphere (cf. Chapter 1).

Molecular Aggregates

As organic chemicals accumulated, perhaps in pools along the oceans' margins, they gradually joined together in more complex associations which happened to be increasingly resistant to the natural tendency for things to become disorganized. The more persistent associations would have been favored by a form of prebiotic natural selection. Molecular aggregates formed and increased gradually until ultimately a self-duplicating cell was produced. Oparin (1938) emphasized the importance of the formation of colloidal solutions. In 1924 he wrote in Russian, "There is no essential difference between the structure of coagula and that of protoplasm" (Bernal 1967, 211). When colloidal solutions developed, their structures would have become more complex and formed gels, and life may have evolved from those substrates.

> The moment when the gel was precipitated or the first coagulum formed, marked an extremely important stage in the process of the spontaneous generation of life. At this moment . . . the transformation of organic compounds into an organic body took place. . . . With certain reservations we can even consider that first piece of organic slime which came into being on the Earth as being the first organism. (Bernal 1967, 229)

Farley (1974, 163) elaborated upon this scenario, using a translation from Oparin (1924):

> Since each bit of slime was an individual that assimilated material at different rates, a "selection of the better organized bits of gel was always going on. . . . Thus, slowly but surely, from generation to generation, over many thousands of years, there took place an improvement of the physico-chemical structure of the gels. Naturally, as the amount of material in the surrounding medium diminished, the more strongly and bitterly the struggle for existence was waged." Finally, only two courses of change were open to the gels, either toward a cannibalistic mode of life or toward autotrophism.

One example of mutual attraction of molecules and the spontaneous increase in complexity of interactions among organic molecules involves movement and the modern muscular system in animals. The two molecules actin and myosin are the basis of muscle and muscular contraction. When they occur in the nonliving condition, they join together to form a muscle-like structure of molecules. What is more, if energy is supplied in a certain form (ATP, or adenosine triphosphate), molecules contract together just as in living muscle. This is the sort of combination of molecules that could have occurred in the primeval earth.

The Evolution of Reproduction

Droplets and Microspheres

Another property of nonliving organic chemicals is that their colloidal protein molecules, when mixed with other compounds (such as gum arabic), tend to clump together like gelatin to form **coacervate droplets,** balls of molecules that float in water. Other spherical particles called **proteinoid microspheres** form when heated amino acids produce proteinoids, or protein-like molecules, and the proteinoids are mixed with water. The coacervate droplets and microspheres absorb some chemicals out of the water automatically (Figure 6-1), and so in the droplet many chemicals in high concentrations are brought into proximity. The chance for chemicals to combine and the speed of the reactions increase greatly. Molecules in the coacervate droplet may well become organized in position instead of being randomly scattered. On the early earth, some molecules may have assembled to form membranes that were even more selective than the original coacervate droplet.

Fox (1976) summarized many perceived improvements in chemical activity in contained microsystems. Some of them follow.

1. Physical protection of organic material
2. Organization of chemical reactions
3. Compartmentalization of functions
4. Hydrophobic zones that are thermodynamically favorable
5. Maintenance of kinetically favorable concentrations
6. Reproduction at the microsystemic level because microspheres bud off new microspheres as they enlarge
7. Adaptive natural selection of individual variants at the microsystem level

8. Screening of macromolecules from diffusable molecules
9. Promotion of dynamic interactions with the environment
10. Juxtaposition of enzymes and subsystemic organized centers equivalent to organelles in a cell
11. Enlargement of metabolic pathways through combination of spheres
12. Favorable spatial relations for coding interactions

Fox (1960, 1965; Fox and Harada 1960; Fox et al. 1959) suggested that amino acids could have been polymerized into long-chain proteinoids on the hot cinder cones of volcanoes. When he experimentally heated a dry mixture of amino acids at 190° C, proteinoids formed. Subsequent flushing with water, to simulate rain, resulted in the formation of microspheres, or microdroplets. The more organized these spheres became, and the more chemicals they absorbed, the larger they grew. Ultimately, physical forces broke them in two, or they budded off new spheres. The more stable droplets would have lasted the longest, always increasing the chances for the production of nucleic acids and proteins that could control the further development of the sphere and ensure that each droplet contained the same organizing mechanisms. This self-replication of spheres would have resulted in living cells because the spheres would have been able to reproduce themselves. At this early stage in the origin of life, the boundary between the living and the nonliving was indistinct. Natural selection may well have started at a prebiotic stage when persistence of coacervate droplets or of more complex spheres was selected for. Fox and Dose (1972) summarized the stages in the evolution of the first cells, emphasizing the importance of water in the process (Figure 6-2).

A debatable topic in the origin of life is the question of which came first—a persistent membrane, such as coacervate droplets or proteinoid microspheres, or a self-replicating molecule. Historically, theorists first emphasized persistent membranes, then shifted their focus to the evolution of molecules that could replicate themselves.

Self-Replicating Molecules

Aggregation of molecules was also probably very important in the evolution of molecules involved with self-replication and the evolution of the genetic material. For example, cytosine and guanine link naturally with three hydrogen bonds, and adenine and uracil do the same with two (Figure 6-3). This complementarity of nucleic acid bases is fundamental to the polymerization that builds more complex molecules.

In ribonucleic acids (RNA), ribose phosphates link the nucleic acid bases in chains without special-

FIGURE 6-1

Cross section of a coacervate droplet with some properties discussed in the text.

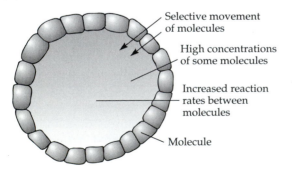

Selective movement of molecules

High concentrations of some molecules

Increased reaction rates between molecules

Molecule

FIGURE 6-2

Possible stages in the evolution of cells, comparing the lack of water on the moon and the importance of water on the earth. *(Based on Fox and Dose 1972)*

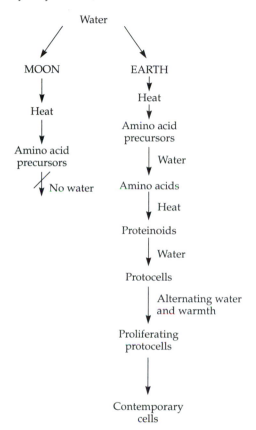

GALAXY

Formaldehyde, ammonia
hydrocyanic acid, carbon monoxide

Water

MOON   EARTH

Heat   Heat

Amino acid   Amino acid
precursors   precursors

No water   Water

Amino acids

Heat

Proteinoids

Water

Protocells

Alternating water
and warmth

Proliferating
protocells

Contemporary
cells

ized catalysts. In a prebiotic world, replication of molecules could have occurred, although quite inaccurately. In fact, part of an RNA molecule may have acted as a template on which another part was built of complementary nucleic acid bases. Such complementarity internal to individual molecules would have resulted in spontaneous folding and a configuration like a hairpin, with a loop and a double-stranded tail held together with hydrogen bonds (Figure 6-4). Such structures would have been stable and could have increased in complexity, resulting in the molecules with alternating loops and complementary bonding seen in RNA (Figure 6-5). Complex molecules like RNA may also have played an important role in early life. They could have handled amino acids directly, as RNA molecules are involved with translation during protein synthesis in cells today, and they could have generated peptide sequences, some of which could become autocatalytic (Loomis 1988), aiding in the synthesis of further such molecules.

Ultimately, some droplet must have contained autocatalytic molecules, and rather complex chemical syntheses evolved. Nucleic acid sequences may have accumulated in the longest-lived droplets, promoting their stability and survival. As more and more sequences coded for more products, the probability of survival would have increased until, perhaps with 50 sequences, a cell could persist and duplicate itself and life began.

Loomis estimated the minimum number of processes, which is broken down in Table 6-3. These 50 sequences would have coded for many enzymes associated with nucleic acid synthesis, protein synthesis, and lipid synthesis plus nucleic acids, proteins, and

FIGURE 6-3

Complementary nucleic acid bases and the hydrogen bonds between them in RNA. Note that the complementary bases cytosine and guanine are bonded with three hydrogen links, and adenine and uracil have two hydrogen bonds. *(From Loomis 1988)*

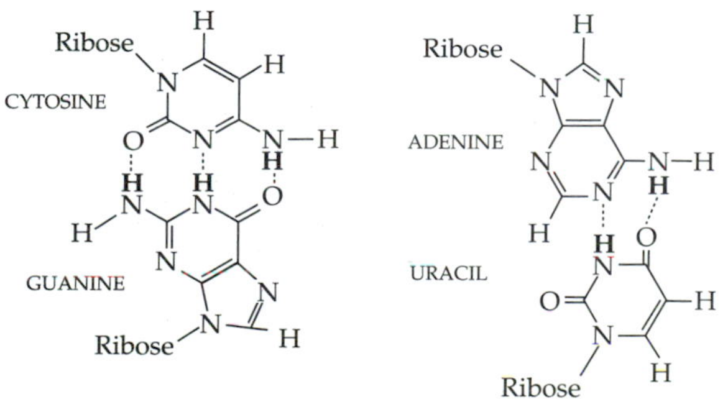

FIGURE 6-4

The spontaneous folding of RNA sequences to match complementary bases results in stable configurations resembling hairpins. The number of lines between bases indicates the number of hydrogen bonds. C, cytosine; G, guanine; A, adenine; U, uracil. *(From Loomis 1988)*

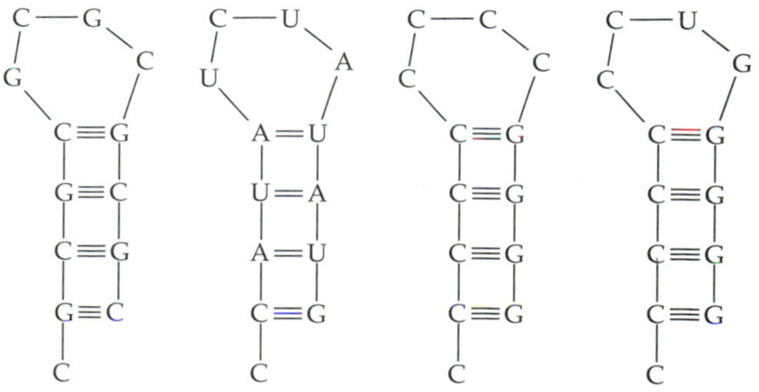

FIGURE 6-5

The two-dimensional structure of 16S ribosomal RNA, the smaller of the two ribosomal RNAs in bacteria. This is a large polymer, about 1500 to 1600 nucleotides in length. The complex three-dimensional shape of such RNA is important in its function. Conservation of sequences and shape and improvement in function would have occurred under stringent natural selection. *(From Woese 1984)*

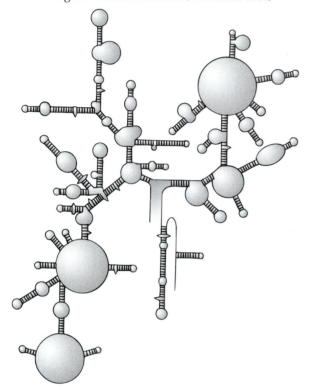

TABLE 6-3 The 50-nucleotide sequence products vital to life[1]

Sequence Product	Number of Sequences Coding for Product
Nucleic acid polymerase	1
Purine synthetases	2
Pyrimidine synthetases	2
Phosphotransferase	1
Peptide elongation factors	2
Ribosomal proteins	3
Ribosomal RNAs	2
Aminoacyl-tRNA synthetases	10
Transfer RNAs	10
Amino acid–metabolism enzymes	4
Oxidation–reduction enzymes	2
Fermentation catalysts	5
Membrane proteins	4
Lipid-metabolism enzymes	2

[1] Identified by Loomis (1988).

lipids themselves. Sequences for catalysts of fermentation would have ensured energy production. Loomis (1988) based his estimate of the minimum kinds and number of sequence products required for survival of early cells on a list of the minimum essential processes in a cell (Table 6-3). Such sequences may have been quite short, with only 30 to 60 bases, and the peptides for which they coded would have been shorter. These weak catalysts would have increased in power as more complexity evolved. Selection for stability was probably strong and populations of droplets very high. Stringent selection in populations with billions of individuals would have resulted in the rapid evolution of more complexity and the origin of the first simple cells: the bacteria.

6.4 ENERGY SOURCES FOR LIFE

Inorganic Molecules

An energy source was required to maintain the organization of the new living cell. Inorganic chemicals could produce chemical energy anaerobically, for there was no free oxygen. For example, the archebacterial genera *Desulfovibrio* and *Desulfatomaculum* contain anaerobic sulfate reducers. The methane bacteria in the genus *Methanobacterium* generate energy and methane by the reduction of carbon dioxide (Woese 1981). Margulis (1981) summarized many of the biochemical cycles in a world without free oxygen.

The methanogenic bacteria make up a large section of the Archaebacteria, many members of which live in extreme environments. These environments may well be refugia maintaining conditions that were much more common near the dawn of life. The methanogens are all anaerobic and live in such habitats as stagnant water, sewage treatment plants, the ocean bottom, hot springs, and the intestinal tracts of ruminants and other animals (Woese 1981).

Another group of Archaebacteria is the extreme halophiles, such as *Halobacterium* and *Halococcus*, but these are aerobic. They live in very salty environments such as the Great Salt Lake, the Dead Sea, and salted fish, and they may produce a red color in salt evaporation ponds. The extreme halophiles have a simple photosynthetic mechanism based on bacterial rhodopsin.

The third group of Archaebacteria, the thermacidophiles, are also aerobic. *Sulfolobus* occurs in hot sulfur springs at 80°C and in very acid conditions with a pH less than 2. The only other genus, *Thermoplasma*, has been found only in smoldering coal piles (Woese 1981; see also Brock 1985)!

In the other major group of bacteria, the Eubacteria, energy acquisition through photosynthesis, either aerobic or anaerobic, became a major strategem. One species of cyanobacterium, *Oscillatoria limnetica,* has facultative anoxygenic photosynthesis (Cohen et al. 1975). The anoxygenic process uses hydrogen sulfide (H_2S) as the electron donor. An anaerobic H_2S-rich layer was found in Solar Lake (near the Gulf of Eilat), in which *O. limnetica* occurs. The cyanobacteria have their beginnings among the oldest living organisms known on this earth, and they are the oldest photosynthesizing group known. They may therefore represent an important link between the anoxygenic photosynthesis typical of bacteria and the oxygenic photosynthesis typical of higher plants (Cohen et al. 1975). Anoxygenic photosynthesis produces no free oxygen because it uses only photosystem I (it is photosystem II that releases oxygen) (cf. Mauseth 1995).

Organic Molecules

Fermentation

Another source of energy for early cells was the organic molecules in the primeval broth. In the absence of oxygen, fermentation is only one possible source of energy from organic molecules. Many kinds of simple carbohydrates were probably common, such as the sugar constituents of nucleic acid, which may have resulted from condensations of formaldehyde (Ponnamperuma and Gabel 1974). Today we can study yeasts as examples of the fermentation of a sugar (glucose). Yeasts ferment sugar without oxygen to produce alcohol plus carbon dioxide and energy:

$$C_6H_{12}O_6 \underset{F}{\rightarrow} 2CO_2 + 2C_2H_5OH + E \ (2)$$

where F denotes fermentation and E is energy in the 2 units of ATP.

This process yields only a small amount of energy because alcohol is still a complex molecule, containing much of the energy originally contained in the sugar. Clearly, fermentation cannot continue indefinitely because the organic foods are used up and the alcohol cannot be used for additional energy. In fact, alcohol becomes toxic at concentrations of 12 to 14 percent. Fermentation thus appears to offer only very limited opportunities for organisms that are restricted to it for their energy acquisition.

Photosynthesis

Before the fermentation process used up all organic food, a new chemical reaction that used the sun's energy evolved: photosynthesis. In photosynthesis, the inorganic molecules carbon dioxide and water, both present in the air, combine to form the organic molecule sugar, and oxygen gas is liberated.

$$CO_2 + H_2O + \text{sun's energy} \rightarrow C_6H_{12}O_6 + O_2 \uparrow$$

Derived from early photosynthesis, the sugar molecules could have served as an energy source to synthesize amino acids in the presence of nitrogen. Amino acids would have been synthesized into proteins, and organisms would have become independent of the organic chemicals accumulated in the primeval broth.

It seems that the oxygen that makes up 21 percent of our atmosphere is derived from the photosynthetic process (see Section 6.5 for evidence). The photosynthesizing oxygenic bacteria had probably evolved by 2.5 to 3.5 billion years ago, and the oxygen in the atmosphere probably reached its present level by about 1.5 billion years ago (Woese 1984) (Figure 6-6). Atmospheric oxygen also gave rise to the ozone layer in the upper atmosphere, which filters out much radiation that is damaging to life, and which is so important in the evolution of life.

Respiration

Originally, it was probably fermentation that broke down organic chemicals. As oxygen increased in the atmosphere because of photosynthesis, and carbon dioxide was used up, the process of respiration evolved. Cold combustion of sugar with oxygen liberated vastly more energy (38 units of ATP per glucose molecule) than did fermentation (2 units of ATP per glucose molecule). This was because cold combustion broke down the molecule more completely into carbon dioxide and water.

$$C_6H_{12}O_6 + O_2 \underset{R}{\rightarrow} CO_2 + H_2O + E \ (38)$$

where R signifies the process of respiration and E is energy in the form of 38 units of ATP.

Thus, a system evolved that was extraordinarily effective in obtaining energy. The nonpolluting, nontoxic end products of the system were recycled in the photosynthetic process:

$$CO_2 + H_2O + S \underset{R}{\overset{P}{\rightleftharpoons}} C_6H_{12}O_6 + O_2 \uparrow$$

where S is the sun's energy, P is photosynthesis, and R is respiration.

The oxygen you breathe may have already been in the leg of a frog, the lung of a spider, or the flower of a lily. In fact, respiration was so efficient that even multicellular organisms could obtain ample energy for rapid growth and reproduction, and photosynthesis was so efficient that primary production increased enormously. The age of microorganisms probably lasted at least from 3.5 billion years ago to about 0.5 billion years ago, if it ever ended (Figure 6-6).

The advent of respiration gave the evolutionary process its impetus. Today's great diversity of conspicuous life has resulted largely from organisms that

FIGURE 6-6

A chronology of events in the evolution of life, showing an earth long dominated by microorganisms and the formation of oxygen in the atmosphere from about 2.2 billion to 1.5 billion years ago. *(Based on Margulis 1981, 1993; Woese 1984)*

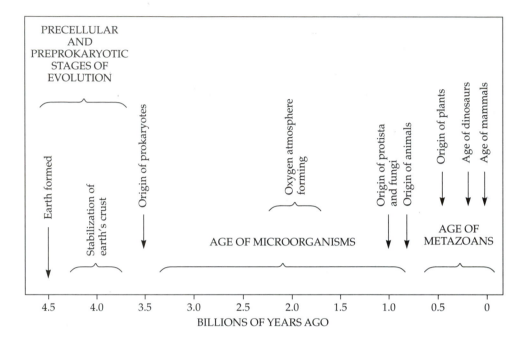

evolved to use oxygen for the release of energy from organic molecules. Organisms that use exclusively fermentation, such as yeasts and bacteria, have remained to this day virtual replicas of early organisms. Margulis (1981) summarized the rich array of biogeochemical cycles in an oxygenated world (Figure 6-7).

When one considers that there are at least 100,000 planets like the earth in our own galaxy, and there are about 100 million galaxies within the range of a powerful telescope, then at least 10 trillion planets like our own probably exist. Life could exist on all these planets, and some of its forms may be similar to life forms on earth. Perhaps our experiences are similar to those of organisms in other parts of the universe. Will we ever be able to communicate with, or meet with, those organisms?

6.5 AN OXYGEN-RICH ATMOSPHERE

Evidence for the timing of an oxygen-rich atmosphere comes from rocks and the forms of iron and sulfur in those rocks. Rocks older than 2.5 billion years usually have iron and sulfur in reduced form—that is, ferrous iron and sulfides. Then, 2.5 billion years ago (earlier in some places, e.g., up to 3.8 billion years ago in Greenland [Walker et al. 1983]), rocks with a rusty color caused by ferric iron started to appear. The

massive deposits mined for iron ore in Australia, around Lake Superior in North America, and in Africa, called the banded iron formations, all date from that time. The bands indicate the deposition of sediments in alternating oxygenated and anaerobic conditions. Thus, any iron implement used today probably contains very old iron that was oxidized with oxygen emitted by photosynthesizing bacteria 2 to 2.5 billion years ago. So much oxygen was being produced at that time that much of the ferrous iron was being oxidized; the banded iron formations may contain up to 15 percent iron. When the oxidizing process was largely complete, free oxygen would have been liberated into the atmosphere and accumulated. The oxidation of ferrous iron probably lasted 700 to 800 million years, because banded iron formations stop and are replaced by red rock beds at about 1.8 billion years ago. This evidence indicates that an oxygen-containing atmosphere formed around 2.2 to 1.8 billion years ago (Margulis 1982, Schopf 1983).

6.6 MICROBIAL MATS AND STROMATOLITES

Photosynthesizing bacteria probably first lived in the interstices of sediments, with both elements forming **microbial mats**.

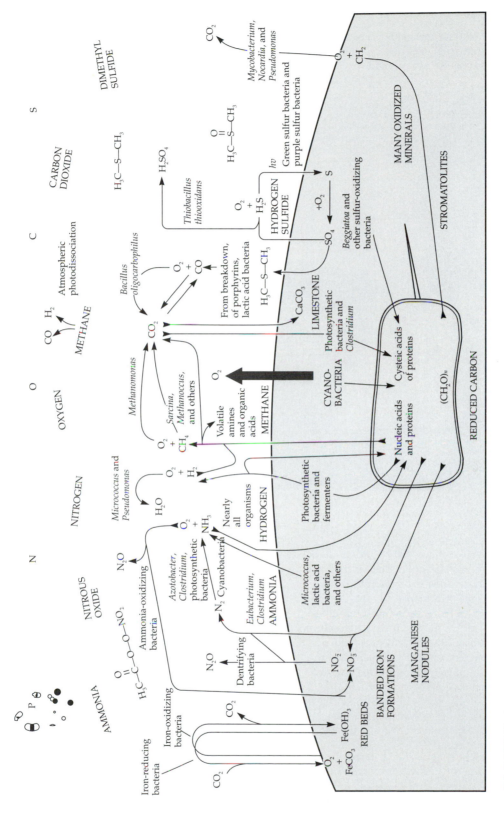

FIGURE 6-7

Biogeochemical cycles on an oxygenated earth, as summarized by Margulis (1981, 1993). The edge of the darkened dome represents the surface of the earth; beneath it, the below-surface processes are indicated.

FIGURE 6-8

Drawings of some stromatolite forms showing their layered-rock appearance. On the right, dark dots in the stromatolite indicate the many microfossils, such as are found in the Gunflint Formation, Ontario, Canada. Such fossils may have grown in an oxygen-rich microenvironment, or "oasis." *(Based on Stearn et al. 1979)*

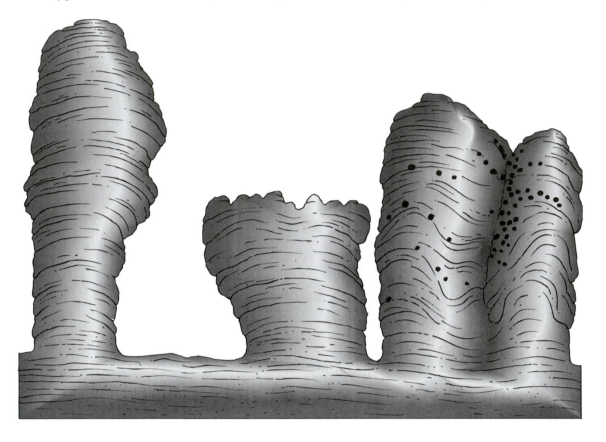

A typical mat looks like mere dirty sand or dark green scum, but one can see upon close inspection that it is fibrous and composed of intertwined filaments. Such a mat traps sediment—clay, mud particles, organic debris. As the sediment accumulates, the microbes either are trapped in it or die or move upward in the light to start a new layer atop the old. If the older layer contains enough sediment and the evaporation rate is high, it gradually turns to stone. As those events repeat themselves periodically, a many-layered stromatolite takes form. Stromatolites are as stable as most other sedimentary rocks, although their structure reveals them to be the work of living things. (Margulis 1982, 65–66)

A stromatolite, or layered rock, may take the form of successive dome-shaped deposits of sediments (Figure 6-8). Present-day stromatolites occur in very saline areas where grazing animals cannot tolerate the salt, such as in Hamelin Pool, Shark Bay, Western Australia. At Shark Bay and in Yellowstone Park, stromatolites and microbial mats, respectively, orient themselves to the sun; they exhibit heliotropism (Awramik and Vanyo 1986), a characteristic seen in fossil-

ized stromatolites. The major kinds of bacteria involved in stromatolite building are the filamentous oscillatoriacean cyanobacteria.

6.7 RELATIONSHIPS AMONG EARLY ORGANISMS

Our own bodies contain vestiges of early forms of life on earth in the form of ribosomal RNA. Ribosomal RNA sequences are very conservative and under strong stabilizing selection, and thus the molecular clock is quite slow (Woese 1984, Wilson 1985), meaning that any change in molecular structure is very rare. Therefore, changes in ribosomal RNA structure can provide evidence about the broad relationships between organisms living today. Analysis of ribosomal RNA sequences showed that the common ancestral bacterium gave rise to three main bacterial groups:

1. The **eubacteria,** or true bacteria, with which we are most familiar, include most pathogenic species

FIGURE 6-9

A universal phylogenetic tree of life on earth, based on rRNA sequence comparisons, by Carl Woese (1992).

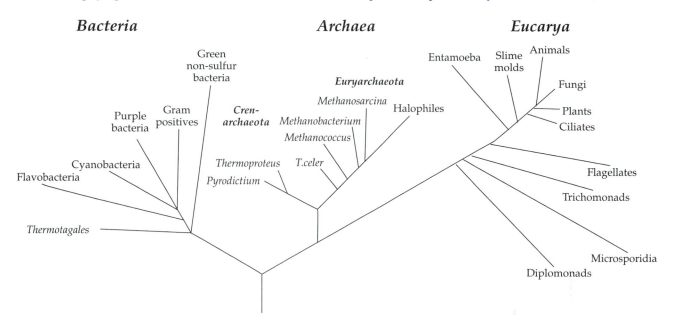

plus the cyanobacteria and other photosynthetic bacteria.

2. The **archaebacteria** include the methanogens and bacteria from extreme environments, such as halophiles and thermophiles.

3. The **urkaryotes** were the ancestral line to the eukaryotes and thus to all forms of higher life on this earth (Figure 6-9). Eukaryotic cells are characterized by complex internal structures with membrane-bound organelles.

How did the complexity of the eukaryotes arise? The theory that the eukaryotic cell is a community of prokaryotic cells has been advanced principally by Margulis (e.g., 1970, 1981, 1982, 1993). A eukaryotic cell is about ten times larger than a prokaryotic cell, and its organelles are about the same size as prokaryotic cells. It now appears that mutualistic relationships developed between prokaryotes, and the eukaryotic cell community developed in a stepwise fashion. This view may be called the **endosymbiotic theory** or the **serial endosymbiosis theory** on the evolution of eukaryotic cells (Taylor 1974).

6.8 THE ENDOSYMBIOTIC THEORY OF EUKARYOTIC CELL EVOLUTION

The first step in the evolution of the eukaryotic cell is hypothesized to have been the invasion of a large cell, such as a mycoplasma, by a purple bacterium.

The bacterium was a respiring mutualist, and so it provided the host cell with respiratory capability. The purple bacterium has retained part of its own genome within the eukaryotic cell, forming what we now recognize as the mitochondria of the cell (Figure 6-10). During cell division, mitochondria are self-replicating so that each new cell receives genetic information from both the mitochondria and the nucleus. Such a cell can be called **digenomic** because it contains the genomes of two organisms.

The second (and more controversial) step is hypothesized to have resulted in the acquisition of undulipodia (cf. Margulis 1993). An **undulipodium** is a cilium or eukaryotic flagellum used for locomotion or feeding. The mutualist in this case is hard to discover, but it could be something like a spirochete that entered into an association with a larger cell, as spirochetes are known to do today. A spirochete has a long flagellum running the length of its body, and the movement of the flagellum would have propelled the larger associated cell. The flagellum bears microtubules reminiscent of those in the cilia and flagella of higher organisms, although not in the same arrangement. Apparently, the spirochete-like cell gradually lost its integrity, but the undulipodium remains a common organelle in eukaryotic cells. (Margulis [1993] provides fascinating support for this exogenous and symbiotic origin of undulipodia, while acknowledging one major alternative hypothesis.) What happened to the spirochete genome is unclear at this time. "Nearly every aspect of spirochete growth and synthesis was

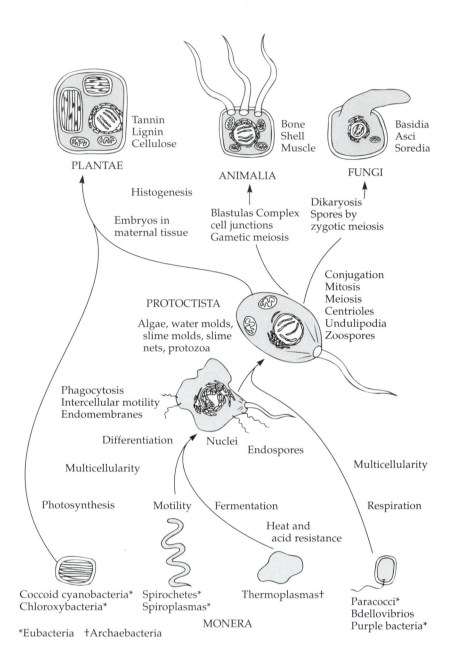

FIGURE 6-10

The stepwise origin of eukaryotic cells by symbiotic mutualism, based on microcosms of tightly integrated prokaryotic organisms. Protoctista is equivalent to Protista. *(From Margulis 1981, 1993)*

relegated to the host genome in the complex act of integration, which, in retrospect, we see was eukaryosis" (Margulis 1993, 300).

In the third step, a bacterium from the photosynthesizing cyanobacteria invaded a eukaryotic cell, providing photosynthesizing capability in cells already capable of respiration. The perfect cycling process, photosynthesis plus respiration, thus established itself. The photosynthesizing bacterium also retained some of its own genome in the host cell; it is now called the chloroplast. Hence, three genomes are involved with the hereditary process in such photosynthesizing eukaryotic cells. That is, cells are **trigenomic,** with nuclei, mitochondria, and chloroplasts dividing, moving to new cells, and carrying their own

DNA. If a vestige of the spirochete genome is present in eukaryotic cells, it remains to be discovered.

The endosymbiotic theory would be incomplete without consideration of other novel structures in the eukaryotic cell: the nuclear membrane and the nucleus. Exogenous, symbiotic origins as well as endogenous, nonsymbiotic origins have been debated as explanations for the evolution of the nucleus. Margulis (1993) points out that the nucleus lacks any mechanism for protein synthesis, which occurs in the cytoplasm of the eukaryotic cell, and therefore a symbiotic origin is unlikely. An alternative hypothesis notes that prokaryotes as a group are rich in a variety of membranous structures that partition activities internal to the cell. Such **endomembrane systems** may have invested "newly synthesized DNA molecules deep inside the cell before the evolution of chromosomes and mitosis" (Margulis 1993, 17–18).

> Although a most problematic aspect of the origin of eukaryotes is this detail of the history of the nucleus, it appears quite likely to me that the tendencies toward membrane-wrapping and engulfment do exist in prokaryotes. Membrane hypertrophy, probably aggravated by infection and predatory attack, was supplemented by hereditary symbiosis, in my opinion. (Margulis 1993, 219–220)

Because the genomes of mitochondria and chloroplasts are conserved and because these members of the cell community synthesize ribosomal RNA, the lineages can be traced accurately (Figure 6-11). All the higher green plants evolved from certain eukaryotes with chloroplasts. All slime molds, protozoa, fungi, and animals evolved from the remaining eukaryotic progenitors (Figure 6-10). From the milestone of the eukaryotic cell, the evolution of higher forms of life seems relatively easy to explain, because all forms are built on the basic plan of multiples of the eukaryotic cell.

Evolution of Mitosis

DNA evolved in the prokaryotes (e.g., see Loomis 1988), but the "dance of the chromosomes," or mitosis, evolved in simple eukaryotic forms—the protists. How did such a complex system develop? Much of the complexity involves microtubules. If we view the cell during mitosis, we can begin to understand the importance of such microtubules (Margulis 1981) (Figure 6-12). Centrioles, kinetosomes, undulipodia, mitotic spindle fibers, and related structures are all composed of microtubules. The microtubules frequently develop from centers such as kinetosomes and centrioles, which have been called, as a general term, microtubule-organizing centers (MCs) and are defined as "structures or loci that give rise to microtubular arrays" (Pickett-Heaps 1974, 1975).

Margulis (1981) developed a five-step scenario for the evolution of mitosis, starting with a eukaryotic cell with an undulipodium and its basal body, or kinetosome (Figure 6-13). In Step I, the MCs served as kinetosomes for undulipodia. The nucleic acid of the MC associated with the undulipodium was responsible for its own synthesis and for replication and synthesis of the undulipodium.

In Step II, MCs were incorporated into the nucleus for segregation of host chromatin, causing the permanent loss of undulipodia. This was a step aside from the main development of microtubule organization. Spirochete-like nucleic acid became associated into or on the surface of the nucleus and functioned as an intranuclear division center, producing naked

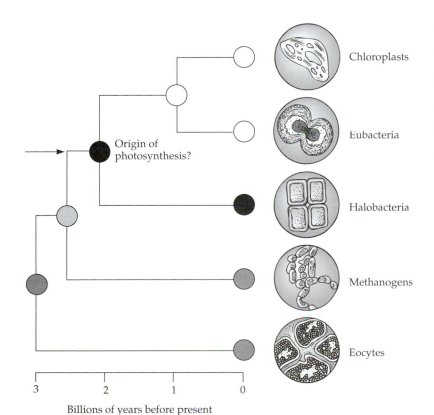

FIGURE 6-11

A phylogenetic tree of early cells, inferred from comparison of the base sequences of ribosomal RNA in the kinds of organisms or organelles shown. The origin of photosynthesis is associated with a common ancestor for all photosynthesizing organisms. Eocytes are the sulfur bacteria. (*Based on Wilson 1985*)

Chloroplasts

Eubacteria

Halobacteria

Methanogens

Eocytes

Origin of photosynthesis?

3   2   1   0

Billions of years before present

FIGURE 6-12

Generalized drawing of a cell during mitosis, showing the locations of microtubules and microtubule organizing centers. *(Based on Margulis 1981, 1993)*

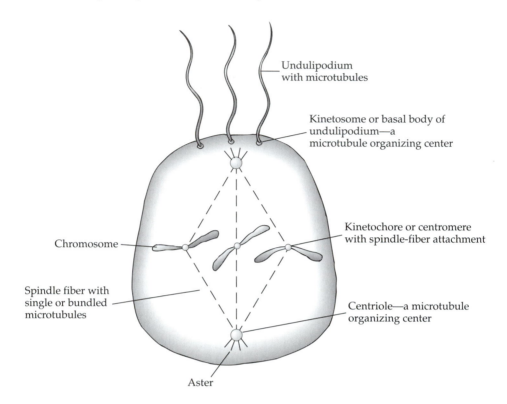

Undulipodium with microtubules

Kinetosome or basal body of undulipodium—a microtubule organizing center

Kinetochore or centromere with spindle-fiber attachment

Chromosome

Centriole—a microtubule organizing center

Spindle fiber with single or bundled microtubules

Aster

amebae, without undulipodia. Such MCs are found in other amebae, cellular slime molds, perhaps microsporidia, amastigote fungi, conjugating green algae, and red algae. Presumably, mitosis evolved in four stages: (1) MCs replicated before chromatin replication and nuclear division; (2) MC–chromatin attachments occurred; (3) MCs functioned as kinetochores or centromeres; and (4) other MCs not attached to chromatin functioned as centrioles.

In Step III, MCs were employed as intranuclear division centers as well as kinetosomes for undulipodia. They performed only one function at a time, however. Protists evolved mitosis, but they lost undulipodia and mobility during cell division. This development resulted in some ameboflagellates: anisonemids or peranemids, some dinoflagellates, and uninucleate metazoans such as *Dimastigamoeba, Mastigella,* and *Mastigina* (Margulis 1981).

In Step IV, a permanent separation of MC function developed. Some MCs formed kinetosomes of undulipodia. Other MCs became permanent inhabitants of the nucleus as intranuclear division centers. This step resulted in such organisms as trypanosomes, chlorophytes, and chrysophytes.

Step V, the final stage, resulted in the evolution of metazoan mitosis. Some products of MC replication differentiate into centrioles, kinetosomes, kinetochores, and other microtubular structures. Other products remain undifferentiated but capable of replication. Cells containing differentiated MCs may not be capable of mitosis in some cell lines.

Evolution of Meiosis

Cleveland (1947) suggested that meiosis developed from the failure of the kinetochores to divide, followed by suppression of kinetochore replication, so that a chromatin/kinetochore ratio of 2:1 became established. Diploidy may have evolved when related cells fused under stress (Margulis 1981), perhaps starting with the cannibalism that is common in protists. With the added genetic diversity of two nuclei, the fused individual survived better than haploid relatives. However, the evolution of meiosis could have deleted the excess chromatin of the new diploid cells. The suppression of all kinetochore replication for one cell division eliminated exactly one haploid number of chromosomes.

FIGURE 6-13

Possible stages in the evolution of mitosis. Steps I through V are indicated on the left. *(From Margulis 1981, 1993)*

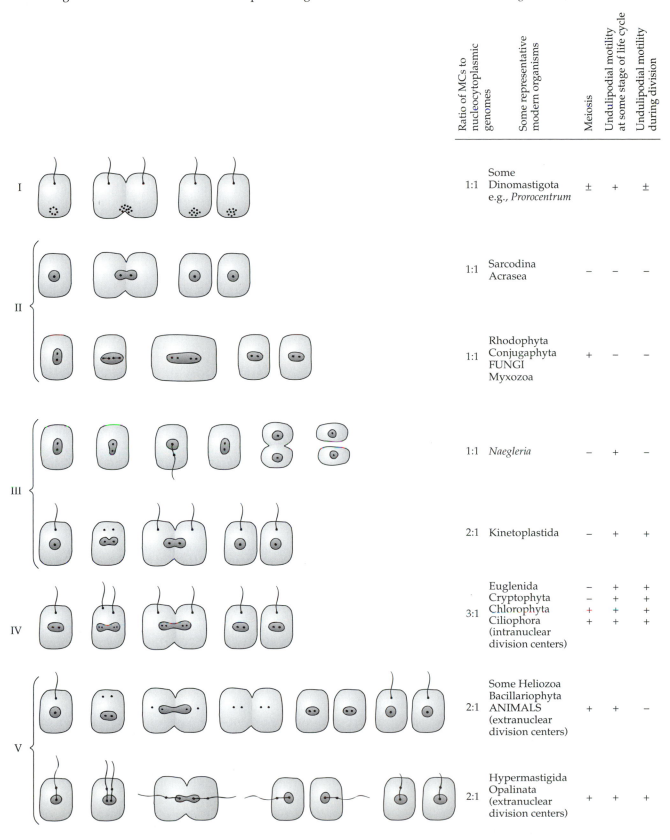

Ratio of MCs to nucleocytoplasmic genomes	Some representative modern organisms	Meiosis	Undulipodial motility at some stage of life cycle	Undulipodial motility during division
1:1	Some Dinomastigota e.g., *Prorocentrum*	±	+	±
1:1	Sarcodina Acrasea	–	–	–
1:1	Rhodophyta Conjugaphyta FUNGI Myxozoa	+	–	–
1:1	*Naegleria*	–	+	–
2:1	Kinetoplastida	–	+	+
3:1	Euglenida	–	+	+
	Cryptophyta	–	+	+
	Chlorophyta	+	+	+
	Ciliophora (intranuclear division centers)	+	+	+
2:1	Some Heliozoa Bacillariophyta ANIMALS (extranuclear division centers)	+	+	–
2:1	Hypermastigida Opalinata (extranuclear division centers)	+	+	+

Key: Characteristic listed is present in
+ all species studied
± some species studied
– no species studied

6.9 BEYOND THE OPARIN–HALDANE HYPOTHESIS

This chapter has presented a general scenario of the evolution of life that can be called the **Oparin–Haldane hypothesis.** A. I. Oparin and J. B. S. Haldane developed their hypothesis in the 1920s to 1940s (Haldane 1929, Oparin 1924, 1938, 1968). Since then, the experimental approaches of Miller and Fox and the endosymbiotic theory on eukaryotic cell evolution championed by Margulis, with their increasing concentration on complex, self-replicating molecules, have made important additions to the Oparin–Haldane hypothesis.

Although the Oparin–Haldane hypothesis is widely accepted as a reasonable scenario, some fascinating questions have emerged that will be hard to resolve, and some modifications may be necessary. As Woese (1984, 29) has said, "Too many accounts imply that we have a basic understanding of the origin-of-life process. We do not." Life may have evolved before oceans formed. Most reactions that result in the molecules of life require the elimination of water, and so such reactions in water would tend to be reversed (Woese 1984). "The origin of life seems inconsistent with fully aqueous chemistry, particularly at high temperature" (Pace 1991, 532). On a hot earth, with only clouds over a merely damp surface, did life evolve in water droplets above the earth? Was the incredibly dynamic early earth an essential environment for the origin of life rather than a condition that had to pass before life could emerge? What was the origin of the process of translating genetic information? How could biological specificity have evolved when large and complex molecular structures define this specificity? Clearly, we still have a fascinating mystery to solve, and probably more questions than Oparin and Haldane envisaged. Pace (1991) clearly presents some of the questions.

The further development of life may well have depended heavily for macroevolutionary innovation on additional symbiotic associations among unrelated organisms (e.g., Margulis and Fester 1991). Land plants may have evolved from the mutualistic association of a semiaquatic green alga as the thallus and an aquatic fungus that became endophytic and essential for the mineral nutrition of the union (Pirozynski and Malloch 1975, Atsatt 1988). Mycorrhizal associations and other symbioses fostered further development of land plants (Malloch et al. 1980, Werner 1992). Radiation of metazoans feeding on land plants has depended on symbiotic associations with microorganisms that have cellulolytic enzymes: termites, wood-boring beetles, and the ruminants, for example (e.g., Hungate 1975, Jones 1984, Martin 1987, Price 1984, 1991). Associated microorganisms mediate many links between the herbivores and their enemies (Price 1991). In fact, in terms of ecology and evolution, the higher organisms seem to be nested within the capabilities evolved long ago by microorganisms (Price 1988, 1991). After the evolution of the eukaryotes, many major breakthroughs in biotic complexity seem to have resulted from symbiotic and mutualistic associations. Microcosms of mutualists have probably provided the basis for macroevolutionary advances beyond the capabilities of gradualism.

The foregoing arguments need further investigation, but the fact is that life has developed by leaps and bounds, building cell upon cell and organism upon organism.

SUMMARY

Since an explosion of pure energy some 15 billion years ago, the cosmos has expanded. Condensation of matter formed the galaxies, among them the Milky Way and, within it, our solar system and planet earth. From a beginning perhaps 4.5 billion years ago, the earth has cooled and the atmosphere has changed, with condensation of water vapor into huge clouds followed by torrential rains and the development of oceans, seas, lakes, and rivers.

Serial events on a prebiotic earth have been suggested as necessary steps toward the formation of living cells. Originally, organic molecules must have been synthesized by abiotic processes, perhaps in extreme heat or with electrical discharges, and such molecules may have aggregated spontaneously into spheres, with intensified internal chemical activity. Such spheres may have been split by impacts or could have budded off duplicates, some persisting longer than others on the basis of their relative stabilities. Abiotic selection of the most stable entities, analogous to natural selection of biotic forms, would then have produced long-lasting lineages of precellular molecular clusters that ultimately became capable of controlled replication. Those clusters are recognizable as the first life on the planet.

The energy sources necessary to maintain the complex organization of living cells and their reproduction developed from anaerobic utilization of inorganic chemicals through fermentation of organic molecules to photosynthesis and respiration. Photosynthesis with the liberation of oxygen became so prevalent on earth that it enriched the atmosphere between 2.2 and 1.8 billion years ago, and the present-day oxygenated environment for life developed.

Photosynthesizing bacteria may well have lived in sediments and formed microbial mats, some with the potential for trapping particles and building up into stromatolites.

According to the endosymbiotic theory of eukaryotic cell evolution, the more complex eukaryotic cell evolved from bacterial cells. Stepwise mutualistic associations between bacteria ultimately resulted in the protists, highly organized cells with mitochondria and microtubule-organizing centers, and some with cilia or flagella and chloroplasts. Mitosis evolved among the protists, with the highly organized division of genetic material orchestrated from microtubule-organizing centers. Fusion of eukaryotic cells, producing diploid organisms with a more robust physiology, may have occurred during stress. Permanent union of two genomes was established when meiosis—the reduction division—created haploid gametes that fused to reinstate the diploid organism.

The Oparin–Haldane hypothesis of the evolution of early life is simply a rational attempt to account for what may have occurred. An objective and empirical understanding of the origin of life is elusive and may remain so.

QUESTIONS FOR DISCUSSION

1. Which do you consider more important in the evolution of the cell, and which probably evolved first—self-replicating genetic material or cell-like membranes?

2. To what extent do you think speculation is a valid approach in science, and is it overused in considerations of the origin of life?

3. Does the recognition that an extraterrestrial impact body (such as a comet or meteor) could introduce new organic molecules from outer space onto the earth help us understand the origin of life?

4. Would you argue that each step in the evolution of the eukaryotic cell involving endosymbiosis was a speciation event that may be regarded as saltational, or an evolutionary leap from one species to another?

5. When new symbiotic associations developed among prokaryotic cells, what kind of symbiosis do you think is likely to have been involved—mutualism, parasitism, or commensalism?

6. Does consideration of the origin of life help with a definition of life? Can you define life?

7. If symbiosis were so central to the evolution of eukaryotes and multicellular species, might we expect that symbiosis continued to be a major source of evolutionary innovation during the radiation of fungi, plants, and animals?

8. In your opinion, what is the prospect of scientists ever being able to develop a living organism—that is, to create life—in the laboratory?

9. Can you think of an exact method of discerning or defining the moment when prebiotic entities became alive?

10. Which environments on earth do you consider to be the most promising research locales in the search for ancient lineages and interactions that may illuminate early life on earth?

REFERENCES

Atsatt, P. R. 1988. Are vascular plants "inside-out" lichens? Ecology 69:17–23.

Awramik, S. M., and J. P. Vanyo. 1986. Heliotropism in modern stromatolites. Science 231:1279–1281.

Banin, A., and J. Navrot. 1975. Origin of life: Clues from relations between chemical compositions of living organisms and natural environments. Science 189:550–551.

Bernal, J. D. 1967. The Origin of Life. Weidenfeld and Nicholson, London.

Brock, T. D. 1985. Life at high temperatures. Science 230:132–138.

Cairns-Smith, A. G. 1985. Seven Clues to the Origin of Life. Cambridge Univ. Press, Cambridge, England.

Cleveland, L. R. 1947. The origin and evolution of meiosis. Science 105:287–288.

Cohen, Y., E. Padan, and M. Shilo. 1975. Facultative anoxygenic photosynthesis in the cyanobacterium, *Oscillatoria limnetica*. J. Bacteriol. 123:855–861.

Farley, J. 1974. The Spontaneous Generation Controversy from Descartes to Oparin. Johns Hopkins Univ. Press, Baltimore.

Fox, S. W. 1960. How did life begin? Science 132:200–208.

———. 1965. The theory of macromolecular and cellular origins. Nature 205:328–340.

———. 1976. The evolutionary significance of phase-separated microsystems. Origins of Life 7:49–68.

Fox, S. W., and K. Dose. 1972. Molecular Evolution and the Origin of Life. W. H. Freeman, San Francisco.

Fox, S. W., and K. Harada. 1960. The thermal copolymerization of amino acids common to protein. J. Amer. Chem. Soc. 82:3745–3751.

Fox, S. W., K. Harada, and J. Kendrick. 1959. Production of spherules from synthetic proteinoid and hot water. Science 129:1221–1223.

Haldane, J. B. S. 1929. The Origin of Life. Rationalist Annual. [Reprinted in Bernal 1967, 242–249]

Haymon, R. M., and K. C. Macdonald. 1985. The geology of deep-sea hot springs. Amer. Sci. 73:441–449.

Hungate, R. E. 1975. The rumen microbial ecosystem. Ann. Rev. Ecol. Syst. 6:39–66.

Jones, C. G. 1984. Microorganisms as mediators of plant resource exploitation by insect herbivores, pp. 53–99. In P. W. Price, C. N. Slobodchikoff, and W. S. Gaud (eds.). A New Ecology: Novel Approaches to Interactive Systems. Wiley, New York.

Loomis, W. F. 1988. Four Billion Years: An Essay on the Evolution of Genes and Organisms. Sinauer, Sunderland, Mass.

Malloch, D. W., K. A. Pirozynski, and P. H. Raven. 1980. Ecological and evolutionary significance of mycorrhizal symbiosis on vascular plants (a review). Proc. Nat. Acad. Sci. 77:2113–2118.

Margulis, L. 1970. Origin of Eukaryotic Cells. Yale Univ. Press, New Haven.

———. 1981. Symbiosis in Cell Evolution: Life and Its Environment on the Early Earth. W. H. Freeman, San Francisco.

———. 1982. Early Life. Van Nostrand Reinhold, N.Y.

———. 1993. Symbiosis in Cell Evolution: Microbial Communities in the Archean and Proterozoic Eons. 2d ed. W. H. Freeman, New York.

Margulis, L., and R. Fester (eds.). 1991. Symbiosis as a Source of Evolutionary Innovation: Speciation and Morphogenesis. MIT Press, Cambridge, Mass.

Martin, M. M. 1987. Invertebrate–Microbial Interactions: Ingested Fungal Enzymes in Arthropod Biology. Cornell Univ. Press, Ithaca.

Mauseth, J. D. 1995. Botany: An Introduction to Plant Biology. 2d ed. Saunders College Publishing, Philadelphia.

Miller, S. L. 1953. A production of amino acids under possible primitive earth conditions. Science 117:528–529.

Oparin, A. I. 1924. Proiskhozdenie zhizny. Izd. Moskovshii Raboehii. [Translated in Bernal 1967, 199–234, as The Origin of Life.]

———. 1938. The Origin of Life. Macmillan, N.Y.

———. 1968. Genesis and Evolutionary Development of Life. Academic, New York.

———. 1976. Evolution of the concepts of the origin of life, 1924–1974. Origins of Life 7:3–8.

Pace, N. R. 1991. Origin of life—facing up to the physical setting. Cell 65:531–533.

Pickett-Heaps, J. D. 1974. Evolution of mitosis and the eukaryotic condition. BioSystems 6:37–48.

———. 1975. Green Algae: Structure, Reproduction and Evolution. Sinauer, Sunderland, Mass.

Pirozynski, K. A., and D. W. Malloch. 1975. The origin of land plants: A matter of mycotrophism. BioSystems 6:153–164.

Ponnamperuma, C., and N. W. Gabel. 1974. The precellular evolution and organization of molecules, pp. 393–413. In M. J. Carlile and J. J. Skehel (eds.). Evolution in the Microbial World. Cambridge Univ. Press, Cambridge, England.

Price, P. W. 1984. Insect Ecology. 2d ed. Wiley, New York.

———. 1988. An overview of organismal interactions in ecosystems in evolutionary and ecological time. Agri. Ecosyst. Environ. 24:369–377.

———. 1991. The web of life: Development over 3.8 billion years of trophic relationships, pp. 263–272. In L. Margulis and R. Fester (eds.). Symbiosis as a Source of Evolutionary Innovation: Speciation and Morphogenesis. MIT Press, Cambridge, Mass.

Schopf, J. W. (ed.) 1983. Earth's Earliest Biosphere: Its Origins and Evolution. Princeton Univ. Press, Princeton, N.J.

Stearn, C. W., R. L. Carroll, and T. H. Clark. 1979. Geological Evolution of North America. Wiley, New York.

Taylor, F. J. R. 1974. Implications and extensions of the serial endosymbiosis theory of the origin of eukaryotes. Taxon 23:229–258.

Walker, J. C. G., C. Klein, M. Schidlowski, J. W. Schopf, D. J. Stevenson, and M. R. Walter. 1983. Environmental evolution of the Archean–early Proterozoic Earth, pp. 260–290. In J. W. Schopf (ed.). Earth's Earliest Biosphere: Its Origin and Evolution. Princeton Univ. Press, Princeton, N.J.

Weinberg, S. 1977. The First Three Minutes. Basic Books, New York.

———. 1985. Origins. Science 230:15–18.

Werner, D. 1992. Symbiosis of Plants and Microbes. Chapman and Hall, London.

Wilson, A. C. 1985. The molecular basis of evolution. Sci. Amer. 253(4):164–173.

Woese, C. R. 1981. Archaebacteria. Sci. Amer. 244(6):98–122.

———. 1984. The Origin of Life. Carolina Biol. Supply, Burlington, N.C.

———. 1992. Archaea, evolution, and the origin of life. Illinois Alumni Newsletter, pp. 3–4.

7

FROM EARLY EUKARYOTES TO FUNGI, ANIMALS, AND PLANTS

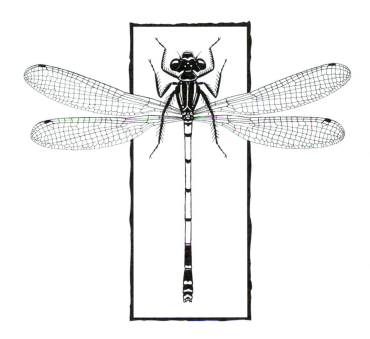

The large red damselfly *(Pyrrhosoma nymphla)*, a modern relative of much larger dragonflies that flew over extensive swamps during the Carboniferous *(Illustration by Stephen Price)*.

7.1 WHAT IS THE PROBLEM?

Once we understand the origins of the prokaryotic cell, "the rest of evolution is relatively easy!" (Woese 1984, 3). However, "the rest of evolution" has been the preoccupation of thousands of scientists for well over 100 years, and there are still interesting and important problems to be resolved, as Woese acknowledged might be the case. Perhaps evolution from prokaryotes to fungi, plants, and animals was "easy" for the organisms involved, but it is certainly not simple for humans to understand.

Woese's point was that, once the first living prokaryotic cell evolved, relatively simple assemblages of cells could account for the subsequent expansion of life on earth: the assemblage of cells into the eukaryotic cell, the assemblage of eukaryotic cells into multicellular organisms, and the assemblage of multicellular organisms into more complex symbiotic units and into ecosystems. The concept is simple—until one wishes to trace actual phylogenetic lineages. What turned into what? What came from what? These are the traditional questions about life that all humans ask from an early age, regardless of level of formal education. Where do babies come from? Where did humans come from? How did the elephant get its trunk? Did birds evolve from dinosaurs? What did the first mammal look like? Are velvet worms the missing link between segmented worms and the jointed-appendage invertebrates or arthropods (Ballard et al. 1992)?

One complication is that any classification of organisms into taxa inevitably must deal with a huge diversity of kinds, many of which are intermediate between the neat compartments we humans envisage. As a result, classification is fluid and hotly debated. Even at the highest, most inclusive categories of taxonomic classification, change and restructuring are commonplace. The **kingdoms** of organisms are often regarded as the most inclusive taxa in the classification of cellular life. Linnaeus recognized two, Plantae and Animalia. But with the advance of knowledge—especially about smaller organisms—it became necessary to recognize five kingdoms (Whittaker 1969). The kingdom Monera contains the prokaryotes, or bacteria. The kingdom Protista contains the unicellular eukaryotic organisms such as protozoa and single-celled eukaryotic algae. The kingdom Fungi includes molds, yeasts, mushrooms, and toadstools. The kingdom Plantae comprises multicellular photosynthesizing organisms, including multicellular algae and, most conspicuously, the terrestrial plants. The kingdom Animalia includes all multicellular animals, from sponges through invertebrates and chordates to the vertebrates.

But if one wishes to group species that are phylogenetically related to each other, rather than simply those appearing to be similar—that is, to strive for a more natural classification—various modifications are necessary as understanding deepens (according to some authorities). For example, a higher category than the kingdom would recognize all prokaryotes as belonging to one group, the **superkingdom Prokaryota,** and all eukaryotes as belonging to the other, the **superkingdom Eukaryota** (Margulis 1993). Another area in need of modification is the unsettling and apparently artificial separation of unicellular and multicellular organisms. Protists obviously aggregated into multicellular organisms and retained stronger phylogenetic ties to protist lineages than to other multicellular organisms with which they were grouped. Multicellular algae, for instance, should be in an expanded kingdom with the protists, for they have direct kinship with photosynthesizing protists but very distant relationships with terrestrial plants. Another case is slime molds and the oomycetes that have been placed with the fungi but are more naturally related to protozoa (see Section 7.4) and fall more logically into an expanded kingdom that includes protists and their closely related multicellular kin. Consequently, Margulis (e.g., 1993) erected the more inclusive kingdom Protoctista. In a third case, authors of textbooks on animal biology refuse to exclude the protozoa—the "unicellular animals," as they are often called (e.g., Hickman et al. 1990). After all, ciliated protozoa often aggregate, and the sponges develop with relatively loose associations of cells and share some characteristics with slime molds (see Section 7.5). The complicating factor is that, even when we know a lot about living organisms, they do not fall into neat categories. Compromises are required, and the compromises are always a good substrate for scientific debate.

Another problem is that much of evolutionary history is hidden because the fossil record is incomplete, and many major bifurcations in the development of life are very ancient. We are grappling with a largely unseen world, trying to imagine what it must have been like. Indeed, the extents to which imagination and the hard, cold, facts of science play roles in our current perception of the evolution of life after the early eukaryotes are open to debate. Certainly, each new discovery of a link between the living and the dead is greeted with enthusiasm—in the scientific community, at least—and each authenticated extension backward in the fossil record of any major taxon receives accolades in the scientific press—e.g., the oldest spider (Shear et al. 1989), the oldest mushroom (Poinar and Singer 1990), the oldest vertebrate (Sansom et al. 1992).

Vestiges of the past in present-day organisms are also important and augment the fossil record. Primitive organisms such as many archaebacteria, slime molds, ginkgo trees, coelacanths, and tuataras exist today much as they may have done in the past. Such species provide many more characteristics to use in phylogenetic analysis than are available in fossils. Mitochondrial structure, protein structure, cell wall composition, the cell cycle with mitosis and meiosis, life history, and DNA and RNA sequences all may contribute to an analysis of ancient phylogenetic links that would otherwise be lost in the mists of time. Indeed, in groups with poor fossil records, living organisms are the predominant mode of investigation of the past; examples are the protists, the fungi, and enigmatic groups such as the velvet worms (onychophorans).

This chapter concentrates on some of the major evolutionary steps from the early eukaryotes to the kinds of organisms with which we are most familiar today—the fungi, plants, and animals. This chapter, like the last, glosses over many uncertainties about the path of evolution and many unresolved debates. The arguments presented here have mostly been developed by major researchers in their fields and are distillations of large bodies of material. Nevertheless, it is conceivable that further scientific debate will result in different opinions and a different general consensus.

In attempts to trace relationships among organisms through time, some groups get left out or left behind in the argument, even though each has its own evolutionary history worthy of consideration. Broader coverage can be found in more specialized volumes—for example, those on fungi (Rayner et al. 1987), plants (Stewart and Rothwell 1993), invertebrates (Ruppert and Barnes 1994), and vertebrates (Colbert and Morales 1991). In this chapter, rather than describing what the fossil record reveals, we emphasize possible mechanistic pathways linking the eukaryotic kingdoms and major groups within those kingdoms.

7.2 THE GEOLOGIC TIME SCALE

As we saw in the last chapter, a good sense of time promotes the understanding of events, and a detailed history of the earth (more detailed than that provided by Figure 6-6 in Chapter 6) will help. However, the geologic time scale of the earth, as recorded in the rocks and the fossils they contain (see Chapters 2 and 9), loses detail as we pass backward in time. Even in what seem to be everlasting rocks we find a certain ephemerality. Whereas each rock stage commonly

represents a few million years or less in the Quaternary and Tertiary periods, tens of millions of years commonly characterize stages in the early Paleozoic and hundreds of millions of years in the Precambrian (Table 7-1). Even this time scale undergoes ongoing revision and refinement. Harland et al. (1990, 11), the authors of the scale presented here, note that

> The refinements possible with existing data, as well as modifications necessitated by new data, entail an indefinitely continuing series of revised numbers. As there is no finality or even temporary stability in this process the moment when the scale is fixed is seemingly arbitrary.... Time scales will be subject to continuing refinement.

Note the book's title, *A Geologic Time Scale 1989*, rather than *The Geologic Time Scale*, signifying that interpretations are open to question and plausible alternatives exist. As Harland et al. (1990, 11) remarked, "initial Cambrian chronometry continued to present distinct problems." This kind of uncertainty is relevant to much of the subject matter of this chapter and should caution the reader against accepting dogmatic statements about the evolution of life on earth.

The long record outlined by the geologic time scale is sufficient to cover the evolution of multicellular organisms on earth and indeed the geologic history of the earth. Fungi probably evolved about 1 billion years ago, and the radiation of animals was well advanced in the Cambrian, although we are missing a prior fossil record for already-complex forms such as trilobites and worms. We have a fossilized Precambrian fauna from more than 570 million years ago, during the Ediacaran Epoch, but its relationships to forms living today seem debatable (cf. Glaessner 1984, Seilacher 1984, Gould 1989).

7.3 MULTICELLULAR LIFE ON EARTH—THE FOSSIL RECORD

After a massive effort to assemble and date fossils, we can map out the major features in the development of multicellular life on earth across a geologic record of time (Figure 7-1, Table 7-2).

The **Ediacaran fauna** was a worldwide fauna of simple, soft-bodied organisms that lived just before the end of the Precambrian. It is named for Ediacara in South Australia (north of Adelaide and south of Lake Eyre), where a rock outcrop yielded many fossil types. The work of Martin Glaessner (e.g., 1984) did the most to reveal the existence of this fauna. Glaessner classified its organisms into known taxa such as soft corals and medusoids (phylum Coelenterata),

TABLE 7-1   A geologic time scale (completed in 1989)

Era	Sub era / Period / Sub period	Epoch	Stage¹ / Tie Point ●	Age² (Ma)	Stage Abbrev.	Intervals (Ma)
Cenozoic (Cz)	Quaternary or Pleistogene	Holocene		0.01	Hol	0.01
		Pleistocene		1.64	Ple	1.63
	Tertiary (TT) / Neogene (Ng)	Pliocene 2 / Pli 1	Piacenzian	3.4	Pia	3.6
			Zanclian	5.2	Zan	
		Miocene 3/2/1 (Mio)	Messinian	6.7	Mes	5.2
			Tortonian	10.4	Tor	
			Serravallian	14.2	Srv	5.9
			Langhian	16.3	Lan	
			Burdigalian	21.5	Bur	7.0
			Aquitanian	23.3	Aqt	(Neogene 22)
	Paleogene (Pg)	Oligocene 2 / Oli 1	Chattian	29.3	Cht	6.0
			Rupelian	35.4	Rup	6.1
		Eocene 3/2/Eoc 1	Priabonian	38.6	Prb	3.2
			Bartonian	42.1	Brt	11.4
			Lutetian	50.0	Lut	
			Ypresian	56.5	Ypr	6.5
		Paleocene 2 / Pal 1	Thanetian	60.5	Tha	4.0
			Danian	65.0	Dan	4.5 (Paleogene 42)
Mesozoic (Mz)	Cretaceous (K)	Gulf / Senonian (Sen)	Maastrichtian	74.0	Maa	9.0
			Campanian	83.0	Cmp	9.0
			Santonian	86.6	San	3.6
			Coniacian	88.5	Con	1.9
		Gul / Gallic (Gal)	Turonian	90.4	Tur	1.9
			Cenomanian	97.0	Cen	6.6
			Albian	112.0	Alb	15.0
			Aptian	124.5	Apt	12.5
			Barremian	131.8	Brm	7.3
		Neocomian (Neo) / K_1	Hauterivian	135.0	Hau	3.2
			Valanginian	140.7	Vlg	5.7
			Berriasian	145.0	Ber	4.9 (Cretaceous 81)
	Jurassic (J)	Malm (J_3 Mlm)	Tithonian	152.1	Tth	6.5
			Kimmeridgian	154.7	Kim	2.6
			Oxfordian	157.1	Oxf	2.4
		Dogger (J_2 Dog)	Callovian	161.3	Clv	4.2
			Bathonian	166.1	Bth	4.8
			Bajocian	173.5	Baj	7.4
			Aalenian	178.0	Aal	4.5
		Lias (J_1 Lia)	Toarcian	187.0	Toa	9.0
			Pliensbachian	194.5	Plb	7.5
			Sinemurian	203.5	Sin	9.0
			Hettangian	208.0	Het	4.5 (Jurassic 62)
	Triassic (Tr)	Tr_3	Rhaetian	209.5	Rht	1.5
			Norian	223.4	Nor	13.9
			Carnian	235.0	Crn	11.6
		Tr_2	Ladinian	239.5	Lad	4.5
			Anisian	241.1	Ans	1.6
		Scythian (Tr_1 Scy)	Spathian	241.9	Spa	0.8 [O]
			Nammalian	243.4	Nml	1.5
			Griesbachian	245.0	Gri	1.6 [I] (Triassic 37)
Paleozoic	Permian (P)	Zechstein (Zec)	Changxingian	247.5	Chx	2.5
			Longtanian	250.0	Lgt	2.5 [D..]
			Capitanian	252.5	Cap	2.5
			Wordian	255.0	Wor	2.5 [G]
			Ufimian	256.1	Ufi	1.1
		Rotliegendes (Rot)	Kungurian	259.7	Kun	3.6
			Artinskian ●	268.8	Art	9.1 [L]
			Sakmarian	281.5	Sak	12.7
			Asselian	290.0	Ass	8.5 (Permian 45)
	Carboniferous / Pennsylvanian	Gzelian (Gze)	Noginskian	293.6	Nog	3.6
			Klazminskian	295.1	Kla	1.5
		Kasimovian (Kas)	Dorogomilovskian	298.3	Dor	3.2
			Chamovnicheskian	299.9	Chv	1.6
			Krevyakinskian	303.0	Kre	3.1
		Moscovian (Mos)	Myachkovskian	305.0	Mya	2.0
			Podolskian	307.1	Pod	2.1
			Kashirskian	309.2	Ksk	2.1
			Vereiskian	311.3	Vrk	2.1
		Bashkirian (Bsh / C_2)	Melekesskian	313.4	Mel	2.1
			Cheremshanskian ●	318.3	Che	4.9
			Yeadonian	320.6	Yea	2.3
			Marsdenian	321.5	Mrd	0.9
			Kinderscoutian	322.8	Kin	1.3 (Pennsylvanian 33)
		C_1 Serpukhovian	Alportian	325.6	Alp	2.8
			Chokierian		Cho	2.7 (40)

Era	Sub era / Period / Sub period	Epoch	Stage¹ / Tie Point ●	Age² (Ma/Ga)	Stage Abbrev.	Intervals (Ma)
Paleozoic	Carboniferous / Mississippian	C_2 Bashkirian	Marsdenian	321.5	Mrd	11.5 (33)
			Kinderscoutian	322.8	Kin	
		Serpukhovian (Spk)	Alportian	325.6	Alp	10
			Chokierian	328.3	Cho	
			Arnsbergian	331.1	Arn	
			Pendleian	332.9	Pnd	
		Visean (Vis)	Brigantian ●	336.0	Bri	17
			Asbian	339.4	Asb	
			Holkerian	342.8	Hlk	
			Arundian	345.0	Aru	
			Chadian ●	349.5	Chd	
		Tournaisian (Tou / C_1)	Ivorian	353.8	Ivo	13
			Hastarian	362.5	Has	(Mississippian 40)
	Devonian (D)	D_3	Famennian	367.0	Fam	15
			Frasnian	377.4	Fra	
		D_2	Givetian	380.8	Giv	9
			Eifelian	386.0	Eif	
		D_1	Emsian	390.4	Ems	22
			Pragian	396.3	Pra	
			Lochkovian	408.5	Lok	(Devonian 46)
	Silurian (S)	S_4 Pridoli (Prd)	Prd	410.7	Prd	2
		S_3 Ludlow (Lud)	Ludfordian	415.1	Ldf	13
			Gorstian	424.0	Gor	
		S_2 Wenlock (Wen)	Gleedonian	425.4	Gle	
			Whitwellian	426.1	Whi	6.5
			Sheinwoodian	430.4	She	
		S_1 Llandovery (Lly)	Telychian	432.6	Tel	
			Aeronian	436.9	Aer	8.5
			Rhuddanian	439.0	Rhu	(Silurian 31)
	Ordovician (O)	Bala (Bal) / Ashgill (Ash)	Hirnantian	439.5	Hir	
			Rawtheyan	440.1	Raw	4
			Cautleyan	440.6	Cau	
			Pusgillian	443.1	Pus	
		Caradoc (Crd)	Onnian	444.0	Onn	
			Actonian	444.5	Act	
			Marshbrookian	447.1	Mrb	
			Longvillian	449.7	Lon	21
			Soudleyan	457.5	Sou	
			Harnagian ●	462.3	Har	
			Costonian	463.9	Cos	
		Dyfed (Dfd) / Llandeilo (Llo)	Late	465.4	Llo_3	
			Mid	467.0	Llo_2	4.5
			Early	468.6	Llo_1	
		Llanvirn (Lln)	Late ●	472.7	Lln_2	7.5
			Early	476.1	Lln_1	
		Canadian (Cnd)	Arenig	493.0	Arg	17
			Tremadoc	510.0	Tre	17 (Ordovician 71)
	Cambrian (€)	Merioneth (Mer)	Dolgellian	514.1	Dol	7
			Maentwrogian	517.2	Mnt	
		St David's (StD)	Menevian	530.2	Men	19
			Solvan ●	536.0	Sol	
		Caerfai (Crf)	Lenian	554	Len	18
			Atdabanian	560	Atb	16
			Tommotian ●	570	Tom	(Cambrian 60)
Sinian	Vendian (V)	Ediacara (Edi)	Poundian	580	Pou	20
			Wonokan	590	Won	
		Varanger (Var)	Mortensnes	600	Mor	20
			Smalfjord	610	Sma	(Vendian 40)
	Z	Sturtian		0.80	Stu	190
	Riphean (Rif)		Karatau	0.80	Kar	250
			Yurmatin	1.05	Yur	300
			Burzyan	1.35	Buz	300
	Animikean			1.65	Ani	550
	Huronian			2.2	Hur	250
	Randian			2.45	Ran	350
	Swazian			2.8	Swz	700
	Isuan			3.5	Isu	300
	Hadean (Hde)		Early Imbrian	3.8	Imb	50
			Nectarian	3.85	Nec	100
			Basin Groups 1-9	3.95	BG1-9	200
		Cryptic		4.15	Cry	410
				4.56		

annelid worms (phylum Annelida), and arthropods (phylum Arthropoda). Seilacher (1984) has seriously questioned the expectation, intrinsic in Glaessner's work, that a very late Precambrian fauna would show direct relationships to the Cambrian period but in simpler form. If the coral-like fossils were really soft corals, why are the apparent branches joined and not open to allow water currents to deliver food to the polyps on the branches? And the organisms of Ediacara seem to have a common body plan: flattened and soft, with sections "matted or quilted together, perhaps constituting a hydraulic skeleton much like an air mattress" (Gould 1989, 312). This plan deviates from all other known phyla and therefore may well represent a group of organisms that went extinct by the end of the Precambrian. The unresolved debate about the Ediacaran fauna raises interesting questions (to be discussed again in Chapter 8): What might the world be like had certain lineages persisted into modern times, and what factors influenced the persistence or demise of certain taxa (see Chapter 9)?

The same questions apply to the many fossils from the early Cambrian Period that make up the **Tommotian fauna,** named after a locality in Russia. The paleontologists' epithet for this group of fossils— the "small, shelly fauna"—illustrates its enigmatic nature (Gould 1989). Fossils are commonly 1 to 5 millimeters long, with shells and skeletons forming plates, tubes and cones, caps and cups, frequently with no obvious relationship to extant phyla (Bengtson 1977, Bengtson and Fletcher 1983).

Then a new fauna emerged quite suddenly in the Cambrian Period. Its complex marine life forms were captured in remarkable detail in "the most precious and important of all fossil localities—the Burgess Shale of British Columbia" (Gould 1989, 13). Sponges, brachiopods, cnidarian polyps, worms, and trilobites were followed by cephalopods, sea lilies, echinoderms, and bryozoans. In the Silurian Period, corals became common while life on land diversified. Arthropods—the scorpions, spiders, millipedes, and insects—colonized terrestrial habitats. The **age of fishes** in the Devonian coincided with the first forests on land and ultimately resulted in the first amphibians, such as *Ichthyostega,* and the colonization of land by the vertebrates. Life in the sea continued to flourish. The bony fishes, the marine reptiles, and eventually the marine mammals emerged. On land, the early reptilian pelycosaurs evolved in the Pennsylvanian and Permian, and the mammal-like reptiles, or theriodonts, developed in the Permian and Triassic. First the insects in the Carboniferous, then the reptilian pterosaurs, and finally birds in the Jurassic colonized the air. Mammals radiated in the Cretaceous and became a dominant group in the Tertiary, after the massive extinctions around the Cretaceous–Tertiary boundary and the demise of the dinosaurs. The primates evolved beginning in the Paleocene, with apes emerging in the Oligocene and humans in the Pleistocene.

Plants proliferated on land from the Ordovician Period to a **Siluro–Devonian explosion.** Vascular plants such as the early *Cooksonia* eventually developed into species occupying the great forests of the Carboniferous. Lycopods, horsetails, tree ferns, and other groups flourished. Seed plants developed in the Upper Devonian, radiating into the seed ferns, the conifers, the cycads, and eventually the flowering plants about 130 million years ago. The heavily vegetated earth was an essential element in the radiation of animal life on land (Figure 7-1).

7.4 EVOLUTION OF THE FUNGI

Missing in Figure 7-1 is any sign of the fungi. No toadstool appears on the ground, and no bracket fungus on a tree trunk. And yet the fungi are conspicuous in nature. What are their origins? In fact, the mycologists' objects of study have diverse and polyphyletic origins, and some scientists argue that they belong to three kingdoms.

The Slime Molds

The slime molds, or Myxomycetes, may be considered to be protozoa and polyphyletic (Cavalier-Smith 1987) or something between protozoa, fungi, and

---

From Harland et al. 1990; the names of eras, periods, epochs, and stages are used throughout this book. This scale represents the geologic history of the earth over 4.56 billion years of earth's existence.

[1] This time scale recognizes stages that differ in a few places from those used later in this book, especially in Chapters 9 and 11. In the Lower Triassic, the stages from Greisbachian through Spathian were divided equally into the Induan (I) and Olenkian (O) stages. In the Permian, the Artinskian, Kungurian, and Ufimian were grouped into the Leonardian (L) Stage. The Wordian and Capitanian were grouped into the Guadalupian (G) Stage. And the Dzhulfian (D) is now considered to overlap half of the Capitanian and all of the Longtanian and Changxingian (cf. Harland et al. 1982, 1990). Each of these stages is indicated in the table with a capital letter to the right of the current scale.

[2] Ages and intervals use the International System of Units. Ma is the abbreviation for **mega annum,** which is 10^6 years, or a million years. Ga stands for **giga annum,** which is 10^9 years, or a billion years, or a thousand million years.

FIGURE 7-1

An artist's interpretation of evolution in marine and terrestrial systems, from the Cambrian Period to the present. Table 7-2 provides a key to numbered organisms. Note that the drawing emphasizes the emergence of novel design in each geological period and does not represent the relative diversity of organisms at any one time. This rendition derives from a mural by D. Roesener at the Field Museum of Natural History in Chicago. After the Cretaceous Period, time periods are indicated by the letters a through f as follows: (a–e) Tertiary: (a) Paleocene, (b) Eocene, (c) Oligocene, (d) Miocene, (e) Pliocene and Quaternary, (f) Pleistocene and Holocene. *Copyright © (1979) by Peter M. Spizzirri, and the Field Museum (1976).*

TABLE 7-2 Key to the organisms numbered in Figure 7-1

CAMBRIAN
1. Pleospongia (sponge)
2. Echinodermata (cystoid)
3. Mollusca (brachiopod)
4. Porifera (Vauxia, a sponge)
5. Scyphozoa (jellyfish)
6. Arthropoda (Hymenocabis, a phyllocarid)
7. Polychaeta (worm)
8. Arthropoda (trilobite)

ORDOVICIAN
9. Arthropoda (Hymenocabis, a phyllocarid)
10. Mollusca (pelecypod or clam)
11. Arthropoda (trilobite)
12. Mollusca (cephalopod or nautiloid)
13. Arthropoda (eurypterid)
14. Echinodermata (crinoid or sea lily)
15. Arthropoda (trilobite)
16. Mollusca (gastropod or snail)
17. Mollusca (brachiopod)
18. Echinodermata (blastoid)
19. Arthropoda (phyllocarid)
20. Ectoprocta (bryozoan)

SILURIAN
21. Cnidaria (coral polyp)
22. Arthropoda (eurypterid)
23. Anaspida (Birkenia)
24. Arthropoda (trilobite)
25. Mollusca (cephalopod or ammonite)
26. Arachnida (scorpion)
27. Heterostraci (Phlebolepis)

DEVONIAN
28. Heterostraci (Pteraspis)
29. Heterostraci (Drepanaspis)
30. Dipnoi (Dipterus or lungfish)
31. Cnidaria (coral polyp)
32. Echinodermata (asteroid or starfish)
33. Amphibia (Ichthyostega)
34. Crossopterygian (Sauripterus)
35. Placodermi (Dunkleosteus, an armored fish)

MISSISSIPPIAN
36. Echinodermata (crinoid or sea lily)
37. Insecta (cockroach)
38. Arthropoda (trilobite)
39. Amphibia (Otocratia)
40. Echinodermata (ophiuroid or brittle star)
41. Insecta (dragon fly)

PENNSYLVANIAN
42. Chondrichthyes (Xenacanthus)
43. Amphibia (Diplovertebron)
44. Hirudinea (Tullimonstrum, a worm)
45. Osteichthyes (Chirodus)
46. Amphibia (Pteroplax)
47. Reptilia (Mesosaurus)
48. Reptilia (Clepsydrops or pelycosaur)

PERMIAN
49. Arthropoda (horseshoe crab)
50. Amphibia (Eryops)
51. Mollusca (cephalopod or ammonite)
52. Amphibia (Diplocaulus)
53. Reptilia (Edaphosaurus)
54. Reptilia (Dimetrodon)
55. Mollusca (cephalopod or nautiloid)
56. Reptilia (Kannemeyeria)

TRIASSIC
57. Reptilia (Placodus)
58. Reptilia (Cynognathus)
59. Osteichthyes (Ceratodus or lungfish)
60. Reptilia (Dicynodon)
61. Reptilia (Mixosaurus or ichthyosaur)
62. Reptilia (Mystriosuchus or phytosaur)
63. Osteichthyes (Perleidus)
64. Chelonia (Proganochelys or turtle)
65. Reptilia (Coelophysis, a dinosaur)
66. Osteichthyes (Dipnoan or lungfish)
67. Amphibia (Metoposaurus)

JURASSIC
68. Reptilia (Ichthyosaurus or icthyosaur)
69. Reptilia (Plesiosaurus or plesiosaur)
70. Reptilia (Allosaurus, a dinosaur)
71. Arthropoda (shrimp)
72. Mammalia (Triconodon)
73. Osteichthyes (Holophagus, a coelacanth fish)
74. Reptilia (Rhamphorhynchus)
75. Reptilia (Diplodocus, a dinosaur)
76. Osteichthyes (Dapedius, a holostean fish)
77. Reptilia (Compsognathus, a dinosaur)
78. Scyphozoa (jellyfish)

79. Aves (Archaeopteryx)
80. Reptilia (Stegosaurus, a dinosaur)
81. Chelonia (Trionyx or soft-shelled turtle)

CRETACEOUS
82. Osteichthyes (Dipnoan or lungfish)
83. Reptilia (Protoceratops, a dinosaur)
84. Reptilia (Tylosaurus, a mosasaur)
85. Amphibia (anuran or toad)
86. Reptilia (Palaeoscincus, an aetosaur)
87. Mollusca (scaphopod)
88. Reptilia (Tyrannosaurus, a dinosaur)
89. Aves (Ichthyornis)
90. Reptilia (Camptosaurus, a dinosaur)
91. Reptilia (Pteranodon)
92. Chelonia (Archelon or sea turtle)
93. Aves (Hesperornis, a diving bird)
94. Reptilia (Triceratops, a dinosaur)
95. Reptilia (Edmontosaurus, a duck-billed dinosaur)
96. Reptilia (Elasmosaurus, a plesiosaur)
97. Reptilia (Coniophis, a burrowing snake)
98. Reptilia (Clidastes, a mosasaur)
99. Aves (Diatryma)
100. Chelonia (Podocnemis or side necked turtle)
101. Mammalia (Barylambda, a pantodont)

TERTIARY & QUATERNARY
102. Insectivora (flying lemur)
103. Osteichthyes (Dipnoan or lungfish)
104. Primates (tree shrew)
105. Cetacea (Basilosaurus or toothed whale)
106. Chelonia (tortoise)
107. Amphibia (anuran or aquatic frog)
108. Aves (Phororhacos)
109. Carnivora (Cynodictis, a primitive dog)
110. Perissodactyla (Epihippus, a primitive horse)
111. Perissodactyla (titanothere)
112. Perissodactyla (Mesohippus, a horse)
113. Lagomorph (rabbit)

TABLE 7-2 Key to the organisms numbered in Figure 7-1 (continued)

114. Artiodactyla (Archaeotherium or giant pig)
115. Perissodactyla (Caenopus, a rhinoceros)
116. Osteichthyes (Dipnoan or lungfish)
117. Chelonia (Dermochelys or leathery turtle)
118. Aves (Mesembriornis)
119. Artiodactyla (Oxydactylus, a camel)
120. Proboscidea (Deinotherium)
121. Perissodactyla (Parahippus, a horse)
122. Primates (monkey)
123. Mollusca (pelecypod or clam)
124. Osteichthyes (surgeon)
125. Arthropoda (hermit crab)
126. Carnivora (Pseudaelurus or false sabertooth cat)
127. Marsupialia (kangaroo)
128. Proboscidea (Mastodon)
129. Marsupialia (koala bear)
130. Chiroptera (bat)
131. Aves (vulture)
132. Osteichthyes (sea horse, a bony fish)
133. Sirenia (Manatus or sea cow)
134. Mollusca (octopus)
135. Carnivora (Smilodon, a sabertooth cat)
136. Edentata (Megatherium, giant ground sloth)
137. Aves (penguin)
138. Cetacea (blue whale)
139. Cnidaria (sea anemone)
140. Primates (Homo or man)

animals (Collins 1987). In either case, a protozoan ancestor and an amoeba-like cell that could move between two phases—flagellated and encysted with no flagellum—are clearly involved. The motile amoeboflagellate stage can aggregate into a plasmodium, a multinucleate mass of protoplasm that feeds and grows by ingesting small pieces of dead organic matter and creeping over the substrate like a macroscopic amoeba, often appearing as a colorful, slimy mass on the soil surface. The mass may then thicken to form a sporangium in which the nuclei divide meiotically; the resulting haploid nuclei with some cytoplasm form resting spores. Under favorable conditions, germination of spores results in the amoeboflagellate stage; cells fuse to form a diploid zygote that undergoes repeated mitotic divisions, forming a small plasmodium, which may fuse with others (Figure 7-2).

This sort of aggregation of protozoan cells into a multinucleated organism was the kind of process that started the evolution of life along a path of increasingly complex design. The abilities of cells to fuse, to cooperate, and to differentiate were essential features. Even now in these ancient organisms, although isogamy is the norm, only haploid cells from different thallus types can fuse to form zygotes in some species. Slime molds are **heterothallic** and outcrossing. Their **heterotrophic nutrition,** i.e., their feeding on organic material originally synthesized by plants, is typical of the true fungi and the animals as well.

The Algal Fungi

Some fungus-like organisms, the oomycetes and other groups, originated from algae after loss of chloroplasts. These organisms resemble green algae in structure and modes of reproduction and are commonly called the algal fungi. They have been classified as fungi, algae, heterokonts, and chromista depending upon which characteristics are emphasized. Cavalier-Smith (1987) places the algal fungi in the kingdom Chromista on the basis of mitochondrial structure.

The True Fungi

This leaves the true fungi in the kingdom Fungi, which may have emerged from stock in common with the kingdom Animalia. Fungi do resemble animals more than plants in six characteristics (Cavalier-Smith 1987).

1. They have a similar mitochondrion design.
2. They have a single posterior cilium on motile cells.
3. Chitinous exoskeletons are common.
4. They store glycogen rather than the starch stored by plants.
5. They lack chloroplasts and therefore are heterotrophic.
6. The mitochondrial code uses UGA to code for tryptophan. (In plants, UGA codes for chain termination.)

Based on ribosomal RNA sequence similarities, the kingdom Fungi appears closer to the animal kingdom than to the plant kingdom (Figure 7-3).

What, then, is the common ancestor of the fungi and animals? One hypothesis is that choanoflagellate protozoa, or the Choanociliata, are the link, on the basis of the two features common to all three groups: (1) the cristae in the mitochondria are flattened and nondiscoidal, and (2) motile cells have a single poste-

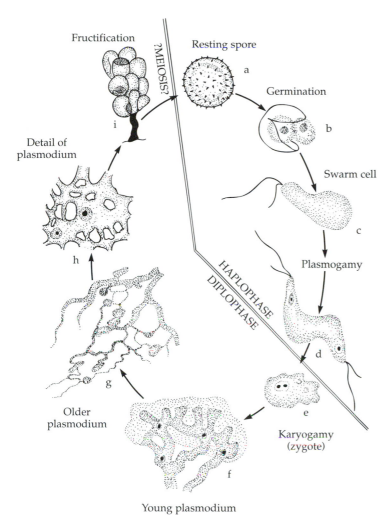

Fructification

?MEIOSIS?

Resting spore

a

Germination

b

Detail of
plasmodium

i

Swarm cell

c

Plasmogamy

HAPLOPHASE
DIPLOPHASE

h

d

g

Older
plasmodium

e

Karyogamy
(zygote)

f

Young plasmodium

FIGURE 7-2

The life cycle of the myxomycete, or slime mold, *Physarum polycephalum*. The plasmodium (f, g) is macroscopic and may be brightly colored in some species. It flows over the substrate and engulfs food particles as it passes. Eventually it coalesces into the spore-producing fructification (i). *(From Alexopoulos 1952)*

rior flagellum (Figure 7-4). Coanociliates use the cilium for wafting onto the collar bacteria, which are consumed by phagocytosis. In a species presumed to be ancestral to the fungi, a choanociliate with a chitinous theca, or enclosing capsule, underwent a mutation that caused it to settle on a substrate upside down and to lose its cilium in that phase of its life cycle (Figure 7-5). Thus, phagocytosis was lost. In its place developed new avenues of nutrition—evaginations from the cell into rhizoids for saprophytic life and penetration of (perhaps) algal cell walls for a parasitic lifestyle. In either case, the theca would become a cell wall and a ciliated dispersal phase would remain. From these hypothetical early developments, the remainder of the true fungi diverged in a series of changes summarized in Figure 7-4. Nobody knows how many fungal species exist on this earth, but a sober estimate is a million or more (Hawksworth 1991). Probably fewer than 5 percent are described.

7.5 EMERGENCE OF THE ANIMALS

Sponges and Other Early Animals

Leadbeater (1983) supported a choanociliate origin of the most primitive animals. The name of a species he studied, *Proterospongia choanojuncta*, a marine member of the choanociliates, tells us a lot: it is the proto-sponge with links to the choanociliates. After all, some choanociliates were already colonial, as *Codosiga* (Figure 7-6) is today, with closely associated, aggregated cells. And it is no accident that the ciliated phagocytosing cells involved with generation of water currents and much of the nutrition in sponges are called choanocytes (Figure 7-6).

An alternative hypothesis is that sponges derive from a hollow-centered, free-swimming flagellate, rather like a chloroplast-free *Volvox*. The inversion of the blastula is seen in sponges and in *Volvox*.

FIGURE 7-3
A phylogenetic tree illustrating relationships among the five kingdoms—Monera, Protoctista, Fungi, Animals, and Plants—based on 16S-like ribosomal RNA sequence similarities. The lengths of lines to named groups indicate genetic distances. Note that the five main eukaryotic assemblages, indicated by triangles, diverged late in the evolution of the eukaryotes, and apparently in a relatively short period of time. *(From Margulis 1993, courtesy of G. Hinkle and M. Sogin, based on an original drawing by Kathryn Delisle)*

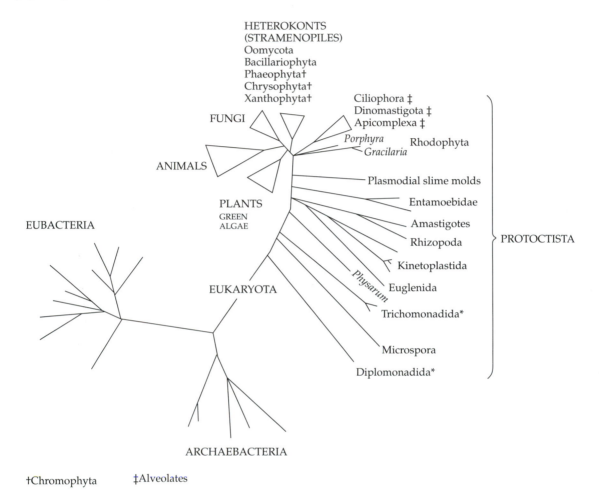

The Cambrian Explosion

Whichever hypothesis is closer to reality, sponges have some characteristics similar to slime molds: single cells can assemble into larger aggregates and form new individuals; cutting up a sponge yields many new sponges because regeneration is so effective; cells can recognize each other and associate; cells can function independently; no organs are present. The sponges developed with a relatively loose association of cells not seen in other metazoans, and so, if the multicellular groups have one common ancestor, the sponges diverged very early from the other lineages.

Mystery still surrounds the emergence of the other primitive metazoans, such as the cnidarians and platyhelminths. A common theme involves a ciliated, free-swimming planula larva that develops from the blastula with such a basic form that almost any body design can be derived from it.

The advent of a fossil record considerably firmed up speculative approaches to the evolution of the early animals. By the end of the Cambrian Period, almost all of the animal phyla with hard parts to fossilize were present, taking part in a **Cambrian explosion** of complex metazoans (Whittington 1985, Gould 1989, Levinton 1992). The Burgess Shale yielded exquisite fossils of many animal phyla: Porifera, Brachiopoda, Cnidaria, Mollusca, Priapulida, Annelida, Arthropoda, Crustacea, Echinodermata, Hemichordata, and Chordata (Whittington 1985). Body plans were already so disparate at the time of the explosion that phylogenies have remained a puzzle (see also Chapter 8). This puzzle may yet be solved. Living species can provide clues to the phylogenetics of the past, as

F I G U R E 7 - 4

The hypothetical relationships of the fungi. Note that mitochondrial characters—specifically those involving the shape of the cristae, the folded membranes of the eukaryotic mitochondrion—are important near the base of the phylogeny. Tubular cristae are associated with the kingdom Chromista in this scheme, discoidal cristae with the Euglenozoa, and plate-like cristae with the main lineage of the fungi and animals from choanociliate ancestors. AAA refers to the α-aminoadipic acid pathway to the biosynthesis of lysine used in fungi and euglenoids; DPA indicates the diaminopimelic acid pathway used in chromists and plants, derived from a cyanobacterial symbiont that developed into chloroplasts. The abbreviation "mt UGA → tryp." indicates that the mitochondrial UGA codon codes for tryptophan in the lineages leading to Fungi and Animalia. TMs indicate mastigonemes on the anterior cilium, and ER indicates chloroplast endoplasmic reticulum. *(From Cavalier-Smith (1987))*

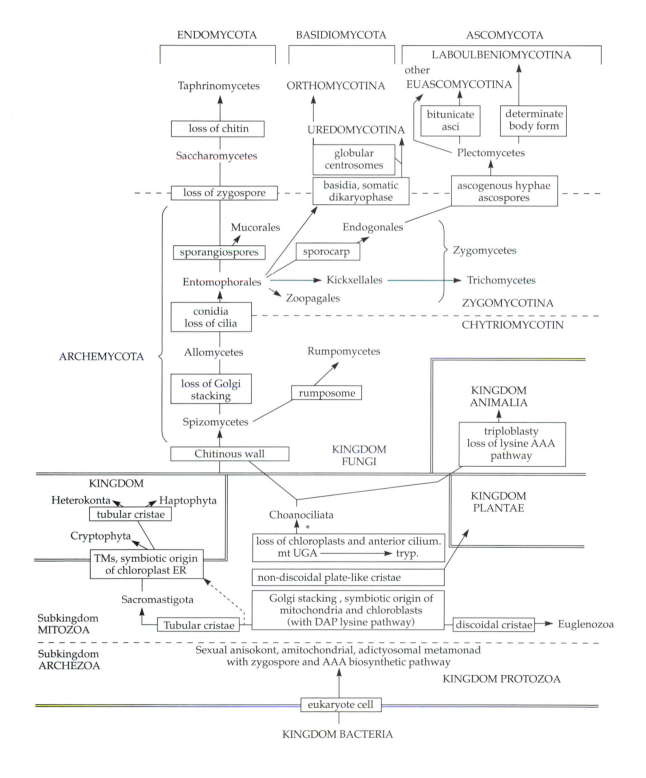

FIGURE 7-5

Hypothetical stages in the evolution of the fungi from a choanociliate protozoan. (a) The life cycle of a choanociliate with (I) a briefly motile dispersal phase and (II) a sessile feeding phase with a chitinous theca. (b) A mutation causes a choanociliate to land with its collar toward the substrate (I). The choanociliate has two ways of surviving. It can develop saprotrophically by growing rhizoids (II); alternatively, it can penetrate an algal cell wall (III) and become parasitic with a complete cell wall (IV). The dispersing phase is preserved (V). *(Based on Cavalier-Smith (1987))*

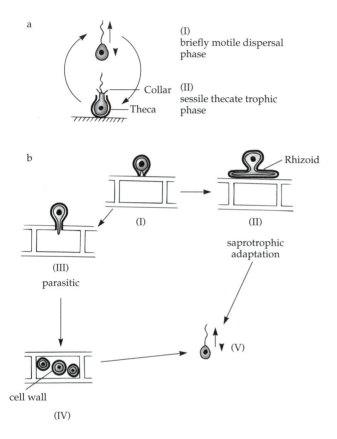

Darwin demonstrated so effectively (Chapters 2 and 12). Where morphological characters are very distinct, as with phyla (because each phylum represents a different body plan), clues may still reside in the molecules. For ancient relationships, slowly evolving, highly conserved molecules and sequences must be used, such as are found in ribosomal RNA.

Annelids, Onychophorans, and Arthropods

One example of molecular detective work was the investigation of the largest phylum of animals, the Arthropoda, and their presumed relatives. Of particular interest was whether the velvet worms, or onychophorans, formed a link between segmented worms, the Annelida, and the Arthropoda (Figure 7-7). The onychophorans are usually placed in their own phy-

lum (e.g., Barnes 1987), and so the detective must investigate phylogenetic relationships mainly among three phyla that diverged more than 500 million years ago.

Figure 7-8 summarizes four hypotheses on the relationships in and around the arthropods that are developed in the literature. Most of the groups (a, b) could have independent origins, making the arthropods polyphyletic—and many scientists skeptical. The origins could be monophyletic (c, d), with primitive ancestors of Onychophora linking annelids and arthropods (c), or the velvet worms could derive from the arthropod lineage and be most closely related to myriapods (millipedes and centipedes) and hexapods (insects in particular). Ballard et al. (1992) used sequences from 12S ribosomal RNA to explore these hypotheses. Their analysis involved taxa from five phyla plus onychophorans: chordates, echinoderms, molluscs, annelids, and arthropods. The results will no doubt generate some debate and stimulate further examination, but under this hypothesis some very interesting relationships emerge in the phylogenetic tree (Figure 7-9).

1. The onychophorans derive from the arthropods and rest in the arthropod clade, and so hypotheses a, b, and c can be rejected.
2. Velvet worms are not a missing link between the arthropods and the annelids.
3. None of the hypotheses in Figure 7-8 is correct, because myriapods appear to be the basal group to the arthropods and may have been the earliest marine arthropods. A fossil from mid-Cambrian marine deposits may be a myriapod-like animal (Robison 1990).
4. However, the arthropods are a monophyletic group, making the results closer to hypothesis *d*.
5. The onychophorans are related to the chelicerates (spiders and scorpions) and to the crustaceans (shrimps, lobsters, and crayfish).

With the increasing use of molecular techniques in phylogenetic analysis, researchers will have many new characteristics to consider in the construction of lineages. Just which characteristics will emerge as the most potent in the developing systematics remains an interesting question. There is no doubt that new techniques will create new controversies. For example, there will be some reluctance to create a new phylum for arthropods and onychophorans, and a new phylum it would have to be, because velvet worms do not have segmented appendages as all the arthropods do. Also, the use of different sequences of RNA, as in studies using 18S ribosomal RNA (e.g., Lake 1990, Turbeville et al. 1991, 1992), yields conclusions different from those by Ballard et al. (1992) on early metazoan divergence patterns. As yet, the

FIGURE 7-6

The similarity in design of choanociliate cells such as those of the gregarious *Codosiga* (a) and the choanocytes of sponges (b, c) suggests the possibility of a common origin. (b) A cutaway view of a sponge shows the cell types and the flow of water maintained by the choanocyte cilium. (c) Magnified choanocytes indicate the flow of food down the collar and phagocytosis in the food vacuole. *(Based on Dorit, Walker, Barnes 1991; Ruppert and Barnes 1994)*

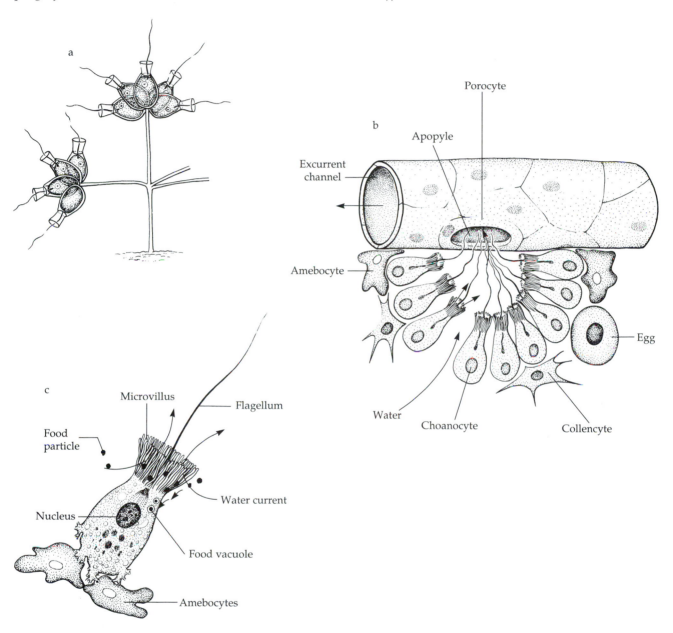

affinities among the arthropods and related groups are by no means clear. The new synthesis of morphological and molecular analyses in systematic arguments promises an engaging and novel perspective on the evolution of life.

The Misty Emergence of the Vertebrates

Except for the chordates and echinoderms, all the animals discussed so far in this chapter lack endoskel-etons. Their bodies depend for support on hydrostatic pressure or on exoskeletons (if the animals are relatively large and not flattened like the platyhelminths and tapeworms). The cephalopods—octopuses, cuttlefish, nautiloids, ammonoids, and squids—were certainly not preadapted for terrestrial life. When arthropods moult their exoskeletons, they are vulnerable in a terrestrial environment, and the early arthropods that were generally confined to land were much smaller than their aquatic counterparts. Hence, a

FIGURE 7-7

(a) A polychaete worm in the phylum Annelida and (b–h) some representatives from the phylum Arthropoda. (b) a millipede, a myriapod in the class Diplopoda; (c) An adult scorpionfly in the class Hexapoda, or Insecta; (d) a larval sawfly in the class Hexapoda, or Insecta; (e) a scorpion in the subphylum Chelicerata, class Arachnida; (f) a crab in the subphylum Crustacea, class Malacostraca; (g) a black widow spider, in the subphylum Chelicerata, class Arachnida; (h) a symphylan in the class Symphyla. (i) A velvet worm, commonly placed in its own phylum, Onychophora. Note the clear segmentation of bodies and appendages in most of the animals illustrated. *(From Barnes 1987, Borror et al. 1989)*

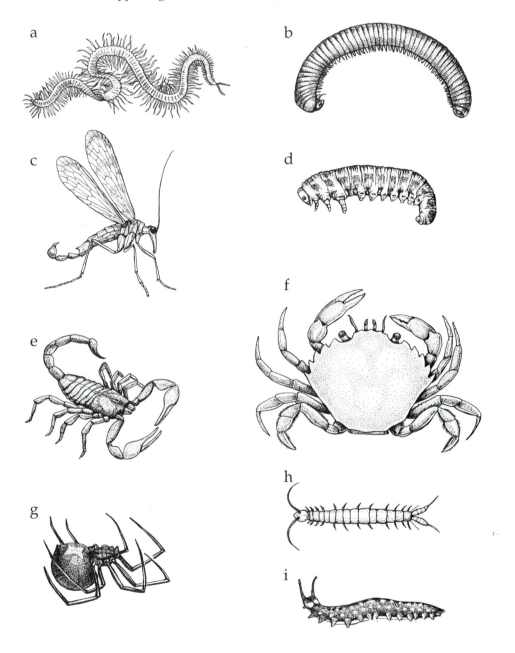

major new body plan, and a new phylum, had to emerge before large organisms such as hyenas, horses, and humans could complete their entire life cycles in a terrestrial environment.

The emergence of vertebrates with endoskeletons in marine environments was the beginning of the **preadaptations** facilitating the colonization of land by lineages that would evolve into the large creatures on which we fasten our attention. (Note that the term "preadaptations" does not imply any preconceived evolutionary plan or preparation for an event in the future, but simply an adaptation in response to natural selection that happens to also become an adaptive advantage in a subsequent environment.) Dinosaurs,

FIGURE 7-8

Four hypotheses tested by Ballard et al. (1992), depicting relationships, or their absence, among arthropods, annelids, and onychophorans. (Figure 7-7 illustrates representatives of major taxa.) (a, b) Polyphyletic origins: (a) Onychophora and Annelida are closely related, more so than to Myriapoda and Hexapoda; (b) onychophorans, centipedes, millipedes, and insects have a common ancestor. (c, d) Monophyletic hypotheses: (c) onychorphorans are primitive and serve as a link between annelids and arthropods; (d) onychorphorans are in the arthropod lineage, closely allied to myriapods and hexapods.

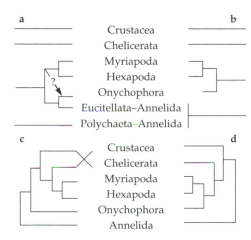

other reptiles, birds, amphibians, and mammals, including humans, all share roots in the beginnings of vertebrate design during the Cambrian explosion more than 500 million years ago. As we hop and skip through the evolution of life, let us light for a moment on the evolution of the vertebrates.

Links with the Echinoderms

The Burgess Shale deposits contained possibly only one fossil species of the phylum Chordata (to which vertebrates belong), *Pikaia gracilens* (Figure 7-10) (Whittington 1985). Or was *P. gracilens* a worm? Echinoderm species were more numerous in the Burgess Shale: crinoids (sea lilies), asteroids (sea stars), and holothurians, represented by a sea cucumber. It is the echinoderms that seem to have embryological characteristics in common with the vertebrates, even though both groups have diverged enormously from their common ancestor in the invertebrates.

The echinoderms evolved with a novel embryological development and are the most primitive deuterostomes. Their dramatic break in design "tradition" was from the **protostomes.** The term "protostome," or "first mouth," is applied because the mouth derives

from or near the blastopore during embryogenesis. In the **deuterostomes** (meaning "secondary mouth"), the blastopore develops into the anus or the anus is associated closely with the blastopore, and the mouth develops secondarily. Other embryological features of deuterostomes include a coelom that buds off from the archenteron, or primitive gut, and a mesoderm derived from the endodermal pouches in the archenteron (Figure 7-11). In the protostomes, mesoderm originates around the blastopore, and the coelom forms as a result of splitting in the mesoderm.

Given these traits of echinoderms in common with the vertebrates, we are still left wondering where vertebrate design originated. The characteristic radial symmetry of body design seen in sea lilies, sea stars, sand dollars, sea urchins, and sea cucumbers could hardly be more different from the bilateral symmetry of the vertebrates.

Primitive Chordates

Again we are left with a break in the evidence for the emergence of the vertebrates, although the primitive chordates fill a large part of the gap. The life cycle of sea squirts (subphylum Urochordata [Tunicata]) in the class Ascidiacea provides a clue to the early development of the vertebrate stock. The ascidian life cycle includes a free-swimming larval form reminiscent of a tadpole and a sedentary adult form that may appear like a little bag, giving rise to the name Ascidiacea. The adult may eject water forcefully from its excurrent siphon, thus earning the common name sea squirt (Figure 7-12). The tadpole larva swims via undulating movements of the tail, and for this it needs a strong axis for muscle attachment. The axis—a breakthrough in animal design—took the form of a flexible, rod-like structure down the back of the tadpole, just below the dorsal nerve cord (notochord). This structure is the fundamental feature of the chordates.

7.6 THE FIRST VERTEBRATES

From Sea Squirts to Fishes?

The tadpole-like larva had a pharynx with gill slits for filter feeding on plankton and presumably evolved into an adult with similar form that could reproduce. Sexual reproduction in an individual that retains larval or immature characteristics is called **neoteny.** Neoteny gave rise to very creative body design, evidently with enormous possibilities for adaptive radiation—such as in the fishes, amphibians, reptiles, and mammals. A modification of the

FIGURE 7-9

Phylogenetic relationships developed by Ballard et al. (1992), based on 12S ribosomal RNA sequences. Major taxa are indicated on the right, and some common names appear on the left. Notice that the onychophorans are in the arthropod clade, close to the myriapods. Also, arthropods are a monophyletic group.

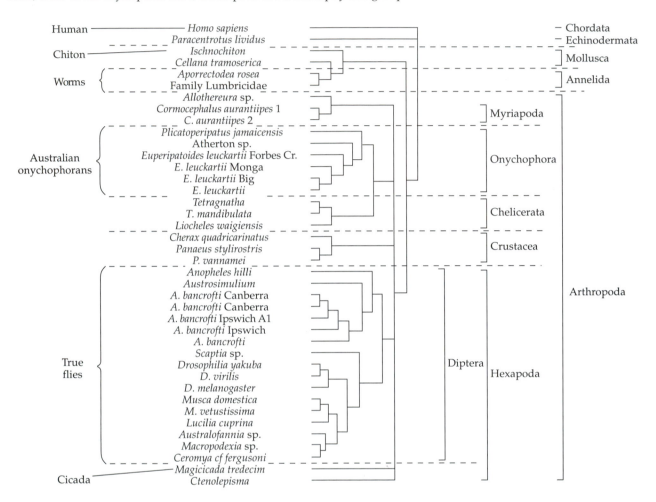

FIGURE 7-10

Pikaia gracilens, the earliest known chordate, was originally described as a worm. However, the fossils show a longitudinal strip along the dorsum, which may well represent a notochord, and the segmented zigzag arrangement of muscles along the body is very fish-like. Shown here are a distinct head, trunk, and tail, with a caudal fin around the tail. *(From Colbert and Morales 1991)*

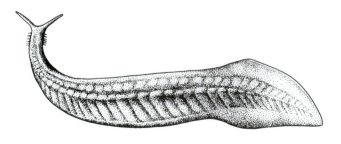

anterior and posterior openings to the alimentary canal as illustrated in Figure 7-12 would yield the beginnings of a fish-like design and perhaps the earliest fish, the ostracoderms. Such fish lacked jaws and were filter feeders, but they had vertebrate features: a brain case, or cranium, and bony plates anticipating the bone of vertebrates. The first fossils appeared in the Ordovician Period, with major radiations occurring in the Silurian and Devonian periods (Figure 7-1). Because the ostracoderms lacked jaws, teeth, and even a bony vertebral column, they remained technically invertebrates.

Jaws

Another breakthrough in design came with the evolution of jaws. The first jawed vertebrates were the

FIGURE 7-11

A major break in development of the embryo occurred from a protostome lineage **(top)** to a deuterostome lineage **(bottom)**. In the protostomes, the mouth is associated with the blastopore; mesoderm develops around the blastopore and splits to form the coelom—hence the term "schizocoelous" for the mollusc. In the deuterostomes, the anus is associated with the blastopore and the mouth develops secondarily. Mesoderm develops as an outpushing of the endoderm in the archenteron, and the coelom originates as space that is enclosed as the endoderm and mesoderm close off the evaginations. The origin of the coelom is enterocoelous; it was first seen in the echinoderms but is also seen in the chordates.

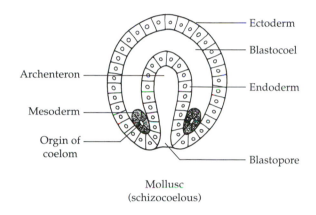

Mollusc
(schizocoelous)

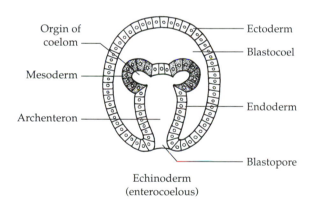

Echinoderm
(enterocoelous)

placoderms, which had plated, armor-like surface protection. The jaw enabled them to prey on large animals, and some placoderms became enormous predators, up to 30 feet long. For example, *Dinichthys* had a head and thoracic shield 8 to 10 feet long with cutting blades on the jaws, and the head and lower jaw articulated to make an enormous gape capable of engulfing any other animal in sight (Figure 7-13).

As we saw in the presumed emergence of the fishes from a neotenic tadpole-like larva, break-throughs in design commonly involve reconstruction or modification of existing structures. This is the case for the evolution of jaws in vertebrates. In the jawless fishes, the gill slits were supported by cartilaginous gill arches, one for each gill. Gradually the anterior gills were lost, and the arches became more and more maneuverable and forward-pointing (Figure 7-14). The muscles in the gill arches also changed, to a more vertical position that enabled them to open and shut the developing jaws. Other gill arches, the hyomandibulars, became supports for articulation of the jaw, and bony denticles in the skin around the mouth became modified into teeth (Figure 7-14). These developments restricted the gill slit to a small spiracle, which remains today in the sharks but is lost in the bony fishes.

Paired Appendages

Paired fins in early fishes were a major advance in propulsion and control in marine environments, with extensive ramifications for further novelties in design. The misty past has not provided a clear picture of how they arose. Embryological studies indicate that fins develop as folds along the sides of the body. In one stage, the continuous fin-like structures along the body resemble those seen in some early ostracoderms. A dorsal fold produces a single dorsal fin. Later in the embryological development of fins, the continuous folds are lost through tissue breakdown, leaving dorsal fins, tail, and paired lateral fins (Colbert and Morales 1991). In evolution, similar fin-like structures were to open up opportunities among the fishes (which we will discuss a little later in this chapter).

From the early fishes in the Devonian came a very successful radiation to the modern cartilaginous and bony fishes (Figure 7-15). The bony fishes, with about 21,000 living species, are the largest vertebrate group today. *Cladoselache* was a primitive shark from the late Devonian, and *Cheirolepis* was an early ancestor of the bony fishes. *Osteolepis* was an early lobe-fin fish in the Devonian. It represented a stock that had limited radiation in the aquatic environment but, as we shall see, that was an ancestor of great radiations.

7.7 FROM WATER TO LAND

Colonization of the land from an aquatic environment required many modifications in body design, physiology, life cycle, and reproduction as well as a novel food supply. Changes in body design involved new methods of locomotion, resistance to abrasion and desiccation, and new methods of oxygen intake from the air instead of from water. Ultimately the life cycle

FIGURE 7-12

The hypothetical origin of ostracoderm fishes through the neoteny of an ascidian tadpole larva, in which the larva becomes reproductive and permanently free-swimming and the notochord is retained throughout life.

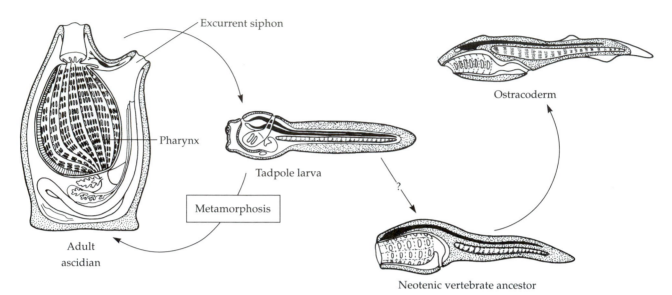

FIGURE 7-13

The head and thoracic shield of the largest Devonian fish, *Dinichthys*. These creatures grew to more than 9 meters (30 feet) in length and could engulf large prey with a hinged head shield and large lower jaw. *(From Colbert and Morales 1991)*

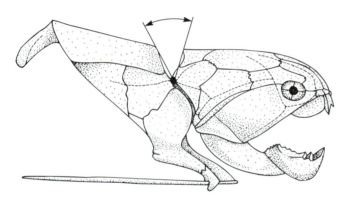

FIGURE 7-14

How the vertebrates got their jaws. The anterior gill bars became modified into jaws and supports for the articulating jaw. This reconstruction pinches the anterior gill slit down to the vestigial spiracle, which is still evident in present-day sharks.

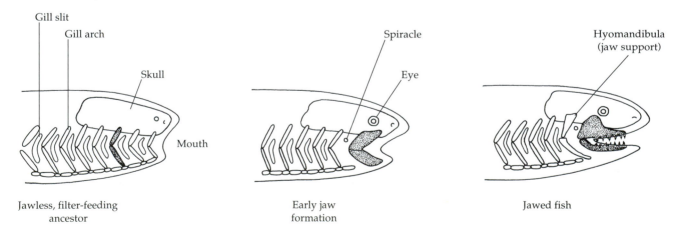

FIGURE 7-15

A general view of the radiation of the fishes into the modern cartilaginous fishes, such as sharks, and the bony fishes, such as perch, pike, catfish, and salmon. Note the early origin of the lineage leading to the lobe-fin fishes, or crossopterygians, indicated on the right. *(Prepared by Lois M. Darling, from Colbert and Morales 1991)*

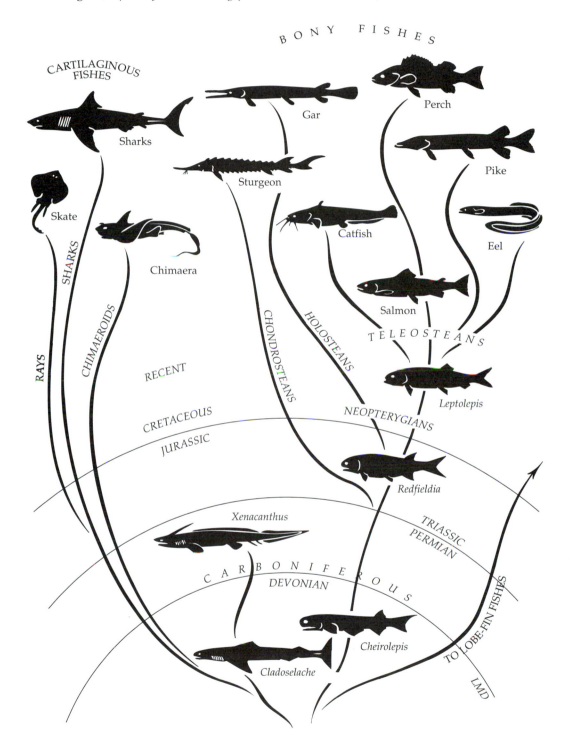

was completed on land with desiccation-resistant eggs or viviparity (live birth). Internal fertilization of eggs thus became mandatory. The colonization of land also involved many adjustments that preserved some semblance of an aquatic environment. For example, animals are still mostly water; delicate surfaces such as the cornea of the eye are still bathed in liquid; sperm still swim; eggs develop in their own little "sea of liquid." The aquatic habitat was never really lost; it became encapsulated.

Devonian Beginnings

The modifications of form that ultimately resulted in the emergence of vertebrates from the water onto land can be argued to have commenced in the Devonian Period, some 400 million years ago. We have already noted the origin of the lobe-fin fishes, or crossopterygians, deep in the Devonian (Figure 7-15). An early lobe-fin fossil is the genus *Osteolepis*. The name "crossopterygian" refers to a tassel-like fin, because the fossils indicate a central skeletal base from which the fin rays radiated, just as the filaments of a tassel hang from a globular base. Like the other crossopterygians, *Osteolepis* had a strong notochord and a well-developed vertebral column, with the skull and jaws formed completely of bone. Notably, the pectoral and pelvic fins were paired, with bones in the bases of the fins and a single bone articulated with the girdles. In fact, in these Devonian fishes we can see the beginnings of terrestrial tetrapod design (Figure 7-16). Directly homologous links probably exist between the proximal bone of these fish and the proximal bones in tetrapod limbs, the humerus and femur. The next bone in a crossopterygian became the radius and ulna of the tetrapod front leg and the tibia and fibula of the hind leg.

Another notable feature of crossopterygians and their relatives, the lungfishes, is that they could breathe air. The bony skeleton of the head was remarkably similar to those of early amphibians, including the paired nostrils close to the anterior tip of the skull.

From Lobe-Fin Fishes to Amphibians

To a remarkable degree, the crossopterygians were preadapted for a life spent at least partially on land. They could breathe air, they could scramble over the ground with their paired lobed fins, and many species dwelled in fresh water. These fish could forage on new sources of food that were abundant on land by this time, they could escape from aquatic predators and parasites, and they could move among water bodies when drought threatened their existence. Little reproductive change was needed, because they could return to water to breed in ways similar to their ancestors.

In fact, already in the late Devonian, the crossopterygians from *Osteolepis* had diverged into a lineage of fishes—of which only the coelacanth, *Latimeria*, remains today—and into the beginning of great radiations of the four-legged animals on land. Recent studies based on comparisons of hemoglobin molecules have concluded that the coelacanth is more closely related to amphibians than are lungfishes or teleost fishes (Gorr and Kleinschmidt 1993). The amphibian *Ichthyostega* is a primitive representative of animals that colonized the land (Figure 7-17).

Food in Early Terrestrial Habitats

By the late Devonian Period, there was plenty to eat on land. Algae, fungi, and plants were thriving, and there is fossil evidence of insects, millipedes, centipedes, spiders, and the related trigonotarbids (Figure 7-18). Terrestrial scorpions from the Lower Carboniferous Period are recorded. Many of the invertebrate groups were carnivorous themselves, suggesting an abundance of detritivores, for herbivores are poorly represented in these early terrestrial assemblages. Centipedes, trigonotarbids, spiders, and scorpions were all predatory arthropods. Ancestors of the amphibians were already carnivorous, probably feasting on aquatic invertebrates as well as small fish, and so the shift to predation on land would have been relatively simple. Even if the last land-adapted lobe-fin fishes or the first amphibians were ungainly, ponderous predators, they were the only vertebrates on land, free from the rapacious carnivores that would eventually specialize in eating tetrapods. The "good life" for the amphibians lasted for a long time! On

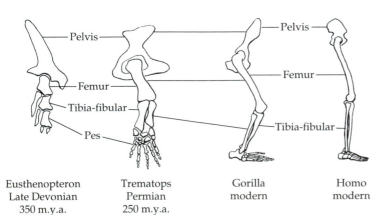

Eusthenopteron
Late Devonian
350 m.y.a.

Trematops
Permian
250 m.y.a.

Gorilla
modern

Homo
modern

FIGURE 7-16

The remarkable similarities in limb structure among a Devonian crossopterygian fish, *Eusthenopteron* (**left**); a Permian amphibian, *Trematops* (**middle**); and modern mammals, the gorilla, *Gorilla*, and human, *Homo*. Hind limbs are not to scale. These are probably homologous structures that demonstrate evolutionary relationships lasting over 350 million years and no doubt well into the future. (*Fish and amphibian limbs from Colbert and Morales 1991; mammal limbs from Napier 1967*)

FIGURE 7-17

A general view of the radiation of lobe-fin fishes into present-day lungfish and coelacanths, now represented by the genus *Latimeria*, and amphibians, represented by the Devonian *Ichthyostega*. Note the great success of the frogs, toads, salamanders, and caecilians today relative to the few remaining descendants of the lobe-fin fishes. Even more terrestrially adapted descendants of the crossopterygians, the reptiles **(right)**, emerged in the Lower Carboniferous Period some 340 million years ago and advanced the rapid colonization of land. *(Prepared by Lois M. Darling, from Colbert and Morales 1991)*

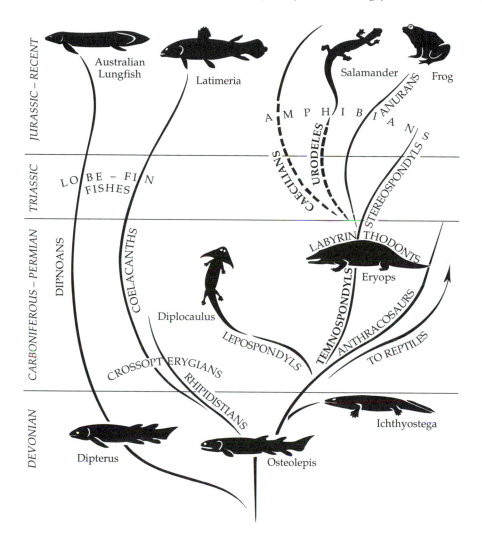

land they faced no vertebrate competitors for food and no vertebrate predators except for other amphibians; in addition, parasite life cycles were still largely confined to the water.

7.8 COMPLETE ADAPTATION TO LIFE ON LAND

However spectacular was the advent of terrestrial tetrapods, they retained important features from their aquatic ancestors. Reproduction was very much an aquatic affair. For most early amphibians, mating probably took place in the water, fertilization of eggs

was in the water, and immature stages remained aquatic. It was the reptiles that made a complete adjustment to terrestrial life.

The reptilian lineage emerged early in the Carboniferous Period, some 340 million years ago. The key to continuous life on land was the amniote egg. The amnion enclosed the developing embryo in an aquatic environment, the yolk sac provided plentiful food for development to an advanced stage, the allantois contained the waste products, and all were sealed against desiccation by a chorion and an outer shell (Figure 7-19). Internal fertilization of the egg and behaviors involved with oviposition into benevolent substrates accompanied the advancing design of the

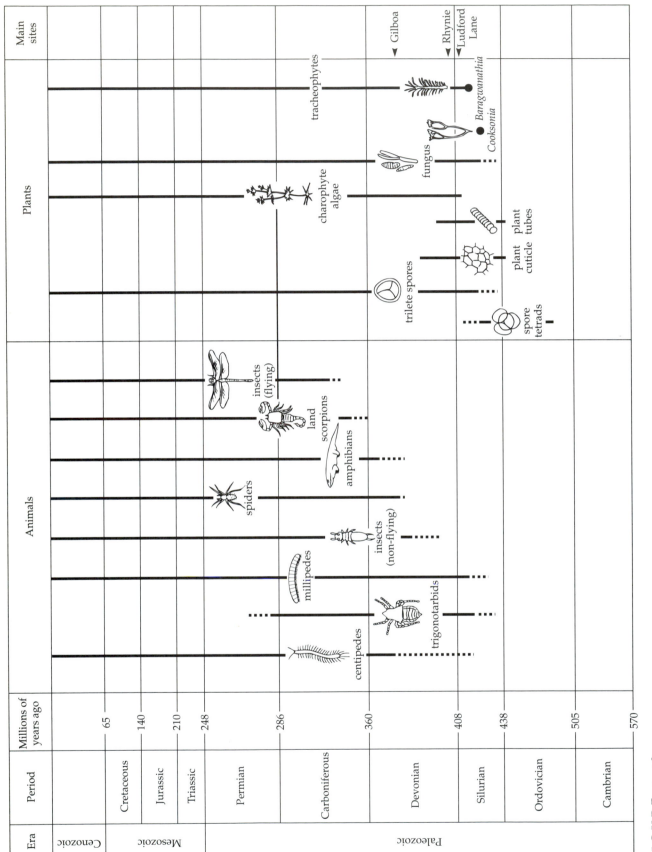

FIGURE 7-18

According to the fossil record, early life on land was enriched rapidly in the Silurian and Devonian periods. Three sites have provided particularly important fossil evidence **(right):** Ludford Lane in Ludlow, England; Rhynie, Scotland; and Gilboa, New York. *Cooksonia* was close to vascular plants, and *Baragwanathia* provides the earliest evidence of a true vascular plant. *(From Gray and Shear 1992)*

FIGURE 7-19

The amniote egg, represented by the egg of a chicken. Note that the embryo is enclosed in the amnion and bathed in amniotic fluid so that early development remains essentially aquatic.

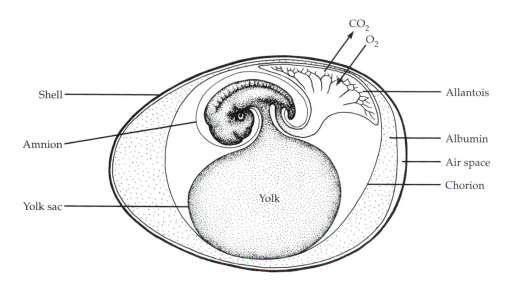

first terrestrial eggs among the vertebrates. How this great increase in complexity of the egg occurred remains lost in the mists of time.

Reptilian Radiations

The breakthrough in egg design opened up enormous potential for adaptive radiation, for lineages were now freed from dependency on persistent water bodies. Very small sources of water, such as raindrops, dew, juicy insects or plants, and small vertebrates, were sufficient to sustain the complete life cycle of the reptile.

Snakes, lizards, and crocodilians are the major groups of terrestrial reptiles today. Even crocodilians lay eggs on land, in carefully constructed nests that maintain a humid environment for the egg. In the Jurassic and Cretaceous periods, the reptiles dominated terrestrial habitats in the forms of saurischian and ornithiscian dinosaurs (Figure 7-20). The first flying vertebrates, the pterosaurs, patrolled the air. Even major radiations back into aquatic habitats resulted in the ichthyosaurs, sauropterygians such as the plesiosaurs, mosasaurs such as *Tylosaurus*, and turtles.

Mammal-like Reptiles

Early in the radiation of the reptiles, during the Pennsylvanian Period, the subclass Synapsida diverged, and by the end of the Triassic the synapsids were so mammal-like that perhaps some species were actually mammals. Hence, the synapsid lineage stopped abruptly at the close of the Triassic Period (Figure 7-20).

Evolutionary trends in the synapsids included increasing dental specialization into incisors, enlarged canines, and lateral cheek teeth; separation of the nasal passage and mouth to enable simultaneous breathing and feeding; advancing skeletal design for more effective locomotion on land, with the legs below the body raising it from the ground; and a more distinctive neck joining the head and body. Along the way, large synapsid reptiles, 6 to 10 feet long with huge sail-like dorsal fins, graced the Permian landscape. *Dimetrodon* and *Edaphosaurus* are probably the most familiar genera of these pelycosaurs (Figures 7-1 and 7-20).

The therapsids diverged early from the other pelycosaurs. They never went extinct, they just evolved into the mammals.

In the order Therapsida, the lineage leading to mammals developed as the suborder Theriodonta, which comprised mostly small to medium-size, rapidly moving carnivores. Some became dog-like in form and size, with teeth that could tear and chew up their prey before swallowing. Scientists have seriously considered the possibility that these predators had hair and a steady body temperature—both mammalian characteristics. But the fossils have not revealed such secrets. Fossils of *Thrinaxodon* have been found curled up, and young have been found nestled against adults, as if conservation of body heat and parental care had evolved in the theriodonts (Colbert and Morales 1991).

FIGURE 7-20

A general view of the major radiations of the reptiles. Of the present-day reptiles, the snakes, lizards, crocodilians, and turtles are the major groups. *Sphenodon*, the tuatara, is a living relic of an ancient lineage. Note the abbreviated record of the Synapsida ending with the therapsids. Did they go extinct or did they evolve into something else? *(Prepared by Lois M. Darling, from Colbert and Morales 1991)*

7.9 THE MAMMALS

Mammals exhibit important differences from reptiles. They have generally larger brain cases, complex cheek teeth, a single lower jawbone, complex ossicles of the ear, a larger ilium in the pelvis, homeothermy or more-or-less constant body temperature, hair instead of scales, live birth with embryonic development in a uterus in the placental mammals, and mammary glands that secrete milk. The first indisputable mammals appeared in the Triassic Period. From the small early mammals throughout the Mesozoic, the lineages radiated in the directions of different modern-day groups even though they remained rather inconspicuous members of a terrestrial fauna dominated by reptiles such as the dinosaurs. Then, during the Cenozoic Era, following the extinction of the dinosaurs at the end of the Mesozoic, the mammals became the dominant group of large animals on land (Figure 7-21).

Unfortunately, many important mammalian features do not become fossilized because they are not

FIGURE 7-21

The radiation of the mammals was well developed in the Mesozoic Era during the Cretaceous Period and continued into the Tertiary Period. Note the emergence of the primates in the early Tertiary. *(Prepared by Lois M. Darling, from Colbert and Morales 1991)*

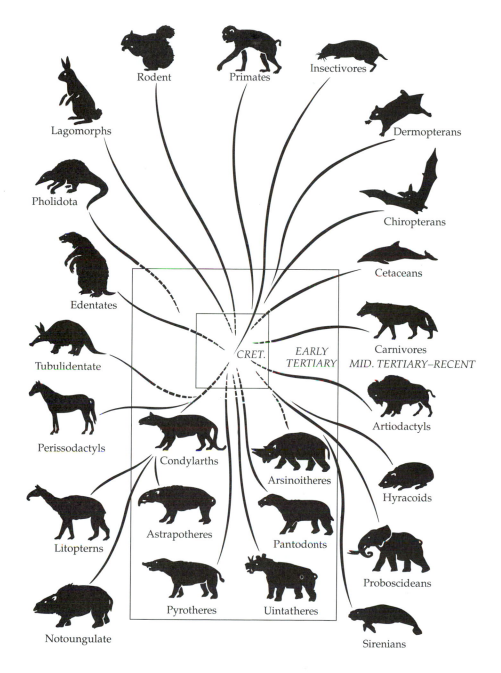

hard and bony. The high metabolic rates and high activity levels of mammals, even on cold nights and in cold climates, are sustained by a particularly good blood and oxygen supply not seen in reptiles today. The four-chambered heart keeps oxygenated arterial blood separate from the deoxygenated venous blood, and the diaphragm increases efficiency of ventilation of the lungs and oxygen supply into the blood. High metabolic rates sustain body temperature well above the ambient temperature in cool climates, making mammals warm-blooded relative to reptiles. Some quantity of hair insulates the warm body. These characteristics permit persistent activity in cold weather and efficient gathering of food to fuel high utilization and production of energy. In the evolution of the more derived mammals, the development of a uterus in which the embryo could remain to an advanced stage and parental care involving feeding the young with milk depended on the efficiencies of food, oxygen, and blood supply. The amniote egg with all its membranes (Figure 7-19) remains intact in mammals, serving with relatively minor modifications the close integration of maternal and fetal vascular systems through the umbilical cord and placenta.

The earliest mammals probably retained the reptilian trait of laying eggs, but their important distinction from reptiles was that they suckled their young from sweat glands on the underside of the body. Those glands developed into the mammary glands of the more derived mammals. The earliest egg-laying mammals are still represented by a few living species in Australia and New Guinea: the duckbilled platypus and two spiny echidnas, or anteaters.

This syndrome of adaptations associated with the high metabolic rates of mammals cannot be regarded as "better" or more "advanced" or more "progressive" than the alternative reptilian design. Reptiles still outnumber mammals almost two to one (there are 7,000 species of reptiles and 4,000 species of mammals extant). Mammals and reptiles simply represent alternative avenues of adaptation to life on land.

7.10 THE PRIMATES

The evolution and radiation of all the mammals is a long and complex story that is best left to books on vertebrate evolution (e.g., Romer 1966, 1971; Halstead 1968, 1978; Savage and Long 1986; Benton 1988; Colbert and Morales 1991). Discussion of the early emergence of the primates not only bypasses much of mammalian radiations but helps to make a direct link to Chapter 12, "Human Evolution." The primates emerged in the early Tertiary Period, probably from the insectivores (Figure 7-21). They had modest beginnings and, as we shall see in Chapter 12, modest success in the design of what many view as the crowning achievement of the evolutionary process: humans. The cloudy nature of primate origins is captured in the first fossil evidence of a primate, which received the name *Purgatorius.* The comparative anatomy of living primates provides a more definite story (Figure 7-22).

The tree shrew, *Tupaia,* is an arboreal, insectivore-like, primate-like mammal with a long snout and a relatively large brain. Life in the trees of tropical forests, gathering insects and fruits, gradually led to a basic set of primate characteristics. Tree shrews are very active, exploratory animals that search for food, check fruits for ripeness, inspect holes for insects, investigate nooks and crannies. Well-developed binocular vision makes them accurate judges of distance in the canopy (see Chapter 12, Sections 12.3

and 12.4). They are agile, with dextrous hands for grasping limbs, handling food, and engaging in social activities such as grooming and caring for young in the trees. The tree shrew's thumb and big toe became somewhat separated from the other digits, allowing better grasping of limbs, and as the digits broadened for grasping, the claws broadened into nails. The dense vegetation of tropical forest probably selected for the development of excellent vocal communication in the treetops, with perhaps two peaks in achievement: the howler monkey and, having carried this trait to the ground, the lineage that resulted in the human species.

It is interesting to note that the major kinds of primates all seemed to diverge from ancestral stock early in its history (Figure 7-22). For example, an early radiation in the Paleocene and Eocene gave rise to plesiadapids, lemuroids, and tarsoids and a second radiation into the New World monkeys, the Platyrrhini, and the Old World monkeys, the Catarrhini, with rapid divergence of the apes and ultimately humans. Thus, good skeletal remains can provide evidence of early primates, such as *Plesiadapis,* but they probably do not provide steps in the direct lineage to humans. A plesiadapid skeleton from the Eocene was already distinctly different from the other lineages that presumably resulted in the apes and humans.

We will pick up the story of the steps in the evolution of humans in Chapter 12. Meanwhile, we have a lot of backtracking to do to catch up on the development of the dense forests in which the primates radiated.

7.11 THE EARLY PLANTS

Figure 7-1 indicates the extent to which we must retrace our steps from Eocene primates back to the origin of plants on land, members of the kingdom Plantae. The earliest known land-plant fossils come from Ordovician deposits about 440 million and more years old. Small relics of spores, cuticles, and tubes have been found in many parts of the world (Gray 1984, 1988; Gray and Shear 1992).

The origin of the land plants is about as "lost in the mists of time" as anything can be, and the mystery has created a fertile arena for debate and conjecture. After all, this major breakthrough in design for terrestrial life provided much of the impetus for the great radiations of animals and fungi we have already discussed. Plants became by far the major autotrophs on land, sustaining the development of complex ecosystems to which the heterotrophic animals and fungi contributed.

FIGURE 7-22

The radiation of the primates with the tree shrews, in the genus *Tupaia*, which are squirrel-sized animals representing a link between the insectivores and the primates. *(Prepared by Lois M. Darling, from Colbert and Morales 1991)*

From Water to Aerial Parts

The first "landed immigrants," as Gray and Shear (1992) put it—i.e., green photosynthesizing organisms—were probably the cyanobacteria. These remarkably adaptable organisms had invaded eukaryotic cells to form chloroplasts (Chapter 6) and had contributed to major marine fossilized rock formations, the stromatolites, as well as freshwater equivalents, before they colonized the land. Today they may be found even colonizing desert soils, forming a crust and stabilizing the ground (Campbell 1979). On the shores of Precambrian oceans where many cyanobacteria lived, no doubt high salinity and the ebb and flow of the tides with regular desiccation were facts of life.

Other landed immigrants probably closer to the basic stock of terrestrial plants were the green algae, or chlorophytes. Chlorophytes have succeeded in fresh water and even on land, and they have been able to reach into the aerial domain a little by associations with fungi. A lichen is a symbiotic assemblage of an algal species and a fungal species. Lichens have proved themselves capable of colonizing extreme environments, from frigid to hot and from moist to dry. Dry rock surfaces can become covered with lichens. Lichens are commonly the first colonizers, and they ameliorate very harsh environments, enabling other life forms, such as mosses and liverworts, to colonize. Lichens possess many traits that were necessary to photosynthesizing organisms in the early colonization of the dry land.

Indeed, the fascinating and controversial suggestion has been made that not lichens but "inside-out lichens" provided the basic stock from which terrestrial plants evolved (Atsatt 1988). That is, the algal associate became the dominant partner and provided a lot of the main plant structure, and the fungus started as a parasite and became an internal mutualist (Pirozynski and Malloch 1975). The argument is based on several fungus-like characteristics of plants and the recognition of their rapid development of complex anatomy. Pollen tubes are fungus-like and frequently parasitic on the recipient; more than one nucleus is present, reminiscent of the heterokaryotic phase in fungi. Many haustoria-like growths occur in plants and resemble those in parasitic fungi. The fungi use these modified hyphae for drawing nutrients from their hosts. In some plants, haustoria develop in the endosperm. So-called idioblasts are peculiar cells that differ markedly from other cells in the same tissue (Esau 1977). Some idioblast cells behave as parasitic fungi, and some parasitic flowering plants grow within host plants just as endophytic fungi do. Many cells in plants grow intrusively into tissues, appearing to behave as parasites.

If such an association developed, then the fungal part of the mutualism could ramify in the soil, absorbing water and nutrients and transporting them to aerial parts, while the algal associate would be photosynthetic and resistant to desiccation. The symbiosis would achieve a nice division of labor, opening up new opportunities for colonizing land.

Divergence of the Bryophytes and Tracheophytes

An alternative, more generally favored hypothesis involves early green algae, the chlorophytes, which share biochemical and structural traits with both the nonvascular bryophytes and the vascular tracheophytes (Stewart 1983, Gensel and Andrews 1987, Gray and Shear 1992). Some green algae do exhibit a division of labor into a multicellular erect thallus and a prostrate rhizoid-bearing portion reminiscent of a root system (Figure 7-23). In addition, some algae produce spores in tetrads (groups of four), and the parental plant retains the fertilized egg. Alternation of generations also occurs in algae, with a haploid phase producing gametes (the gametophyte generation) and a diploid phase producing spores. Thus, we can readily assign some living algae to a group to which the progenitor of plants may well have belonged. These algae have an erect thallus, rhizoids, spores in tetrads, retention of the zygote by the parental thallus, and alternation of generations. Because the mosses and liverworts, or bryophytes, share these traits with plants with vascular tissues, the tracheophytes, biologists suspect a common ancestor. In addition, in both groups spores are invested in a wall composed of sporopollenin, which provides a durable casing that preserves well in the fossil record (cf. Figure 7-18). Both groups develop an embryo from the zygote, and together they are called embryophytes.

The steps in the early evolution of the land plants may have progressed as follows. A hypothetical algal ancestor had alternation of generations, with diploid and haploid phases represented as similar thallus

FIGURE 7-23

The green alga *Fritschiella,* showing an erect multicellular thallus and a prostrate, rhizoid-like extension from its base. Species living today still represent this genus. *(From Stewart 1983)*

types with a simple dichotomous branching pattern. As explained in Figure 7-24, the alga had a **diplohaplontic life cycle** with **isomorphic sporophyte and gametophyte generations.** This ancestral state then diverged into (1) the dominance of the gametophyte as the dichotomizing, photosynthetic thallus in the bryophytes and (2) the large sporophyte with a diminished gametophyte generation in the tracheophytes (Figure 7-25). Both produced spore tetrads with durable sporopollenin coats. As the tracheophytes developed, the dichotomous sporophyte thallus tended toward a division of labor among reproductive functions and photosynthetic functions (Figure 7-26). Leaf-like structures retaining dichotomous veins developed.

It does not take much imagination to see how these early land plants could leave fossil evidence, as in the genus *Cooksonia* from the Silurian, 428 million years ago (Figure 7-27). The fossil record has not yet revealed whether *Cooksonia* had vascular tissues, but certainly the rhyniophytes were soon vascularized (Figure 7-28). The type genus was designated as *Rhynia* (Figure 7-27) because of its discovery in the Rhynie (Scotland) chert beds from the Lower Devonian (cf. Figures 7-18 and 7-28). Note in *Rhynia major* the Y-shaped dichotomous branching pattern, with sporangia capping some stems, and the spores in tetrads. This species grew to 50 cm tall, with a creeping rhizome anchored with rhizoids. Presumably it

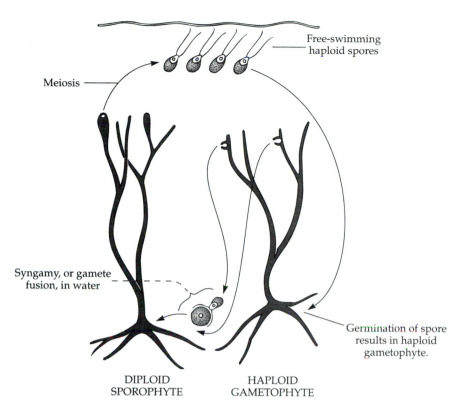

Free-swimming
haploid spores

Meiosis

Syngamy, or gamete
fusion, in water

Germination of spore
results in haploid
gametophyte.

DIPLOID
SPOROPHYTE

HAPLOID
GAMETOPHYTE

FIGURE 7-24

A hypothetical algal precursor to green terrestrial plants, showing the isomorphic alternation of generations and a dichotomous branching pattern. **(Left)** The spore-producing diploid sporophyte. **(Right)** The haploid gametophyte, which produces male and female gametes. This diplohaplontic life cycle was a precursor to the life cycle of plants that diverged into the bryophytes and the tracheophytes. *(Based on Stewart 1983; see also Stewart and Rothwell 1993)*

FIGURE 7-25

Early terrestrial plants diverged into (a) the bryophytes, with a relatively large gametophyte thallus and a smaller, dependent sporophyte, and (b) the tracheophytes, with a large sporophyte and a much smaller gametophyte. Note, in both cases, the sedentary female gamete, the motile male gamete, and spores that are produced in tetrads. *(Based on Stewart 1983; see also Stewart and Rothwell 1993)*

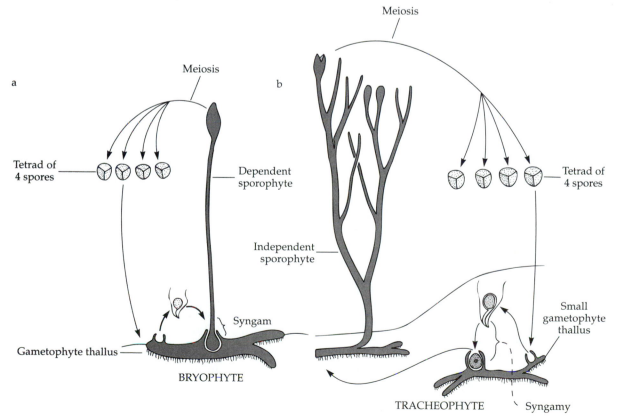

Meiosis

a   b

Meiosis

Tetrad of
4 spores

Dependent
sporophyte

Independent
sporophyte

Tetrad of
4 spores

Small
gametophyte
thallus

Syngam

Gametophyte thallus

BRYOPHYTE

TRACHEOPHYTE   Syngamy

FIGURE 7-26

A hypothetical image of an early vascular plant. Dichotomous branching is retained in the large, erect sporophyte generation, which shows division of functions into fertile, sporangium-bearing fronds; vegetative branches with leaf-like extensions; a rhizomatous stem enabling vegetative reproduction; and rhizoids stabilizing the plant in the soil. *(From Stewart 1983)*

had green photosynthesizing stems, because its stomata are clearly preserved. Stewart (1983) discusses the interesting possibility that *Rhynia* may have retained the primitive isomorphic sporophyte and gametophyte life cycle.

The Lycophytes

The sequence of events in the evolution of land plants is complicated by a *Lycopodium*-like fossil, *Baragwanathia*. Note its record from the Upper Silurian in Figure 7-18. Its advanced design included cauline leaves and stems up to 1 meter long with sporangia in leaf axils, probably with rhizomatous stems creeping along the ground (Figure 7-27). "In essence, the ?Silurian–Devonian genus *Baragwanathia* is simply a very large version of shining club moss (*Lycopodium lucidulum*), a living lycopod" (Gensel and Andrews 1987, 484). The question mark acknowledges debate on the date of the fossil beds in which *Baragwanathia* has been found. Even so, the existence of such a large, complex plant close to the Siluro–Devonian border

suggests an already long history of life on land or else remarkably rapid evolution (see Chapter 18). The similarity between living plants and fossils from more than 400 million years ago is remarkable evidence of the early origins of very successful design for life on land. From these early beginnings, the lycopod-like plants, or lycophytes, radiated to some of the dominant land plants, reaching a climax in the Upper Devonian and the lush, swampy forests of the Carboniferous (Figure 7-28).

Rhyniophytes to Trimerophytes

By the end of the Lower Devonian Epoch, plants had met the challenges of life on land and diversification of major groups was well in progress. Scientists have recognized 28 genera of vascular plants in the Lower Devonian. One of the major advances was the development of a strong central axis with lateral, spirally arranged branches allocated to photosynthetic autotrophy, some bearing sporangia. The genus *Psilophyton* (Figure 7-27) is an example of a lineage that may have given rise to the major groups of plants extant today: ferns, gymnosperms, and angiosperms. The three-dimensional branching pattern was achieved with dichotomous branching retained, but one branch became much stronger and more central and the other became lateral—a growth pattern called **pseudomonopodial branching.**

The type genus for the Trimerophytopsida is *Trimerophyton*, which has a branching pattern that includes an inverted tripod-like design—trimeric branching. *Psilophyton* retained a more dichotomous branching plan, but all the trimerophytes had a more centralized main stem, more vascular tissue in the stems, richer branching patterns with separate vegetative and reproductive functions, and paired sporangia in clusters (Figure 7-27). *Psilophyton* had dehiscent sporangia; some had leaf-like appendages. After closely resembling the rhyniophytes at the beginnings of their divergence, the trimerophytes had achieved a major new feature in design—the central stem that characterizes all the derived plant groups (Figure 7-28).

7.12 THE EMERGENCE OF SEED-BEARING PLANTS

Cambial Activity Producing Secondary Xylem

In the evolution of land plants, a main axis brought with it a clear limitation on size until additional strength was available to make a developing trunk more rigid. The production of more woody tissue,

FIGURE 7-27

Some examples of early land plants. (a) *Cooksonia*. (b) *Rhynia major*, with an inset of a spore tetrad. (c) *Asteroxylon* with an aerial stem, and a rhizome with root-like extensions, found in the Lower Devonian Rhynie chert. (d) A small section of stem from *Baragwanathia*, showing a thick stem 1 to 2 centimeters in diameter, microphyllous leaves, and axillary sporangia. (e) The aerial axis of a *Psilophyton* species, with sterile branches below and fertile branches with sporangia above. *(From Stewart 1983)*

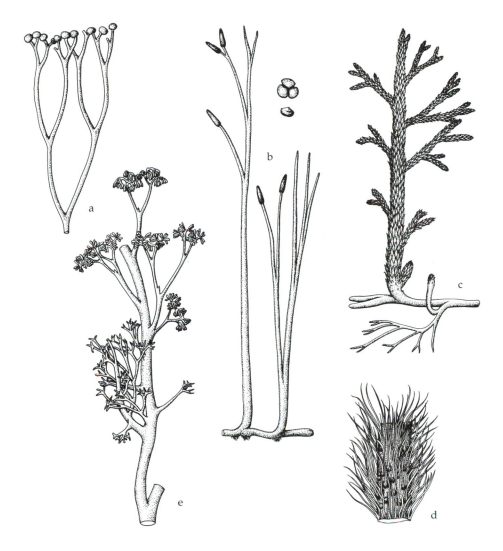

or xylem, in the main stem during growth became possible with the evolution of a cambial layer—a meristem producing secondary xylem and the corky bark on the exterior. The fossil record contains 370-million-year-old evidence of secondary xylem that had evolved by the Middle Devonian (Figure 7-28). Trees were in evidence in the Upper Devonian, and by the Lower Carboniferous there were stems 45 centimeters (1.5 feet) in diameter.

The Gymnosperms

The progymnosperms seem to have evolved from the trimerophytes (Figure 7-28). The traits in common include dichotomous and pseudomonopodial branching, terminal parts of branches appearing to form the beginnings of fronds, sporangia clustered on the ends of fertile branches, and dehiscent sporangia. The basic advance of the progymnosperms was secondary growth with a cambium down the length of the main stem.

From the progymnosperms came a major breakthrough in the reproductive system—the development of seeds and the evolution of the gymnosperms. Evolution of seeds is generally regarded as one of the most important events in the evolution of vascular plants because it is associated with the divergence of the plant groups that dominate landscapes today. A

FIGURE 7-28

The course of time and the development of terrestrial plants as summarized by Gensel and Andrews (1987). The major groups of plants are illustrated as lenses to indicate origins, maximum richness, and decline. Major stages in the development of plants are illustrated; for example, the first records of heterospory, secondary growth, arborescence, seeds, and large megaphyllous leaves. Note that the Silurian and Devonian periods are expanded on the time scale to capture this Siluro–Devonian explosion of land plants in more detail.

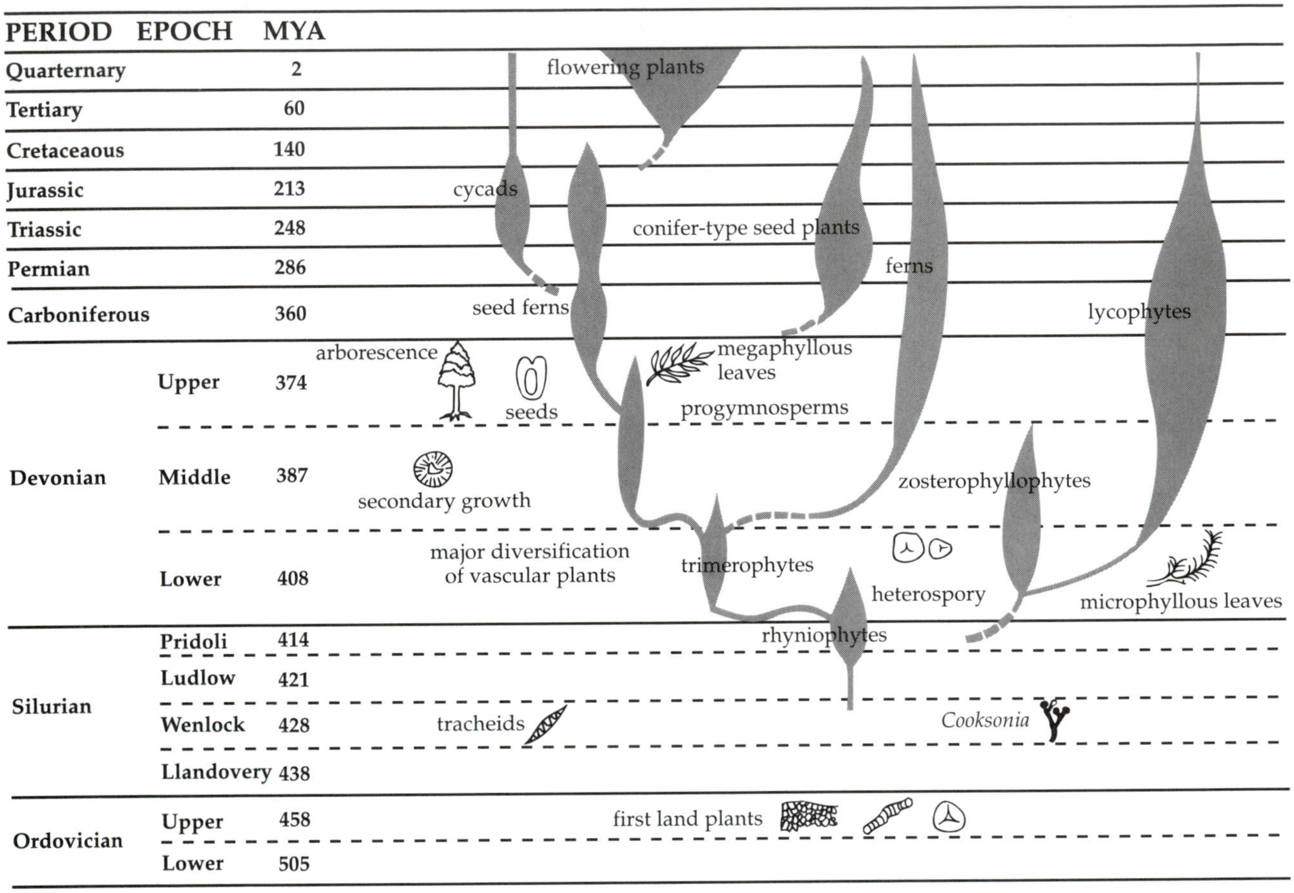

PERIOD	EPOCH	MYA
Quarternary		2
Tertiary		60
Cretaceaous		140
Jurassic		213
Triassic		248
Permian		286
Carboniferous		360
Devonian	Upper	374
Devonian	Middle	387
Devonian	Lower	408
Silurian	Pridoli	414
Silurian	Ludlow	421
Silurian	Wenlock	428
Silurian	Llandovery	438
Ordovician	Upper	458
Ordovician	Lower	505

Figure labels: flowering plants; cycads; conifer-type seed plants; ferns; seed ferns; lycophytes; arborescence; seeds; megaphyllous leaves; progymnosperms; secondary growth; zosterophyllophytes; major diversification of vascular plants; trimerophytes; heterospory; microphyllous leaves; rhyniophytes; tracheids; *Cooksonia*; first land plants

seed is not only multicellular; it contains an embryo, usually considerable stored nutrients, and a protective coat. Seeds can be dispersed great distances by wind, water, or animals and exhibit a remarkable range of adaptations to the growing conditions typically encountered. They can disperse in time through complex controls of dormancy, persisting in the soil for years or decades.

The seed plants that are conspicuous on modern landscapes are the coniferous gymnosperms—such as pines, spruces, junipers, and firs—and the angiosperms, or flowering plants, which include broad-leaved trees, grasses, cacti, daisies, lilies, orchids, and many others. Many angiosperms provide humans with food such as grains, wheat, corn and rice, the onion family, the cabbage family, the potato family, the family containing the pulses, peas, beans, lentils, and many other crop plants. The fruits we buy at the market are all from angiosperms, and many herbs, spices, and drugs are of angiosperm origin. Strangely enough, the conifers have provided little in the way of human nutriments: pine nuts are widely utilized, spruce gum was chewed during the Second World War, and the cambium of some conifers is eaten.

We should not forget that a seed usually becomes an independent living organism once it is released from the parental plant. It is diploid, with a haploid ovule having been fertilized by a haploid nucleus from a pollen grain. Thus, the evolution of seeds also depends on the evolution of pollen, because only with a pollen tube can the ovule nucleus be reached in its concealed and protected position. Internal fertilization without the need for free water, with a fertilized ovule encased in a protective "shell" released from the parent plant, provided the same kind of breakthrough for terrestrial plants as did the amniote egg for vertebrates.

Stewart (1983) recognized three stages in the evolution of seeds.

1. Plants produced both small microspores and large megaspores and released them from dehiscent sporangia.

2. The parental plant retained megaspores in the indehiscent megasporangium and released microspores from a dehiscent microsporangium. Microspores released sperm cells that had to swim in water to reach and fertilize megaspores.

3. The indehiscent megasporangium and the dehiscent microsporangium were retained. The megasporangium became invested in an integument that would form the seed coat, and a pollen tube transported sperm cells to ovules in the absence of free water.

Clearly, the evolution of different spore types, or **heterospory,** was critical to the ultimate development of small pollen, large ovules, and seeds. Again, it is fossils from the Middle Devonian that yield evidence of this novelty, at a time when secondary growth emerged as a major breakthrough. Perhaps the development of trees and the retention of megaspores reduced the probability of sperm meeting egg in a liquid medium, and airborne microspores compensated for difficulties with reproduction high in the windy tops of tall plants. The earliest record of heterospory is in the genus *Chauleria* in the progymnosperms (Stewart and Rothwell 1993). Sporangia contained both microspores and megaspores, conveniently illustrating a transitional phase between isospory and the evolution of separate micro- and megasporangia.

The first seed-bearing plant to be named in the fossil record was assigned a new genus name, *Archaeosperma*, which alludes to the ancient origin of this seeded plant from the Upper Devonian (Figure 7-28). The fossils indicate the presence of a single functional megaspore, probably surrounded by a nucellus, and an integument, all invested in sterile branches (Figure 7-29). Dichotomous branching was retained in the

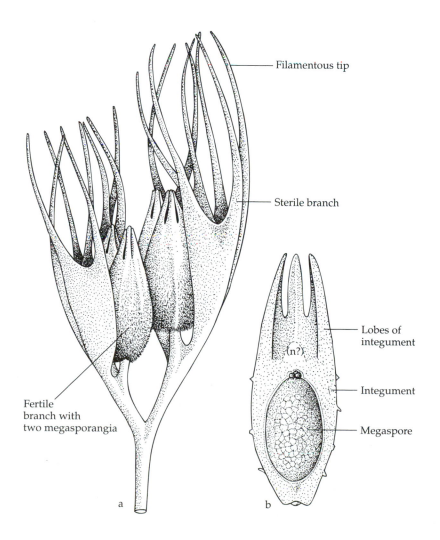

Filamentous tip

Sterile branch

Fertile branch with two megasporangia

Lobes of integument

Integument

Megaspore

(n?)

a

b

FIGURE 7-29

(a) A reconstruction of the first named seed-bearing plant, *Archaeosperma arnoldi*. The megasporangia have become invested in an integument and appear to be protected by dichotomous, bract-like, sterile branches with filamentous tips. (b) A cutaway section of the megasporangium. Note the one large functional megaspore and the three sterile products of meiosis, the probable existence of the nucellus (n?), and a lobed integument that has not closed in to form the micropyle seen in later plants. *(Based on Stewart 1983)*

paired ovules and sterile branches. In fact, this genus provided an opportunity to see the transition from open sporangia to megasporangia invested in sterile tissue (Figure 7-30). The surrounding sterile tissue eventually evolved to become more closely associated with the megasporangium and form the integument of the ovule, with the micropyle providing access to pollen grains (Figure 7-31).

From these early plants the seed-producing gymnosperms radiated into the major groups of conifers with simple leaves, needles, or scale-like leaves and the major groups of plants with large frond-like leaves—the seed ferns and the cycads (Figure 7-28). Notice that the true ferns had long before diverged from the trimerophytes, and the fern-like fronds of ferns, cycads, and seed ferns are the result of convergence in design.

FIGURE 7-30

Possible stages in the evolution of the seed. (a) Dichotomous branching with sporangia. (b) The development of heterospory, with spores produced in micro- and megasporangia. (c) The protection of the megasporangium by sterile branches. (d) Sterile branches have fused to form an integument around the megasporangium. *(From Stewart 1983)*

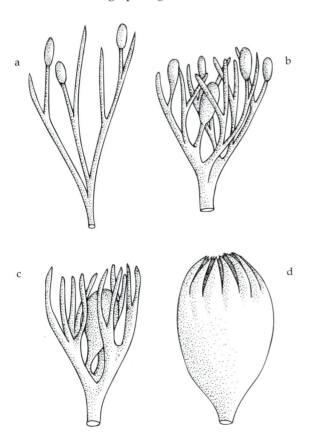

7.13 THE FLOWERING PLANTS

Characteristics of Angiosperms

The flowering plants no doubt beguiled the first members of the human race with their beauty. Their nutritional properties had been the basis of whole ecosystems for millions of years. Their pharmacological properties had mediated host and parasite interactions for as long. The flowering plants radiated to the greatest number of plant species and today possibly represent the richest flora that ever existed on earth. Figure 7-28 suggests this richness with the width of the clade representing present-day flowering plants relative to the other plant groups. Biological classification of these plants is in flux. Between 300 and 400 families and more than 250,000 species have been recognized.

For most biologists the most distinctive features of the angiosperms relate to their mode of reproduction. Flowers commonly are composed of whorls of sepals, petals, stamens, and central carpels in which ovules develop and seeds ripen. The carpel surrounds the ovule, precluding direct access of pollen to the nucellus, and pollen grains germinate on a stigmatic surface above the carpel. Once fertilization is complete, the carpel and seed or seeds form a fruit. True fruits may be dry, like the nuts and the so-called "seeds" of grasses and the Aster family, or fleshy and succulent, like the fruits we eat—e.g., tomatoes, cucumbers, apples, plums, oranges, and pomegranates. Dry fruits may also be associated with fleshy plant parts, as in strawberries, figs, and cashews. Such dry and juicy fruits have certainly delighted the human eye, palate, and stomach as much as the beauty of angiosperm flowers.

Although the angiosperms have many other distinctive features, in this chapter we explore only the evolution of the flower—in particular, the carpel.

Angiosperm Origins

Although the source of the angiosperms probably lies among the gymnosperms, it remains enigmatic. As we have seen already in this chapter, the fossil record often provides unsatisfactory clues about major breakthroughs in design and function. This dearth of clues may well be the result of very rapid transitions from one condition to another, caused by new evolutionary impetus, that leave scant fossil evidence or extant intermediates. The transition from gymnosperms to angiosperms is a possible example. The phenomenon of rapid change appears to be common in the evolutionary history of life; Chapter 18 will discuss it more extensively.

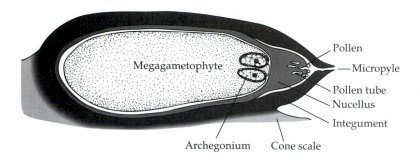

A section through a modern pine seed in the genus *Pinus*, a gymnosperm and a conifer. This median longitudinal view shows the large megagametophyte, the archegonia, the integument, the nucellus, pollen tubes growing through the nucellus, pollen, the micropyle, and the cone scale with which the integument is closely associated. *(Based on Stewart 1983)*

As Figure 7-28 indicates, early clearly angiospermous fossils have appeared in the Lower Cretaceous, Barremian–Aptian stages (cf. Table 7-1), dating back to 112 million years ago. Leaves and pollen were preserved, but wood fossils are rare. Fruits were diverse and well formed. A dig in Upper Cretaceous deposits revealed an amazingly advanced flower with whorls of sepals, petals, anthers, and carpels—all with five parts, which are typical of modern dicotyledonous angiosperms (Figure 7-32). Until more intermediary stages between gymnosperms and angiosperms enter the fossil record, we are left in a rather speculative mode for discussion of the origins of the flower.

Some evidence suggests that sepals and petals derive from leaves modified to protect the flower bud and to "advertise" the flower to pollinating insects. Anthers and ovaries clearly derive from micro- and megasporangia. The carpel may have derived from the merger of megasporangia within a folding leaf, and the seed ferns, or pteridosperms, seem to provide the pathway to the angiosperm carpel (cf. Figure 7-28). Stewart (1983) formulated a six-stage scenario for this breakthrough in floral design and reproductive function.

Stage 1: Grouping of ovules. Many pteridosperms had ovules grouped and protected by cup-like umbrellas, or cupules.

Stage 2: The fertile axis. The several ovules plus cupule became grouped on fertile axes, derived from fertile fronds of seed ferns (Figure 7-33).

Stage 3: Association of the fertile axis with a leaf. The fertile axis develops in the axil of a leaf (Figure 7-33a).

Stage 4: Fusion of the fertile axis and leaf midrib. The stem of the fertile axis fuses with the midrib of the leaf, leaving two rows of ovules and cupules associated with the leaf surface (Figure 7-33b).

Stage 5: Origin of the single ovule with two integuments. The number of ovules per cupule is reduced to one (Figure 7-33d) and the cupule fuses with the ovule's integument to produce a second integument. This structure folds down toward the axis of the plant (Figure 7-33f, g).

Stage 6: Carpel formation. The lamina of the leaf folds around the ovules to form the carpel, it fuses, and the distal part of the leaf becomes a stigma on which the pollen germinates (Figure 7-33h). Protection of the ovule was complete. But whether this is what really happened remains to be seen. Perhaps more than one lineage of gymnosperms developed carpels and the angiosperms are polyphyletic.

Figure 7-1 illustrates many of the plant types discussed in this chapter. Although we have considered the evolutions of fungi, plants, and animals as separate concerns, Figure 7-1 reminds us of the simultaneous development of all kingdoms, in the presence also of the protists and bacteria. Rich biotic interactions between all these kinds of organisms shaped

A reconstruction of a flower from the Upper Cretaceous. Note the clear whorls of five sepals, five petals, five anthers, and five carpels, with one sepal almost concealed in the lower left part of the flower. *(Based on Stewart 1983)*

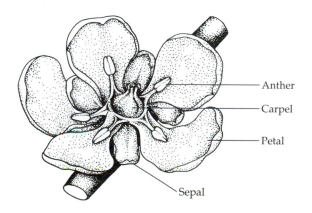

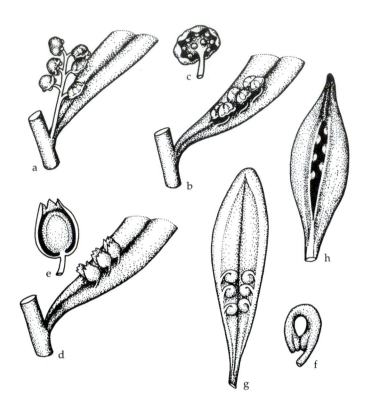

FIGURE 7-33

The stages in the evolution of the carpel from seed ferns to angiosperms. (a) Several cupules with multiple ovules become grouped on a fertile axis, and the axis develops in the axil of a leaf (Stages 2 and 3 in text). (b) The stem of the fertile axis fuses with the midrib of the leaf, producing two rows of cupules with ovules (Stage 4 in text). (c) The underside of a cupule with multiple ovules. (d) The ovule number is reduced to one per cupule (Stage 5 in text). (e) The cupule encloses a single ovule more completely and more closely. (f) The cupule fuses with the integument of the ovule to form a double integument of the seed. (g) The leaf lamina begins to fold around the ovules. (h) Complete enclosure of the ovules results in a carpel with a stigmatic surface at the leaf extremity (Stage 6 in text). *(From Stewart 1983)*

the evolution of life and the evolution of ecosystems and landscapes in ways that are still being unraveled.

Now we can consider an earth clothed in plants and populated by plants, animals, fungi, protists, and bacteria, all interacting in complex ways. Organisms entered into symbioses, they defended themselves against attack from other organisms, and they dispersed as land masses changed on a global scale. Mutually induced evolutionary change among interacting species resulted in coevolution (Chapters 19 and 20). Great adaptive radiations (Chapter 8) and mass extinctions (Chapter 9) became parts of life on earth. Eventually, humans evolved that were capable of studying all of these developments and unraveling as best they could the complex interactions involved in the evolution of life on earth.

To end this chapter, let us summarize approaches to the analysis of relationships among organisms. (Chapter 13, "Biological Classification," will take a closer look at how we systematize and order relationships.)

7.14 THE COMPARATIVE METHOD IN DECIPHERING EVOLUTIONARY HISTORY

The comparative method is a potent component of the scientific method. It may take several different forms and use several different kinds of evidence, depending on the available data and the difficulty of the problem at hand. In the virtual absence of opportunities for experimentation on the evolutionary relationships among organisms through time, comparison among living organisms is a crucial tool. Let us review the comparative methods used in this chapter and the kinds of organisms to which they were applied.

A. **Comparison of fossils** (vertical comparisons through time, from lower, older strata up into more recent strata)
　1. Using comparative anatomy. Examples:
　　(a) Evolution from fishes to amphibians, amphibians to reptiles, reptiles to mammals
　　(b) Evolution of land plants—emergence of tracheophytes, rhyniophytes to trimerophytes, seed-bearing plants, and the gymnosperms

B. **Comparison of living organisms**
　1. Using comparative anatomy (vertical comparisons). Examples:
　　(a) Relationships between bacteria and protists and the eukaryotic cell
　　(b) Relationships between protists and fungi and the animals
　　(c) Relationships among early mammals, primates, and humans
　　(d) Evolution of plants

2. Using embryology (horizontal comparisons using contemporaneous organisms). Examples:
   (a) Relationships among deuterostomes
   (b) Evolution of paired fins in fishes
3. Using molecular and biochemical techniques (horizontal comparisons). Examples:

   (a) Relationships of worms, onychophorans, and arthropods
   (b) Relationships of the major groups of organisms from bacteria to plants, fungi, and animals
   (c) Relationships between bryophytes and tracheophytes

S U M M A R Y

Unraveling the pathways along which life evolved and tracking relationships are major challenges for the evolutionary biologist. The fossil record is short and incomplete relative to the history of life on Earth, and major innovations may have occurred rapidly, leaving scant evidence. In addition to fossils, however, living organisms can provide views into the past, and in this endeavor the comparative method becomes an important and potent aspect of the scientific method.

The geologic time scale gives a sense of the passing of the ages and provides some details of the organisms whose remains are trapped in the rocks, back to almost 600 million years ago. We can map out the major features in the development of life since then in a timed sequence of events, and we can explore major breakthroughs in design in terms of mechanisms, pathways, and opportunities. But the Cambrian explosion of marine invertebrates, the Siluro–Devonian explosion of terrestrial plants, the colonization of land, and the full proliferation of marine and terrestrial life all pose problems yet to be solved, and the current secrets of evolution will fascinate us well into the future.

The fungi that mycologists study, so important as decomposers and parasites, seem to have evolved from three different lineages of protists. The slime molds can be regarded as amoeboflagellated protists that aggregate to form a plasmodium and sporangium. The algal fungi derive from algae that have lost chloroplasts to become heterotrophic like the true fungi. The true fungi share several characteristics with the animals and seem to have origins in common with them, among the choanociliate protists.

Animals may have their origins in the choanociliates, with cell aggregation having led to the sponges and an early offshoot from this stock having led to all other animal groups. By the time of the Cambrian explosion, animals of advanced design represented many phyla. Even a chordate may have been present in the Burgess Shale deposits. In the absence of a fossil record of intermediate types, scientists can use living animals to search for phylogenetic relationships. Researchers have employed molecular techniques to address longstanding questions on the relationships among living annelid worms, velvet worms, and arthropods. Embryological evidence played a role in the search for vertebrate ancestors among the invertebrates, with the advent of the echinoderms marking the major shift in development from protostomes to deuterostomes. Deuterostomes developed a notochord as a flexible axis on which muscles were attached for swimming. The neotenic tadpole-like larva of the sea squirt may have been a link to the vertebrates. The first fishes retained a notochord and lacked jaws, but strengthened gill arches supported gills and enabled filter feeding in the ostracoderms. The first jawed vertebrates, the placoderm fishes, evolved by the gradual conversion of anterior gill arches to upper and lower jaws, and some evolved relatively rapidly into large predatory species. Paired fins may have arisen from a pair of fin folds along the body followed by tissue breakdown between the persistent fins.

The vertebrates' colonization of the land resulted from a lineage of fishes that could breathe air, with strongly boned pectoral and pelvic fins enabling locomotion over land. These beginnings of tetrapod design led to the amphibians that could persist on land, although they reproduced in the aquatic medium. Several advantages of a more terrestrial existence included new and abundant food sources, escape from aquatic predators and parasites, and, for freshwater species, escape from drying lakes and rivers. Evolution of a persistently terrestrial life cycle and the colonization of dry habitats occurred in reptiles with designs for encapsulating aquatic life in water-holding membranes: a skin, or integument, covered with scales; internal fertilization of the egg; embryonic development in the amnion surrounded by a chorion and egg shell. Major adaptive radiations resulted in reptiles that became very large—the dinosaurs—and reptiles that persist in diverse forms today—snakes, lizards, crocodiles, tortoises, and turtles.

Among this radiating group was a lineage of synapsids that came close to being mammals. The therapsids developed into carnivores, some with a dog-like design including teeth adapted for tearing and chewing flesh. The transition from reptile to mammal cannot be pinpointed at this time. Mammalian characteristics include a relatively large brain, homeothermy, hair to insulate the body, live birth, and the suckling of young accompanied by extended parental care. Among the mammals, primates probably evolved from small insectivores that took to the trees and fed on insects and fruits. Active, inquisitive, and nimble, they adapted to an arboreal life in tropical forests with increasing binocular vision, dexterity, sociality, and vocalization.

The fossils record the evolution of plants, starting in the Ordovician about 440 million years ago. The green algae, or Chlorophyta, are likely to be associated with the colonization of land, and whether or not they became associated with fungi to form green plants has been debated. Some green algae developed a multicellular thallus and a prostrate rhizoid system. Alternation of generations between a gametophyte and a sporophyte was a characteristic that accompanied the colonization of land and the divergence of early plants into the bryophytes and the tracheophytes. In the tracheophytes, the sporophyte became dominant, with an early dichotomizing rhyniophyte, *Cooksonia*, from 428 million years ago being followed by the genus *Rhynia*, with upright photosynthesizing stems, a creeping rhizome, and rhizoids. Early lycophytes appeared at the same time with strong similarities to today's *Lycopodium* species.

A major advance in plant design involved the evolution of the trimerophytes, with a central axis and branches that were spirally arranged and divided between the sterile and the fertile. That advance was followed by the development of cambial meristems and the production of secondary wood, which greatly strengthened plants that were increasing in size—the progymnosperms. Among the progymnosperms was a lineage that resulted in the first seed-bearing plants, the gymnosperms. Heterosporic plants began to retain large megaspores in indehiscent megasporangia, and an integument formed the seed coat. *Archaeosperma*, the first named genus of seeded plants, dates from the Upper Devonian. Gymnosperms radiated into the very successful conifers and into the seed ferns and cycads.

The seed ferns seem to have provided the basis for evolution of the angiosperms. With flowers and fruits, the angiosperms form a distinctive group with uncertain links to progenitors. Sepals and petals are likely to have derived from leaves, and anthers and ovaries arose from micro- and megasporangia. Development of the carpel, which protects the seed, may have involved a gradual investment of a fertile branch by a leaf, with the leaf tip becoming the stigmatic surface receptive to pollen.

This chapter used the comparative method to explore relationships on the bases of the anatomy, embryology, and molecular and biochemical characteristics of living organisms and the anatomy of fossils. This comparative approach is an essential tool in the scientific method applied to biological evolution.

QUESTIONS FOR DISCUSSION

1. Do you agree with the argument that the most complex evolutionary step ever to occur on this earth was the origin of the prokaryotic cell, and all evolution since has been simple, relatively speaking?

2. If you were to create new, higher taxa, such as new kingdoms, given the species known to exist on earth, how would you identify the new taxa, and of what would they be composed?

3. Do you think that the Cambrian explosion resulted from an incredibly creative wave of evolutionary diversification, or is it more likely that the fossil record is incomplete and a more gradual diversification of life preceded the apparent explosion?

4. At the time of the colonization of land by invertebrates and vertebrates, can you suggest other directions in which evolution might have gone, such that the resultant faunas would be considerably different in structure from those in existence today? What particular design differences would be possible?

5. The amniote egg is regarded as a major adaptive breakthrough to life on land. What alternative avenues of complete adaptation to a terrestrial existence were realistically possible?

6. The evolution of homeothermy, or warm-bloodedness, which resulted in the mammals, had heavy costs and strong benefits. Can you list some of each? When you consider the status of many mammals today, and their small radiations compared with those of fishes and reptiles, do you think that homeothermy was too costly an adaptation to permit long-term success in a changing world?

7. Would you argue that the development of terrestrial life was more complex and unlikely in the lineage leading to plants than in the lineage resulting in the reptiles and their descendents?

8. Why do you think the life cycles of most terrestrial plants have lost the separate gametophyte and sporophyte forms?

9. Do you find it remarkable, given the struggle for existence, that some terrestrial plants living today are very similar to those that existed some 400 million years ago? What does this situation suggest in terms of natural selection and evolution?

10. Considering all groups of plants, what characteristics do you think have made the angiosperms so numerous, with about 250,000 species?

REFERENCES

Alexopoulos, C. J. 1952. Introductory Mycology. Wiley, New York.

Atsatt, P. R. 1988. Are vascular plants "inside-out" lichens? Ecology 69:17–23.

Ballard, J. W. O., G. J. Olsen, D. P. Faith, W. A. Odgers, D. M. Rowell, and P. W. Atkinson. 1992. Evidence from 12S ribosomal RNA sequences that onychophorans are modified arthropods. Science 258:1345–1348.

Barnes, R. D. 1987. Invertebrate Zoology. 5th ed. Saunders College Publishing, Philadelphia.

Bengtson, S. 1977. Early Cambrian button-shaped phosphatic microfossils from the Siberian platform. Palaeontology 20:751–762.

Bengtson, S., and T. P. Fletcher. 1983. The oldest sequence of skeletal fossils in the Lower Cambrian of southwestern Newfoundland. Can. J. Earth Sci. 20:525–536.

Benton, M. J. (ed.) 1988. The Phylogeny and Classification of the Tetrapods. 2 vols. Clarendon, Oxford.

Borror, D. J., C. A. Triplehorn, and N. F. Johnson. 1989. An Introduction to the Study of Insects. 6th ed. Saunders College Publishing, Philadelphia.

Brown, J. H., and A. C. Gibson. 1983. Biogeography. C. V. Mosby, St. Louis.

Campbell, S. E. 1979. Soil stabilization by a prokaryotic desert crust: Implications for Pre-Cambrian land biota. Origins of Life 9:335–348.

Cavalier-Smith, T. 1987. The origin of fungi and pseudofungi, pp. 339–353. In A. D. M. Rayner, C. M. Brasier, and D. Moore (eds.). Evolutionary Biology of the Fungi. Cambridge Univ. Press, Cambridge, England.

Colbert, E. H., and M. Morales. 1991. Evolution of the Vertebrates. Wiley–Liss, New York.

Collins, O. R. 1987. Reproductive biology and speciation in Myxomycetes, pp. 271–283. In A. D. M. Rayner, C. M. Brasier, and D. Moore (eds.). Evolutionary Biology of the Fungi. Cambridge Univ. Press, Cambridge, England.

Esau, K. 1977. Anatomy of Seed Plants. Wiley, New York.

Gensel, P. G., and H. N. Andrews. 1987. The evolution of early land plants. Amer. Sci. 75:478–489.

Glaessner, M. F. 1984. The dawn of Animal Life. Cambridge Univ. Press, Cambridge, England.

Gorr, T., and T. Kleinschmidt. 1993. Evolutionary relationships of the coelacanth. Amer. Sci. 81:72–82.

Gould, S. J. 1989. Wonderful Life: The Burgess Shale and the Nature of History. Norton, New York.

Gray, J. 1984. Ordovician–Silurian land plants: The interdependence of ecology and evolution. Palaeontol. Assn. Special Pap. Palaeontol. 32:281–295.

———. 1988. Land plant spores and the Ordovician–Silurian boundary. Bull. Brit. Mus. (Natur. Hist.), Geol. 43:351–358.

Gray, J., and W. Shear. 1992. Early life on land. Amer. Sci. 80:444–456.

Halstead, L. B. 1968. The Pattern of Vertebrate Evolution. W. H. Freeman, San Francisco.

———. 1978. The Evolution of the Mammals. Eurobook, London.

Harland, W. B., R. L. Armstrong, A. V. Cox, L. E. Craig, A. G. Smith, and D. G. Smith. 1990. A Geologic Time Scale 1989. Cambridge Univ. Press, Cambridge, England.

Harland, W. B., A. V. Cox, P. G. Llewellyn, C. A. G. Pickton, A. G. Smith, and R. Walters. 1982. A Geologic Time Scale. Cambridge Univ. Press, Cambridge, England.

Hawksworth, D. L. 1991. The fungal dimension of biodiversity: Magnitude, significance, and conservation. Mycol. Res. 95:641–655.

Hickman, C. P., L. S. Roberts, and F. M. Hickman. 1988. Integrated Principles of Zoology. 8th ed. Times Mirror/Mosby College Publishing, St. Louis.

———. 1990. Biology of Animals. 5th ed. Times Mirror/Mosby College Publishing, St. Louis.

Lake, J. A. 1990. Origin of the metazoa. Proc. Natl. Acad. Sci. USA 87:763–766.

Leadbeater, B. S. C. 1983. Observations on the life history and ultrastructure of the marine choanoflagellate Proterospongia choanojuncta. J. Mar. Biol. Assoc., U.K. 63:135–160.

Levinton, J. S. 1992. The big bang of animal evolution. Sci. Amer. 267(5):84–91.

Margulis, L. 1993. Symbiosis in cell evolution: Microbial communities in the Archean and Proterozoic Eons. 2d ed. W. H. Freeman, New York.

Napier, J. 1967. The antiquity of human walking. Sci. Amer. 216(4):56–66.

Pirozynski, K. A., and D. W. Malloch. 1975. The origin of land plants: A matter of mycotrophism. BioSystems 6:153–164.

Poinar, G. O., and R. Singer. 1990. Upper Eocene gilled mushroom from the Dominican Republic. Science 248:1099–1101.

Rayner, A. D. M., C. M. Brasier, and D. Moore (eds.). 1987. Evolutionary Biology of the Fungi. Cambridge Univ. Press, Cambridge, England.

Robison, R. A. 1990. Earliest-known uniramous arthropod. Nature 343:163–164.

Romer, A. S. 1966. Vertebrate Paleontology. 3d ed. Univ. of Chicago Press.

———. 1971. The Vertebrate Story. Univ. of Chicago Press.

Ruppert, E. E., and R. D. Barnes. 1994. Invertebrate Zoology. 6th ed. Saunders College Publishing, Philadelphia.

Sansom, I. J., M. P. Smith, H. A. Armstrong, and M. M. Smith. 1992. Presence of the earliest vertebrate hard tissues in conodonts. Science 256:1308–1311.

Savage, R. J. G., and M. R. Long. 1986. Mammal Evolution: An Illustrated Guide. British Museum (Natural History), London.

Seilacher, A. 1984. Late Precambrian Metazoa: Preservational or real extinctions? pp. 159–168. In H. D. Holland and A. F. Trendall (eds.). Patterns of Change in Earth Evolution. Springer, Berlin.

Sepkoski, J. J., and D. M. Raup. 1986. Periodicity in marine extinction events, pp. 3–36. In D. K. Elliott (ed.). Dynamics of Extinction. Wiley, New York.

Shear, W. A., J. M. Palmer, J. A. Coddington, and P. M. Bonamo. 1989. A Devonian spinneret: Early evidence of spiders and silk use. Science 246:479–481.

Stewart, W. N. 1983. Paleobotany and the Evolution of Plants. Cambridge Univ. Press, Cambridge, England.

Stewart, W. N., and G. W. Rothwell. 1993. Paleobotany and the Evolution of Plants. 2d ed. Cambridge Univ. Press, Cambridge.

Turbeville, J. M., K. G. Field, and R. A. Raff. 1992. Phylogenetic position of Phylum Nemertini, inferred from 18S rRNA sequences: Molecular data as a test of morphological character homology. Mol. Biol. Evol. 9:235–249.

Turbeville, J. M., D. M. Pfeifer, K. G. Field, and R. A. Raff. 1991. The phylogenetic status of arthropods, as inferred from 18S rRNA sequences. Mol. Biol. Evol. 8:669–686.

Whittaker, R. H. 1969. New concepts of kingdoms of organisms. Science 163:150–160.

Whittington, H. B. 1985. The Burgess Shale. Yale Univ. Press, New Haven, Conn.

Woese, C. R. 1984. The Origin of Life. Carolina Biological Supply, Burlington, N.C.

8

ADAPTIVE RADIATION

The soldier fish *(Holocentrus rubrum)*, a species from Indo-Pacific Oceans that colonized the Mediterranean Sea after construction of the Suez Canal was completed in 1869. *(Illustration by Stephen Price)*

8.1 ADAPTIVE RADIATION AND PROGRESSIVE OCCUPATION

Mayr (1963, 663) defined **adaptive radiation** as "evolutionary divergence of members of a single phyletic line into a series of rather different niches or adaptive zones." An adaptive zone is simply a set of similar ecological niches (Futuyma 1986). Earlier, Simpson 1953, 223) had made an admittedly vague distinction between adaptive radiation and **progressive occupation:**

> So far as adaptive radiation can be distinguished from progressive occupation of numerous [adaptive] zones, a phenomenon with which it intergrades, the distinction is that adaptive radiation strictly speaking refers to more or less simultaneous divergence of numerous lines all from much the same ancestral adaptive type in different, also diverging adaptive zones [Figure 8-1]. Progressive occupation of such zones is not simultaneous and usually involves in any one period of time the change of only one or a few lines from one zone to another, with each transition involving a distinctly different ancestral type. Theoretically, at least, the whole of the diversity of life is

explicable by these two not sharply distinct processes plus the factor of geographic isolation which may permit essential duplication of adaptive types by different organisms in different regions.

Therefore, after speciation, the processes of adaptive radiation and progressive occupation are fundamental to our understanding of the increase in organismal diversity and are central to our knowledge of the mechanisms resulting in the richness of life on earth.

Simpson was right in asserting that adaptive radiation and progressive occupation are not clearly distinct processes. In fact, they result largely from scientists' attempts to pigeonhole parts of nature's continuum of change. On what scale do we measure "more or less simultaneous divergence"? Does progressive occupation in real time come to look like adaptive radiation when we collapse the record into geological strata? Questions of rapid or slow rates of change are always debated (as Chapter 18 will discuss). In this chapter, then, while we concentrate on the more celebrated cases of adaptive radiation, keep in mind that similar processes on the continuum from adaptive radiation to progressive occupation have resulted in the rich species assemblages around

FIGURE 8-1

The concepts of adaptive radiation and progressive occupation, representing the extremes of a continuum of patterns occurring in nature. Both processes can result in the same ultimate diversity—in this example, ten species in five adaptive zones—but in adaptive radiation from a recent common ancestor, evolution into these zones is rapid, whereas in progressive occupation with common ancestors much more distant in the phylogeny, evolution into the zones is gradual. Adaptive zones may represent major differences in food resources, such as nectar, seeds, fruits, and insects for birds. The ecological niches occupied by single species may subdivide an adaptive zone on the basis of food size (such as large, medium, or small seeds), accessibility (such as insects under bark, on leaves, or flying), food location (such as seeds on plants versus seeds on the ground), or combinations of these.

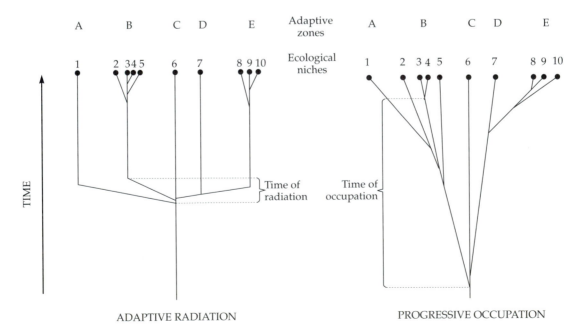

this globe. In addition, although studies of archipelagoes have dominated the literature on adaptive radiation, much more diversification of life has occurred on mainlands and continental shelves, albeit with some habitats distributed in island-like patches.

8.2 GENERALIZATIONS

By 1963 Mayr could present a general consensus on aspects of adaptive radiation:

1. A shift into a new ecological niche is most likely at the edge of the species range because of the increased chances of genetic reconstruction in peripheral populations. Also, in these suboptimal habitats, a shift in ecology may match the needs of new genotypes better than those of old ones.
2. Lack of competitors facilitates adaptive radiation. A population finds it difficult or impossible to enter a new niche if that niche is already occupied by another species. Even on continents and in seas, however, there is evidence of vacant niches in rich biotas. The impressive success of transfers of animals and plants affirms this point (Elton 1958, Drake et al. 1989). The destruction of faunal barriers also indicates the presence of empty niches. After the Suez Canal was opened in 1869, Red Sea fish species colonized the Mediterranean, and some became abundant—presumably filling vacant niches, because decreases in abundance of Mediterranean species were not evident. No Mediterranean fish colonized the Red Sea (Kosswig 1950). By 1971, the number of fish species along the Mediterranean coast of Israel that were derived from Red Sea species had increased to 30, or 10.6 percent of the fauna (Ben-Tuvia 1971) (Figure 8-2). For radiations to be extensive, then, it appears that not only must empty niches be available, but extensive sets of adaptive zones must be free for colonization.

FIGURE 8-2

Some fish species that have invaded the Mediterranean Sea from the Red Sea via the Suez Canal, listed in Ben-Tuvia (1971): *Hemirhamphus far* (**top left**), *Tylosurus choram* (**top right**), *Siganus rivulatus* (**bottom right**), *Holocentrus rubrum* (**bottom left**), and *Epinephelus tauvina* (**background center**). What has made these species successful colonizers? Did they occupy vacant niches in the Mediterranean? If so, why did Mediterranean groups not radiate into these niches before the Suez Canal was opened? Is competition really an important force preventing radiation, if so, how many vacant niches persist in communities? *(Illustration by Stephen Price)*

3. Islands provide many empty niches, but continents provide few. Empty niches are particularly numerous on islands with much habitat heterogeneity. The Hawaiian honeycreepers (Drepanididae) were early colonists of the Hawaiian Islands and evolved into rather varied adaptive zones that on the mainland would typically have been occupied by finches, honey-eaters, creepers, and woodpeckers (Amadon 1950). Later colonists such as thrushes, flycatchers, and honey-eaters did not radiate to the same extent even though they were preadapted for exploiting ecological niches occupied by drepaniids—presumably because of competitive exclusion in evolutionary time. Mayr noted that insects on the Hawaiian Islands have patterns similar to those of the drepaniids (Zimmerman 1948), as do Darwin's finches on the Galapagos Islands (Lack 1947).

4. Predators may prevent radiation; specifically, they may prevent colonization of empty niches by maladapted colonists. The frequency of successful shifts into radically new niches on oceanic archipelagoes (such as the Galapagos and the Hawaiian Islands) in the *absence* of predators supports this generalization.

5. The occupation of a new adaptive zone, which Mayr (1963, 617) considered to be a "breakthrough," is followed by two events. (A) The rate of evolution and speciation increases radically, and a period of adaptive radiation, which in retrospect appears to be synchronous, ensues. (B) Some major types develop, many minor evolutionary pathways are exploited. Examples of event A include the evolution of bats (Chiroptera) in the Paleocene, but the first fossils were basically modern-style bats. Also, the lungfishes in the middle and late Devonian evolved more than twice as much in about 30 million years as they did in the following 250 million years (Westoll 1949) (Figure 8-3). About 75 percent of new characteristics in the lungfishes had been expressed by the end of the Devonian, and only about 20 percent of new characteristics have evolved in the last 250 million years. Tertiary birds are all of modern lineages, and so most of their radiation must have taken place in the late Jurassic to early Cretaceous.

6. Speciation and opportunities for adaptive radiation define the size of taxonomic categories.

> If a group speciates actively . . . , without much adaptive radiation, it will have many species per genus. If a group radiates actively . . . without much speciation, it will develop many monotypic genera and even families. The independence of these two processes is the main reason for the so-called "hollow curve" of taxonomists. (Mayr 1963, 619)

FIGURE 8-3

The rate of change in characteristics of lungfishes, and the rate of change in genera from the early Devonian to the present. Note the rapid origination and extinction of new genera (bars indicate the time ranges of main genera) that accompanied the rapid acquisition of new characteristics (sigmoid curve) from 300 to 280 million years ago, followed by 250 million years of long-lived genera and the development of relatively few new characteristics. (*After Westoll 1949*)

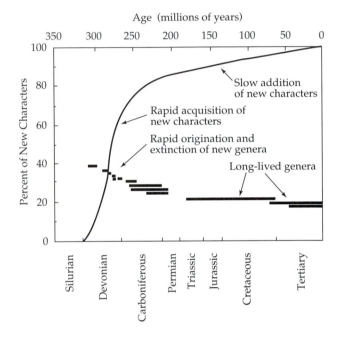

If Mayr's analysis is correct, then perhaps we can glimpse the relative contributions of adaptive radiation and progressive occupation. At the left of the hollow curve, much diversification may result from progressive occupation, whereas much of the taxonomic diversity in the right-hand tail of the distribution may be a consequence of adaptive radiation (Figure 8-4). Alternatively, patterns in lineage diversification, if assumed to be generated by chance events of speciation and extinction, would yield similar ranges of taxon size (see Chapter 11, Sections 11.1 and 11.2).

8.3 EXAMPLES OF RADIATIONS FROM THE FOSSIL RECORD

Ammonite Phylogeny and Radiation

The empirical data from the fossil record certainly support the patterns Mayr (1963) recognized, but whether the mechanisms are essentially correct is

FIGURE 8-4

The commonly observed hollow curve of distribution of lower taxa in a higher taxon. Note that the concave, or hollow, distribution always results from high taxon riches in a small number of the largest groups followed by a rapid decline to more numerous taxa with low taxonomic richness. These examples illustrate the number of species per phylum with the kingdom Animalia (**top left**), in which the arthropods are the largest taxon; the number of familes per order within the fishes (**bottom left**), in which the perches predominate; the number of species per family within the Monocotyledonae (**top right**), with orchids as the largest group; and the number of families per suborder within the Australian beetles (**bottom right**), in which the Polyphaga is the largest group. The organism depicted in each distribution is a member of the largest taxon in the higher taxonomic unit. *(From Dial and Marzluff 1989)*

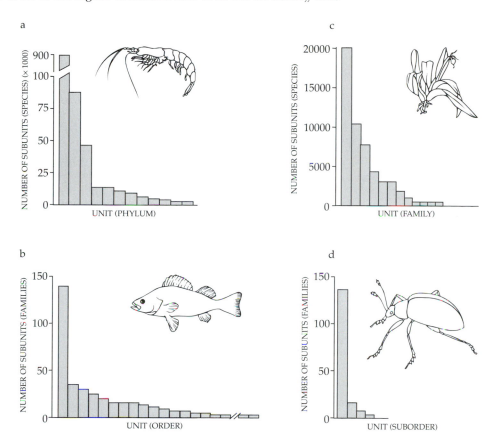

open to some debate. Let us explore ammonite phylogeny and radiation (Lehmann 1981) as an example. The ammonites were a group of extremely numerous shelled cephalopod molluscs; Chapter 9 discusses their extinctions. Ammonite shells have fossilized well, leaving a rich record of adaptive radiation and, eventually, complete extinction. The class Cephalopoda is divided into two infraclasses, Ectocochlia and Endocochlia. The subclasses within the Ectocochlia include the following:

Subclasses:

1. Nautiloidea
2. Endoceratoidea
3. Actinoceratoidea
4. Bactritoidea
5. Ammonoidea

Orders:

(a) Anarcestida ⎫
(b) Clymeniida ⎬ "Goniatites"
(c) Goniatitida ⎭ "Palaeoammonoidea"

(d) Prolecantida ⎫ "Ceratites"
(e) Ceratitida ⎬ "Mesoammonoidea"

(f) Phylloceratida ⎫
(g) Lytoceritida ⎬ "Ammonites"
(h) Ammonitida ⎮ "Neoammonoidea"
(i) Ancyloceratida ⎭

The phylogeny of Ammonoidea shows that, during each radiation after a major extinction episode, it was often a new order that radiated from a rather small, unimpressive representation of species in an earlier period (Figure 8-5). In the Devonian, the Goniatitida radiated extensively while the Prolecanitida

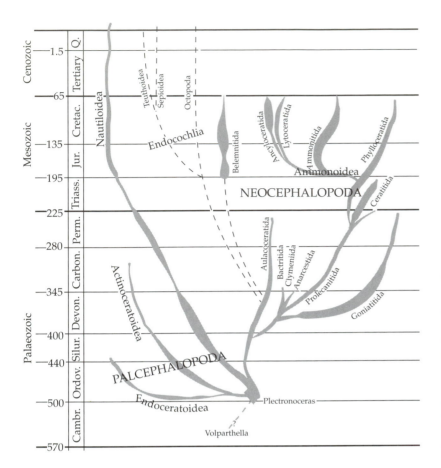

FIGURE 8-5

Adaptive radiation of ammonites, showing successive radiation events from rather unimpressive lineages. The thickness of the line indicates the relative richness of taxa in each lineage through time. Note, for example, that the Goniatitida was the richest taxon in the Devonian, but the comparatively poor relatives in the Prolecanitida provided the stock for major radiations in the Jurassic. (*After Lehmann 1981 and Moore 1957*)

remained a relatively small group. At the end of the Permian, it was the Prolecanitida that radiated while the Goniatitida went extinct. Then the Ceratitida radiated in the Triassic but went extinct, and another lineage underwent a massive and rapid radiation in the early Jurassic to produce the four orders of the Neoammonoidea.

One might easily infer from this pattern that the radiation of one group precludes the radiation of another, just as Mayr explained. Competitive interactions set a carrying capacity on taxa and limit radiations to a small subset of their potential expansions. But this argument is hard to reconcile with the pattern of abundance of ammonoid genera from the Devonian to the Cretaceous (Figure 8-6). There is no sign of any equilibrium number of genera being reached, and about 60-plus genera were extant for continuous periods of millions of years. Thus, the following questions arise. (1) Why did no new radiations occur in the Carboniferous when 30 or fewer genera were present? (2) If 100 genera can be supported for periods of time, why did only one taxon, the Ammonitida, radiate extensively in the early Jurassic, when fewer than ten genera survived the Triassic–Jurassic extinc-

tions? (3) How could the Ammonitida suppress radiation of other groups when diversity of genera was below 100 for most of the Jurassic and yet 140 to 160 genera coexisted at other times? (4) How could new genera reach a peak of almost 160 in the Albian, just when continuous genera were at their maximum of about 35? If one considers only the ammonoids, the evidence for some kind of carrying capacity for genera is not impressive. The record for the ammonoids appears to exhibit a strongly stochastic element involved with radiation and extinction events. Alternatively, we must consider the role of other taxa in competition for habitat and resources, the role of extinction of other groups (Chapter 9), the nature of survival through extinction episodes (Chapter 10), and insights about chance or stochastic events in the fossil record gleaned from models (Chapter 11).

Other patterns in the ammonites agree with Mayr's generalizations.

1. At any one time, a small number of large taxonomic groupings and a large group of smaller taxa are producing the hollow-curve distribution of taxa (Figure 8-5).

FIGURE 8-6

The number of genera of ammonoids through the fossil record. Note that all extant ammonites went extinct at the end of the Cretaceous. Hatching indicates continuous genera, and the more erratic stippling represents the record of new genera. *(From Lehmann 1981 and Moore 1957)*

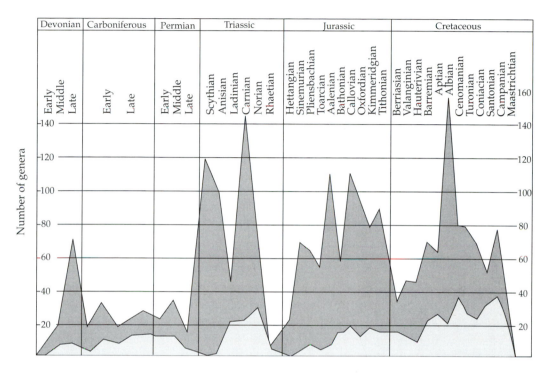

2. Adaptive radiation was rapid after an extinction episode, as seen in the heteromorph ammonites after the Jurassic–Cretaceous and Albian extinctions (Figure 8-7).
3. The opportunities for radiation coincided with the availability of vast new expanses of continental shelf for colonization.

Wiedmann (1973) argued that, as sea level rose and fell, up onto and then off continental shelves, the taxa of ammonites also waxed and waned (Figure 8-8). Deep-water species with smooth shells were largely responsible for maintaining the continuity of the ammonites. When continental shelves became available again, during transgression of seas, species radiated from the deep-basin forms, became progressively more ribbed, and went extinct during the next regression of seas.

Echinoid Radiation

Other taxa exhibit patterns similar to those of the ammonites, with large radiations from a single lineage and usually with no obvious superiority over other lineages, which either remain small or go extinct. For example, the echinoids (including sea urchins and sand dollars) originated in the Ordovician and underwent a small radiation through the Silurian and Devonian, but all except one lineage went extinct by the end of the Permian (Smith 1984) (Figure 8-9). The major echinoid groups radiated rapidly in the Triassic and early Jurassic and have largely persisted into the present with no further radiation equivalent in magnitude to the Triassic–Jurassic radiation. In the echinoids there is an apparent plateau in the number of species per million years; about 900 species are extant (Smith 1984). In total, there have been 124 Paleozoic species, 3,672 Mesozoic species, and 3,250 Cenozoic species. It is interesting to note the relative stability of the echinoids after the Jurassic in contrast with the wide fluctuations of the ammonoids (Figure 8-10).

Radiation of Brachiopods and Foraminiferans

The brachiopods, or "lampshells" (like early Roman oil lamps), also exhibit patterns of one lineage radiating while others are almost static (e.g., early Devonian mutationellinid brachiopods; Boucot 1975). Foraminiferans are protozoans, related to the amebas, that secrete a shell-like calcareous test. They have left a

FIGURE 8-7

Rapid adaptive radiation of heteromorph ammonites (those with unconventional coiling) after the extensive extinctions around the Jurassic–Cretaceous boundary. *(From Lehmann 1981 and Wiedmann 1969)*

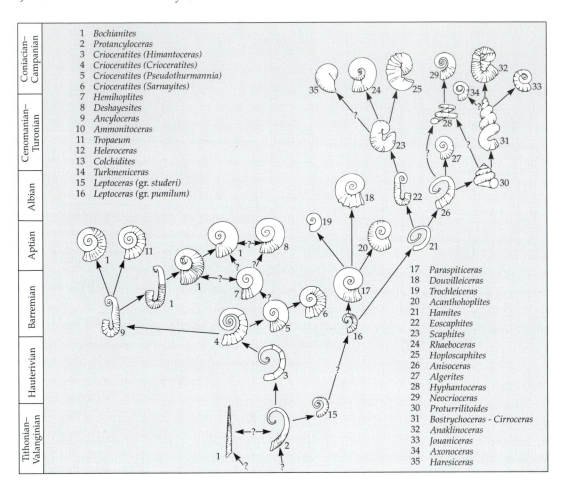

1	*Bochianites*
2	*Protancyloceras*
3	*Crioceratites (Himantoceras)*
4	*Crioceratites (Crioceratites)*
5	*Crioceratites (Pseudothurmannia)*
6	*Crioceratites (Sarnayites)*
7	*Hemihoplites*
8	*Deshayesites*
9	*Ancyloceras*
10	*Ammonitoceras*
11	*Tropaeum*
12	*Heleroceras*
13	*Colchidites*
14	*Turkmeniceras*
15	*Leptoceras (gr. studeri)*
16	*Leptoceras (gr. pumilum)*

17	*Paraspiticeras*
18	*Douvilleiceras*
19	*Trochleiceras*
20	*Acanthohoplites*
21	*Hamites*
22	*Eoscaphites*
23	*Scaphites*
24	*Rhaeboceras*
25	*Hoploscaphites*
26	*Anisoceras*
27	*Algerites*
28	*Hyphantoceras*
29	*Neocrioceras*
30	*Proturrilitoides*
31	*Bostrychoceras - Cirroceras*
32	*Anaklinoceras*
33	*Jouaniceras*
34	*Axonoceras*
35	*Haresiceras*

very rich fossil record of tests, usually made of calcium carbonate. In the Foraminifera, two genera, *Globigerina* and *Globorotalia*, radiated rapidly into several subgenera in the Late Oligocene and Early Miocene while others, such as *Cassigerinella* and *Globoquadrina* (Figure 8-11), did not radiate at all (Kennett and Srinivasan 1983). The fossil record of these groups reinforces the perception that survival through episodes of extinction and subsequent radiation contain a strong element of either chance or as-yet-undetected governing forces.

8.4 ARE RADIATIONS PREDICTABLE?

Gould (1986) broached the question of whether there is anything predictable in the kinds of taxa that radiate and the kinds that do not. If we rewound the tape of life and played it again, would the results be the same? If they were, we would deduce a strongly de-

terministic evolutionary sequence and would find ourselves in our present shape once again. But what if the replay produced a totally different assemblage? From this result we could assume one of two alternatives to determinism: (1) chaotic evolutionary sequences or lineages, i.e., a random series of radiations and extinctions, or (2) evolutionary sequences that are unpredictable but, in retrospect, rational. Alternative pathways to dominant groups were not clearly inferior to the lineages that radiated. Life could have evolved, at any phase in adaptive radiation, in several alternative directions.

The Burgess Shale Fauna: Wiwaxiids and Clams

Gould used three scenarios to support the last-mentioned possibility: unpredictable results with rational explanations. During the great Cambrian explosion of early metazoans more than 500 million years ago,

FIGURE 8-8

The movement of seas over continental shelves (transgressions) and the corresponding adaptive radiation of ammonites derived from long-lived deep-basin species, followed by lowering of sea levels, regression of seas off the continental shelves, and extinction of all but the deep-basin species. In each phase of radiation, the newly available shelf must be similar to a remote archipelago on which many vacant niches are available to new colonists. Shading indicates the marine environment that is continuous in the deep basins but increasingly interrupted from the outer to the inner continental shelf. *(From Wiedmann 1973)*

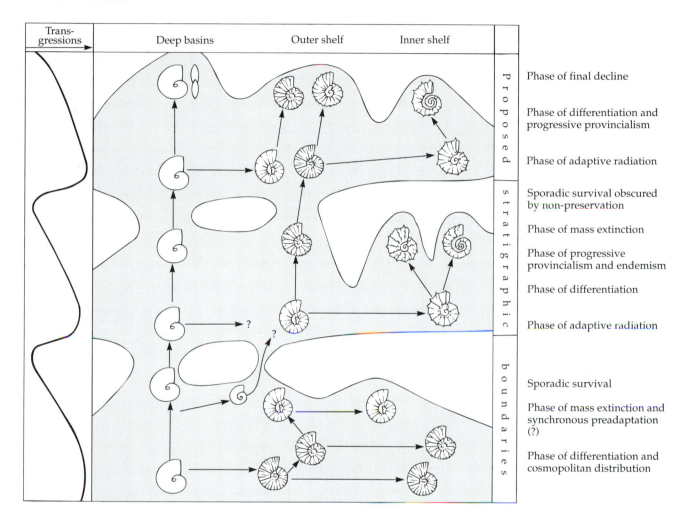

some extraordinary creatures were present that ended up preserved in the Burgess Shale (Whittington 1985). *Wiwaxia* was a genus of worm-like organisms covered with plates, or sclerites, and two rows of spines (Figure 8-12) (Conway Morris 1985, Gould 1985, 1989). *Opabinia* had five large, stalked eyes and, extending beyond them, a highly mobile tube fringed at its end by vicious teeth (Figure 8-12). To modern eyes, *Hallucigenia* is a strange misfit; at first sight it seems to have walked on seven pairs of spines and had tentacles on its upper surface; but perhaps it stood the other way up (Figure 8-12) (see also Briggs and Whittington 1985)? (Such might be the hallucination of a tired paleontologist sifting through piles of

rock.) Further examination of these fossils indicates locomotory function of the paired "tentacles" and presumably a defensive role for the spines, as depicted in Figure 8-12 (Ramsköld and Xianguang 1991, Ramsköld 1992). These authors have proposed a possible affinity with the onychophorans.

Some taxa present in the Cambrian are readily linked to modern forms, such as the clams and snails. However, whereas wiwaxiids may have had six genera in the Cambrian, there were only two genera of small clams. Therefore, had we been able to observe the Burgess community and predict the future radiations of animals, we would almost certainly have expected wiwaxiids to radiate and clams to remain

FIGURE 8-9

The pattern of echinoid radiations and extinctions. *(From Smith 1984)*

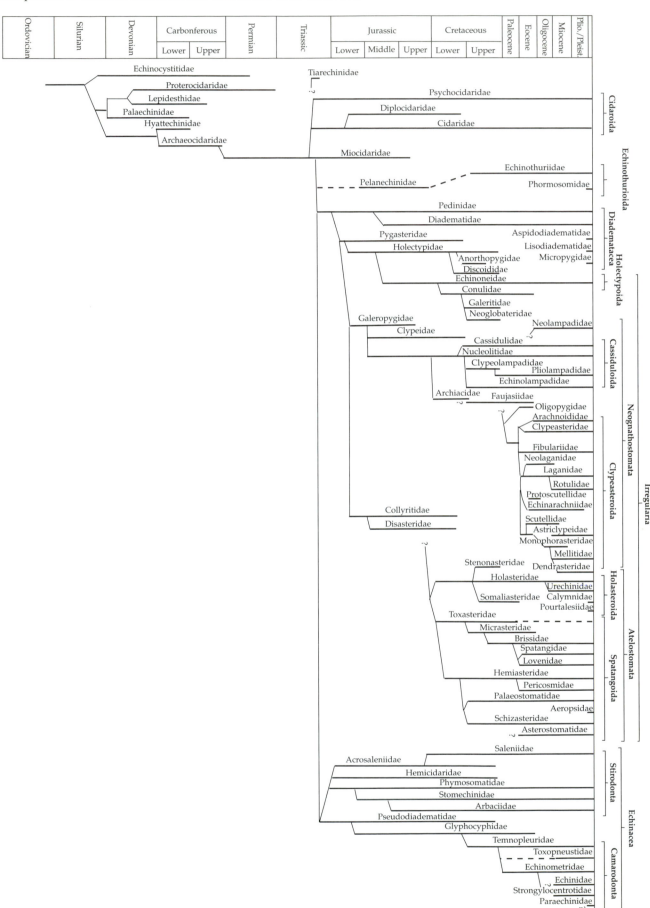

FIGURE 8-10

The number of species of echinoids per million years. *(From Smith 1984)*

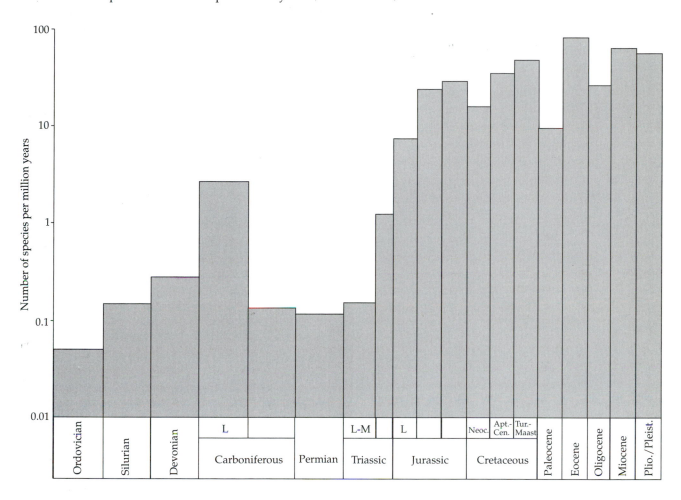

depauperate in species or go extinct. Quite the reverse took place, for clams underwent a major radiation (although very much later, in the Triassic), and the wiwaxiids went extinct without leaving evidence for a rational explanation.

The Burgess Shale Fauna: Priapulids and Polychaetes

Gould's second scenario concerning Burgess Shale organisms compares the polychaete worms and the priapulids, or "little penis worms." In the Burgess fauna the two were equally represented in number of genera, but the priapulids were abundant and the polychaetes were rare (Table 8-1). And yet it was the polychaetes that radiated spectacularly, while priapulids became a very insignificant component of any fauna. No convincing reason for this turn of events is available.

Predaceous Birds or Mammals?

The third scenario suggests even more convincingly that, at any time of radiation, alternative pathways are equally viable, at least for a few million years. When the large carnivorous dinosaurs died out in the Cretaceous, there was clearly room for a radiation of large carnivores from a lineage that had survived the Cretaceous–Tertiary extinctions. Four different taxa radiated, two in the north and two in the south. In the north, the canids and felids radiated, as well as huge predaceous birds represented by *Diatryma gigantea*, found in Wyoming Eocene deposits (Figure 8-13).

The appearance of this great bird at a time when mammals were, for the most part, of very small size (the contemporary horse was the size of a fox terrier) suggests some interesting possibilities—which never materialized. The great reptiles had died off, and

FIGURE 8-11

Planktonic foraminiferans in groups that radiated and groups that did not radiate in the late Oligocene and early Miocene. Why did some genera radiate while others remained narrow phylogenetic lineages? *(From Kennett and Srinivasan 1983)*

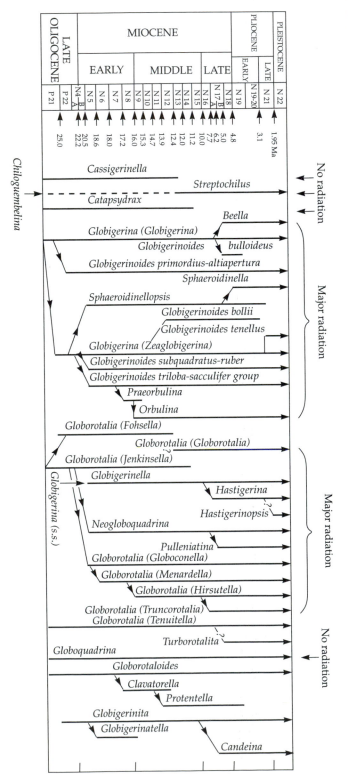

the surface of the earth was open for conquest. As possible successors there were the mammals and the birds. The former succeeded in the conquest, but the appearance of such a form as *Diatryma* shows that the birds were, at the beginning, rivals of the mammals. (Romer 1945, 270–271)

Gould points out that the "tape" ran again in isolation in South America, and in this case the giant ground birds, or phororhacids, probably did dominate for a while (Figure 8-14). There the potential mammal contenders were marsupials—the borhyaenids. Simpson (1980, 147, 150) explains:

It has sometimes been said that these and other flightless South American birds . . . survived because there were long no placental carnivores on that continent. That speculation is far from convincing. Rheas still survive although there have been placental carnivores in their communities for at least 2 million years. Most of the phororhacids became extinct before, only a straggler or two after, placental carnivores reached South America. Many of the borhyaenids that lived among these birds for many millions of years were highly predaceous. . . . The phororhacids . . . were more likely to kill than to be killed by mammals.

Conway Morris, Simpson, and Gould make a convincing case that the radiation of groups is largely unpredictable. Perhaps the best we can do is to predict that, when a large adaptive zone becomes available, some taxon or other will radiate into it.

8.5 TYPES OF ADAPTIVE RADIATION

Despite the case for unpredictability, some general patterns have emerged in the mechanisms involved with adaptive radiation. If we define an **adaptive zone** as "a set of ecological niches that may be occupied by a guild of organisms" and a **guild** as "a group of species that exploits the same resources in a similar manner" (Root 1967), we can identify the following types of adaptive radiation.

Type 1: General Adaptation

A new general adaptation opens up new adaptive zones. One example is the evolution of flight in birds, which opened up possibilities for utilization of prey in the air; colonization of islands, cliff faces, and tall trees; and extensive migration. The result has been about 8,600 extant species. Bats have radiated to about 900 extant species.

Another possible example is the bark lice (Psocoptera), which colonized birds, became parasitic (Mallophaga), and radiated across bird species and into different microhabitats on the bird's body, with some 1,500 species of bird lice (Philopteridae) now described (Price 1980). In a similar way, agromyzid

TABLE 8-1   Change in relative abundance of little penis worms (priapulids) and polychaete worms

	Priapulids	Polychaetes
Burgess fauna	6 genera	6 genera
Abundance in Burgess	Very abundant	Much rarer
Present world fauna	9 species	87 families, 1,000 genera, 8,000 species
Habitats	Live in unusually harsh environments	Sea floor, brackish and fresh water, and moist soil in terrestrial habitats

FIGURE 8-12

Some of the strange organisms from 500 to 600 million years ago found in the Burgess Shale. Some animals do not fit any currently recognized phyla: *Hallucigenia sparsa* (**bottom left**), *Opabinia regalis* (**center**), and *Wiwaxia corrugata* (**top center**). *Aysheaia* (**left center**, on sponge), *Sidneyia* (**right middle ground**), and *Leanchoilia* (**top right**) are arthropods of unknown affinity. *Pirania* (**bottom center**) and *Vauxia* (**top center**, two species) belong in the phylum Porifera. *Ottoia* (**bottom right**) is a priapulid worm. (Wiwaxia *from Conway Morris 1985; other species from Whittington 1985) (Illustration by Tad Theimer)*

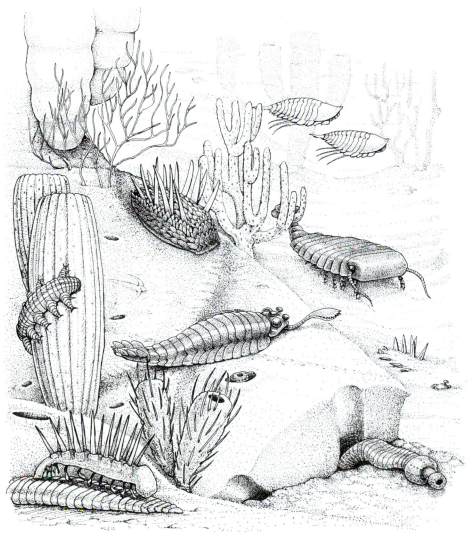

FIGURE 8-13

A reconstruction of *Diatryma* based on skeletons found in North America. This bird reached up to 9 feet in height and lived in Eocene grasslands. It is pictured chasing an oreodont artiodactyl, whose short legs were hardly a match for the great bird's speed. *(Based on Špinar 1972) (Illustration by Tad Theimer)*

FIGURE 8-14

The large predaceous bird *Phororhacos* as it may have looked during the Tertiary in South America. *(Based on Špinar 1972) (Illustration by Stephen Price)*

flies acquired the general adaptation of mining and feeding in the leaf tissue of angiosperms. They radiated across plant species and into different plant parts such as leaves, seeds, and stems—the adaptive zones that guilds of flies may occupy.

These examples represent major radiations on continents.

Type 2: Environmental Change

Environmental change opens new adaptive zones. The ammonites should fulfill this principle, for they radiate as newly flooded continental shelves opened up new habitats. Likewise, as the savannas opened and developed with climatic drying, the ruminants radiated. As the climate in South America became wetter, tropical forests spread from small refugia, or expanded extensively from centers of distribution, and many plant and animal species radiated.

Type 3: Archipelagoes

We may regard remote, geographically isolated locales with heterogeneity as archipelagoes, whether they are oceanic island groups or habitat islands, such as mountaintops in a sea of desert or rock patches in a large freshwater lake. On an archipelago, adaptive zones remain open in places remote from potential colonists, and the first colonists radiate. This type of adaptive radiation encompasses the classic cases that have received the most attention: the honeycreepers on the Hawaiian Islands (Amadon 1947, 1950) (Figure 8-15); Darwin's finches on the Galapagos Islands (Lack 1947); and the genus *Drosophila* (Carson et al. 1970, Ringo 1977, Templeton 1979), the beetles (Zimmerman 1948), and the silverswords (Carlquist 1974, Carr and Kyhos 1986, Carr et al. 1989) on the Hawaiian Islands.

Hawaiian Honeycreepers

Amadon (1950) emphasized the following processes in the radiation of the honeycreepers.

1. Quantum evolution or rapid divergent evolution
2. Isolated biotas with many families absent
3. Plentiful empty ecological niches
4. Absence of competing species outside the radiating taxa
5. Importance of environmental heterogeneity: "The Drepaniidae could not have evolved on the bleak Aleutian Islands" (241)
6. Depauperate predators and parasites that do not "inhibit adaptive radiation by eliminating variant individuals" (241)
7. Importance of competition between radiating species and the resultant character displacement: "On

the islands where but one species of this group (*Loxops virens*) occurs, it has generalized feeding habits. On Kauai where two species are present as a result of a double invasion, they have, so to speak, divided the ecological niche of the parental species between them (and extended it)" (246). *Loxops virens stejnegeri* feeds on insects beneath bark while *Loxops parva* forages on leaves and shallow flowers. "Thus divergence and incipient specialization have resulted from competition between two similar species" (246).

Lack's (1947) scenario for Darwin's finches was essentially the same.

Many habitat archipelagoes occur on continents; these include mountaintops (Brown 1971, Vuilleumier 1970), caves (Vuilleumier 1973, Culver 1982), refuges of vegetation such as tropical forest in Amazonia during the Pleistocene (Prance 1981), and patchy distributions of hosts for parasitic species (Price 1980). Continental archipelagoes have received less attention than oceanic archipelagoes in considerations of adaptive radiation even though they are the sites of much more diversification of species.

Hawaiian Silverswords

Relatively recent study of the plant group known as the Hawaiian silverswords suggests a complex array of interacting factors in the process of adaptive radiation. The original tarweed progenitor from North America colonized Kauai less than 6 million years ago, and adaptive radiation has resulted in 28 species in three genera: *Argyroxiphium*, *Dubautia*, and *Wilkesia*. Radiation has occurred particularly in growth form and habitat exploitation. The group includes cushion plants, rosette shrubs, mat-forming woody plants, erect shrubs, trees, and lianas. Species occur in almost all habitats on the Hawaiian Islands, from sea level to 3,750 meters, and occupy places with less than 400 millimeters (about 16 inches) of rainfall per year as well as some of the wettest places in terrestrial environments on earth, with 12,300 millimeters (about 500 inches) of rainfall (Carr et al. 1989). Figure 8-16 illustrates the diversity of leaf forms among the 28 species and one of the most spectacular species.

The mechanisms involved with adaptive radiation in the silverswords probably include interactions among the following factors (Carr et al. 1989).

1. Great diversity of habitats existed within the islands, with close local heterogeneity, as Amadon emphasized.
2. Allopatry and ecological isolation occurred between and within islands.

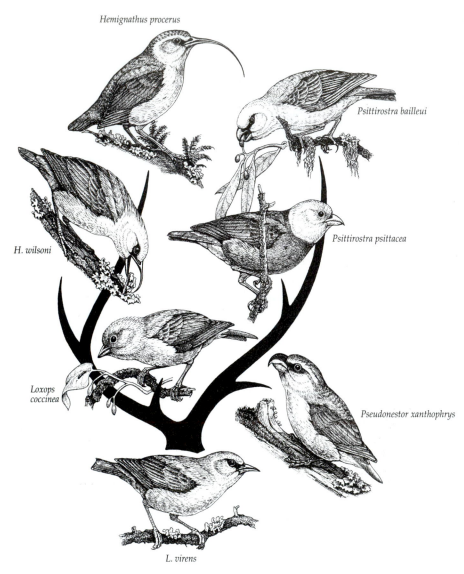

Hemignathus procerus

Psittirostra bailleui

Psittirostra psittacea

H. wilsoni

Loxops coccinea

Pseudonestor xanthophrys

L. virens

FIGURE 8-15
Adaptive radiation of the Hawaiian honeycreepers in the family Drepanididae, exhibiting great diversification in beak morphology and feeding habits. *(From Lewin 1982) (Illustration by Tad Theimer)*

3. Reciprocal chromosomal translocations produced reproductive isolation.
4. Major adaptive shifts occurred when long-distance dispersal between islands resulted in founder effects.
5. Novel physiological traits enabled radiation into new habitats; for example, an aneuploid with 13 pairs of chromosomes instead of the ancestral 14 extended the silversword lineage into much drier sites, although the detailed mechanisms are not understood.
6. Self-incompatibility, outcrossing, and hybridization between genotypes adapted to different conditions may have accelerated the radiation process.
7. Volcanic activity repeatedly opened new areas for colonization by novel hybrids.

Using Carson's (1983, 1984) overviews on *Drosophila*, Carr et al. (1989) noted the similarities between the adaptive radiation of the silverswords and that of the picture-winged *Drosophila*. The founding events on new islands in the two lineages were similar; great diversity of form has ensued during radiation in both cases; many habitats have been colonized; both groups are obligate outcrossers; and chromosomal evolution has been extensive.

The silverswords "appear to be the most spectacular example of adaptive radiation in the plant kingdom" (Carr et al. 1989, 95). The exciting prospect is

FIGURE 8-16
Above: An example of Haleakala, *(Argyroxiphium sandwicense)*, one of the most striking silverswords, growing on volcanic ash, with an immature plant on the right and a plant in flower on the left *(Illustration by Tad Theimer).* **Below:** Leaf silhouettes of the 28 species of Hawaiian silverswords all at a scale of ×.25, showing the diversity of leaf shape in this radiation. *(From Carr et al. 1989)*

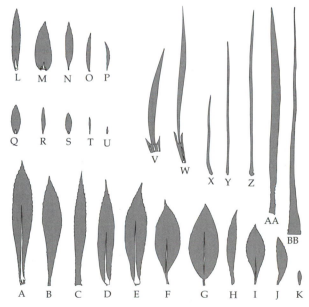

that, with the help of broadly ranging studies focused on variation (in form, morphology, and anatomy), floral morphology and breeding systems, cytogenetics and hybridization, enzyme polymorphism, and adaptive physiology, we will probably come to understand in much greater detail the complexity of interacting factors in adaptive radiation. A detailed mechanistic view will shed new light on the whole diversification of life on earth.

Type 4: Combination of Environmental Change on Archipelagoes

A combination of Types 2 and 3 is necessary to accommodate the fish species flocks in African lakes (McKaye and Gray 1984) (Figures 8-17, 8-18). The lakes change dramatically in water level. When they rise, they flood new areas and open up new habitat. Then, as water levels decline, populations become isolated on a rather local geographic scale and speciate. For example, Lake Nabugabo became cut off from Lake Victoria 4,000 years ago and now harbors five endemic cichlid species (Greenwood 1965).

In addition, for the rock-dwelling cichlids, any lake houses an archipelago of rocky islands, separated by wide expanses of sandy and weedy bottoms, that are available for colonization (McKaye and Gray 1984). Islands are so far apart, and fish so tied to rocky outcrops, that geographic isolation becomes important.

Finally, changes in water level create new rock islands, make some deeper or shallower, or unite formerly isolated populations, alternating habitats for species with shallow-water and deep-water preferences (Figure 8-19). The effect is to change archipelago arrangements and structures quite rapidly, even in a decade. This kind of change opens and closes new sites for colonization and brings new species together, which may then evolve together, resulting in character displacement and finely and narrowly defined ecological niches (Liem 1978, 1980, Liem and Osse 1975).

Very impressive adaptive radiations have resulted from this combination of dynamics. Greenwood (1984) estimated that in the Nile, the source of the colonists of Lake Victoria, 8 percent of fish were cichlids (10 species). In Lake Victoria, however, the cichlids radiated to represent 84 percent of fish species, with about 200 species (Table 8-2). The "cichlid lakes" in Africa contain an amazing diversity of species—some 580 in total.

It is not clear why Lakes Turkana and Albert have remained relatively depauperate in cichlids. Albert may have been part of another drainage system dissociated from the Nile (Greenwood 1974b). Low

FIGURE 8-17

Part of the adaptive radiation in the Lake Victoria genus *Haplochromis*, showing the diversification of feeding ecologies from an *H. bloyeti*-like ancestor. Species with asterisks are illustrated in Figure 8-18. *(From Greenwood 1974a)*

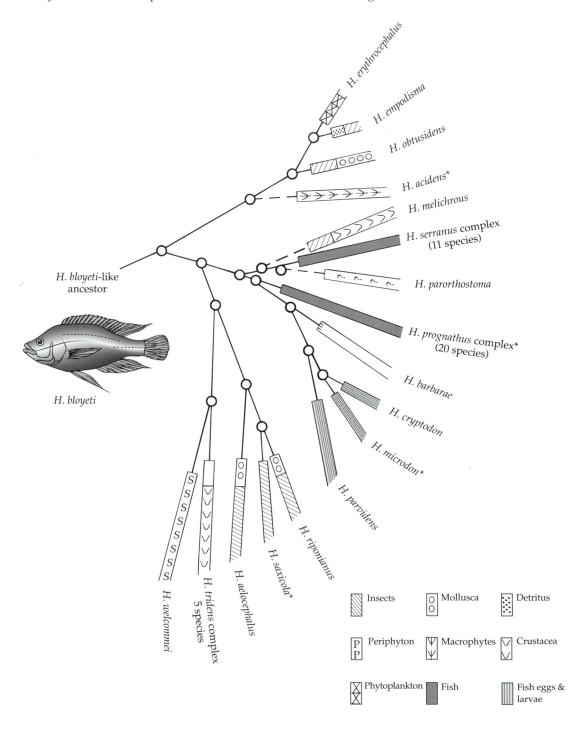

diversity in other lakes, such as Lake Turkana, implicates drying periods in the middle Pleistocene. The whole Great Rift Valley system of East Africa has been so dynamic that geologists are still working out the connections and drainages.

8.6 BEYOND THE ADAPTIVE MODEL OF RADIATION

Amadon (1950) and Mayr (1963) both emphasized the importance of natural selection in adaptive radiation.

FIGURE 8-18

Four species of *Haplochromis* from Lake Victoria, in the lineage depicted in Figure 8-17, showing major diversification in ecology without very dramatic changes in morphology. *H. acidens* (**upper right**), about 6 inches long, is a macrophyte feeder usually found in dense stands of plants; it occupies an almost vacant niche in Lake Victoria. *H. prognathus* (**upper left**) is a fish predator, about 10 inches long, that occupies areas above hard substrates in exposed and sheltered localities. *H. microdon* (**lower left**) is about $5\frac{1}{2}$ inches long; it may feed on fish eggs and larvae and tends to occupy littoral zones over firm or soft substrates. *H. saxicola* (**lower right**) is about $5\frac{1}{2}$ inches long and feeds on insect larvae on the lake bottom. This species may be found on exposed sand or shingle or in dense stands of plants. *(From Greenwood 1959, 1967, 1974a)* *(Illustration by Tad Theimer)*

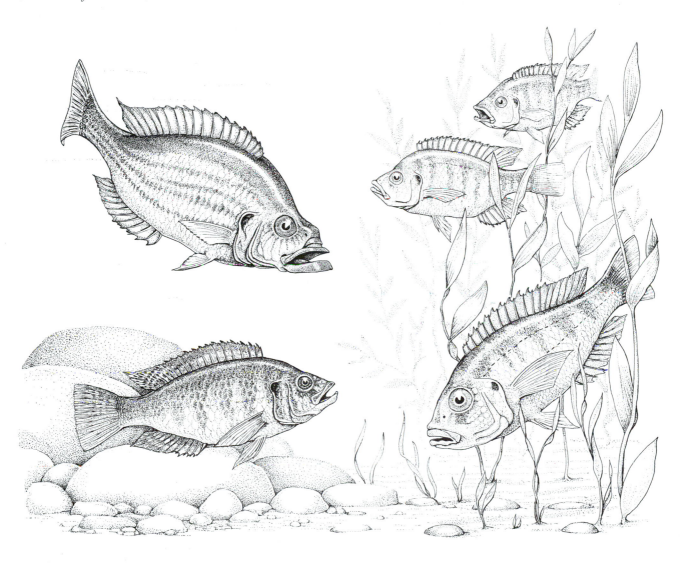

This emphasis may be called the adaptive model of radiation; the term itself implies that selection plays a central role. Other hypotheses have been proposed as well, especially in relation to the impressive radiation of more than 300 species in the genus *Drosophila* on the Hawaiian Islands. That is 25 percent of the genus worldwide, occurring on six small islands. Ringo (1977) and Templeton (1979) summarized four hypotheses.

The Adaptive Model

The adaptive model of radiation refers to the general pattern discussed in Section 8.2, in which mutation and natural selection play the central role in shifts to new niches and adaptive zones (Dobzhansky 1972, 1976). This model does not explain why Hawaiian species of *Drosophila* are so distinct from each other morphologically and behaviorally. For any level of

FIGURE 8-19

The effects of fluctuating lake levels on the division of habitats for cichlid fishes associated with rocks. In each case (a, b, c), lake water levels rise, with the following consequences. (a) Two populations become united. (b) One population is divided into two. (c) A rock providing habitat for shallow-water species becomes deeply submerged, and new rocky shore habitat becomes available for colonization. *(Based on McKaye and Gray 1984)*

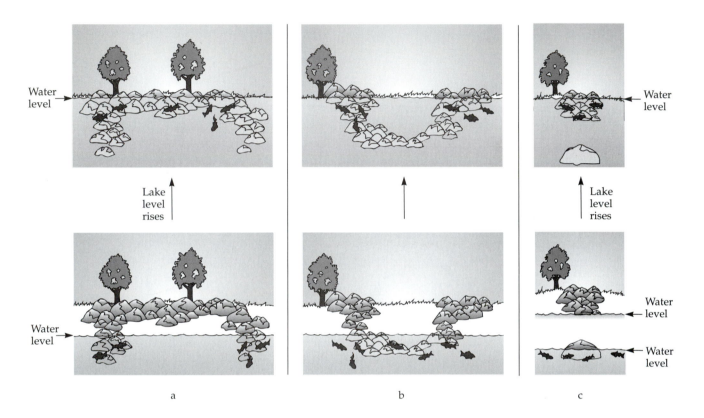

TABLE 8-2 Fish species in African lakes and rivers

Geographic Location[1]	Total Fish Species	Total Cichlid Species	Percent Cichlid Species
Rivers: Zambezi	110	20	18%
Zaire	690	40	6%
Niger	134	10	7%
Nile	115	10	8%
Lakes: Turkana	39	7	18%
Victoria[2]	238	200	84%
Edward[2]	57	40	70%
Albert	46	9	20%
Tanganyika[2]	247	136	55%
Malawi[2]	242	200	83%

Data from Greenwood 1984.

[1] Brackets (not inclusive) connecting rivers and lakes show faunal relationships.

[2] Cichlid lakes.

FIGURE 8-20

Some species of Hawaiian *Drosophila*, showing the diversification of head structure and picturesque wings so different from those of the small mainland species. Species 4 and 5 below are about the size of the common mainland species, *D. melanogaster*. *(Based on Lawrence 1992, and Yoon 1989)*

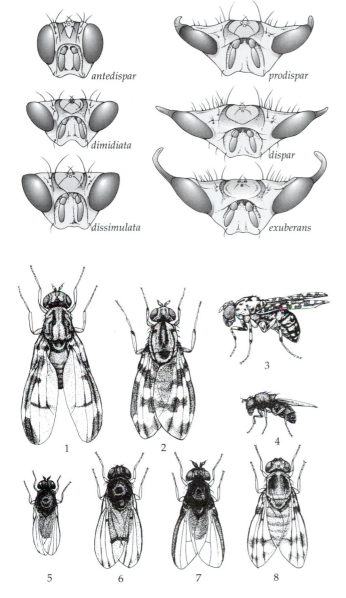

more extreme than those of continental species. Simple adaptation cannot account for these differences between Hawaiian and continental *Drosophila* species.

The Genetic Drift Model

Carson (1970, 1975) proposed that island hopping by founding individuals, followed by genetic drift, could account for the extensive speciation of *Drosophila* on the Hawaiian Islands. However, this process is unlikely to result uniformly in significant phenotypic differences between closely related species. It seems as if character displacement really must play a role beyond the effects of drift.

The Sexual Selection Model

Spieth (1974a, 1974b) argued that predation at oviposition sites (which were conspicuous to the predators) was strong enough to select for lekking behavior away from those sites, and lekking fostered sexual selection between males. Ringo (1977) envisaged a two-step scenario.

A. Speciation starts with colonization of new islands, which produces geographic isolation.
B. "Strong directional intrasexual selection acts as the heritable variation in male reproductive structures and behavior and in female preference, causing rapid changes in the frequencies of alleles responsible both for the preferred behavior and for the preference" (695).

Templeton (1979) criticized Ringo's model, arguing that sexual selection is strongly stabilizing. In place of selection for deviant males, female choice forces male morphology and behavior into a narrowly defined range of acceptable males and thus drives sexual selection.

The Drift Plus Sexual Selection Model

Templeton (1979) argued that both drift and sexual selection are essential in a speciation model, to account for the patterns in genetic variation and behavioral and morphological differences. The actual speciation event occurs as in Carson's genetic drift model. Sexual selection reinforces this speciation event in two ways.

1. "With a founder event, the release of genetic variability in the mate recognition system coincides with a new external selective environment that frees the species from its old constraints and may even actively select for a new mate recognition system—i.e., the sexual selection may become temporarily directional, rather than stabilizing in the founding population. These conditions are optimal

genetic similarity, they are more different phenetically than continental species. In morphology, they have distinct wing patterning, prominent bristles, stalked eyes, and strange tarsi, and the sexes are distinct (Figure 8-20). Behaviorally, in many species males are territorial. They frequently aggregate in leks away from oviposition sites. There they attract females, and copulation occurs. Males also engage in intricate displays and complex courtship patterns, all for causing a rapid reorganization of the mate recognition system that could serve as the basis of speciation" (516).

2. The development of strong premating reproductive isolating mechanisms, and the potential in their rapid shifts through directional selection, reduce the chances of hybridization with sympatric populations and preserve the speciation event.

The important point differentiating Templeton's model from the sexual selection model just discussed is that, during the founding effect, genetic drift causes a shift from the parental adaptive peak, and sexual selection then establishes a new stabilized mating system that may be substantially different from the parental core adaptations (Carson and Templeton 1984).

8.7 SOURCES OF RADIATION IN DOMINANT GROUPS

Scenarios for adaptive radiation in the archipelago-like arenas are fairly well developed. Radiations on large continental landmasses or in the ocean, while contributing in a major way to biotic diversity, have undergone less conceptual scrutiny. Eisenberg's (1981) book *The Mammalian Radiations* provides little that is generalizable, despite its name. Much earlier, Darlington (1959) asked where dominant groups of animals evolved and what factors permitted some groups to expand geographically and radiate. Dominant groups must possess adaptations that are of general utility, i.e., that are adaptive in every environment. Brown (1958) contrasted the general adaptations of dominant species with adaptations specific to a particular set of environmental conditions. General adaptations may increase an animal's efficiency in food utilization, rate of development, or efficiency of locomotion. To occupy a large area, a group must have qualities such as superior competitive ability, effective colonizing ability, and traits enabling exploitation of many different environments. The group must possess numerous general adaptations.

Darlington examined the evidence for the origins of several dominant groups of animals. The family Bovidae is the most recent family of artiodactyls to evolve and has radiated extensively into the cattle, antelopes, sheep, and goats. The Bovidae are most numerous and diverse in Africa, where they probably originated. The earliest fossils of the elephant–mastodont group were found in Egypt, and Simpson (1940) argued that these animals radiated from North Africa. The murid rodents, or Old World rats and mice, are most diverse in the Old World tropics and are dominant almost everywhere they have dispersed. Humans probably also evolved in the Old World tropics.

In fact, evidence suggests that many groups evolved in the Old World tropics, radiated there, and then dispersed to other parts of the globe. Much less movement has occurred in the other direction (Figure 8-21). The mechanisms associated with this pattern, identified by Darlington, are as follows:

1. Dominant groups originate in the largest areas.
2. The largest areas also have the most favorable climates.
3. Because many other species are present, species evolve strong competitive ability, an important general adaptation. They do not evolve to tolerate stressful abiotic environmental conditions such as exist at higher latitudes.
4. In warmer climates, generation time is reduced, and so evolutionary rates are higher than in colder climates.
5. Large areas contain many populations. The larger the area, the more numerous the populations, the greater the maintained genetic diversity, and the greater the evolutionary potential of the species and of the group as a whole (see also Brown 1957 on centrifugal speciation).

FIGURE 8-21

Darlington's view of the major movement patterns of radiating groups of organisms. Large land area in a favorable climate results in many populations per unit area (indicated by large squares) and evolution of new generally adapted species. Such species then disperse, principally in the directions indicated by long arrows, with little dispersal in the opposite direction, indicated by short arrows. Megagea is the landmass composed of Africa in the south, Eurasia, and North America. *(From Darlington 1959)*

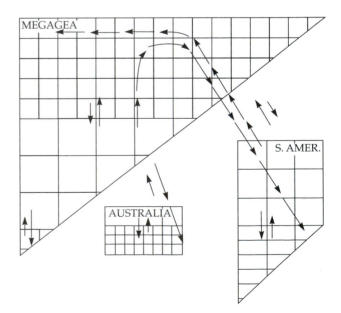

6. Because larger areas contain more species, the probability of species moving from species-rich areas is greater than the probability of the reverse. Recent studies on the faunal relationships across Beringia and across the Isthmus of Panama (Hopkins et al. 1984, Kontrimavichus 1985 on Beringia; Marshall et al. 1982, Stehli and Webb 1985 on the Isthmus of Panama) still support this grand, global pattern, which contrasts with findings of more narrowly focused studies.

   The processes of adaptive radiation have been central issues in the development of evolutionary theory, but much remains to be discovered. What were the driving forces for the fish species flocks in Africa's cichlid lakes? Do we have any power to predict the approximate extent to which adaptive radiation can proceed? Is the process of radiation over large areas, such as mainlands and continental shelves, essentially the same as that on isolated archipelagoes? What really makes Hawaiian *Drosophila* species so different from mainland species? This fascinating subject has captured many an evolutionist's imagination, from Darwin to the present, and continues to be a stimulus for research and debate.

SUMMARY

Adaptive radiation is the relatively rapid evolutionary divergence of members of a single phyletic line, more or less simultaneously, into a series of rather different ecological niches or different sets of similar niches. Generally, such radiation occurs in a new ecological setting, free from competition and natural enemies and often involving impoverished archipelagoes. The fossil record illustrates many such radiations, with small lineages erupting into diverse clades. Ammonites, echinoids, brachiopods, and foraminiferans exemplify the probably unpredictable nature of cladogenesis, with one lineage radiating extensively while others diverge narrowly and slowly, showing a pattern of **progressive occupation.**

   Direct examination of the predictability and repeatability of radiations indicates that alternative pathways are generally viable and outcomes are not foreseeable. Wiwaxiids and clams, and priapulids and polychaete worms, all from the Burgess Shales, illustrate the unpredictability of success in the long-term fossil record. Had humans been able to evaluate the situation after the demise of the predaceous dinosaurs, they would not have anticipated the ultimate predominance of predaceous mammals over large avian predators.

   Adaptive radiation may be classified into four main types.

1. General adaptive breakthroughs may open up new adaptive zones, as occurred in the evolution of wings in lineages leading to birds and bats. Development of the parasitic habitat resulted in greatly enriched chances of radiation across animals and plant host species.
2. Environmental change opened up new adaptive opportunities—for instance, newly flooded continental shelves for ammonite radiation and extensive savannas in which the ruminants radiated.
3. On isolated and therefore depauperate archipelagoes, early colonists may diversify into adaptive zones well beyond the extents of their mainland progenitors. Honeycreepers, *Drosophila* species, and silverswords have all radiated on the Hawaiian Islands, and Darwin's finches on the Galapagos Islands are another classic case. The silversword alliance provides evidence for the complex interplay of factors in adaptive radiation: habitat heterogeneity, allopatric isolation on different islands and ecological isolation within islands, chromosomal translocations, founder effects, aneuploidy, and volcanic activity.
4. A combination of environmental changes and archipelagoes may best explain the impressive radiations of some fishes in the cichlid lakes in the Great Rift Valley system of East Africa. Islands of rock in these large lakes form archipelagoes, and shifts in water level change the configurations of these habitats for many cichlid species, creating an ever-changing set of geographically isolated "islands."

   The special case of radiation in Hawaiian *Drosophila* species raises questions about the adequacy of the adaptive model of radiation. Divergence of morphology and behavior appears to be more extreme than that predicted by the adaptive model, stimulating debate on alternative hypotheses of the process of radiation. The genetic drift model recognizes the potential importance of frequent founding events on new islands and subsequent rapid drift in small populations. The sexual selection model invokes strong selection among males and rapid divergence from the parental type on a new island. A third model combines the elements of drift and sexual selection.

   The dominant taxa that radiate on continents have commonly originated in the Old World tropics, where landmass area is large, climate is favorable, and population sizes and species richness are high.

QUESTIONS FOR DISCUSSION

1. Diversification of species has been much more extensive on large landmasses and in the sea than on oceanic archipelagoes. Do you think that archipelago-like topography on the continents and in the oceans was important or essential in this diversification?

2. The ammonites were an extremely diverse group of organisms, showing several major radiations in the fossil record, but are now extinct. Does this suggest, in your opinion, that groups with explosive evolutionary potential have habitats or biological traits that predispose them to catastrophic declines in diversity and even extinction?

3. Does the evidence for adaptive radiation suggest that, given the right physical conditions, any lineage is likely to radiate, or must the potential for radiation include special biological traits, such as common reciprocal chromosomal translocations in the silverswords?

4. What research would you undertake to convincingly establish the argument that a group has actually undergone adaptive radiation rather than progressive occupation?

5. Do you agree that a special case needs to be made for the adaptive radiation of the genus *Drosophila* on the Hawaiian Islands?

6. Given that the specific examples of adaptive radiation discussed in this chapter come from tropical archipelagoes and lakes, can we infer that most adaptive breakthroughs have occurred in tropical latitudes, and northern latitudes have been populated by novel forms that subsequently dispersed from the tropics?

7. Do you think that the panglossian paradigm is too prominent in arguments for the processes involved with adaptive radiation?

8. If it is true that the silverswords represent a most spectacular case of adaptive radiation in the plant kingdom but number only 28 species, must we conclude that most diversification of plants has involved progressive occupation?

9. Generally speaking, after episodes of adaptive radiation in a lineage on an archipelago, recolonization of the mainland source of the lineage has not occurred. What factors would you argue are important for an understanding of this phenomenon, and how would you partition factors among characteristics of the archipelagoes themselves, traits of the organisms involved, and conditions on the mainland?

10. If you received relevant information on a given set of habitats, such as the Galapagos Islands, and the characteristics of a named invading species, do you think you could predict the course of adaptive radiation in the resulting lineage? Would several independently developed scenarios be similar, in your opinion?

REFERENCES

Amadon, D. 1947. Ecology and the evolution of some Hawaiian birds. Evolution 1:63–68.

———. 1950. The Hawaiian honeycreepers (Aves, Drepaniidae). Bull. Amer. Mus. Nat. Hist. 95:151–262.

Ben-Tuvia, A. 1971. Revised list of the Mediterranean fishes of Israel. Israel J. Zool. 20:1–39.

Boucot, A. J. 1975. Evolution and Extinction Rate Controls. Elsevier Sci. Pub., Amsterdam.

Briggs, D. E. G., and H. B. Whittington. 1985. Terror of the trilobites. Natur. Hist. 94(12):34–39.

Brown, J. H. 1971. Mammals on mountaintops: Nonequilibrium insular biogeography. Amer. Natur. 105:467–478.

Brown, W. L. 1957. Centrifugal speciation. Quart. Rev. Biol. 32:247–277.

———. 1958. General adaptation and evolution. Syst. Zool. 7:157–168.

Carlquist, S. 1974. Island Biology. Columbia Univ. Press, New York.

Carr, G. D., and D. W. Kyhos. 1986. Adaptive radiation in the Hawaiian silversword alliance (Compositae–Madiinae), [Part] II: Cytogenetics of artificial and natural hybrids. Evolution 40:959–976.

Carr, G. D., R. H. Robichaux, M. S. Witter, and D. W. Kyhos. 1989. Adaptive radiation of the Hawaiian silversword alliance (Compositae–Madiinae): A comparison with Hawaiian picture-winged *Drosophila*, pp. 79–97. *In* L. V. Giddings, K. Y. Kaneshiro, and W. W. Anderson (eds.). Genetics, Speciation and the Founder Principle. Oxford Univ. Press, New York.

Carson, H. L. 1970. Chromosome tracers of the origin of species. Science 168:1414–1418.

———. 1975. The genetics of speciation at the diploid level. Amer. Natur. 109:83–92.

———. 1983. Chromosomal sequences and interisland colonizations in Hawaiian *Drosophila*. Genetics 103:465–482.

———. 1984. Speciation and the founder effect on a new oceanic island, pp. 45–54. *In* F. J. Radovsky, P. H. Raven, and S. H. Sohmer (eds.). Biogeography of the Tropical Pacific. B. P. Bishop Museum Special Publication 72.

Carson, H. L., and A. R. Templeton. 1984. Genetic revolutions in relation to speciation phenomena: The founding of new populations. Ann. Rev. Ecol. Syst. 15:97–131.

Carson, H. L., D. E. Hardy, H. T. Spieth, and W. S. Stone. 1970. The evolutionary biology of the Hawaiian Drosophilidae, pp. 437–543. *In* M. K. Hecht and W. C.

Steere (eds.). Essays in Evolution and Genetics in Honor of Theodosius Dobzhansky. Appleton–Century–Crofts, New York.

Conway Morris, S. 1977. Fossil priapulid worms. Special Papers in Paleontology No. 20. Paleontological Assoc., London.

———. 1985. The middle Cambrian metazoan *Wiwaxia corrugata* (Matthew) from the Burgess Shale and Ogygopsis Shale, British Columbia, Canada, pp. 507–582. Phil. Trans. Royal Soc., London B 307.

Culver, D. C. 1982. Cave Life: Evolution and Ecology. Harvard Univ. Press, Cambridge, Mass.

Darlington, P. J. 1959. Area, climate, and evolution. Evolution 13:488–510.

Dial, K. P., and J. M. Marzluff. 1989. Nonrandom diversification within taxonomic assemblages. Syst. Zool. 38:26–37.

Dobzhansky, T. 1972. Species of *Drosophila*. Science 177:664–669.

———. 1976. Organismic and molecular aspects of species formation, pp. 95–105. *In* A. J. Ayala (ed.). Molecular Evolution. Sinauer, Sunderland, Mass.

Drake, J. A., H. A. Mooney, F. di Castri, R. H. Groves, F. J. Kruger, M. Rejmánek, and M. Williamson (eds.). 1989. Biological Invasions: A Global Perspective. Wiley, New York.

Eisenberg, J. F. 1981. The Mammalian Radiations. Univ. of Chicago Press.

Elton, C. S. 1958. The Ecology of Invasions by Animals and Plants. Methuen, London.

Futuyma, D. J. 1986. Evolutionary Biology. 2d ed. Sinauer, Sunderland, Mass.

Gould, S. J. 1985. Treasures in a taxonomic wastebasket. Natur. Hist. 94(12):22–33.

———. 1986. Play it again, Life. Natur. Hist. 95(2):18–26.

———. 1989. Wonderful Life: The Burgess Shale and the Nature of History. Norton, New York.

Greenwood, P. H. 1959. A revision of the Lake Victoria *Haplochromis* species (Pisces, Cichlidae), Part III. Bull. Brit. Mus. Natur. Hist. (Zool.) 5:179–218.

———. 1965. The cichlid fishes of Lake Nabugabo, Uganda. Bull. Brit. Mus. Natur. Hist. (Zool.) 12:315–357.

———. 1967. A revision of the Lake Victoria *Haplochromis* species (Pisces, Cichlidae), Part VI. Bull. Brit. Mus. Natur. Hist. (Zool.) 15:29–119.

———. 1974a. The cichlid fishes of Lake Victoria, East Africa: The biology and evolution of a species flock. Bull. Brit. Mus. Natur. Hist. (Zool.) Suppl. 6:1–134.

———. 1974b. The *Haplochromis* species (Pisces: Cichlidae) of Lake Rudolf, East Africa. Bull. Brit. Mus. Natur. Hist. (Zool.) 27:139–165.

———. 1984. African cichlids and evolutionary theories. pp. 141–154. *In* A. A. Echelle and I. Kornfield (eds.). Evolution of Fish Species Flocks. Univ. of Maine Press, Orono.

Hopkins, et al. (eds.). 1984. Paleoecology of Beringia. Academic, New York.

Kennett, J. P., and M. S. Srinivasan. 1983. Neogene Planktonic Foraminifera. Hutchinson Ross, Stroudsberg, Penn.

Kontrimavichus, V. L. (ed.). 1985. Beringia in the Cenozoic Era. Int. Publ. Serv. Accord, Mass.

Kosswig, C. 1950. Erythräische Fische in Mittelmeer und an der Grenze der Agäis, pp. 203–212. Syllegomena Biol. Festschr. Kleinschmidt, Wittenberg.

Lack, D. 1947. Darwin's Finches. Cambridge Univ. Press, Cambridge, England.

Lawrence, P. A. 1992. The Making of a Fly: The Genetics of Animal Design. Blackwell Scientific Publications, Cambridge, England.

Lehmann, U. 1981. The Ammonites. Cambridge Univ. Press, Cambridge, England.

Lewin, R. 1982. Thread of Life. Smithsonian Books, Washington, D.C.

Liem, K. F. 1978. Mandibulatory multiplicity in the functional repertoire of the feeding mechanism in cichlid fishes, Part 1: Piscivores. J. Morph. 158:323–360.

———. 1980. Adaptive significance of intra- and interspecific differences in the feeding repertoires of cichlid fishes. Amer. Zool. 20:295–314.

Liem, K. F., and J. W. M. Osse. 1975. Biological versatility, evolution and food resource exploitation in African cichlid fishes. Amer. Zool. 15:427–454.

Margulis, L. 1981. Symbiosis in Cell Evolution. W. H. Freeman, San Francisco.

Marshall, L. G., S. D. Webb, J. J. Sepkoski, and D. M. Raup. 1982. Mammalian evolution and the great American interchange. Science 215:1351–1357.

Mayr, E. 1963. Animal Species and Evolution. Belknap Press of Harvard Univ. Press, Cambridge, Mass.

McKaye, K. R., and W. N. Gray. 1984. Extrinsic barriers to gene flow in rock-dwelling cichlids of Lake Malawi: Microhabitat heterogeneity and reef colonization, pp. 169–183. *In* A. A. Echelle and I. Kornfield (eds.). Evolution of Fish Species Flocks. Univ. of Maine Press, Orono.

Moore, R. C. (ed.) 1957. Treatise on Invertebrate Paleontology, Part L. Univ. of Kansas Press, Lawrence.

Prance, G. T. 1981. Discussion, pp. 395–405. *In* G. Nelson and D. E. Rosen (eds.). Vicariance Biogeography: A Critique. Columbia Univ. Press, New York.

Price, P. W. 1980. Evolutionary Biology of Parasites. Princeton Univ. Press, Princeton, N.J.

Ramsköld, L. 1992. The second leg row of *Hallucigenia* discovered. Lethaia 25:221–224.

Ramsköld, L., and H. Xianguang. 1991. New early Cambrian animal and onychophoran affinities of enigmatic metazoans. Nature 351:225–228.

Ringo, J. M. 1977. Why 300 species of Hawaiian *Drosophila*? The sexual selection hypothesis. Evolution 31:694–696.

Romer, A. S. 1945. Vertebrate Paleontology. 2d ed. Univ. of Chicago Press.

Root, R. B. 1967. The niche exploitation pattern of the bluegray gnatcatcher. Ecol. Monogr. 37:317–350.

Simpson, G. G. 1940. Mammals and land bridges. J. Wash. Acad. Sci. 30:137–163.

———. 1953. The Major Features of Evolution. Columbia Univ. Press, New York.

———. 1980. Splendid Isolation: The Curious History of South American Mammals. Yale Univ. Press, New Haven, Conn.

Smith, A. 1984. Echinoid Palaeobiology. George Allen and Unwin, London.

Spieth, H. T. 1974a. Courtship behavior in *Drosophila*. Ann. Rev. Entomol. 19:385–405.

———. 1974b. Mating behavior and evolution of the Hawaiian *Drosophila*. pp. 94–101. *In* M. J. D. White (ed.). Genetic Mechanisms of Speciation in Insects. D. Reidel, Dordrecht, Holland.

Špinar, Z. V. 1972. Life Before Man. Thames and Hudson, London.

Stehli, F. G., and S. D. Webb (eds.). 1985. The Great American Biotic Interchange. Plenum, New York.

Templeton, A. R. 1979. Once again, why 300 species of Hawaiian *Drosophila*? Evolution 33:513–517.

Vuilleumier, F. 1970. Insular biogeography in continental regions, [Part] I: The northern Andes of South America. Amer. Natur. 104:373–388.

———. 1973. Insular biogeography in continental regions, [Part] II: Cave faunas from Tessin, southern Switzerland. Syst. Zool. 22:64–76.

Weidmann, J. 1969. The heteromorphs and ammonoid extinction. Biol. Rev. 44:563–602.

———. 1973. Evolution or revolution of ammonoids at Mesozoic system boundaries? Biol. Rev. 48:159–194.

Westoll, T. S. 1949. On the evolution of the Dipnoi, pp. 121–184. *In* G. L. Jepsen, E. Mayr, and G. G. Simpson (eds.). Genetics, Paleontology, and Evolution. Princeton Univ. Press, Princeton, N.J.

Whittington, H. B. 1985. The Burgess Shale. Yale Univ. Press, New Haven, Conn.

Yoon, J. S. 1989. Chromosomal evolution and speciation in Hawaiian *Drosophila*, pp. 129–147. *In* L. V. Giddings, K. Y. Kaneshiro, and W. W. Anderson (eds.). Genetics, Speciation, and the Founder Principle. Oxford Univ. Press, New York.

Zimmerman, E. C. 1948. Insects of Hawaii. Univ. of Hawaii Press, Honolulu.

9

EXTINCTION

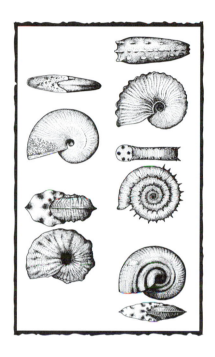

Ammonites (shown here in side and front views) represent a group on which mass extinction events have had a strong impact, with termination of the lineage at the end of the Cretaceous.

9.1 CURRENT THREATS TO SPECIES

Extinction as a process has affected every species that has ever existed on this planet, and it will have its impact on the human race. Whether we like it or not, our species is vulnerable to the same kinds of forces of nature that caused the demise of trilobites, ammonites, dinosaurs, mastodonts, and the dodo. "As dead as the dodo" is a definitive expression that will ultimately be applicable to all species living today. When and how will extinction occur? In this chapter, we explore extinction as a biological phenomenon with processes in need of further understanding. Extinction is not simply a failure of evolutionary processes and a consequence of events beyond the coping capacities of species. It is not simply bad genes or bad luck (cf. Raup 1991). Even now, habitat destruction, clearcutting of forests, clearing of land for agriculture, and use of pesticides and other polluting agents threaten thousands of species to unusual degrees. Serious declines in neotropical migrating birds in northern latitudes affect perhaps 650 species in North America alone, with steep reductions in some species—such as the bay-breasted warbler, the scarlet tanager, and the wood thrush—in the past 10 years.

Ironically, human impact on wild species reflects the potential for mass destruction of the human species itself. Nuclear arsenals around the world are sufficient to cause the extinction of humans. During nuclear winter, planet earth would be a cold place. According to Sagan (1984), temperatures could plummet to −45°C in the first 75 days after a war involving 10,000 megatons of nuclear bombs used in a major confrontation. Temperatures would probably remain below freezing for more than a year (Figure 9-1).

On the biological effects of a large-scale (say, 10,000-megaton) nuclear war, Ehrlich (1984, 62) concluded that, "if the atmospheric effects did spread over the entire planet, then we cannot be sure that *Homo sapiens* would survive." Primary production would cease in the whole of the Northern Hemisphere away from the coasts, and the cold would be particularly damaging during the growing season, when plants are not acclimated to the cold. Many species would abruptly cease to exist.

Although nuclear bombs seem to be less of a threat now than they once were, they are equivalent in power to natural disasters such as asteroid impacts, which may be frequent enough in geological time to make humans uncomfortable. It is estimated that asteroids with individual destructive impacts equivalent to that of the atomic bomb dropped on Hiroshima, Japan (August 6, 1945), pass through the edge of the earth's atmosphere almost once per year, on average. (The Hiroshima bomb was equivalent to 20,000 tons of TNT.) And an estimated 1 percent likeli-

FIGURE 9-1

Some estimates of temperature changes on Northern Hemisphere landmasses, away from oceans, after a hypothetical nuclear war. Ambient temperature is averaged across latitudes and seasons. The figure next to each curve is megatons (MT) of nuclear warheads in the exchange. A megaton is the explosive equivalent of a million tons of trinitrotoluene (TNT), an explosive used as the bursting charge for conventional bombs, grenades, and shells. The cases differ in the kinds of targets attacked and the conservatism of estimates, as follows: **100 MT**—only cities are attacked; smoke from burning cities prevents penetration by the sun, and temperatures decline; soot settles rapidly. **3,000 MT**—missile silos are the main targets; the dust produced circulates in the stratosphere and settles very slowly. Fires and smoke are excluded from this estimate because cities and forests are not involved. **5,000 MT (a)**—a case similar to the combined effects of the 100-MT and 3,000-MT examples. The rapid but rather short temperature decline results from the burning and smoke caused by 1,000 MT devoted to cities. The long-term cooling effect results from the stratospheric dust produced by 4,000 MT used against missile silos and so on. **5,000 MT (b)**—an exchange concentrated on missile silos in which the (estimated) fine dust produced per megaton increases and much more dust reaches the stratosphere as a result. **10,000 MT**—both cities and silos are attacked in a major exchange that produces both smoke and dust. *(After Sagan 1984)*

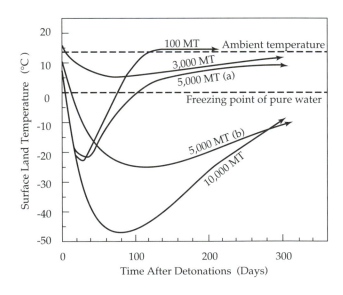

hood exists that an asteroid will hit the earth with an impact equivalent to a 1,000-megaton nuclear explosion in the next 100 years (Adams 1994). This magnitude of impact would be locally devastating. In the recent past, such an asteroid passed by the earth at less than twice the moon's distance from earth. Obviously, there is a real, albeit very low, probability that in the next century an impact with the energy of a 10,000-megaton nuclear explosion will occur, making the estimates in Figure 9-1 relevant for the near future. Chapman and Morrison (1994) express the danger in another, perhaps more persuasive, way: "Each typical person [in an industrialized country] stands a similar chance of dying in an asteroid impact as in an aeroplane crash or in a flood."

Extinction, then, is not only an academic subject but is of real concern to human beings in the present and the near future. We must consider, in addition to our own fate, the fates of many species of plants and animals that have been pushed to extinction in the recent past or are in jeopardy of extinction. Based on estimates for birds and mammals, current extinction rates are 4 to 40 times greater than the normal, or background, extinction rate in geological time and may be 40 to 400 times above normal before long (Ehrlich 1986) (see Section 9.2 for estimates on background extinction rates). Wilson (1992, 1993) lists some particularly alarming cases.

1. About 20 percent of bird species in the world have been pushed to extinction in the past 2,000 years, mainly after human colonization of islands. About 11 percent of the surviving bird species are endangered. Without the extinctions there might have been 11,000 bird species, but in the near future we may have 30 percent fewer, or 7,700 species.
2. Populations of migratory songbirds in the eastern United States have declined by 50 percent from the 1940s to the 1980s, and many species are extinct locally.
3. Among freshwater fish species, some 20 percent are extinct or near extinction (see also Minckley and Deacon 1991).
4. Freshwater molluscs are vulnerable to habitat loss and change. Loss of local populations is common, and in the United States some 32 percent of mussel species have become extinct or are endangered. On some islands, native snail extinctions are likely to reach 100 percent because of exotic predatory snails introduced by humans.
5. Among plant species in the United States, about 4 to 5 percent may have become extinct by the year 2000.
6. In Europe, estimates of threatened or endangered species of invertebrates vary from 17 percent in England to 34 percent in western Germany.
7. Among the European fungi, total extinction of species is hard to evaluate, but at least local decline in sampled areas has reached 40 to 50 percent of the species that were present 60 years ago.

Human-caused habitat loss, introduction of exotic predators, and pollution are increasing, with ubiquitous effects on almost all taxa. The estimated extent of future declines in biodiversity is even more like a doomsday scenario (e.g., De Blij 1988, Ehrlich and Wilson 1991, Wilson 1992, 1993). Certainly, we must learn to judge human impacts on biodiversity. Because many more species of organisms live in the tropics than in the higher, extratropical latitudes, and habitat loss through clearing of tropical forests is rapid, calculations have related area loss of habitat to an estimated correlated loss of species, as Chapter 11 (Sections 11.1 and 11.3) discusses. Data from oceanic islands show a decrease in number of species within a taxon (such as birds or reptiles) with island area. Similarly, we can view habitats such as forests as continental islands of varied sizes and thus make predictions about the decline of species in any one taxon as habitat destruction continues. The following passage from Ehrlich and Wilson's (1991, 759–760, references excluded) article on biodiversity, with more detail provided by Wilson (1992, 1993), illustrates the process of estimating and predicting.

Biodiversity reduction is accelerating today largely through the destruction of natural habitats. Because of the latitudinal diversity gradient, the greatest loss occurs in tropical moist forests (rain forests) and coral reefs. The rate of loss of rain forests, down to approximately 55% of their original cover, was in 1989 almost double that in 1979. Roughly 1.8% of the remaining forests are disappearing per year. By the most conservative estimate from island biogeographic data, 0.2 to 0.3% of all species in the forests are extinguished or doomed each year. If two million species are confined to the forests, surely also a very conservative estimate, then extinction due to tropical deforestation alone must be responsible for the loss of at least 4,000 species annually.

But there may well be 20 million or more species in the forests, raising the loss tenfold. Also, many species are very local and subject to immediate extinction from the clearing of a single habitat isolate, such as a mountain ridge or woodland patch. The absolute extinction rate thus may well be two or three orders of magnitude greater than the area-based estimates given above. If current rates of clearing are continued, one quarter or more of the species of organisms on Earth could be eliminated within 50 years—and even that pessimistic estimate might be conservative.

Clearly, we exist in a time of catastrophic extinction. In the marine fossil record, mass extinctions have been marked by extinction rates about four times greater than background levels at most (Raup and Sepkoski 1982). Against this backdrop, life on earth is unusually threatened now and in prospect.

Yet we must also emphasize the constancy and normality of extinction. Some attrition of species and higher taxonomic groups is continual, as if extinction continues at a regular rate over great expanses of geological time (see Chapter 11, Section 11.6). While a view of broad expanses of time may appear to reveal a simple, linear pattern, a dissection of extinction rates using more limited time scales reveals variation (see also Chapter 17, Section 17.7). If cognizant organisms were to be present on this planet a few million years from now, it would be interesting to see, from that vantage point, how our current extinctions compare with the background level of extinction and whether we are leaving a record sufficient to reveal the mass extinctions in progress now.

We do have a global laboratory for the study of the evolutionary consequences of extinction. Although we may regard extinction as a nonevolutionary event, that would be a remarkably narrow view, because an extinction inevitably affects the evolutionary biology of the survivors. The scientific community has yet to treat current extinctions as a research opportunity to help with predictions of the evolutionary consequences of extinction and of the post-extinction ecological and evolutionary processes among survivors. The next chapter will emphasize that there always have been survivors. There are patterns in survival, although we have only a crude understanding of the mechanisms involved. Opportunities to study the mechanisms abound.

9.2 EXTINCTIONS IN THE HISTORY OF LIFE

Background and Mass Extinction

Extinction is a prominent part of the history of life on earth. Whether catalogued at the species, genus, family, or higher taxonomic level, taxa have been going extinct throughout geologic history, producing a normal, or background, level of extinction. For marine invertebrates and vertebrates, the normal extinction rate has averaged about two to five families per million years (Figure 9-2). Times of mass extinction, when rates were prominently greater than normal, have punctuated this background level (Raup and Sepkoski 1982, Jablonsky 1991) (Table 9-1). Note that the names and dates in Table 9-1 and in the following examples differ in some cases from the more recently published geologic time scale in Table 7-1 (Chapter 7).

1. Late Ordovician (Ashgillian), 435 million years ago, with an extinction rate of 19 families per million years and a loss of 12 percent of families, 61 percent of genera, and 85 percent of species
2. Late Devonian (Givetian, Frasnian, Famennian), 345 million years ago, with an extinction rate of 10

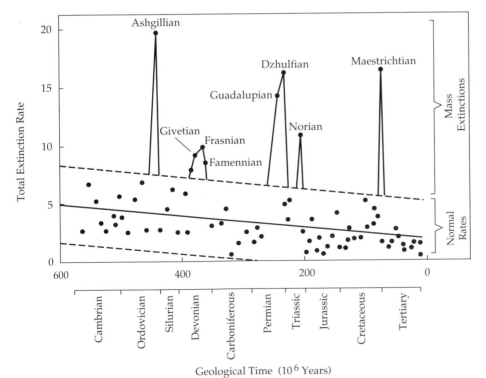

FIGURE 9-2

Total extinction rates (number of families going extinct per million years) for marine invertebrates and vertebrates over the last 600 million years. The regression line and the 95 percent confidence intervals (dashed lines) identify normal (background) levels of extinction. The five prominent departures from normal extinction rates—i.e., mass extinctions—are labeled with the relevant geological stage(s). (*After Raup and Sepkoski 1982*)

families per million years and a loss of 14 percent of families, 55 percent of genera, and 85 percent of species

3. Late Permian (Guadalupian, Dzulfian), 253 million years ago, with an extinction rate of 14 families per million years and a loss of 52 percent of families, 85 percent of genera, and 96 percent of species

4. Late Triassic (Norian, Rhaetian), 213 million years ago, with an extinction rate of 16 families per million years and a loss of 12 percent of families, 47 percent of genera, and 76 percent of species

5. Late Cretaceous (Maestrichtian), 65 million years ago, with an extinction rate of 16 families per million years and a loss of 11 percent of families, 47 percent of genera, and 76 percent of species

Later studies of the last 270 million years of geologic time identified, in addition to the three major mass extinctions just listed (3–5), five minor peaks of extinction: one each in the early Jurassic (Pliensbachian), late Jurassic (Tithonian), and middle Cretaceous (Cenomanian) and two in the Tertiary (Upper Eocene and Middle Miocene) (Raup and Sepkoski 1984, 1986, Sepkoski and Raup 1986, Jablonski 1991). We will discuss the pattern of mass extinctions later in this chapter.

In response to originations of taxa, radiations, and background and mass extinctions, the diversity of taxa on earth has changed dramatically since the Precambrian (Figure 9-3). After each mass extinction, radiation was relatively rapid and usually compensated for the loss of taxa.

Notable for its absence in the marine mass extinctions is a Pleistocene extinction. In fact, survival of marine taxa was remarkably high in the Pleistocene, with a loss of only one family (0.1 percent) per million years (Sepkoski and Raup 1986). Contrasting with this are the catastrophic extinctions of large mammals during the Pleistocene, reaching 73 percent of genera in North America and 80 percent of genera in South America (Martin 1984a). Such differences in survival and extinction of taxa in a given geologic stage have fostered the development of idiographic approaches to extinction—scenarios that invoke the special features of the times coupled with the characteristics of the particular taxa involved.

Pattern and Mechanism

Part of the process of science is to first recognize a pattern in nature and subsequently attempt an explanation of the processes or mechanisms producing the pattern. Once the mechanisms are generally validated, then science can move from a hypothetical explanation to scientific theory (cf. Chapter 1). Studies

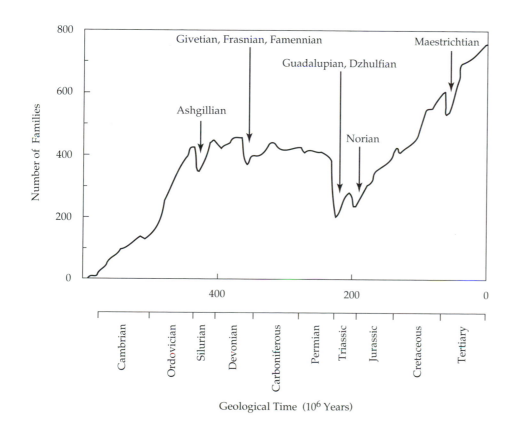

FIGURE 9-3
Numbers of families of marine invertebrates and vertebrates from the Cambrian to the present, with five conspicuous episodes of extinction. (*After Raup and Sepkoski 1982*)

T A B L E   9 - 1   Major extinction events during earth's history

Extinction Event (and Geological Boundary Involved)	Millions of Years Before Present (Top of Stage)	Environments of the Epochs and Periods	Organisms in Marine Environments
Pleistocene (Pleistocene–Recent)	0.011	Ice age occurred in north, sea levels dropped, land bridges formed between continents.	Marine organisms show no major extinctions.
Middle Miocene (Middle–Upper Miocene)	11	Tectonic activity forced India into Asia; Himalayas formed. Rockies and Andes formed. Grasslands spread extensively.	Extinction among foraminifers, dinoflagellates, coccolithoporids, and silicoflagellates.
Upper Eocene (Eocene–Oligocene [E/O])	38	Climates changed from warm or mild to cool.	Major extinctions in planktonic foraminifers, dinoflagellates, coccolithoporids, and silicoflagellates. Larger marine animals were also affected generally.
Maestrichtian (Cretaceous–Tertiary [K/T])	65	Climates tended to cool. Shallow seas on continental shelves retreated, exposing new land in inland North America, Africa, and Australia.	Last of the ammonoids and plesiosaurs, ichthyosaurs and mosasaurs went extinct. Familial extinctions among other cephalopods, bivalves, gastropods, bryozoans, echinoids, sponges, bony fishes, and marine reptiles. Major declines in planktonic foraminifers, dinoflagellates and coccolithoporids; 67 percent reduction in oceanic microplankton.
Cenomanian (Cenomanian–Turonian in Cretaceous)	91	Large, shallow seas resulted in extensive chalk deposits.	Extinctions among sponges, echinoids, cephalopods, bony fishes, planktonic dinoflagellates, and globigerinids.
Tithonian (Jurassic–Cretaceous)	145	Gondwanaland was breaking up, and shallow seas had developed on continental shelves.	Extinctions among bivalves and cephalopods (including ammonoids). Planktonic dinoflagellates were heavily affected.
Pliensbachian (Pliensbachian–Toarcian in Jurassic)	187	Climates were warm and moist.	Extinctions among bivalves and cephalopods (including ammonoids).
Norian–Rhaetian (Triassic/Jurassic)	208	Pangaea was splitting into Laurasia and Gondwanaland. Vast deserts occupied inland areas.	Brachiopods, bivalves, gastropods, cephalopods (including ammonoids), and marine reptiles were heavily affected. Total extinction of conodonts and conulariids.
Guadalupian–Dzhulfian (Permian–Triassic)	245	Assembly of one landmass, Pangaea, and one ocean, Panthalassa.	Fifty percent of marine families and >95 percent of marine species went extinct.

TABLE 9-1   Major extinction events during earth's history (continued)

Extinction Event (and Geological Boundary Involved)	Millions of Years Before Present (Top of Stage)	Environments of the Epochs and Periods	Organisms in Marine Environments
			Brachiopods, crinoids, corals, and cephalopods (including ammonoids) all lost 20 or more families. Total extinction of trilobites, rostroconchs, and blastoids.
Givetian–Frasnian–Famennian (Devonian–Carboniferous)	367	Warm climates with development of extensive shallow seas.	Decline of corals, brachiopods, jawless fishes (agnathans, including ostracoderms), and placoderms.
Ashgillian (Ordovician–Silurian)	439	Extensive, warm, shallow seas.	Decline of graptolites, trilobites, and some cephalopod and brachiopod taxa.

Sources: Lambert 1985, Lipps 1986, Martin and Klein 1984, Raup and Sepkoski 1986, Sepkoski and Raup 1986.

of extinction have revealed many patterns, although more detailed studies are continually modifying earlier perceptions. Explanation of patterns is still in a highly debatable state, with several alternative hypotheses frequently competing. Quite possibly, alternative hypotheses are viable for the same extinction episode for different organismal taxa. However, it is important to recognize in this chapter that science frequently follows a path from pattern discovery to the understanding of mechanisms driving pattern. Use of the fossil record is thus a difficult path to follow because a chasm of missing credibility and scientific knowledge yawns between the two endeavors, pattern description and mechanism description.

This nature of science, to proceed from pattern to mechanism, is evident in the study of extinction. Throughout this chapter, bear in mind the clear difference between pattern and process. Both aspects of science can be debated. New information makes our understanding an almost fluid property—as we saw in the case of the geologic time scale, which is a fundamental pattern (Chapter 7, Section 7.2)—and makes the field of evolution vibrant and stimulating. One of the most enduring fascinations of science is this ongoing debate, this tension, this dialectic and rapid movement. The strength of science is in its fluidity. Science cannot be dogmatic. It must be open to alternative views. The more vigorous the debate, the more rapid the progress.

9.3 HISTORICAL PERSPECTIVE ON EXTINCTIONS

Idiographic views, or special-case scenarios, of extinction dominated paleontologists' perspectives until the early 1970s (see Chapter 11). It is therefore valuable to review the development of ideas on extinction, although we should recognize the heavy influence of data from the Pleistocene extinctions—the first mass extinction to be studied in detail.

Grayson (1984) discusses early views on extinction. Until the 1790s, the concept of the "Great Chain of Being" dominated concepts related to extinction. The Maker preserved every link in the chain, or else the chain would be broken. Therefore (the logic went), extinction was probably not a reality. The Creation was perfect, and the Creator conserved this perfection.

Fossils turned up during the 18th century, but they were assumed to represent animals and plants still living on some other part of the globe. When Cuvier (1796a, 1796b) started discovering and describing fossils of giant ground sloths (*Megatherium*) (Figure 9-4) and elephants (*Mammuthus*) (Figure 9-5) and demonstrated clearly that no living specimens of these animals existed, he established the reality of extinction.

The first scenarios for the Pleistocene extinctions were catastrophic. Theorists invoked rapid geologic

FIGURE 9-4

Cuvier named the giant ground sloth, *Megatherium*, in 1795. It was 6 meters long and probably browsed tree foliage as shown, standing semierect and supported by its strong tail. *(Modified from figures in Bates 1964 and Lambert 1985) (Illustration by Stephen Price)*

FIGURE 9-5

Mammuthus columbi in Pleistocene times lived in what is
now North America. Here it
encounters saber-toothed tigers (*Smilodon*), a predator
that exploited large animals
and probably killed by stab-
bing its long canine teeth into
vital body areas. (*Modified from
figures in Špinar 1972, Anderson
1984, and Carrington 1963*). (*Illustration by Tad Theimer*)

changes that revolutionized life on earth (Cuvier
1812). Many scientists thought global phenomena
were involved, but Cuvier emphasized more local
events, such as rapid cooling that "refrigerated" Siberian mammoths and local flooding that changed land
into sea, causing extinction of terrestrial quadrupeds.

Lyell countered these scenarios by taking the uniformitarian approach, positing slow, natural changes
on the globe. In the second volume of his *Principles
of Geology* (1832, 141), he argued as follows:

> The possibility of the existence of a certain species in
> a given locality, or of its thriving more or less therein,
> is determined not merely by temperature, humidity,
> soil, elevation, and other circumstances of like kind,
> but also by the existence or nonexistence, the abun-
> dance or scarcity, of a particular assemblage of other
> plants and animals in the region.

If we show that both these classes of circumstances, whether relating to the animate or inanimate
creation, are perpetually changing, it will follow that
species are subject to incessant vicissitudes and if the
result of these mutations, in the course of ages, be so
great as to materially affect the general conditions of
stations, it will follow that the successive destruction
of species must now be part of the regular and constant order of nature.

Extinction was a uniformitarian phenomenon.

Where Lyell focused on specific extinctions, he
first emphasized the long duration over which individuals in a species died, as in the case of the woolly
mammoths "refrigerated" in Siberia, then asserted
that the change causing extinction was a matter of
degree rather than kind. He used as an example the
changes from climates in which winter and summer

were uniformly cold to climates with strongly contrasting seasons (Lyell 1830). Lyell emphasized gradualism in the extinction process.

An alternative hypothesis for Pleistocene extinctions was what is now called Pleistocene overkill: caused by man's hunting large animals to extinction. Although Martin coined the term in 1967, the hypothesis had been discussed in the 18th century. We will examine this hypothesis after we treat another idiographic approach—the effects of competition.

9.4 THE ROLE OF COMPETITION

Scientists frequently invoke the mechanism of competition to explain extinction: a newly evolving group competes with a less well adapted existing group and drives it to extinction. Because the actual mechanisms of competition, the limiting resources, and the similarity of exploitation patterns are seldom discussed such appealing explanations are unconvincing. For example, Colbert (1980, 82) wrote,

> The most important event in the history of the vertebrates during middle Devonian times was the appearance of the Osteichthyes, and it is probable that the sudden influx of the new and comparatively advanced bony fishes was instrumental in leading to the decline and final disappearance of the more primitive vertebrates such as the ostracoderms and most of the placoderms [Figure 9-6].

In a similar vein, Simpson (1953), with more recent support from Krause (1986), argued that competition from rodents probably caused the extinction of multituberculates (Figure 9-7). The powerful jaw, strong incisors, and molars adapted for grinding suggest that the multituberculates were the first herbivorous mammals (Colbert and Morales 1991).

Certainly, equilibrium concepts of faunal assemblages have prevailed during the modern synthesis, with the implication that radiation of a new group is not possible without the extinction of an ecologically equivalent group. This view always stretches the imagination as to how every member of an extant group is competitively inferior to members in the radiating taxon. And the matter of which is cause and which is effect is largely a matter of taste. Do radiations occur because extinction has vacated ecological niches, or do radiations cause extinction?

Colbert's (1980, 228) view was that the demise of the reptiles permitted the radiation of the mammals:

> As soon as the dinosaurs and other ruling Mesozoic reptiles vanished, the mammals came into their own. Although the first mammals appeared during the Triassic Period, and although modern placental mammals inhabited the Cretaceous world, these vertebrates remained as small, minor members of the mid-

FIGURE 9-6

Devonian fishes. (a) *Hemicyclaspis* was a bottom-dwelling agnathan (jawless fish) and ostracoderm (shell-skinned fish) living in the late Silurian to Devonian and was about 13 centimeters long. (b) *Bothriolepis* was a placoderm (an armored or plate-skinned fish about 20 to 30 centimeters long. (c) *Osteolepis*, about 17 centimeters long, was a very early representative of the Osteichthyes, or bony fishes. It was a rhipidistian, or fleshy-finned fish, and was in a lineage that probably gave rise to the amphibians. (*Modified from figures in Lambert 1985, Špinar 1972, and Colbert 1980*) (*Illustration by Stephen Price*)

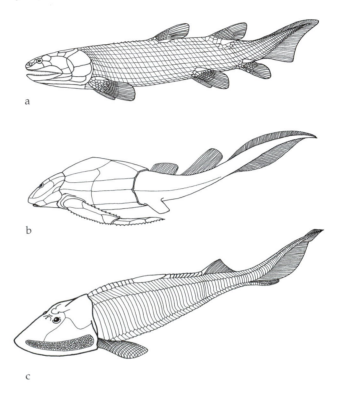

dle and late Mesozoic faunas. It would seem that the varied reptiles of the Mesozoic "held down" the first mammals, and it was not until the dominant reptiles had vanished, to vacate numerous ecological niches, that the mammals enjoyed their first great burst of evolutionary adaptation. By the opening of the Paleocene epoch the world was abundantly inhabited by the mammals, and from that time until today the mammals have reigned supreme.

9.5 ALTERNATIVE EXPLANATIONS

Biology and Life History

Belief in tight complementarity of taxa in an equilibrium world is losing its hold as more detailed stratigraphic records enable more exact timing of radia-

FIGURE 9-7

Numbers of species in the mammal orders Multituberculata and Rodentia during the early Cenozoic. Data are from mid-20th-century studies of Rocky Mountain deposits where the last multituberculates and earliest rodents then known could be found. Fossils indicate a possible competitive replacement of multituberculates by rodents. However, the coincidence of the early decline from 17 to 11 multituberculate species in the middle to late Paleocene with the emergence of only one rodent species is hard to reconcile with a competitive cause. (a) The skull and jaw of *Taeniolabis*, a Paleocene multituberculate about 15 centimeters long. (b) The skull and jaw of *Paramys*, a primitive Paleocene and Eocene rodent about 9 centimeters long. *(Data from Jepsen 1949 and Simpson 1953; after Colbert 1980)*

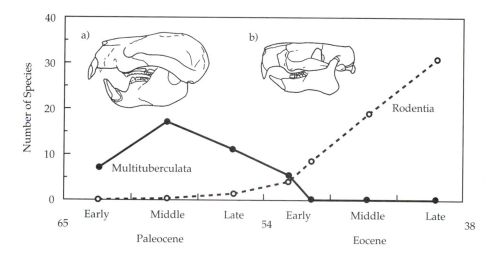

tions and extinctions (e.g., Bakker 1983), which—it turns out—are often not coincident. Also, researchers increasingly take into account the biology of the species. For example, many have taken the view that clams (Mollusca, Bivalvia, and Pelecypoda) replaced brachiopods (Brachiopoda) through competition over the past 400 million years (Gould and Calloway 1980) (Figure 9-8). But the major shift in relative abundance was during the Permian extinctions, when brachiopods were affected heavily and clams to only a minor degree. A fundamental difference between the two groups is that brachiopods brood their young and spend little time in the plankton as larvae, whereas clams have planktonic larvae and do not brood. A general pattern seems to exist in which brooders have small local populations and thus rapid speciation and extinction rates (Valentine and Jablonski 1982). Figure 9-8 reflects this dynamic situation. By contrast, species with planktonic larvae have large, widely dispersed populations that are resistant to extinction but are also slow to speciate. Stanley (1968) pointed out that the post-Permian radiation of bivalves resulted largely from a lineage with mantle fusion and siphon formation that exploited new infaunal ecological niches by burrowing into substrata. The epifaunal brachiopods never achieved exploitation of infaunal habitats, and bivalves achieved it only after the Permo-Triassic extinctions. Thus, the idiosyncratic characters of the involved species seem to be much

more important and tangible than the possible effects of competition.

Another case, long regarded as competitive displacement between the order including horses, the Perissodactyla, and the order including horned and antlered animals, the Artiodactyla, has not withstood careful examination by Cifelli (1981). With increasing precision of the fossil record and emphasis on the biologies of individual species, the competition hypothesis will probably weaken and be frequently falsified. As Jablonski (1989, 124) argued,

> The rise to dominance of previously unimportant groups therefore does not require an adaptive breakthrough and competitive superiority. . . . A group might have suffered losses along with the rest of its contemporaries during a mass extinction, but undergone a more rapid pre-emptive diversification than the other survivors.

Environmental Change

Environmental change is a common theme in arguments about the causes of extinction. We can divide environmental changes into climatic change, usually cooling events; sea-level changes, usually declines; and other changes in the sea, such as salinity and nutrients. As we shall see, it is frequently difficult or unrealistic to separate these types of changes. For example, plate tectonic activity, involving the shifting

FIGURE 9-8

Numbers of brachiopod and clam genera from the Cambrian to the present. Note the dramatic shift in relative abundances at the end of the Permian. The five major extinctions from Figures 9-2 and 9-3 are shown: A, Ashgillian; GFM, Givetian, Frasnian, Famennian; GD, Guadalupian, Dzhulfian; N, Norian; M, Maestrichtian. (a) The carboniferous brachiopod *Spirifer striatus*, which was about 50 millimeters wide (from British Museum 1969). (b) A member of the modern genus *Mya*, displaying its siphon and its ability to burrow into the substrata (from Stanley 1968). *(Data from Gould and Calloway 1980)*

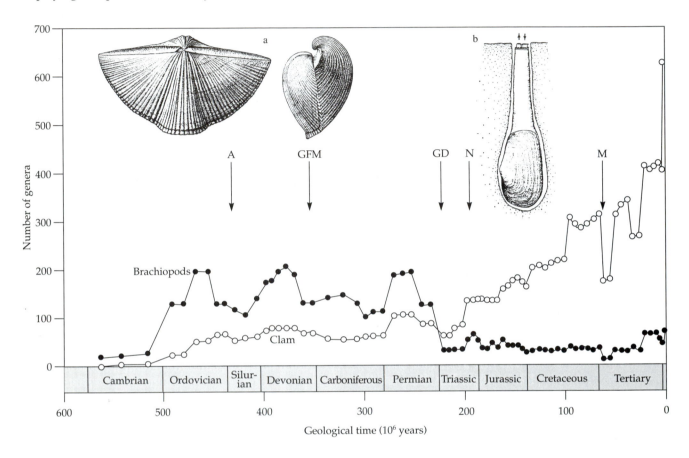

of plates over the earth's crust, changes the relationships of continents to one another, which influence sea level, and volcanoes may erupt. When continents assembled into the supercontinent Pangaea in the late Permian Period, the massive inland areas became very dry and sea level declined (as Chapter 11, Section 11.3, discusses). In Table 9-1, several cases of extinction are associated with tectonic events and climate change, making it unrealistic to focus on one factor as the cause of a mass extinction. The locations in which species survive extinction events (as discussed in Chapter 10) also indicate the importance of ocean surface temperature and nutrient supply in marine systems and arid periods in terrestrial environments. Kauffman (e.g., 1977, 1984, 1985, 1986) has repeatedly emphasized multiple associated events in relation to the Western Interior Cretaceous Seaway in North America. This massive sea once spread through what are now the states of Arizona and Texas up to Utah

and Colorado and into Wyoming, Nebraska, and Kansas. Plate tectonics, sea-level change, climate, and salinity all had strong effects on this epicontinental sea, far more so than in the more stable deep-sea zones. Vermeij (1987) has also cautioned against emphasis on only one driving force, while noting that various authors have championed particular causative agents.

The critical reader should consult the proponents of particular environmental changes as influential in major extinction events to obtain evidence supporting their arguments. Stanley (1984a, 1984b) made an impressive case for the role of especially cold conditions. Rapid changes in salinity of the sea, caused by flooding of the sea into highly saline inland basins or large-scale upwelling of highly saline seawater, is a recurring argument (e.g., Holser 1977, Thierstein and Berger 1978, Degens and Stoffers 1976, Wilde and Berry 1984). Kauffman (1986) recognized plate tectonics as a central force in extinction, although he emphasized its multiple effects and additional influences, such as shifts in climatic belts. He called major mechanisms causing extinction, external to the system under consideration, **allocyclic forcing mechanisms.**

> In summary, the sedimentologic and biologic records of the Western Interior Cretaceous Seaway of North America are demonstrably controlled by regional and global allocyclic forcing mechanisms such as plate and regional tectonics, large scale volcanism, oceanography, and climate. (Kauffman 1986, 297)

As noted in Section 9.2, we can record the pattern of extinction, but discovering the mechanisms is more problematic. An assertion that cooling climates are a mechanism causing extinction is unconvincing unless selective extinction and geographical patterns are clearly documented. The correlation of a comet impact with an extinction event is insufficient as a mechanistic explanation without some knowledge of climatic, atmospheric, and oceanographic changes that influence organisms directly. To attain a mechanistic understanding of extinction in this chapter and Chapter 11, and survival in Chapter 10, keep in mind the differences between pattern recognition and the interpretation of the pattern.

9.6 THE PLEISTOCENE OVERKILL HYPOTHESIS

The Patterns

The best-developed idiographic explanation of an extinction is the Pleistocene overkill hypothesis. Martin (1984a) noted that late Pleistocene extinctions were not synchronous; most groups of organisms escaped mass extinction in the late Pleistocene. Those most heavily affected by extinction were large mammals on continents—"large" meaning animals weighing 44 kilograms (100 pounds) or more, i.e., the megafauna.

("Small" would mean less than 44 kilograms.) Table 9-2 summarizes megafaunal extinctions of the late Pleistocene. Impacts were very different on different continents (note that the African megafauna was the least severely affected). Impacts and timing for small and large mammals also differed (Table 9-3). Small mammals tended to go extinct mostly through the late Pliocene and early Pleistocene, but extinction of the megafauna was heavily concentrated in the late Pleistocene in both North and South America.

Climatic effects on extinction could thus be excluded as a cause, because they would have had largely synchronous impact around the globe, and climates of the late Pleistocene had also occurred earlier in the Pleistocene without causing extinction. Also, because extinctions were not synchronous and were concentrated on large animals, they were unlikely to have resulted from asteroid or comet impact with catastrophic global effects as in the model of Alvarez et al. (1980).

The competition hypothesis could also be excluded, because extinctions occurred without replacement. No new potentially competitive species invaded megafaunal ranges; no mixing of faunal groups took place. Extinctions created many vacant ecological niches that remain largely unfilled to this day. The flourishing of feral populations of certain human-introduced animals—such as horses and burros in the American Southwest—illustrate that such niches continue to exist.

Human Hunters

Martin argued that human movements around the globe can best account for highly selective, asynchronous, and patchy extinction. Much of the megafaunal extinction in Africa occurred in the Lower Pleistocene before 700,000 years ago (21 genera); less occurred in the Middle and Upper Pleistocene (respectively, 700,000 to 130,000 years ago [nine genera] and 130,000 years ago to the present [seven genera]). As highly

---

T A B L E 9 - 2 Estimates of extinct and living genera of megafauna—species weighing 44 kg (100 lb) or more

Continent	Total Genera	Extinct Genera	Living Genera	Percent Extinction
Africa	49	7	42	14%
North America	45	33	12	73%
South America	58	46	12	79%
Australia	22	19	3	86%

After Martin 1984a.

TABLE 9-3 Timing of extinction of small and large mammals, compared for North and South America

North America

Epoch	Late Pliocene			Pleistocene					
Stage	Blancan			Irvingtonian			Rancholabrean		
	E	M	L	E	M	L	I	S	W
Million Years Ago	3.5			1.7			0.7		
Small mammals (n = 46)	10	10	8	6	3	3	1	1	4
Large mammals (n = 53)	3	0	7	3	2	2	2	1	33

South America

Stage	Chapadmalalan	Uquian	Ensenadan	Lujanian
Million Years Ago	3.0	2.0	1.0	0.3
Small mammals (n = 36)	17	14	5	0
Large mammals (n = 70)	14	7	3	46

E = Early, M = Middle, L = Late, I = Illinoian, S = Sangamon, W = Wisconsin.
Braces indicate roughly equivalent stages in North and South America.
Data from Martin 1984a.

efficient hunters invaded new continents and islands from Africa, they came into contact with a naive fauna unadapted to this new form of predation. The result was overkill of the megafauna spread through the late Pleistocene, closely matching the arrival of humans. Mass extinctions occurred from more than 15,000 to 30,000 years ago in Australia and New Guinea, 10,000 to 12,000 years ago in North and South America, and only 1,000 to 6,000 years ago in the Greater Antilles, Mediterranean islands, Madagascar, and New Zealand. Archaeological sites in Australia date back to 30,000 years ago, and those in North America, to 14,000 years ago. Until seafaring and exploration increased, survival of species into historic times occurred only on remote islands with little or no human traffic. Examples of island survivors include the giant tortoises on Aldabra, the Seychelles, and the Galapagos; Stellar's sea cow on the Commander Islands (which may have been the original mermaids [Bruemmer 1986]); and the dodo on Mauritius.

Differential Impact

Several factors may explain the differential impact of humans in Africa and the other continents.

1. Humans evolved in Africa, and so, as skills in hunting developed, animals had time to learn avoidance skills and evolve evasive behaviors. Humans were skillful hunters when they invaded other parts of the world, and the local megafaunas were naive to such predation.

2. In Africa, because many large herbivores were fleet of foot and many cursorial predators (especially Carnivora) were part of the fauna, the fauna was harder to hunt. Other parts of the world were home to sluggish herbivores such as the giant sloth in North America and the large kangaroos (and even larger herbivores) in Australia (Figure 9-9). Martin (1984a, 377) commented,

One remarkable feature of the Australian Pleistocene megafauna is the relative scarcity of carnivores. They are limited to three species; the fiercely clawed large "lion," *Thylacoleo;* the large "dog," *Tylacinus;* and the giant varanid lizard, *Megalania.* Only the thylacine could have functioned as a cursorial predator. Presumably because the Australian large herbivores did not coevolve with an array of fleet predators like the canids, hyaenids, and felids of Africa, most of them, at least the extinct ones, appear to have been at best sprinters rather than distance runners.

FIGURE 9-9

Reconstructions of some extinct Australian marsupials: (a) the largest true kangaroo (*Macropus ferragus*) with the largest extant species, by comparison, half its weight, (b) the red kangaroo (*Macropus rufus*), (c) the largest of all extinct Australian marsupials, *Diprotodon optatum*, and (d) the "marsupial lion," or "giant killer possom" (*Thylacoleo carnifex*). All are drawn to scale, with *M. ferragus* standing at 2.5 meters. *(After Murray 1984)*

3. Disease in Africa may also have prevented large human populations, and hence overexploitation of large animals, from developing. Sleeping sickness, malaria, and other tropical diseases would have kept human groups small and prevented settlement in some areas. Such diseases were not as devastating in novel human locations in America and Australia.

Testing Extinction Dates

The best-dated last remains of extinct megafauna in North America coincide at 11,000 years ago, at the time of Clovis human activity in the Southwest (Mar-

tin 1984a, 1984b). Dates for the shasta ground sloth, *Nothrotheriops shastensis*, in the Southwest are particularly reliable because the dry caves of that region conserve sloth dung in excellent condition for radiocarbon dating. Martin (1984b) provided locations and dates for all known records, and they concentrate strongly around 11,000 years ago. These data are particularly valuable because they enable us to postulate that the extinction time of ground sloth in the West Indies was later than 11,000 years ago, since humans arrived later. The exact dates of sloth extinction and human arrival are yet to be determined. We have a clear prediction and a strong test of the Pleistocene overkill hypothesis.

9.7 GENERAL APPROACHES TO THE STUDY OF EXTINCTION

Much more recent than idiographic arguments are nomothetic approaches to the study of extinction, which search for general patterns in extinction without invoking special cases. Simulation of cladograms by Raup et al. (1973) was an early attempt at a general treatment of extinction (see Chapter 11). This simulation permitted the identification in the real fossil record of extinctions with very low probabilities of occurrence, meaning that very unusual events must have been important. Rare and unusual events are difficult to study, and they make the discovery of pattern especially challenging.

Periodicity of Mass Extinctions

Raup and Sepkoski (1982, 1984, Sepkoski and Raup 1986) undertook another nomothetic approach and identified periods of major extinctions. As described earlier in this chapter, they identified eight episodes of extinction in the last 270 million years. Their remarkable discovery was highly significant periodicity in the timing of extinctions. The cycles had a. period length of 26.2 million years (Sepkoski and Raup 1986). This periodicity depended on the recognition of two

cycles among the eight in which no detectable extinction occurred (Figure 9-10).

Sepkoski and Raup (1986) concluded from such periodicity in Mesozoic–Cenozoic extinctions that mass extinctions are interrelated and caused by a single "ultimate forcing agent," or causative force. No cycles of this length are known on earth or in the solar system, suggesting that forces beyond the solar system drive the cycle.

Large-Body Impacts with the Earth

The search for cyclical events that could cause extinction was under way. One of the first possibilities explored was the impact of large bodies, such as meteorites, on the earth. If such an impact occurred, the explosion into the atmosphere of dust, debris, and perhaps smoke from fires would cause significant cooling of the earth, as in nuclear war. Alvarez and Muller (1984) studied the ages of 11 large impact craters on the earth. All the craters were more than 10 kilometers in diameter and had occurred within the last 5 million to 250 million years, plus or minus 20 million years. Alvarez and Muller found significant periodicity in the timing of impacts, with a mean period length of 28.4 million years. This estimate was very close to the extinction periodicity, and high impact frequency has coincided with high extinction

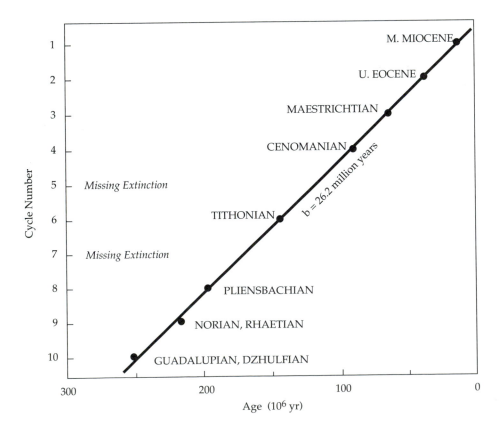

FIGURE 9-10

The pattern of extinction episodes in the last 270 million years, with missing extinctions indicated at cycles 5 and 7. The slope, *b*, of the regression is 26.2 million years. Figures 9-2 and 9-3 plot some of these extinctions, and Table 7-1 gives the stages in the context of the geologic time scale on earth. (*After Sepkoski and Raup 1986*)

for the last 150 million years. Observations of "cycle slippage" before that time may arise purely from problems of timing crater ages and geological ages, which grow worse with the passage of time. Rampino and Stothers (1984) used 42 crater dates and estimated a 31-million-year periodicity. Sepkoski and Raup (1986) excluded three craters used by Rampino and Stothers that were older than 210 million years and therefore had uncertain ages. They obtained a revised estimate of cratering periodicity of 27 to 28 million years, although this must be regarded as a tentative figure until additional, more accurately calculated crater dates are available.

If asteroids or comets are involved with mass extinctions, then the mechanisms may fall into two main types (Sepkoski and Raup 1986):

1. Catastrophic atmospheric changes—resulting, for example, from a single, large impact as at the Cretaceous–Tertiary boundary (e.g., Alvarez et al. 1980)—that cause global cooling.
2. Cumulative deleterious effects of many small impacts in a period of 1 or 2 million years, which may cause cooling of climates (Stanley 1984a, 1984b), and changes in stable isotope ratios (Kauffman, reported by Weisburd 1986). We will examine these alternatives in more detail later in this chapter.

The Ultimate Forcing Agent

The identity of the "ultimate forcing agent" is even more problematic and cannot yet be determined. What regulates the extraterrestrial clock, if it really exists? Sepkoski and Raup (1986, 26–27) briefly reviewed four astronomical mechanisms.

1. **Transit of the solar system through the spiral arms of the galaxy.** The solar system passes through the spiral arms of the Milky Way galaxy every 50 million years, increasing asteroid bombardment of the earth. Sepkoski and Raup (1986) reject this hypothesis because the periodicity is two times longer than the extinction periodicity.
2. **Vertical oscillation of the solar system about the plane of the galaxy.** The solar system oscillates vertically through the galactic plane about every 33 million years. Two effects may be evident. (a) At the oscillatory extremes, increased X-rays and ultraviolet radiation disturb the ionization balance of the upper atmosphere, causing climatic changes that induce mass extinction. (b) As the solar system approaches the galactic plane, tidal forces from large molecular clouds may perturb the Oort Cloud and inner cometary reservoir, producing comet showers that last up to several million years. One large or several small comets may hit the earth. Again, the 33-million-year period is not close

enough to the 26-million-year extinction period to allow us to accept this hypothesis.

3. **Precession of an undetected tenth planet.** Precession means the slow gyration of the rotational axis of a spinning body—in this case, a planet (a spinning top shows the same pattern of movement). If Planet X beyond Pluto exists, its orbit through the planetary plane may produce comet showers, and some comets may hit the earth. The lack of evidence for this planet, and therefore the lack of any estimate of its periodic precession, leaves this idea in the realm of imagination.
4. **Orbital dynamics of an unobserved solar companion.** The sun may be part of a binary star system, its companion being of low mass and luminosity. This companion star would enter the inner comet reservoir, producing a shower of 10^9 comets in the inner solar system, and 10 to 200 (mean of 25) could hit the earth over 100,000 to 1 million years. Davis et al. (1984) named the sun's companion Nemesis, but it has yet to be discovered, so this hypothesis has a problem.

Effects of Large-Body Impact

A massive comet impact producing a crater 10 kilometers or more in diameter would cause global and largely synchronous extinction. Small comets in succession or climatic changes, which may be coupled, would cause more graded extinctions. To cover the continuum of possibilities, Erle Kauffman (reported by Weisburd 1986) developed three categories of extinction as a heuristic stimulus for testing among ideas: (1) catastrophic extinction, (2) stepped mass extinction, and (3) graded mass extinction (Figure 9-11).

Catastrophic extinction follows large-body impact on the earth (Alvarez et al. 1980, 1984). Abrupt and global changes cause synchronous extinctions at a geologic boundary, with few survivors from which new originations and radiations can develop, if a lineage persists at all. **Stepped mass extinction,** a hypothesis Kauffman favors, involves groups of species that are sensitive to change going extinct first and more robust groups going extinct in subsequent episodes, with generally elevated rates of extinction across a geologic boundary. **Graded mass extinction** was the hypothesis generally supported by paleontologists through the 1970s, with climatic cooling commonly invoked; it remains viable (e.g., Clemens 1986). The rate of extinction increases beyond the background level, straddling a geologic boundary. The appearance of a largely synchronous extinction results from a poor record across the boundary.

Let us examine some evidence for these alternative hypotheses.

FIGURE 9-11

FIGURE 9-11

The three hypotheses of mass extinction examined in the text. Note that the hypotheses and predictions could not be distinguished if a continuous fossil record across a boundary were not available. Think of each represented hypothesis as a column of rock deposits crossing a major geologic boundary. Older fossils are embedded lower in the column, and more recent species occur higher in the column. The numbered notes on the right therefore outline a chronological sequence of events across a major geologic boundary, moving from lower rock strata to higher strata. (*After Kauffman as reported by Weisburd 1986, Kauffman 1986*)

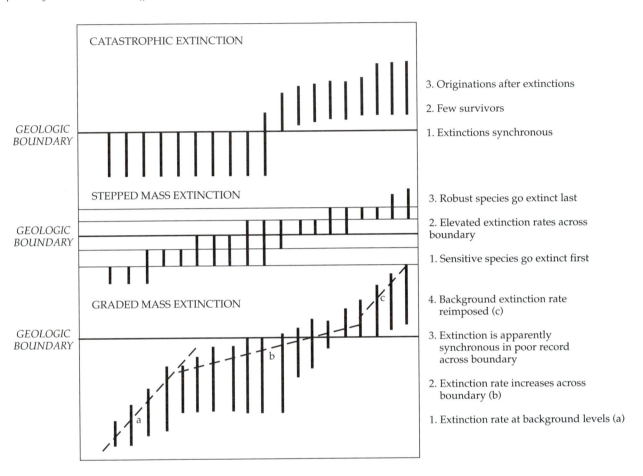

9.8 TESTING ALTERNATIVE HYPOTHESES FOR MASS EXTINCTIONS

Catastrophic Extinction

Alvarez and coworkers (e.g., Alvarez et al. 1980, 1984) favor catastrophic extinction. The Cretaceous–Tertiary (K/T) boundary, at 65 million years ago, is defined by a very rapid change in calcareous microplankton in limestone sequences. In 1978, researchers discovered unusual levels of iridium precisely at this boundary (e.g., Alvarez et al. 1979), and further research revealed high levels of iridium in 48 out of 50 sections through the K/T boundary (Alvarez et al. 1984). This iridium anomaly was thought to result

from the impact of an asteroid or comet measuring about 10 kilometers in diameter. Such an impact would cause darkness and a rapid change in temperature. A large comet would also shock-heat the atmosphere, causing nitrogen and oxygen to form oxides of nitrogen, which dissolve in water and form nitric acid and acid rain. High concentrations of oxides would absorb sunlight, asphyxiate terrestrial animals, kill plants, and decalcify the oceans, dissolving animal shells containing calcium carbonate. A strong candidate for a massive impact has left a 300-kilometer-diameter basin, known as the Chicxulub multiring impact basin, in the northern Yucatán, Mexico (Sharpton et al. 1993).

Alvarez et al. (1984) examined the fossil record of ammonites, bryozoans, brachiopods, and bivalves across the K/T boundary. They used the best possible stratigraphic sections, mainly at Stevns Klint, South of Copenhagen in Denmark. Such sections were important because (1) there is no discontinuity in stratigraphy across the boundary; (2) a rich fossil record is present; (3) the researchers used high-resolution stratigraphy with modern fossil recovery methods; and (4) an independent published literature existed and was used in the analysis.

For the ammonite record, Alvarez et al. found considerable extinction of short-lived species through the Lower and Upper Maestrichtian but at most only three synchronous extinctions of species (Figure 9-12). In contrast, nine species went extinct at the end of the Upper Maestrichtian, and none was found after that time. Those nine extinctions occurred at the iridium anomaly layer. Not only that, but three long-lived taxa went extinct at that time, whereas none had gone extinct in the Lower or Upper Maestrichtian before the boundary. Ammonites had suffered major extinctions several times, but those extinctions involved short-lived taxa, while the surviving long-lived species staged new radiations. With the sudden extinction of long-lived taxa, however, the ammonites were doomed.

Thus, in interpretation of the fossil record, a subtle shift in emphasis occurred. The conventional wisdom regarded extinction of short-lived ammonites as the beginning of a trend that would involve all ammonites, i.e., the gradual extinction of a clade. Alvarez et al. emphasized that the significant and unpredictable event was the extinction of the long-lived species, which resulted in the extinction of the whole group and was caused by a massive impact.

A large extinction, involving 60 out of 71 species (85 percent), occurred in the bryozoans. Eleven species remained into the Lower Tertiary. This event must be regarded as a major extinction, and it occurred just below the iridium anomaly.

In the brachiopods, 20 species existed in the Upper Maestrichtian. Extinction was abrupt and involved at least 75 percent of the species. Up to the boundary there was no gradual decline in species diversity, and no early extinction of specialized groups. The extinction coincided with the nearly

FIGURE 9-12

The ranges of ammonite species in Denmark through Maestrichtian times. Note, marked with asterisks, the three very long-lived species that went extinct at the K/T boundary. *(After Birkelund 1979)*

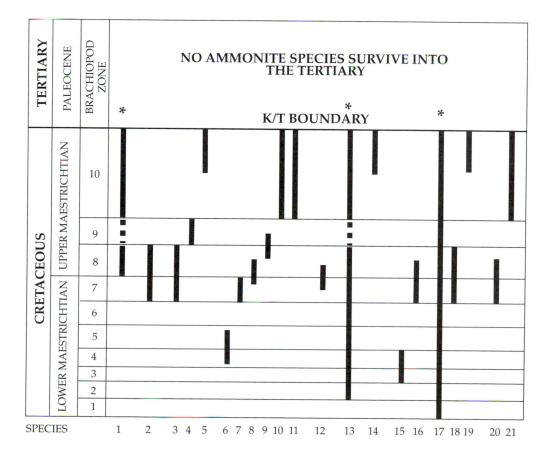

complete extinction of coccoliths and planktonic fora-minifera as well as with the extinction of the bryozo-ans. These synchronous extinctions in several taxa suggest a common cause, and correlation with the iridium anomaly implies that extraterrestrial impact was that cause.

Bivalve shells may or may not be preserved, de-pending on their compositions. Most shells are made of aragonite, which preserves poorly in chalk. Some bivalves, such as oysters and scallops, have shells made largely of calcite, which preserves well. Among the calcite-shelled taxa, 4 genera out of 12 went extinct at the iridium anomaly.

The just-mentioned four groups show significant and synchronous extinctions at the K/T boundary, coincident with the iridium anomaly. Alvarez et al. urge us to undertake much more detailed analysis of the very best available sections on other invertebrate groups that are generally regarded as being little af-fected by the K/T extinction event: the gastropods, corals, and echinoderms.

Stepped Mass Extinction

Kauffman favors stepped mass extinction as a model. Alvarez et al. (1984) used the record we discussed for ammonites, bryozoans, brachiopods, and bivalves to document a synchronous mass extinction at the K/T boundary. Under closer scrutiny, however, the same data indicate that some species went extinct before the boundary, especially ammonites (Figure 9-12), and in other taxa, species persisted through the boundary and went extinct later. A stepwise model with sequential episodes of extinction may provide a better description of events than a synchronous mass extinction model.

The short-lived ammonites went extinct early, several in the early Upper Maestrichtian. Among the bryozoans, 11 species remained into the Lower Ter-tiary. In the brachiopods, about five species persisted into the Lower Tertiary, and in the bivalves, eight genera survived. Kauffman suggested that specialists go extinct first and generalists last, as seems to have been the case for the ammonites.

Destabilization of the Earth's Atmosphere and Oceans

In the stepped mass extinction model, the central cause of extinction is major destabilization of the earth's atmosphere and oceans, as evidenced by the stable isotope record—the change in the ratio of oxy-gen isotopes in deep-sea sediments. Fluctuations in stable isotopes correlate with extinction periods. At some geological boundaries, major impact or multiple small impacts may cause such destabilization, but impact is not an essential feature of mass extinction.

Major destabilization in the earth's atmosphere might cause large (2 to 5°C), rapid drops in tempera-ture and changes in ocean chemistry. Such changes would disrupt the ecology of species, particularly those adapted to stable conditions. Therefore, more extinction would occur in tropical than in temperate regions. A collapse in phytoplankton would mean a collapse of the whole food chain in the oceans. For example, large amounts of nitric acid produced at impact would cause decalcification of the ocean. In fact, 99.2 percent of the world's marine plankton be-came extinct. In the Stevns Klint K/T boundary sec-tion, deposition of calcium stopped for a brief geologi-cal period, and fish clay in which the iridium anomaly occurs was deposited.

The sulfur isotope ^{34}S has undergone occasional sharp rises in the surface waters of the world's oceans (Holser 1977). One episode occurred roughly at the end of the Devonian, and another in the Lower and Middle Triassic, possibly influencing the extinctions of marine organisms in late Devonian and late Trias-sic times. The sharp changes indicate catastrophic mixing of salt-laden water from evaporating basins with surface ocean waters. Such waters entering the oceans would probably be rich in salts, sulfide ions, ^{34}S, and nutrients but would also be high in oxygen demand, dramatically changing the ecology of the surface waters into which they flowed.

An Example Using Pelagic Microfossils

Lipps (1986) provided evidence in support of the stepped mass extinction hypothesis. Pelagic microfos-sils provide abundant data for accurate description of distributions in time and space. Three major mass extinctions have occurred in the last 150 million years in the pelagic marine environment, at the Cenoman-ian–Turonian boundary (91 million years ago), at the Cretaceous–Tertiary boundary (65 million years ago), and near the Eocene–Oligocene boundary (36.6 mil-lion years ago). Lipps noted common elements in the environmental changes and consequent biological changes that occurred at those times. The ocean envi-ronment changed in the following ways: marine cir-culation reduced, upwelling intensity declined, and the thermocline and/or halocline became diffuse. The net result was greatly reduced habitat heterogeneity, both horizontally across oceans and vertically in the water column, with severe declines in high-quality habitats rich in nutrients. Loss of vertical heterogene-ity caused extinction of the smaller plankton that seg-regate along this environmental gradient. Phyto-plankton could not survive the strongly reduced nutrients as upwelling declined. Reduced phyto-plankton undermined the whole trophic system of the oceans, and larger pelagic organisms, including the large carnivores, became extinct. For example, the enormous marine reptilian mosasaurs and the fish-eating birds in the genera *Ichthyornis* and *Hesperornis* all went extinct at the end of the Cretaceous.

In summary, environmental changes at these boundaries had major repercussions in the biotas of marine environments. (1) Before extinction, the pelagic biota was very diverse, and after extinction it was much reduced. (2) Planktonic foraminifera were distributed in latitudinal provinces before extinction, but surviving species became cosmopolitan after the extinction episode (Figure 9-13). (3) In tropical waters, foraminifera had complex morphologies, but after the extinction event, all species had simple morphology. (4) Large vertebrate consumers were absent after the extinction.

These patterns demonstrate the selective nature of extinction emphasized by Lipps and probably also demonstrate sequential episodes of extinction starting with the most sensitive microplankton in tropical seas. Such selective extinctions force us to search for rather subtle causes. Sluggish ocean circulation may initiate changes in oceanographic patterns throughout the world, but the causes are unclear. Asteroid impact would increase turbulence and nutrients in the ocean—effects opposite those observed. Subtler influences act over longer periods with less obvious and more selective effects and with the expectation of stepped mass extinctions. Whether such effects would be so attenuated as to register across geological time spans is open to debate.

FIGURE 9-13

Latitudinal provinces in modern planktonic foraminifera. **Right:** modern species. **Left:** species from the late Cretaceous in tropical waters and from the early Paleocene in colder waters. The latter, however, would have been distributed across provinces after the K/T extinctions. Note that in tropical waters species with complex morphologies are common, but in colder seas species exhibit much simpler structures. The three aspects of two late Cretaceous tropical fossil species, A and B, illustrate characteristics of complex morphology: elongated chambers and ridged surfaces. Simple morphology involves more spherical chambers arranged in spirals. *(Based on Bradshaw 1959 and modified from Lipps 1986)*

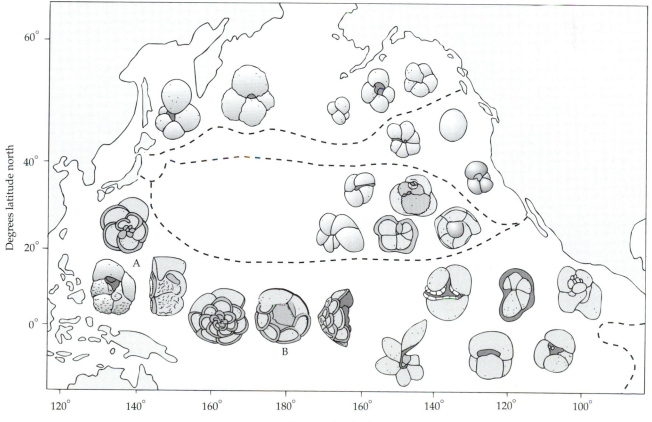

Graded Mass Extinctions

Terrestrial Vertebrates

Graded mass extinction is a logical alternative hypothesis positing gradualism and the slow influence of global changes in climate, accompanied by changes in sea level and continental landmasses (see Figure 9-11). On the basis of the fossil record of terrestrial vertebrates across the K/T boundary, Clemens (1986) has advocated a view of highly selective extinction spread across geological time. He used the best fossil sites available, in Montana, Wyoming, and Alberta. High selectivity of extinction is evident in the total extinction of dinosaurs (Saurischia and Ornithischia) and pterosaurs, 66 percent extinction of marsupial families, only 25 percent loss of placental mammal families, and no loss of families in some reptiles (turtles, crocodiles, and lizards) (Table 9-4). In addition, among ornithiscian dinosaurs a significant decline in taxa—from 20 genera to 14 genera, mainly in the families Hadrosauridae and Ceratopsidae—occurred well before the K/T boundary. However, although local extinction was evident in the north (Montana and Alberta), some genera, such as *Hadrosaurus*, persisted farther south (New Mexico) until the end of the Cretaceous, indicating a southward shift in distribution, presumably in response to changing climatic conditions.

From a search for general patterns among the taxa that went extinct, Clemens (1986) concluded the following:

1. Species of large adult body size tended to go extinct more frequently than small-bodied species.
2. No general pattern occurred among ectotherms and endotherms. Among warm-blooded animals the marsupials were heavily affected, but the placentals were not. Several cold-blooded taxa such as lizards and turtles were less affected than marsupials.

Clemens (1986) argued that evidence independent of the vertebrate fossil record indicates a large recession of shallow epicontinental seas toward the end of the Cretaceous, which may well account for much of the marine invertebrate extinction around the K/T boundary. Such recession would remove geographic barriers to the dispersal of terrestrial species; floras and faunas would merge; and climate would change with modification of oceanic circulation patterns and greatly enlarged continental land-

T A B L E   9 - 4   Estimates of percentage extinctions of some vertebrates with adequate fossil records across the K/T boundary

Taxon	Common Name	Total Genera	Total Families	Extinct Genera (%)	Extinct Families (%)
Fish					
Chondrichthyans	Cartilaginous fish (sharks, skates, rays, etc.)	5	4	60	25
Osteichthyans	Bony fish	13	10	38	10
Amphibia		12	8	33	13
Reptilia					
Chelonia	Turtles	18	4	11	0
Crocodilia	Crocodiles	4	1	25?	0
Lacertilia	Lizards	15	7	27?	0
Serpentes	Snakes	2	1	0?	0
Saurischia	Lizard-hipped dinosaurs	8	5	100	100
Ornithischia	Bird-hipped dinosaurs	14	8	100	100
Pterosauria	Flying reptiles	?	?	100	100
Mammalia					
Multituberculata	Multituberculates	11	8	36	25
Marsupialia	Marsupials	4	3	75	66
Placentalia	Placentals	9	4	11	25

Data are from localities that were coastal lowlands in what is now Montana, Wyoming, and Alberta.
Data from Clemens 1986.

masses. Global climates cooled, and the seasons became more distinct on the large continental landmasses. These long-term changes were perhaps sufficient to account for the observed southerly distribution of some ornithiscians and a rather gradual extinction of species in the late Cretaceous. (See also Sloan et al. 1986 for similar views, with the additional influences of a resultant major deterioration of the flora and diffuse competition from newly evolving herbivorous mammals.) "The possibility remains that extinction of these lineages may not prove to be extraordinary in rate or magnitude when the overall pattern and tempo of change in diversity through the Phanerozoic is adequately assessed" (Clemens 1986, 75). But were gradual climatic changes sufficient to account for all extinctions during Cretaceous/Tertiary times?

Terrestrial Plants

Using primary producers as evidence, Hickey (1981) supported a scenario of gradual extinction across the K/T boundary. Land plants, which form the base of the terrestrial food chain, are particularly important in any analysis of mass extinction, for we need to know whether extinctions are based on food-chain collapse or on other events more specific to certain kinds of animals. Using studies in Montana and Wyoming, Hickey found moderate levels of land-plant extinction at the end of the Cretaceous, but extinctions occurred largely about 50,000 to 90,000 years after the extinction of the dinosaurs.

Nichols et al. (1986) took a contrary view. Studying the K/T boundary in Saskatchewan, they found a sudden increase in fern spore abundance relative to angiosperm pollen, coincident with and immediately following the iridium anomaly (Figure 9-14). In the Maestrichtian, angiosperms dominated assemblages, with their pollen representing 42 to 92 percent of the total pollen-plus-spore record. Spores of pteridophytes, mainly ferns, represented 5 to 46 percent, and gymnosperm pollen was low in abundance, 3 to 12 percent. At the iridium anomaly, a large proportion of fern spores developed, making up almost 100 percent of the pollen–spore assemblage. This fern-spore spike has also been observed in Montana and New Mexico. It represents a low number of fern species but a high density of the species present. Researchers interpreted this pattern as a new recolonization of the terrestrial surface after a devastating event at the K/T boundary.

Nichols et al. (1986) estimated that 30 percent of the palynoflora, mostly angiosperm taxa, went extinct at the K/T boundary in Saskatchewan. Most plant taxa survived the K/T extinction, however, and some palynological evidence indicates a fairly stable climate across the K/T boundary. "If paleoclimate was altered, it may have undergone only a brief change consistent with a 3 to 6 month period of darkness, after which it returned to the previous frost-free condition" (Nichols et al. 1986, 716). Even thermophilic palms and screw pines survived into the Tertiary, suggesting that no persistent and drastic change in climate occurred.

More recent estimates of taxon richness indicate rapid and high extinction at the K/T boundary. For the western North American flora, from New Mexico through Alberta and Saskatchewan up to the North Slope of Alaska, Nichols et al. (1990) concluded that a mean of 45 percent of all plants went extinct at the K/T boundary: 25 percent of pteridophytes, 36 percent of gymnosperms, and 51 percent of angiosperms.

9.9 FUTURE STUDIES

The argument by Nichols et al. is interesting in the light of the studies by Hickey and Clemens. Could climates change enough to cause animal extinctions but not many plant extinctions? Could climatic

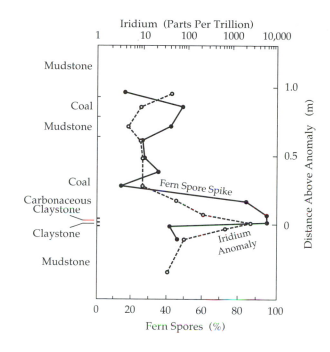

FIGURE 9-14

Fern spore abundance across the K/T boundary at Morgan Creek, Saskatchewan, and iridium concentrations, showing a large increase in relative spore abundance—a fern-spore spike—coincident with the iridium anomaly. *(After Nichols et al. 1986)*

changes or large-body impact have altered the relative abundance of plants enough so that vertebrate herbivores ran out of food? And of what relative importance are the processes—large-body impact, destabilization of the earth's atmosphere and oceans, changes in temperature, salinity, acidity, calcium content or nutrient status, and sea-level changes? And will we eventually be able to isolate single causes, or will several changes acting in concert escalate the magnitude of impact?

Probably, no one scenario can explain all the major episodes of extinction. Several authors (e.g., Keller 1983, Keller et al. 1983, Lipps 1986, Sepkoski and Raup 1986) accept a stepwise extinction event that lasted for perhaps 8 to 12 million years in the Eocene–Oligocene. The Cretaceous–Tertiary extinctions were more sudden and persisted for about 3 million years (Smit 1982, Lipps 1986). In the end, depending on the extinction event and the taxon studied, all of Kauffman's alternative hypotheses may be supported.

The views and questions raised by the controversial literature significantly change the perspective on mass extinctions. We should not be as impressed by the brutal extirpation of life as by the subtlety and selectivity of extinction. Emphasis on differential impact on the basis of different biological attributes is creating a fertile climate in which to study the mechanistic biology of extinction. Both idiographic and nomothetic approaches seem essential to the testing of extinction hypotheses, and the mechanistic biology of extinction should be a fertile field of study for many years to come.

SUMMARY

Extinction is a process that ultimately affects all species, from fossil organisms through the species that live around us to humankind itself. Indeed, the human species may be causing the next round of mass extinction.

Background rates of extinction have been interspersed with episodes of mass extinction, with the balance of originations and extinctions resulting in a generally increasing diversity of life on earth. Historically, scientists have regarded extinctions as special events caused by catastrophes, gradual environmental change, competition from newly emerging groups, and overexploitation by humans as they spread around the world.

The Pleistocene extinctions of predominantly large animals are particularly interesting because of their relative recency (they occurred in the last 50,000 years) and the rich fossil record available for analysis. The Pleistocene overkill hypothesis is worth examining in some detail because it accounts for differential impact on large versus small vertebrates, differences among continents, and differences in timing of extinction.

General approaches to the study of extinction have identified the periodicity of mass extinctions and have revealed the need to find and understand forcing factors beyond the solar system, on the scale of the Milky Way galaxy. The existence of a solar companion, Nemesis, is but one of several speculations in this uncertain territory. Large-body impact on the earth from massive comets, meteorites, or asteroids could cause mass extinction, but alternative hypotheses may be more relevant to certain groups of the earth's biota.

When we test alternative hypotheses for mass extinction, we can envision three basic processes: catastrophic mass extinction, stepped mass extinction, and graded mass extinction. A catastrophic extinction would be the result of a single major event such as an asteroid or comet impact and would include rapid and almost simultaneous extinctions, such as those recorded in marine groups of ammonites, bryozoans, brachiopods, and bivalves. Even in those cases, however, extinction was stepwise, possibly involving sensitive species first and broadly adapted species later. Pelagic microfossils such as the Foraminifera offer an example of stepped mass extinction: specialized tropical species succumbed first, because of reduced heterogeneity of tropical waters, and subsequently undermined the whole trophic structure of pelagic food webs in the ocean; the more tolerant, cold-adapted, generalized species formed the basis for radiations after mass extinction. Turning to terrestrial environments, evidence from vertebrates and some higher plant assemblages supports the hypothesis of graded mass extinction, which may involve cooling climates and increasing variation in seasons.

For all these hypotheses conflicting evidence exists, and the search for less controversial, more solid evidence is in progress.

QUESTIONS FOR DISCUSSION

1. While recognizing that competition between species occurs locally, do you think global patterns of extinction in major groups can result from competition? Can you provide sound evidence of resource use and distribution overlap to make a strong case for the role of competition in extinction?

2. Given that many clades are expanding or contracting over similar periods of time, do you think that invoking an interactive connection between clades is akin to the adaptationist program—i.e., convenient but not rigorous?

3. How would you plan a research program to develop a critical test of the Pleistocene overkill hypothesis?

4. In your opinion, would our understanding of the mechanisms involved with extinction progress more rapidly if more emphasis were placed on testing alternative hypotheses simultaneously? For any particular event, which hypotheses would you wish to see tested?

5. Extinction appears to be highly selective at most geological boundaries, with major loss of species. How can this observation help to resolve debates on the causes of extinction?

6. Do you think scientists should place more emphasis on the individual biologies of the groups involved with extinction and survival, with a balanced examination of habitat, food, and life history traits, including the possibility of competition if the evidence permits?

7. In your opinion, does the resolution of much of the debate on causes of extinction depend on the discovery of many more continuous stratigraphic sections that enable researchers to make a detailed examination of extinctions and originations across a geological boundary?

8. Do you consider the bases of predictions of human-caused future extinctions to be adequate, given the fact that in nature and in human economics compensatory influences are frequently important?

9. Would you regard extinction as simply a result of destruction in nature, without relevance to evolutionary studies, or is there a tight mechanistic link between extinctions and originations, of vital interest to the evolutionist?

10. The fossil records for marine species and for angiosperms are two of several indications that diversity of taxa is higher now than ever before. Is it conceivable that diversity will continue to rise with humans present, or will further increases have to wait until after the human species goes extinct?

REFERENCES

Adams, R. McC. 1994. Smithsonian horizons. Smithsonian 24(12):10.

Alvarez, W., L. W. Alvarez, F. Asaro, and H. V. Michel. 1979. Experimental evidence in support of an extraterrestrial trigger for the Cretaceous–Tertiary extinctions. Eos 60:734.

———. 1980. Extraterrestrial cause for a Cretaceous–Tertiary extinction. Science 208:1095–1108.

Alvarez, W., E. G. Kauffman, F. Surlyk, L. W. Alvarez, F. Asaro, and H. V. Michel, 1984. Impact theory of mass extinctions and the invertebrate fossil record. Science 223:1135–1141.

Alvarez, W., and R. A. Muller. 1984. Evidence from crater ages for periodic impacts on the Earth. Nature 308:718–720.

Anderson, E. 1984. Who's who in the Pleistocene: A mammalian bestiary, pp. 40–89. In P. S. Martin and R. G. Klein (eds.). Quaternary Extinctions. Univ. of Arizona Press, Tucson.

Bakker, R. T. 1983. The deer flees, the wolf pursues: Incongruencies in predator–prey coevolution, pp. 350–382. In D. J. Futuyma and M. Slatkin (eds.). Coevolution. Sinauer, Sunderland, Mass.

Bates, M. 1964. The Land and Wildlife of South America. Time, New York.

Birkelund, T. 1979. The last Maastrichtian ammonites, pp. 51–57. In T. Birkelund and R. G. Bromley (eds.). Cretaceous/Tertiary boundary events symposium. Vol. 1. Univ. of Copenhagen, Copenhagen.

Bradshaw, J. S. 1959. Ecology of living planktonic foraminifera in the north and equatorial Pacific. Contrib. Cushman Found. Foraminif. Res. 10:25–64.

British Museum (Natural History). 1969. British Paleozoic Fossils. British Museum (Natural History), London.

Bruemmer, F. 1986. How the mermaid perished. Int. Wildlife 16(1):24.

Carrington, R. 1963. The Mammals. Time, New York.

Chapman, C. R., and D. Morrison. 1994. Impacts on Earth by asteroids and comets: Assessing the hazard. Nature 367:33–40.

Cifelli, R. L. 1981. Patterns of evolution among the Artiodactyla and Perissodactyla (Mammalia). Evolution 35:433–440.

Clemens, W. A. 1986. Evolution of the terrestrial vertebrate fauna during the Cretaceous–Tertiary transition, pp. 63–85. In D. K. Elliott (ed.). Dynamics of Extinction. Wiley, New York.

Colbert, E. H. 1980. Evolution of the Vertebrates. Wiley, New York.

Colbert, E. H., and M. Morales. 1991. Evolution of the Vertebrates. 4th ed. Wiley–Liss, New York.

Cuvier, G. 1796a. Notice sur le squelette d'une très-grande espèce de quadrupède inconue jusqu'à présent, trouvé au Paraguay, et déposé au Cabinet d'Histoire naturelle de Madrid. Mag. Encyclopédique 2me annee 1:303–310.

———. 1796b. Memoire sur les espèces d'elephans tant vivantes que fossiles. Mag. Encyclopédique, 2me annee, 3:440–445.

———. 1812. Recherches sur les ossemens fossiles des quadrupèdes, où l'on rétablit les caractères de plusieurs espèces d'animaux que les révolutions du globe paroissent avoir détruites. Deterville, Paris.

Davis, M., P. Hut, and R. A. Muller. 1984. Extinction of species by periodic comet showers. Nature 308:715–717.

De Blij, H. J. (ed.). 1988. Earth '88: Changing Geographic Perspectives. National Geographic Society, Washington, D.C.

Degens, E. T., and P. Stoffers. 1976. Stratified waters as a key to the past. Nature 263:22–27.

Ehrlich, P. R. 1984. The biological consequences of nuclear war, pp. 41–71. In P. R. Ehrlich, C. Sagan, D. Kennedy, and W. O. Roberts (eds.). The Cold and the Dark: The World After Nuclear War. W. W. Norton, New York.

———. 1986. Extinction: What is happening now and what needs to be done, pp. 157–164. In D. K. Elliott (ed.). Dynamics of Extinction. Wiley, New York.

Ehrlich, P. R., and E. O. Wilson. 1991. Biodiversity studies: Science and policy. Science 253:758–762.

Gould, S. J., and C. B. Calloway. 1980. Clams and brachiopods—ships that pass in the night. Paleobiology 6:383–396.

Grayson, D. K. 1984. Nineteenth-century explanations of Pleistocene extinctions: A review and analysis, pp. 5–39. In P. S. Martin and R. G. Klein (eds.). Quaternary Extinctions. Univ. of Ariz. Press, Tucson, Arizona.

Hickey, L. J. 1981. Land plant evidence compatible with gradual, not catastrophic, change at the end of the Cretaceous. Nature 292:529–531.

Holser, W. T. 1977. Catastrophic chemical events in the history of the ocean. Nature 267:403–408.

Jablonsky, D. 1989. The biology of mass extinction: A palaeontological view. Phil. Trans. R. Soc. Lond. B 325:357–368.

———. 1991. Extinctions: A paleontological perspective. Science 253:754–757.

Jepsen, G. L. 1949. Selection, "orthogenesis," and the fossil record. Proc. Amer. Phil. Soc. 93:479–500.

Kauffman, E. G. 1977. Geological and biological overview: Western Interior Cretaceous Basin, pp. 75–99. In E. G. Kauffman (ed.). Cretaceous Facies, Faunas, and Paleonenvironments Across the Western Interior Basin. Mountain Geologist 14(3/4).

———. 1984. Paleobiogeography and evolutionary response dynamic in the Cretaceous Western Interior Seaway of North America, pp. 273–306. In G. E. G. Westermann (ed.). Jurassic–Cretaceous Biochronology and Paleogeography of North America. Geol. Assoc. Can. Special Paper No. 27.

———. 1985. Cretaceous evolution of the Western Interior Basin of the United States, pp. iv–xiii. In L. M. Pratt, E. G. Kauffman, and F. B. Zelt (eds.). Fine-Grained Deposits and Biofacies of the Cretaceous Western Interior Seaway: Evidence of Cyclic Sedimentary Processes. Soc. Econ. Paleont. Mineral., Field Guidebook 4.

———. 1986. High-resolution event stratigraphy: Regional and global Cretaceous bio-events, pp. 279–335. In O. H. Walliser (ed.). Global Bio-events: A Critical Approach. Springer-Verlag, Berlin.

Keller, G. 1983. Biochronology and paleoclimatic implications of middle Eocene to Oligocene planktic foraminiferal faunas. Marine Micropaleontol. 7:463–486.

Keller, G., S. D'Hondt, and T. L. Vallier. 1983. Multiple microtektite horizons in Upper Eocene marine sediments: No evidence for mass extinctions. Science 221:150–152.

Krause, D. W. 1986. Competitive exclusion and taxonomic displacement in the fossil record: The case of rodents and multituberculates in North America. Cont. Geol. Univ. Wyoming, Special Paper 3:95–117.

Lambert, D. 1985. The Cambridge Field Guide to Prehistoric Life. Univ. of Cambridge Press, Cambridge, England.

Lipps, J. H. 1986. Extinction dynamics in pelagic ecosystems, pp. 87–104. In D. K. Elliott (ed.). Dynamics of Extinction. Wiley, New York.

Lyell, C. 1830. Principles of Geology, Being an Attempt to Explain the Former Changes of the Earth's Surface by Reference to Causes Now in Operation. Vol. 1. J. Murray, London.

———. 1832. Principles of Geology, Being an Attempt to Explain the Former Changes of the Earth's Surface by Reference to Causes Now in Operation. Vol. 2. J. Murray, London.

Martin, P. S. 1967. Pleistocene overkill. Natur. Hist. 76:32–38.

———. 1984a. Prehistoric overkill: The global model, pp. 354–403. In P. S. Martin and R. G. Klein (eds.). Quaternary Extinctions. Univ. of Arizona Press, Tucson.

———. 1984b. Catastrophic extinctions and late Pleistocene blitzkrieg: Two radiocarbon tests, pp. 153–189. In M. H. Nitecki (ed.). Extinctions. Univ. of Chicago Press.

Martin, P. S., and R. G. Klein. 1984. Quaternary Extinctions. Univ. of Arizona Press, Tucson.

Minckley, W. L., and J. E. Deacon (eds.). 1991. Battle Against Extinction: Native Fish Management in the American West. Univ. of Arizona Press, Tucson.

Murray, P. 1984. Extinctions down under: A bestiary of extinct Australian late Pleistocene Monotremes and Marsupials, pp. 600–628. In P. S. Martin and R. G. Klein (eds.). Quaternary Extinctions. Univ. of Arizona Press, Tucson.

Nichols, D. J., D. M. Jarzen, C. J. Orth, and P. Q. Oliver. 1986. Palynological and iridium anomalies at Cretaceous–Tertiary boundary, South-Central Saskatchewan. Science 231:714–717.

Nichols, D. J., R. F. Fleming, and N. O. Frederiksen. 1990. Palynological evidence of effects of the terminal Cretaceous event on terrestrial floras in western North America, pp. 351–364. In E. G. Kauffman and O. H.

Walliser (eds.). Extinction Events in Earth History. Springer–Verlag, Berlin.

Rampino, M. R., and R. B. Stothers. 1984. Terrestrial mass extinctions, cometary impacts and the sun's motion perpendicular to the galactic plane. Nature 308:709–712.

Raup, D. M. 1991. Extinction: Bad Genes or Bad Luck? Norton, New York.

Raup, D. M., S. J. Gould, T. J. M. Schopf, and D. Simberloff. 1973. Stochastic models of phylogeny and the evolution of diversity. J. Geol. 81:525–542.

Raup, D. M., and J. J. Sepkoski. 1982. Mass extinctions in the marine fossil record. Science 215:1501–1503.

———. 1984. Periodicity of extinctions in the geologic past. Proc. Natl. Acad. Sci. U.S.A. 81:801–805.

———. 1986. Periodic extinction of families and genera. Science 231:833–836.

Sagan, C. 1984. The atmospheric and climatic consequences of nuclear war, pp. 1–39. In P. R. Ehrlich, C. Sagan, D. Kennedy, and W. O. Roberts (eds.). The Cold and the Dark: The World After Nuclear War. W. W. Norton, New York.

Sepkoski, J. J., and D. M. Raup. 1986. Periodicity in marine extinction events, pp. 3–36. In D. K. Elliott. Dynamics of Extinction. Wiley, New York.

Sharpton, V. L., K. Burke, A. Camargo-Zanoguera, S. A. Hall, D. S. Lee, L. E. Marín, G. Suárez-Reynoso, J. M. Quezada-Muñeton, P. D. Spudis, and J. Urritia-Fugugauchi. 1993. Chicxulub multiring impact basin: Size and other characteristics derived from gravity analysis. Science 261:1564–1567.

Simpson, G. G. 1953. The Major Features of Evolution. Columbia Univ. Press, New York.

Sloan, R. E., J. K. Rigby, L. M. Van Valen, and D. Gabriel. 1986. Gradual dinosaur extinction and simultaneous ungulate radiation in the Hell Creek Formation. Science 232:629–633.

Smit, J. 1982. Extinction and evolution of planktonic Foraminifera after a major impact at the Cretaceous/Tertiary boundary. Geol. Soc. Amer. Special Paper 190:329–352.

Špinar, Z. V. 1972. Life Before Man. Thames and Hudson, London.

Stanley, S. M. 1968. Post-Paleozoic adaptive radiation of infaunal bivalve molluscs—a consequence of mantle fusion and siphon formation. J. Paleontol. 42:214–229.

———. 1984a. Marine mass extinctions: A dominant role for temperature, pp. 69–117. In N. H. Nitecki (ed.). Extinctions. Univ. of Chicago Press.

———. 1984b. Temperature and biotic crises in the marine realm. Geology 12:205–208.

Thierstein, H. R., and W. H. Berger. 1978. Injection events and ocean history. Nature 276:461–466.

Valentine, J. W., and D. Jablonski. 1982. Larval strategies and patterns of brachiopod diversity in space and time. Geol. Soc. Am. Abstr. 14:241.

Vermeij, G. J. 1987. Evolution and escalation: An ecological history of life. Princeton Univ. Press, Princeton, N.J.

Weisburd, S. 1986. Extinction wars. Sci. News 129:65–80.

Wilde, P., and W. B. N. Berry. 1984. Destabilization of the oceanic density structure and its significance to marine "extinction" events. Palaeogeogr. Palaeoclimatol., Palaeoecol. 48:143–162.

Wilson, E. O. 1992. The Diversity of Life. W. W. Norton, New York.

———. 1993. The Diversity of Life. Special edition. Harvard Univ. Press, Cambridge, Mass.

10

THE SURVIVORS

The Moorish idol, a tropical Indo-Pacific fish found in coral reefs, represents a geographic area apparently somewhat buffered against mass extinction events.

(Illustration by Stephen Price)

The human tendency toward overemphasis of catastrophic extinction is perhaps understandable, because we are good at focusing on disaster. Nevertheless, it is unrealistic to concentrate on extinction without also giving attention to survivors and the postextinction events that repopulate depauperate biotas. Thus, just as Wilson (1988, 73) liked to deal with the dual phenomena of "extinction and renewal," so we should try to understand how the budget of life is balanced.

How does the restocking of species after catastrophes proceed? The recovery of biotic diversity is remarkable on a geological time scale. For example, consult Figure 9-3 (Chapter 9) and note the recovery of the number of marine families in the Jurassic, Cretaceous, and Tertiary periods. In fact, at the end of the Tertiary (according to the fossil record, at least), the marine fauna was the richest ever recorded. During a similar period, a massive and rapid evolutionary radiation of terrestrial flowering plants occurred. "In short, global biological diversity is now at or close to its all-time high. Another way of putting it is that mankind came into existence in the most interesting time in which any sentient species could live" (Wilson 1988, 74).

Despite the inevitable dual nature of extinction–renewal phenomena, "analyses of selectivity during mass extinctions are still scarce" (Jablonski 1991, 754), and the mechanisms permitting survival are poorly understood. Once survivors are established after a period of extinction, then the processes of speciation (Chapter 5) and radiation (Chapter 8) presumably proceed as already discussed. We scientists should be searching for the mechanisms involved in survival and renewal. The predictive science of survival after catastrophic global events (see Figure 9-1) is in its infancy.

In earlier chapters of this book, we caught glimpses of the likely survivors after mass extinction: the deep-basin ammonites; the clams with their planktonic larvae, and hence wide geographical distribution, coupled with mantle fusion and siphon formation; the morphologically simple but cold-adapted species of Foraminifera; and species with a long history of exposure to a threat, such as the African megafauna (compared with other megafaunas). But are there any patterns? Are predictions of the likely survivors possible? Must we persist with an idiographic view of extinction and survival, or are nomothetic principles at work?

10.1 LINKAGE BETWEEN SPECIATION AND EXTINCTION

A factor that complicates the study of extinction and renewal is that rapid evolutionary rates inevitably result in frequent originations (or speciations) but also high extinction rates, as we saw in the case of the lungfishes. Thus, an apparently volatile evolutionary history, such as that of the ammonites, may not produce as viable a long-term pattern as, say, the history of the "imperturbable gastropods" (Jablonski 1986, 210).

An interesting case of linkage between speciation and extinction rates concerns some African mammal lineages that have living members and are recorded in fossils (Vrba 1987). Researchers can use the mammals' present ecological relationships to food and habitat to evaluate whether a lineage has been rather more specialized or more generalized in its utilization of resources and how that factor influences geographic distributions. Vrba calls species and lineages with typically narrow resource exploitation **stenobiomic organisms** (just as a *steno*haline species has narrow tolerances to salt concentration). Thus, a specialized grazer such as a waterbuck (*Kobus*) or a specialized browser such as a giraffe (*Giraffa*) would be a stenobiomic species. Conversely, species and lineages with broad food and habitat tolerances and utilization patterns, spreading their foraging over more than one biome, are called **eurybiomic organisms.** For example, African buffaloes (*Syncerus* and *Ugandax*) are both grazers and browsers and move between grassland and woodland biomes.

Vrba noted the first appearance in the fossil record of each lineage, or clade, of African mammals as time *t,* in millions of years before present, and added the numbers of extinct species and living species for the total species, *N,* produced since the first record. The number of speciations in the time available could be calculated as N/t, and the speciation rate (S) was estimated as $S = \ln N/t$. Therefore, the speciation rate could be calculated, and habitat and food utilization could be estimated, for each lineage.

A clear pattern emerged that showed a correlation between high speciation rates and stenobiomic lineages (Figure 10-1). This pattern was consistent with Vrba's resource-use hypothesis, which predicted that through evolutionary time stenobiomic clades would pass through stages of dramatic decline in biome area, large populations would become subdivided, and repeated episodes of allopatric population structure would result in high speciation rates. Low rates of speciation would be observed in clades of eurybiomic organisms, because dramatic habitat change would have less impact on more generalized species. At the same time, the number of extinct species strongly influenced the total species in a clade, indicating a link between originations and extinctions.

Vrba's analysis perhaps creates a paradox in juxtaposition with other patterns to be considered in this chapter. We will see that geographically widespread

FIGURE 10-1

A slightly simplified depiction of the realtionship between the feeding characteristics of mammalian lineages in Africa and their speciation rates, according to Vrba's (1987) analysis. The stenobiomic species on the left exhibit the more specialized types of food and habitat utilization. On the right are the more generalized exploiters: the eurybiomic species, which utilize at least two biomes, grassland and woodland, and omnivores such as bushpigs and baboons that range over several biomes. A representative genus with living species and a common name or names for the group appear in each rectangle. Note the strong negative correlation between speciation rate and the gradient from stenobiomic to eurybiomic organisms.

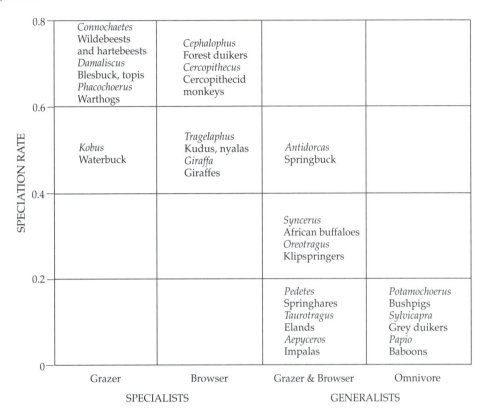

species survive extinctions better than species with narrower distributions. But it may be the more steno-biomic lineages that speciate more rapidly after an extinction event and contribute more to the recovery of numbers of species and higher taxa.

10.2 PATTERN OF EVENTS AFTER EXTINCTION

The fossil record at any one locality reveals a series of predictable events after mass extinction: deposits almost free of fossils, the beginnings of a recovery in species number, diversification of species and ecological niches utilized, a diverse biota, and then another extinction event (Figure 10-2). The Western Interior Cretaceous Basin of North America, already discussed in Chapter 9 (Section 9.5), provides an example of this cycle on a regional scale (Kauffman 1986). For various reasons, such as increased stratification and reduced mixing of the water column, anoxic (low-oxygen) environments cause massive extinction of benthic, or bottom-dwelling, organisms. As oxygenation of the benthic region increases, a few species of detritus-feeding burrowers appear in the fossil record. Next comes colonization by burrowing species living in domicilia—established living quarters—such as the burrowing shrimps in the decopod crustacean superfamily Thalassinoidea. A third stage includes surface-dwelling, or epifaunal, species tolerant of relatively low oxygen levels, such as bivalve mollusks, members of the Inoceramidae. Finally, a diverse assemblage of species has formed (Figure 10-2). Such local observations, with colonization from relatively nearby sites, provide an accelerated image of the restocking of a biota, involving colonization and radiation from surviving lineages, after a global extinction event.

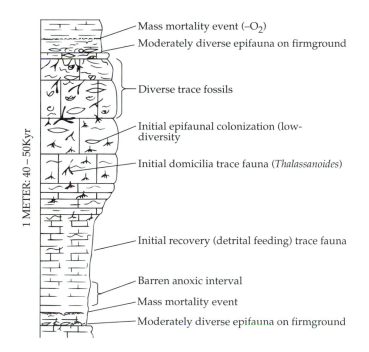

1 METER: 40 – 50Kyr

Mass mortality event (–O_2)

Moderately diverse epifauna on firmground

Diverse trace fossils

Initial epifaunal colonization (low-diversity)

Initial domicilia trace fauna (*Thalassanoides*)

Initial recovery (detrital feeding) trace fauna

Barren anoxic interval

Mass mortality event

Moderately diverse epifauna on firmground

FIGURE 10-2

A typical pattern of species recovery after an extinction episode in a shale–limestone deposit about 1 meter deep and representing 40,000 to 50,000 years. The extinction was regional rather than global, with moderately rapid colonization of vacated habitat. *(From Kauffman 1986)*

Harries and Kauffman (1990) have reconstructed the events following a global extinction at the end of the Upper Cretaceous. Again a "dead zone" appears just above the Cretaceous boundary (Figure 10-3). Surviving taxa include rare pelagic (open-ocean) ammonites, stress-resistant oysters, and inoceramid mollusks such as *Mytiloides.* Surviving species may appear very rarely in the dead zone or may be totally absent from it, and yet they are clearly not extinct, because the same species reappear in the fossil record in higher strata. This sequence—loss of species and their reappearance later in the fossil record—is called the **Lazarus effect** after the biblical Lazarus, who rose from the dead. The **Lazarus species** that reappear may have become generally rare for a period, with low probability of fossilization, or may have "weathered the storm" in certain kinds of refuges from extinction (discussed later in this chapter), then returned to colonize. As chemical and thermal conditions in the ocean stabilize, originations increase, based on Lazarus lineages, and ecological and morphological novelties and innovations follow. At first, originations of new species may be rapid, or "punctuated" (as discussed in Chapter 18), as in a radiation of new forms. This phase is followed by more gradual evolution of new species until a more balanced community with normal species richness is reestablished.

A multitude of traits in lineages may happen to provide buffers against extinction (Figure 10-4) (Harries and Kauffman 1990). Some of them follow.

1. **Ecological generalism.** Taxa with broad resource exploitation—Vrba's eurybiomic taxa (Section 10.1).
2. **Opportunism.** Taxa with high reproductive rates (*r*-selected species), usually exposed to stressful environments—prolific reproducers with short generation times.
3. **Disaster species.** Species adapted to intermittent stress, with population "blooms" after highly stressful conditions.
4. **Migration to alternative habitat.** A species capable of major shifts in habitat may escape the most stressful locations and persist in a secondary habitat. After the extinction event, the lineage may persist in the secondary habitat or in the original, or a speciation event may lead to occupation of both habitats.
5. **Refugia species.** Taxa whose ranges include refugia from stressful episodes (discussed in Section 10-4).
6. **Widely ranging taxa.** Taxa represented by many populations in wide geographic ranges may have a few populations that survive otherwise heavy impact—perhaps especially marginal populations frequently exposed to stressful conditions.
7. **Neotenic taxa.** With the potential for neoteny, a reproductively youthful form may have broader environmental tolerances than a form with mature traits, or shortened generation time may allow a more opportunistic existence (see item 2).
8. **Preadapted taxa.** Taxa exposed to repeated stress,

FIGURE 10-3

A model of recovery of a biota after a mass extinction event at the Cretaceous–Tertiary boundary, developed by Harries and Kauffman (1990). The so-called Lazarus window is variable, depending on the sedimentation and fossilization across the extinction boundary. If these processes are extensive, Lazarus species will be few; reduction of marine deposits increases the probability that a Lazarus effect will be evident.

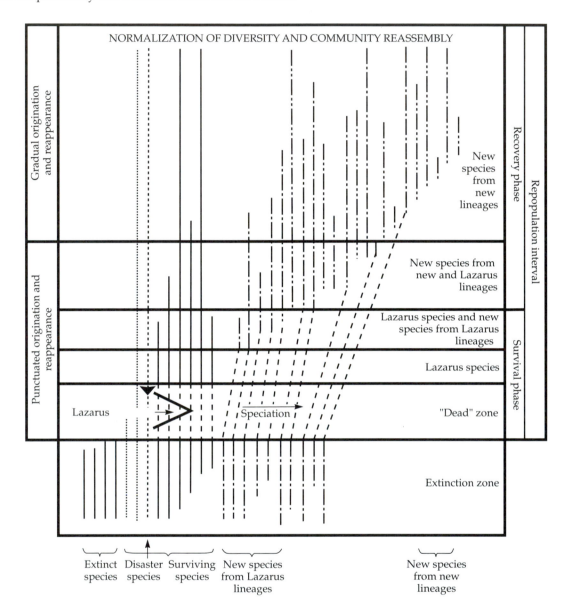

like those adapted to extreme environments, are likely to have relatively high resilience.

9. **Taxa with dormancy.** If the life cycle includes a prolonged dormant phase, that phase may coincide with a period of major stress and thus permit survival.

10. **Taxa with low critical population size.** Taxa that retain high reproductive potential with low population size may be more durable through high stress than others, such as inbreeding species.

11. **Rapid evolutionary rates.** Taxa with high evolutionary rates may be capable of adapting to changing conditions.

12. **Chance.** There is always the possibility that "good luck" or chance plays a role in survival through extinction events.

This categorization focuses on the biological traits of organisms. With such a focus, we need to know the specifics of each taxon to understand its persistence through an extinction episode.

FIGURE 10-4

Models of possible reasons for survival through mass extinction events, summarizing some of the types recognized by Harries and Kauffman (1990). The line *B* represents a mass extinction boundary. The width of each shaded area indicates the relative abundance of a taxon.

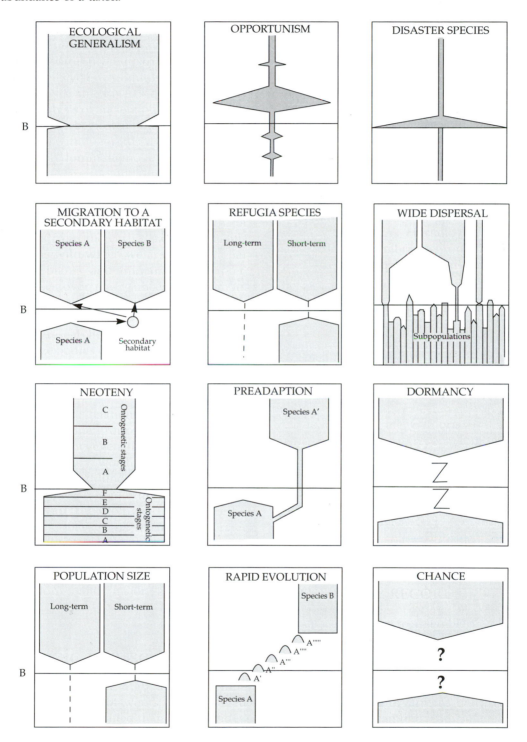

10.3 GENERAL PATTERNS IN TYPES OF SURVIVORS

Jablonski (1986; see also 1991) sought generalities about survivors, especially after mass extinctions.

Among marine organisms, tropical species have always been more susceptible to extinction, and extinctions have been especially concentrated in reef communities. Such communities were slow to recover, taking millions of years to regain lost diversity. For example, in the case of the Permo-Triassic extinctions in articulate brachiopod families, only 25 percent of tropical families survived while 44 percent of nontropical families persisted (Table 10-1). The discrepancy was wider in bivalves and gastropods after the Cretaceous–Tertiary extinctions. The forcing factor for tropical extinction of higher taxa and survival outside the tropics may well be the higher endemicity, and hence reduced buffering against change and the effects of stochastic events, of taxa in the tropics. Recent analysis of marine bivalve K/T extinctions has revealed that high extinction in the tropics is specific to the rudists (Raup and Jablonski 1993). The shells of rudists, highly specialized filter feeders, had formed extensive reefs in tropical waters, but rudists were uncommon outside the tropics. When we ignore the extinction of rudists, extinction of other bivalve mollusks appears to be uniform around the globe, arguing against any specific regional or latitudinal effects.

A nonintuitive generalization by Jablonski was that, in general, the species of globally species-rich taxa do not survive much better than those of species-poor taxa (Table 10-1). At issue is the coupling of high origination and high extinction.

> It is probably not species per se that is a liability during mass extinctions, but the covariation of biological attributes that determine a clade's extinction rate with traits that affect its speciation rate. For example, among bivalves and gastropods, low larval dispersal capability tends to heighten speciation rates and is generally accompanied by restricted geographic ranges [higher endemicity] and narrow environmental tolerances, imparting high extinction rates as well. Conversely, species with high dispersal capability, and thus low speciation rates, tend to have broader geographical ranges and environmental tolerances, which in turn impart extinction resistance to those species. (Jablonski 1986, 206–207)

These data suggest that mass extinctions are qualitatively different from background extinction, for in the latter we should predict that stochastic events usually eliminate species-poor taxa before species-rich taxa.

The patterns Jablonski identified are supported by a direct test comparing geographically widespread genera with those that are endemic (Table 10-1). Widespread genera represented 96 percent of surviving bivalve genera and 89 percent of surviving gastropod genera. Jablonski (1989) emphasized that it is widespread *genera*, not species, that have a high probability of survival. In a North American assemblage of mollusks through the Cretaceous–Tertiary extinctions, widespread genera survived significantly better than endemic genera. For bivalves, the difference was 55 percent survival for widespread genera versus 9 percent for endemic, and for gastropods, 50 percent versus 11 percent.

Among terrestrial vertebrates, small-bodied species have tended to survive better than large-bodied members of the same taxon (Diamond 1984, Jablonski 1991). Small-bodied species have a syndrome of characters that foster population resilience in times of stress.

1. Population density and total number are usually greater than those of large-bodied species.
2. Home ranges are relatively small.
3. Generation times are shorter.
4. Trophic requirements are smaller.

Small species generally have higher evolutionary potential, with less exacting environmental requirements, than their larger counterparts.

With a narrower focus on studies in southern Australia, Diamond (1984) detected three characteristics of survivors.

TABLE 10-1 Generalizations about survival through mass extinctions

Comparison	Taxon	Extinction Event	Percent Survival
Tropical/nontropical	Brachiopod families	Permo-Triassic	25%/44%
Tropical/nontropical	Bivalve families	Cretaceous–Tertiary	14%/89%
Tropical/nontropical	Gastropod families	Cretaceous–Tertiary	17%/91%
Species rich/species poor	Bivalve genera	Cretaceous–Tertiary	42%/58%
Species rich/species poor	Gastropod genera	Cretaceous–Tertiary	33%/67%
Endemic/widespread	Bivalve genera	Cretaceous–Tertiary	3%/96%
Endemic/widespread	Gastropod genera	Cretaceous–Tertiary	11%/89%

Based on Jablonski 1986.

1. Herbivores survived better than carnivores because they are lower on the trophic chain from plant to herbivore to carnivore.
2. Small carnivores survived better than large carnivores.
3. Habitat generalists survived better than habitat specialists.

In general, we have a syndrome of characteristics associated with taxa predisposed toward mass extinction: endemic, tropical, rapidly speciating, and diverse; in addition, the large species in a geographic area are more vulnerable than the small species. The most compelling characteristic is the endemicity (cf. Table 10-1). As a corollary, the survivors are geographically widespread. Jablonski (1991, 755) noted the stronger persistence of widespread and common species in the fossil record and related it to present-day biotas in which extinction is frequent:

> In the face of ongoing habitat alteration and fragmentation, this implies a biota increasingly enriched in widespread, weedy species—rats, ragweed, and cockroaches—relative to the larger numbers of species that are more vulnerable and potentially more useful to humans as food, medicines and genetic resources.

Are humans themselves a "weedy species?"

10.4 REFUGES FROM EXTINCTION

The next question, addressed by Vermeij (1986), becomes not "What survives?" but "Where does survival occur?" This is not so easy to answer, for researchers must identify refuges from extinction with a continuous record of occupation through a catastrophic event while simultaneously documenting range contraction. Refuges may be small and thus have a low probability of discovery. In fact, many higher taxa, such as families, seem to go extinct during a mass extinction, only to "rise from the dead" in demonstration of the Lazarus effect (Jablonski 1986). Evidence suggests that refuges sustained some taxa

for millions of years, although the locations of those refuges have not been discovered.

Geographic Refuges

Despite the difficulties of discovery, Vermeij (1986) reached some generalizations about the locations of refuges during biotic crises. Although survival in the tropics was generally low, there must have been refuges for many taxa limited to the lower latitudes, such as corals. A major **geographic refuge** was the tropical Indo-West Pacific region, which seems to be resistant to the cooling trends in sea-surface temperatures often associated with marine mass extinctions. Average sea-surface temperatures during the Pleistocene glaciations, for example, are estimated to have been 7 to 8°C lower than current temperatures in the Caribbean and 5 to 6°C lower in the open equatorial Atlantic, but only 3 to 4°C lower in the Pacific. The Atlantic Ocean is comparatively small and is surrounded by large continents with cold interiors during a glaciation; thus it is less buffered against temperature change than the larger oceans (Vermeij 1978). Apparently as a result, the Indo-West Pacific region appears to have repeatedly supported taxa that have contracted into this area (Table 10-2). For example, among the reef-building scleractinian corals, 27 genera were widely distributed in the Eocene, but 29 percent of these are now restricted to the Indo-West Pacific.

Another aspect of this important refugium may well be the nutrient-rich conditions created by runoff from continents and oceanic islands and upwelling of nutrients around many of the large Pacific islands and continental coasts. Persistent high-nutrient waters may well buffer threatened and small populations against extinction, ensuring effective reproduction and recruitment from larvae dependent on plankton feeding. A possible present-day example concerns the crown-of-thorns starfish, *Acanthaster planci*, which seems to have been erupting in numbers on coral reefs, such as the Great Barrier Reef, for at least the last 8,000 years (Walbran et al. 1989). Such

---

TABLE 10-2 The Indo-West Pacific region as a geographic refuge for marine organisms

Taxon	Former Distribution and Epoch	Refuge in Indo-West Pacific
Scleractinian corals	27 genera in East and West Pacific, Eocene	8 genera in Recent (29%)
Scleractinian corals	10 genera in East and West Pacific, Pliocene	3 genera in Recent (30%)
Gastropods	62 subgenera in tropical America, Pliocene	11 subgenera in Recent (18%)

Based on Vermeij 1986.

epidemics appear to occur after an increase of nutrients in the sea, resulting from very heavy rains that transport nutrients from land into coastal waters (Birkeland 1982, Vermeij 1986).

Oceanic islands may sustain large proportions of the shallow-water families of bivalves (80 percent), gastropods (97 percent), echinoids (76 percent), asteroids (82 percent), and scleractinian corals (100 percent) (Jablonski 1986). Such faunas should be resistant to extinction, because their habitat does not decline during marine regression. In fact, as sea level falls, the area of a cone-shaped island that is habitable by shallow-water species increases (Figure 10-5). In contrast, the faunas of continental shelves, such as the ammonoids, are extinction-prone during sea-level decline off the shelves. In general, however, islands do not appear to be adequate refugia for many species. If islands generally acted as refuges during marine regression, we would anticipate a loss of only about 13 percent of families of shelf biota—the mean of the abovementioned families not represented on oceanic islands. During the Permo-Triassic extinctions, about 52 percent of families were lost. Generally, there are fewer species per family on islands than on continental shelves, making families more vulnerable to stochastic events on islands. "Relict taxa appear to have continental distributions" (Vermeij 1986, 239), and no cases are known of marine mollusks that have survived on islands but not on continental shelves. As we shall see in the next chapter, the best predictor of taxon richness and taxon extinction through the Permo-Triassic is global area of continental shelf.

Ecological Refuges

Vermeij (1978, 1986) also recognized the importance of **ecological refuges** from extinction. For many marine groups, the deep sea, especially in polar regions, has acted as a refuge. But do relict taxa in extreme environments "inherit the earth"? "Most refuges represent evolutionary backwaters—'old folks' homes'—that receive and protect many species but export very few" (Vermeij 1986, 240); [ecological refuges are] "museums for the adaptively obsolete" (Vermeij 1987, 102). So where do radiations after mass extinctions come from? Do they originate only in productive refuges with high populations and high colonizing ability, involving species adapted to withstand strong biotic selection from competitors or predators? Or does the unpredictable sometimes (rarely) happen? Deep-basin ammonoids provided the seeds for subsequent radiation on continental shelves. Simple, cold-water-adapted foraminiferans survived and radiated into exuberant tropical forms. Clams evolved well before the Permian with a remarkable and unpredictable general adaptation, the siphon, and radiated into all kinds of hitherto unavailable ecological niches after surviving the Permo-Triassic extinctions. Are bottlenecks and founder events critical for the creatures that inherit the earth? Do we know anything about the population structure of survivors? These are perplexing questions.

Analysis of the fossil record of bottom-dwelling marine invertebrates since the early Mesozoic (lower Triassic), when recovery from the Permo-Triassic extinctions was generally rapid (cf. Figure 9-3), has revealed a strong pattern. Jablonski and Bottjer (1991) recorded the sites of first appearance (originations) in the fossil record for 26 well-preserved orders, on a water-depth gradient from the shallow nearshore habitats to the continental shelf slope and into deep ocean basins. To simplify general conclusions, they divided the gradient into onshore and offshore environments. Originations of orders were strongly biased toward the onshore habitats close to the shore:

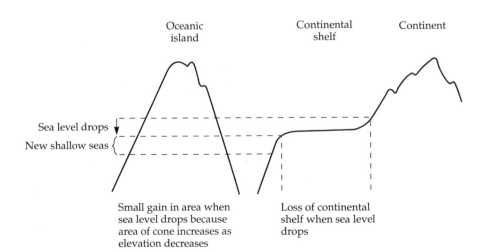

Oceanic island

Continental shelf

Continent

Sea level drops

New shallow seas

Small gain in area when sea level drops because area of cone increases as elevation decreases

Loss of continental shelf when sea level drops

FIGURE 10-5
The differential effect of lowered sea level on oceanic islands and continents. On islands, areas in shallow water increase slightly, whereas on continents they decrease dramatically.

"77% of the well-preserved orders first appear in relatively disturbed, onshore environments" (Jablonski and Bottjer 1991, 1831).

Our discussion has focused on origins of high (order-level) taxonomic rank, which involve major innovations in design. Clearly, the environment must select for new or derived characteristics that depart from the ancestral stock and are then recognized by systematists as worthy of new ordinal status. Highly variable and heterogeneous onshore environments may promote selection for novelty. High-nutrient onshore waters could maintain high populations and act as local refugia during extinction episodes (cf. Section 10.2). This kind of analysis promises to yield more insight into not only survival after extinction but the identities of the groups that radiate to colonize and replenish depauperate biota.

Another tack on the understanding of survivors is to study modern animals and plants that clearly are members of lineages that have survived extinctions in the past. One example is the red kangaroo.

10.5 SURVIVAL OF THE RED KANGAROO

If we take a more idiographic approach, we ask more specifically what modern species are doing today compared to what their extinct relatives did that got them into trouble. Take the red kangaroo. How did it survive? It is good to eat and large, and so how did it escape human overkill? Many species in the same genus, *Macropus*, went extinct, and other bipedal marsupials in the genera *Protemnodon*, *Sthenurus*, and *Procoptodon* also did not survive (Murray 1984).

Thirty thousand years ago a biologist studying Australian mammals would have considered red kangaroos to be very much the poor relations of the large mammal fauna: sheltering to conserve water and therefore forced to subsist on poor-quality food, or endlessly moving to follow the rains to obtain water and good-quality food, and existing in low numbers in the harshest part of Australia. (Horton 1984, 677)

Horton's explanation for the red kangaroo's survival is radically different from Martin's (1984) scenario of Pleistocene overkill. It had to be, because Martin did not account for the survival of large mammals. Horton argued that no mass extinction coincided with the arrival of humans in Australia; instead, the mass extinctions resulted from an arid period 26,000 to 15,000 years ago (Figure 10-6). Most of the megafauna consisted of woodland dwellers that depended on local water supplies as the focus of their range. In the arid period, woodlands declined, water supplies dried up, focal sites were overgrazed, and large mammals died. Smaller species with lower water requirements survived. As the woodlands shrank, the dry zone, where the red kangaroo lived, expanded. This species was preadapted to an arid period and thrived. Human distribution probably shrank back toward coastal regions during the arid period and impeded the red kangaroo's range extensions little if at all.

FIGURE 10-6

The coincidence of extinctions of large species with the arid period from 26,000 to 15,000 years ago in Australia, and the absence of mass extinctions soon after aborigines arrived on the continent. Note that mass extinctions since European colonization have involved mainly small species. *(Based on Horton 1984)*

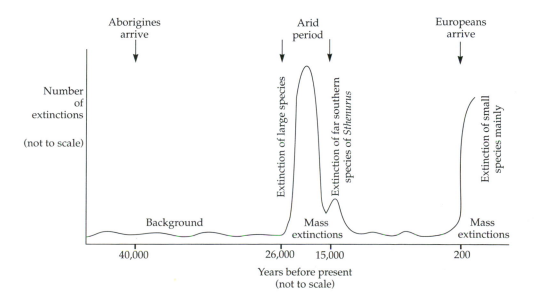

Aborigines probably concentrated on small game rather than large species, anyway.

Beyond addressing the welfare of the red kangaroos, this example may illuminate a major point made by Martin on differential impact of extinction on continents. Differential levels of megafaunal extinction and survival may result from special regional geographies. The concentric distribution of climates in Australia, for instance, leaves no refugia from a strong drying trend and thus promotes heavy extinction (86 percent) and low survival (cf. Table 10-2). In North America, there was little escape from Pleistocene cooling as ranges were pushed toward the south, and survival was low. But in Africa, megafaunas could retreat into large tropical havens, and survival of genera was high at 86 percent. (Why, then, was survival so low in South America at 21 percent of genera?)

Horton's scenario for the red kangaroo helps to focus our attention on the crux of the survival issue. Species are the basis on which we must understand survival through catastrophic extinctions, for if species survive, genera, families, and higher taxa survive. Talking about extinction of genera and families largely removes the debate from the mechanisms involved. If we agree to focus on species, then it is interesting to ask whether we understand modern assemblages well enough to predict species' future survival through biotic crises. One example from a relatively well-understood assemblage (discussed in the next section) suggests that biotic interactions and innate species attributes create a complex mosaic that leaves us with relatively little power to interpret the long-term survival of species, in the absence of detailed studies on living organisms. The study of small assemblages on small archipelagoes has little bearing on our understanding of major extinctions, but such research does raise our awareness of the importance of the real biologies and life histories of the species involved (cf. Chapter 9, Section 9.5).

10.6 RED MANGROVE ANT COMMUNITIES

On red mangrove (*Rhizophora mangle*) islands in the Florida Keys, ants are the dominant arthropods. The major ant species are members of five genera and three subfamilies on small islands (Table 10-3). All nest within dead hollow twigs. Each species has its own minimum island size requirement, and as islands increase in size, more species coexist (Figure 10-7). The inhospitable nature of very small islands buffeted by wind and waves defined the lower limit of island size for the primary species. For the secondary species, competition from *Crematogaster* or *Xenomyrmex* precluded occupation until an island became large enough to support two species. The resident dominant species always resisted invasion by another primary species, and so occupation depended on the species that arrived and established itself first. When primary species arrived simultaneously, *Crematogaster* always prevailed, its individuals being slightly larger than *Xenomyrmex* and therefore winning in one-to-one encounters.

In the ants we see the drama of survival and extinction played out on a very small scale. The primary and secondary species have different pathways for staying in the system (Figure 10-8). On each island, the ants stage a confrontation that determines survival or extinction for each participating species. This confrontation may mirror interactions over mainland species ranges, especially during crises when ranges are small and habitats become more uniform. Based on Cole's detailed studies, we could predict the order

TABLE 10-3 Major species of ant on small mangrove islands in the Florida Keys

Ant Species	Subfamily	Minimum Island Size for Habitation[1]	Basis for Minimum Size	Species Status	Colonizing Ability— Rank
Crematogaster ashmeadi	Myrmicinae	0.3 m³	Abiotic stress	Primary	2
Xenomyrmex floridanus	Myrmicinae	1.2 m³	Abiotic stress	Primary	1
Pseudomyrmex elongatus	Pseudomyrmicinae	5.1 m³	Competition	Secondary	3
Zacryptocerus varians	Myrmicinae	12.6 m³	Competition	Secondary	4
Camponotus (Colobopsis) sp.[2]	Formicinae	25.4 m³	Competition?	Secondary?	5

[1] Island volume was estimated as length × width × height.

[2] Results for *Camponotus* are less certain because of small sample sizes.

Based on Cole 1983a, 1983b.

FIGURE 10-7

The distribution of ant species in the Florida Keys in relation to island size. Bars indicate the range of island sizes on which ant species were found. The left-hand limit of each bar indicates the minimum land-volume requirement for a species, as listed in Table 10-3. Solid circles indicate the presence of *Crematogaster ashmeadi*. Open circles with no ant species represent sampled islands. Open circles with ant species present show the presence of *Xenomyrmex floridanus*. Note that *C. ashmeadi* and *X. floridanus* were never discovered on the same island. *(Based on Cole 1983a)*

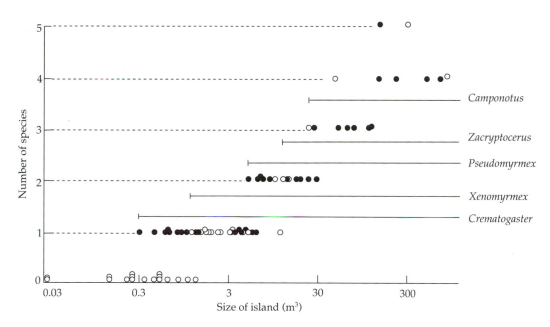

FIGURE 10-8

Mechanisms in the assembly of mangrove ant communities in the Florida Keys, for primary and secondary species. *(Based on Cole 1983a)*

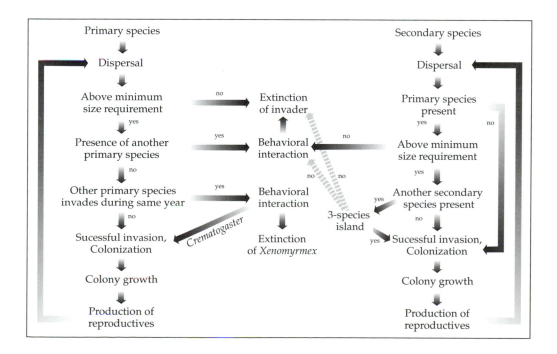

of survival probability with reasonable confidence, taking into account both competitive and colonizing abilities. The system involving ants, islands, colonization, and competition does have strong deterministic properties. The predicted order would match the listing in Table 10-3. We would have no basis for making predictions if the abiotic environment became the most stressful aspect of a crisis.

The question we should contemplate is the extent to which anyone could predict survivors in this assemblage on the basis of the fossilized remains of all five coexisting species, unearthed 10 million years from now. Those examining the fossils would lack all the behavioral information gathered by Cole and all the detailed distributional records. They would probably not know that all the ant species used dead hollow twigs as nesting sites and thus were forced into strongly competitive relationships. Predictive ability would be very weak for all five species and all five genera.

With increasing interest in the survivors of mass extinctions and a growing linkage between modern ecological relationships and paleobiology, the field of extinction and survival promises to develop rapidly. As we learn more about "assembly rules" in modern coexisting groups of species, perhaps we can begin to apply those rules to earlier times, as has been done with species–area relationships and the theory of island biogeography, discussed in the next chapter.

SUMMARY

An interesting relationship exists between rates of extinction and rates of speciation; it is a link between the potential for destruction and creation. An analysis of extant and extinct African mammals showed that the relatively specialized lineages experienced higher extinction and speciation rates than the more generalized lineages. Thus, the nature of survivors is somewhat paradoxical: rapid recovery of taxon richness after an extinction event might involve components of both the survival of evolutionarily phlegmatic generalists and the subsequent adaptive radiation of lineages with relatively high speciation rates.

Patterns in recovery of biota after regional and global extinction events show a predictable trend, from early recolonization by hardy species frequently exposed to stress to a rich fauna with a variety of lifestyles, exploiting many habitats. In an incomplete fossil record, we find a **Lazarus effect**—the disappearance of a species from the strata immediately after an extinction, followed by its reappearance, intact, later in the stratigraphic record. Such survivors, benefiting from a range of multiple traits that enable them to persist through stressful times, may contribute to a rapid, saltational, or punctuated episode of radiation followed by more gradual expansions of lineages.

Researchers have searched for pattern in the organisms that survive best after extinction events. Among the marine invertebrates, extratropical and widespread taxa are generally more resistant to extinction. In the tropics, an important **geographic refuge** appears to be the Indo-West Pacific marine environment, which has nutrient-rich shallow seas and less extreme cooling of surface waters than other regions during glaciations. Islands in general make poor refuges during mass extinctions. **Ecological refuges** in marine environments include deep-sea habitats, but whether such refuges are simply "old folks' homes" or the sources of new radiations after extinction remains an elusive issue.

Another approach to studying the survivors after extinction is to examine special cases such as the red kangaroo, which survived the Pleistocene extinctions of large animals in Australia. This species lived in dry, harsh environments, in contrast with its relatives, and was thus preadapted for surviving the arid period from 26,000 to 15,000 years ago. The red kangaroo expanded its range in the Pleistocene as its relatives contracted their ranges and went extinct.

Studies of present-day species assemblages, such as ant communities on mangrove islands, alerts us to the complexities involved with understanding colonization, survival, and extinction. Each species has unique properties, and only careful study reveals the reasons for species presence, absence, and persistence in the face of competition from other species. However, the comparative method, using living assemblages in efforts to understand the dynamics of extinction and survival, holds considerable promise for the future.

QUESTIONS FOR DISCUSSION

1. Is the increased use of modern groups with a good fossil record a valuable approach for understanding the processes involved with the survival of taxa through episodes of mass extinction, in your opinion?

2. To gain a stronger biological understanding of groups that are vulnerable or resistant to extinction, which particular taxa would you select for comparative study?

3. What kinds of ecological studies would you plan or explore to determine the resource exploitation patterns and habitat utilization that are relevant to vulnerability or resistance to extinction?

4. If extinction can be attributed to bad genes or bad luck, can survival after mass extinction be encapsulated as resulting from good genes or good luck?

5. Lazarus species and coelacanth-like species may be argued to have simply had extraordinarily good genes or good luck. If that is true, we should not expect to see any patterns among these kinds of species. Do you think there are some generalities to be found in the study of Lazarus species and species exhibiting the coelacanth effect?

6. What research would you undertake on Australian marsupials to further investigate the survival of the red kangaroo and the extinction of many other species in the megafauna?

7. Prediction of local survival and extinction in the present-day ant communities on the Florida Keys depends heavily on a detailed understanding of behavioral traits and social organization that are lost in the fossil record. Do you think that this dependence is an argument for despair that we will ever understand the nature of survivors in an extinction event?

8. Scientists have acknowledged competition as a possible factor in extinction events. They have also recognized refuges from extinction that maintain high biotic diversity and thus have high potential for interspecies competition. Can you resolve the paradox wherein competition is emphasized in one case and ignored in the other?

9. If a large proportion of bottom-dwelling marine invertebrates originated in disturbed, heterogeneous, shallow-water environments, which also show high extinction rates, do you think evolutionary rates after extinction events are greatly accelerated in these environments?

10. Can you think of groups that are particularly valuable for high-detail combined studies on the ecology of living species and their paleobiology?

REFERENCES

Birkeland, C. 1982. Terrestrial runoff as a cause of outbreaks of *Acanthaster planci* (Echinodermata: Asteroidea). Mar. Biol. 69:175–185.

Cole, B. J. 1983a. Assembly of mangrove ant communities: Patterns of geographical distribution. J. Anim. Ecol. 52:339–347.

———. 1983b. Assembly of mangrove ant communities: Colonization abilities. J. Anim. Ecol. 52:349–355.

Diamond, J. M. 1984. "Normal" extinctions of isolated populations, pp. 191–246. *In* M. H. Nitecki (ed.). Extinctions. Univ. of Chicago Press.

Harries, P. J., and E. G. Kauffman. 1990. Patterns of survival and recovery following the Cenomanion–Turonian (late Cretaceous) mass extinction in the Western Interior Basin, United States, pp. 277–298. *In* E. G. Kauffman and O. H. Walliser (eds.). Extinction Events in Earth History. Springer–Verlag, Berlin.

Horton, D. R. 1984. Red kangaroos: Last of the Australian megafauna, pp. 639–680. *In* P. S. Martin and R. G. Klein (eds.). Quaternary Extinctions: A Prehistoric Revolution. Univ. of Arizona Press, Tucson.

Jablonski, D. 1986. Causes and consequences of mass extinctions: A comparative approach, pp. 183–229. *In* D. K. Elliott (ed.). Dynamics of Extinction. Wiley, New York.

———. 1989. The biology of mass extinction: A palaeontological view. Phil. Trans. Royal Soc. London B. 325:357–368.

———. 1991. Extinctions: A paleontological perspective. Science 253:754–757.

Jablonski, D., and D. J. Bottjer. 1991. Environmental patterns in the origins of higher taxa: The post-Paleozoic fossil record. Science 252:1831–1833.

Kauffman, E. G. 1986. High-resolution event stratigraphy: Regional and global Cretaceous bio-events, pp. 279–335. *In* O. H. Walliser (ed.). Global Bio-events: A Critical Approach. Springer–Verlag, Berlin.

Martin, P. S. 1984. Prehistoric overkill: The global model, pp. 354–403. *In* P. S. Martin and R. G. Klein (eds.). Quaternary Extinctions. Univ. of Arizona Press, Tucson.

Murray, P. 1984. Extinctions down under: A bestiary of extinct Australian late Pleistocene monotremes and marsupials, pp. 600–628. *In* P. S. Martin and R. G. Klein (eds.). Quaternary Extinction: A Prehistoric Revolution. Univ. of Arizona Press, Tucson.

Raup, D. M., and D. Jablonski. 1993. Geography of end-Cretaceous marine bivalve extinctions. Science 260: 971–973.

Vermeij, G. J. 1978. Biogeography and adaptation. Harvard Univ. Press, Cambridge, Mass.

———. 1986. Survival during biotic crises: The properties and evolutionary significance of refuges, pp. 231–246. *In* D. K. Elliott (ed.). Dynamics of Extinction. Wiley, New York.

———. 1987. Evolution and Escalation: An Ecological History of Life. Princeton Univ. Press, Princeton, N.J.

Vrba, E. S. 1987. Ecology in relation to speciation rates: Some case histories of Miocene–Recent mammal clades. Evol. Ecol. 1:283–300.

Walbran, P. D., R. A. Henderson, A. J. T. Jull, and M. J. Head. 1989. Evidence from sediments of long-term *Acanthaster planci* predation on corals of the Great Barrier Reef. Science 245:847–850.

Wilson, E. O. 1988. The diversity of life, pp. 68–78. *In* H. J. De Blij (ed.). Earth '88: Changing Geographic Perspectives. National Geographic Society, Washington, D.C.

11

MODELING THE FOSSIL RECORD

Coelacanths, which represent a remnant from a Devonian group of fishes, are still living today—hence the term "coelacanth effect." *(Illustration by Tad Theimer)*

The union of paleontology with evolutionary biology to form paleobiology has led to creative avenues of research and synthesis. Much of the progress in this field has relied on the application of the results of work with living systems to the fossil record. A close relationship has developed between ecology and paleontology, in particular (cf. Gould 1981). Some examples described in this chapter illustrate the ways in which integration is achieved, the kinds of thinking involved, and some of the interesting results that have unfolded.

There are two basic types of phylogenetic development: anagenesis, or the linear progression of species through time, and cladogenesis, the branching of lineages into subgroups, or clades.

11.1 THE EQUILIBRIUM THEORY OF ISLAND BIOGEOGRAPHY

Ecological Development

We can assume that cladogenesis continues to occur until the available living spaces have been filled and communities and ecosystems are saturated with species. (That such extensive radiation has actually taken place is a debatable point, but the assumption leads to interesting avenues of thought.) Then we may apply the **equilibrium theory of island biogeography** (MacArthur and Wilson 1967), which predicts the number of species present in local equilibrial conditions in ecological time. The theory was founded on the empirical observation that, as island area increases, the number of species in a taxon increases. The area–species curve has a positive slope and is frequently linear on a log–log plot (Figure 11-1). The general form of the relationship is $\hat{S} = kA^z$, where $\hat{S}$ is the equilibrium number of species on an island, A is the area of the island, k is a constant that depends on the taxon and region under study, and z is a constant that tends to vary between 0.20 and 0.35 for oceanic islands and between the lower values 0.12 and 0.17 for habitat islands within larger landmasses, such as continents.

MacArthur and Wilson argued that the balance of **colonization** by new species and local **extinction** of resident species maintained the number of species on each island at equilibrium. Colonization of a vacant island was rapid at first but declined as slower dispersers arrived and as fewer new species were available for colonization, producing a concave **immigration curve.** The opposite force, extinction, was weak initially, because populations of colonizers were small relative to food supply, competition was weak, and natural enemies might not have colonized. As the number of species increased, so did population

FIGURE 11-1

The equilibrium theory of island biogeography used patterns such as the number of species of amphibians and reptiles on West Indian islands **(top)** to develop a mechanistic scenario explaining the prevalence of such equilibrial conditions **(bottom)** and relating the rates of immigration and extinction to the maintenance of an equilibrium number of species ($\hat{S}$). I is the maximum immigration rate, and P is the number of species in the mainland pool from which colonization occurs. (*After MacArthur and Wilson 1967*)

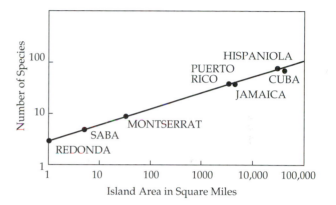

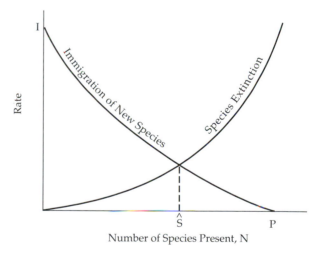

sizes, competition, predation and parasitism, and probability of extinction, yielding an exponentially increasing rate of extinction (Figure 11-1).

Application to Phylogenetic Patterns

In a gross but heuristic extrapolation, it is possible to extend the equilibrium theory to global phylogenetic patterns through evolutionary time. Implicit is the assumption that different major taxa have different enough ecologies that they are unlikely to preempt available space from other taxa and thereby inhibit the development of equilibrium within a taxon.

Raup, Gould, Schopf, and Simberloff (1973) undertook to apply the equilibrium model to paleontological data to see if an equilibrium pattern of phyletic radiation existed and what it might be. Their approach is important because, like the island biogeographic model, it may lead to the discovery that very few variables are significant in regulating the diversity of organisms. If, however, the simple interaction of these variables fails to account for some real-life examples, then special causes may be sought. It becomes a relatively simple matter to sort out the ordinary from the extraordinary. Raup et al. took a nomothetic approach to paleontology and searched for general patterns, in contrast with many paleontologists, who studied individual events in isolation (i.e., took the idiographic approach). As a result, the argument Raup et al. developed is both taxon-unspecific and unspecific in time.

The phylogenies of several major fossil groups have been well documented yet are difficult to interpret. Some lineages have been much more successful than others in terms of numbers of species or ecological diversity in ways of life. Some groups diversified early and became extinct suddenly; others radiated less but over long periods of time with no sudden extinction. In almost all cases, researchers could not predict such patterns. Raup et al. attempted to simulate phylogenies as an equilibrium process by adopting the simplest possible position: aspects of phylogeny such as speciation and extinction act as random, or stochastic, variables. They do not imply that any evolutionary event is random but that the timing of events may be simulated as if they were occurring at random. With this assumption, one can use computers to generate **artificial phylogenies** based on random events and regulated by a small number of constraints (Figure 11-2). If these phylogenies bear the same characteristics as real phyletic lines, then simple universal models may be assumed to apply. If, on occasion, the real phylogenies appear to be very different from the randomly generated clades, then the researcher should seek special causes. But the possibility remains that a large proportion of the patterns may be adequately simulated, just as the theory of island biogeography correlates with, and may account for, much of the diversity of organisms on islands.

Raup et al. controlled extinction and branching events, or speciation, in the hypothetical cladogram stochastically, using computer-generated random numbers. They assigned certain probabilities of branching and extinction (Figure 11-3) and extrapolated from island biogeographic theory to assume the existence of an equilibrium diversity. When the assigned equilibrium was reached, the researchers damped oscillations in diversity around this equilib-

FIGURE 11-2

An imaginary phylogeny that develops through time by branching into new lineages, which ultimately all go extinct. On the right is a summary of the diversity of the lineage, or the number of lineages for each time interval, to provide a clear visual impression of the "shape" of the clade. (*From Raup et al. 1973*)

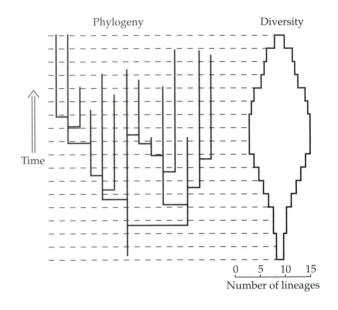

rium by changing probabilities of branching and extinction (for example, see Simberloff's [1974] equations described later in this chapter). The computer program also contained automatically determined times when major divisions took place, equivalent to the evolution of a new genus, family, or class of organisms.

11.2 COMPUTER-GENERATED PHYLOGENIES

The Results

The computer cladogenesis produced a great variety of clade shapes (Figure 11-4). Clade A-16 radiated rapidly and then declined gradually to extinction, just as the trilobites did. Clade B-3 shows the reverse pattern, with a mass extinction just like that of the dinosaurs. Some clades, such as B-4 and C-7, passed through a phase close to extinction but survived to develop a new phase of high diversity, as did the ammonites. In some cases, a clade shows a decline while a related clade rapidly increases in number, such as A-2 and A-3, C-18 and C-19, and D-3 and D-4. And some groups of clades, such as C-19, C-20,

FIGURE 11-3

Results of a partial computer-simulated cladogram showing two clades, the origin of the second clade, and the units of time over which originations and extinctions occurred. By adding, for each time interval, all taxa extant at that time, we can develop clade shapes as in Figure 11-2. For example, at time *X* three taxa were present in clade 2, at time *Y* four taxa, and at time *Z* only two. *(After Raup et al. 1973)*

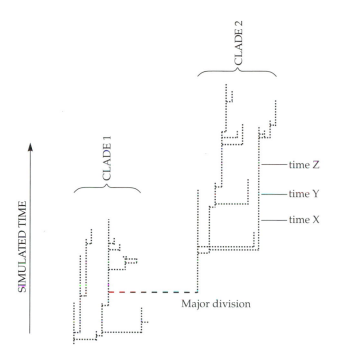

In 24 computer runs, generating 481 clades, the diversity of clades increased rapidly, tended to level off, and then declined as a few clades increased in dominance, but an equilibrium diversity was maintained for many units of computer time (Figure 11-5).

Comparisons with the Fossil Record

Raup et al. (1973) were understandably interested in how their phylogenetic trees and diversity patterns compared with cladograms derived from the fossil record, beyond the rather superficial comparisons already discussed. They constructed cladograms from the reptile record to compare with computer cladograms. The similarities and differences are interesting (cf. Figures 11-4 and 11-6):

1. Many of the clades in both simulated and real examples are spindle-shaped.
2. Origins are concentrated early in the sequence.
3. Some real clades are smaller than simulated clades and others are larger, as a result of more stringent limits on the computer program than apparently exist in nature.
4. The simulated clades did not show a mass extinction of the magnitude seen in reptiles at the Cretaceous–Tertiary boundary. That extinction affected half of the clades then living and must be recognized as an unusual event worthy of a special explanation.
5. Very rapid radiation occurred in the therapsid reptiles (clade 17), but it has a very low probability of occurrence in the simulated cladograms.
6. The **coelacanth effect,** in which one survivor of an ancient clade persists for a long time, as the coelacanth has done, does not occur to the same degree in simulated clades. The reptile clade (7) consists of primitive lepidosaurs; its single survivor is the tuatara (*Sphenodon*), now found only on a few islands near the coast of New Zealand (Figure 11-7). Some simulated clades approach this extreme, such as C-16, D-5, and D-18, but the coelacanth effect is well beyond the pattern expected for chance originations and extinctions.

When we compare reptile clade diversity (Figure 11-8) with simulated clade diversity (Figure 11-5), considerable agreement is evident. An equilibrium level of diversity was maintained from the late Triassic to the late Cretaceous, lasting more than 100 million years.

Validity of a Simple Stochastic Model

Raup et al. concluded that a simple stochastic model can produce phylogenetic diversity patterns similar to those seen in the fossil record. The general similar-

and C-21, seem to assume dominance at the expense of other groups, such as C-5, C-6, C-7, and C-10. Commonly, scientists finding such a pattern in the fossil record would have accounted for it by invoking apparent competition from new radiating groups as the cause of extinction of formerly successful clades, but this explanation is not tenable in regard to computer-generated scenarios.

In fact, many characteristics of the simulated cladograms are relevant to interpretations of the real fossil record. Faced with differences between clades, paleontologists have tended to assume that the organisms involved were inherently different, that they had different evolutionary potentials. But the simulation modeling shows that groups with exactly the same "evolutionary potential" can show very different diversity patterns.

Another way of using the simulated data is to examine the number of clades created and extinguished per unit of time, because the balance determines the **diversity of clades** present at any one time.

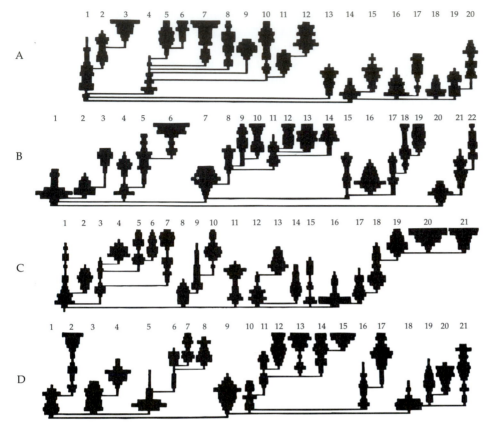

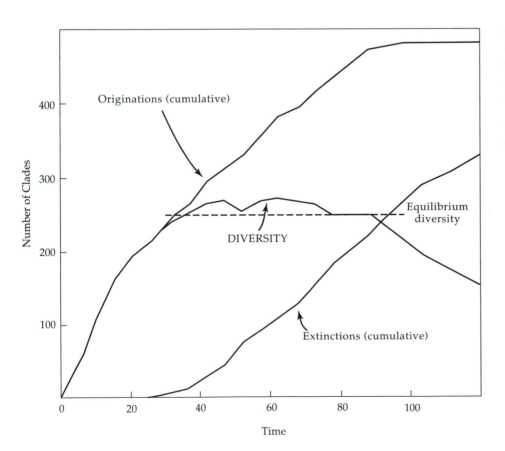

FIGURE 11-4

Results of four computer runs (A–D), showing randomly generated phylogenies and clades, with differences in clade shape resulting from chance events. Each computer run was terminated in the midst of high diversity of several clades; the termination does not represent extinction but is perhaps more like the end of a geological period. *(From Raup et al. 1973; see also Raup 1977, Gould 1981)*

FIGURE 11-5

The development of an equilibrium number of clades in about 60 units of computer time, resulting from the balanced effects of originations and extinctions. *(After Raup et al. 1973)*

FIGURE 11-6

The shapes of 17 reptilian clades derived from the fossil record. *(After Raup et al. 1973)*

FIGURE 11-7

The tautara (*Sphenodon punctatum*) is an example of the coelacanth effect, being the sole modern-day survivor of an ancient lineage of primitive lepidosaurs that diversified in the Triassic Period, about 200 million years ago. *(Illustration by Tad Theimer)*

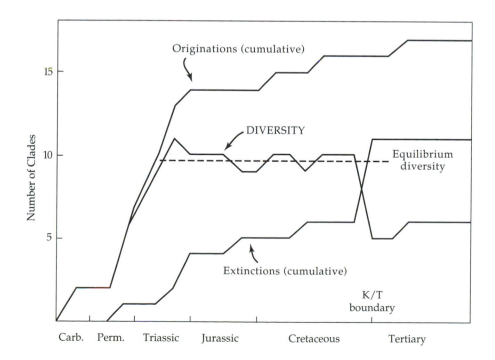

FIGURE 11-8

Change in diversity of reptilian clades. An equilibrium number of clades lasted more than 100 million years but was terminated by the Cretaceous–Tertiary mass extinctions. Only the extinctions caused a substantial deviation from the pattern developed by simulation in Figure 11-4. *(After Raup et al. 1973)*

ity exists for two important reasons. First, the researchers simulated phylogenies assuming equilibrium conditions of diversity, and it appears that in the real world those conditions also hold. Second, the taxonomic system used by paleontologists imposes constraints on the shapes of clades: in real lineages, if a clade becomes very large, it is likely to be subdivided; if one species or a small group diverges from a larger taxon, it is likely to be lumped with the larger taxon anyway. These conventional limits are similar to those in the otherwise stochastic simulation. It is therefore important to recognize procedural biases and distinguish between their results and real biological differences in phylogeny.

Thus, computer simulations help with interpretation of the biological factors involved with diversity in phylogenies. For example, clades B-7 and B-8 (Figure 11-4) existed at the same time under the same conditions for five units of computer time after the origin of B-8. Then, in the sixth unit of time, B-7 became extinct while B-8 did not, even though the probability of extinction for B-8, because of its lower diversity, was higher than that for B-7. It seems that in real life the causes of extinction are so diverse and probabilistic that over a long period all biological groups are equally prone to extinction, and this complexity produces a random array of extinctions (see also the discussion of taxonomic survivorship curves later in this chapter).

The stochastic model provides another biological insight into extinctions. In Figure 11-4, clades B-4, B-15, and B-17 all went extinct at the same time. B-4 was taxonomically very distantly related to the others. These clades represent 20 percent of clades existing at that time, and so the extinction was significant. Whereas paleontologists have tended to search for a single cause to explain all such synchronous extinctions, the stochastic model provides an alternative explanation: extinctions had wholly independent causes, and their coincidence in time was due to chance.

11.3 MODELING THE PERMO-TRIASSIC EXTINCTIONS

Perhaps the greatest challenge to an understanding of pattern in the fossil record is what has been called paleontology's outstanding dilemma: the massive extinctions in the late Paleozoic Era, between the Permian and Triassic periods, about 230 million years ago. Some of the major extinctions concern families of marine invertebrates, and the classical view has been that an enormous drop in sea level, which greatly reduced the extents of epicontinental seas

(shallow seas on continental shelves), caused massive extinctions of marine invertebrates.

Schopf (1974) took a uniformitarian approach to understanding environmental changes by studying known relationships between area and diversity as well as aspects of the Permian and Triassic in relation to processes going on now. He combined plate tectonics with an understanding of species–area curves from the theory of island biogeography.

The Patterns

First, Schopf plotted the pattern of extinctions of families of marine animals occupying shallow seas during the late Permian and early Triassic. The diversity of families during this period was reduced to half (Figure 11-9), and the many affected taxa included brachiopods, bivalves, foraminifers, ostracods, ammonoids, and nautiloids.

Second, Schopf reasoned that an exponential function of area adequately describes faunal diversity. Therefore, an estimate of changes in area of epicontinental seas due to plate tectonic movement of landmasses during this period may reveal a relationship between area and diversity. Schopf constructed maps (Figure 11-10) and estimated the percentage of continental area under shallow water. At the beginning of this period, Pangaea had not yet fully formed. Shallow seas shrank from a coverage of 43 percent of their possible distribution in the early Permian to 13 percent coverage in the very late Permian, then expanded to 34 percent coverage in the Lower Triassic.

The Mechanisms

Area reduction and family diversity reduction are clearly correlated. If there is a real cause-and-effect relationship, what might the mechanisms be?

1. Suturing of continental masses during assembly of Pangaea would amalgamate previously distinct faunas, with the potential for new competitive, predatory, and parasitic interactions, greatly increasing the probability of extinctions.
2. Reduction of sea level would add to this effect by reducing the width of shallow continental shelves.
3. The near-shore area rich in nutrients, over which primary production is highest, would diminish, reducing food supply and increasing the probability of extinction.

The reduction of shallow seas would also have a substantial effect on terrestrial climate, leading to harsher conditions in continental interiors. Perhaps sea-level changes indirectly affected the Permo-Triassic extinctions of land organisms.

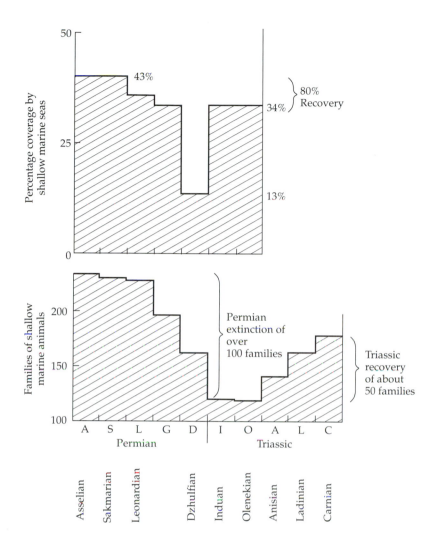

FIGURE 11-9

Changes in percent coverage of continents by shallow seas and number of families of shallow marine animals in the Permian and Triassic. *(After Schopf 1974)*

In the early Triassic the area of shallow seas returned to 34 percent coverage, which was 80 percent of the coverage in the early Permian (Figure 11-9). New adaptive radiations proceeded to replace some of the extinct families. The Triassic recovery amounted to about 50 families of shallow marine invertebrates by the Carnian, and then the Norian–Rhaetian extinctions occurred.

Schopf concluded that the Permo-Triassic extinctions were another example of the relationship between faunal diversity and area, although on a larger-than-usual scale. Area changed because of continent suturing and sea-level changes, both under tectonic influence. Active sea-floor spreading in early Permian times involved an increase in volume of oceanic ridges, which caused sea levels to rise. The dramatic decline in sea level in the Dzhulfian coincided with reduced activity of oceanic ridges. After all, Pangaea had assembled by the Guadalupian Epoch (Figure 11-10), and so there was considerable resistance to further sea-floor spreading.

The Model

Simberloff (1974) modeled the relationships Schopf had studied. First Simberloff established that the relationships between area and numbers of families and genera, like the species–area relationship, were linear on a log–log plot. He then plotted Schopf's data on logarithmic scales (Figure 11-11). If the responses of family originations to area were instantaneous, we would expect the point for the Lower Triassic to be on the line for the family–area relationship of the Permian. The origin of new families is *not* instantaneous and must lag behind the areal extension of epicontinental seas. In fact, faunal diversity did not recover before the impact of the Norian–Rhaetian extinctions (Figure 9-3), and levels of familial diversity similar to that in the early Permian were not seen again for about 100 million years! Diamond (1973) coined the term **relaxation time** for the time period required to reestablish the equilibrium number after a perturbation. But equilibrium was never

Changes in area of shallow marine seas (shaded) in the Permian and Lower Triassic, showing a large decline in the Dzhulfian and a dramatic recovery in the Lower Triassic (Induan and Olenekian). *(After Schopf 1974)*

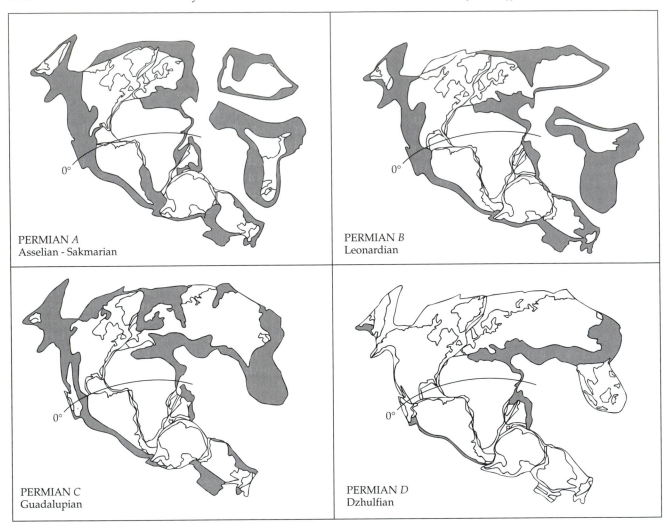

PERMIAN *A*
Asselian - Sakmarian

PERMIAN *B*
Leonardian

PERMIAN *C*
Guadalupian

PERMIAN *D*
Dzhulfian

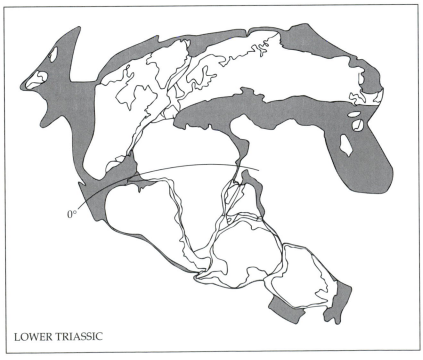

LOWER TRIASSIC

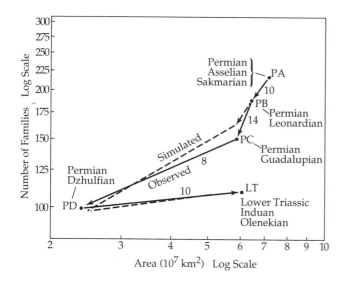

FIGURE 11-11

Number of families and area relationships during the Permian and Lower Triassic, for shallow marine invertebrates studied by Schopf (1974) and Simberloff (1974). (*After Simberloff 1974*)

reestablished! Since the end of the Triassic, the number of families seems to have been increasing, far surpassing the equilibrium that had prevailed from Ordovician to Permian times—about 380 families for 200 million years (Figure 9-3).

Simberloff's model took the form

$$E(F_{t+1}) = E(F_t) + [ORIG - EXT + DEV \cdot E(F_t) - (\hat{F}_t)^2] \cdot E(F_t)$$

if the number of families was lower than equilibrium $E(F_t) < (\hat{F}_t)$. The symbols are defined as follows:

$E(F_t)$ = expected number of families at time t.

$\hat{F}$ = equilibrium number of families at time t, which is a function of area through the species–area equation ($\hat{F} = k'A^{z'}$) (see MacArthur and Wilson 1967).

F = starting number of families.

$ORIG$ = a constant probability of origination of a new family.

EXT = a constant probability of extinction of a family.

DEV = an additional probability of family origination when the area is undersaturated, assumed to be proportional to the square of the deviation of F from $\hat{F}$, which operates only when F is less than $\hat{F}$. The probability is arbitrarily set equal to the additional probability of extinction when F is greater than $\hat{F}$.

When the number of families was above equilibrium $(E(F_t) > \hat{F}_t)$, the equation remained the same except that a $-DEV$ term replaced the $+DEV$ to represent

an additional probability of family extinction that is equal and opposite to the $+DEV$ term.

Simberloff used the square of the deviation for nonequilibrium states because of the exponential rates of immigration and extinction originally conceived by MacArthur and Wilson (1967). The same reasoning applies here to origination of families instead of immigration of species, and extinction of families instead of extinction of species.

With these equations, Simberloff calculated the expected number of families at each stage in the Permian and Lower Triassic, given the starting number in the Asselian and values for the unknowns: $ORIG$, EXT, DEV, k', and z'. One of the stochastic simulations produced results as in Figure 11-11, using the probabilities $ORIG = EXT = 0.01$, $DEV = 0.00002$, $k' = 0.000000364$, and $z' = 1.109$. The text of Simberloff's paper does not justify the use of these values, but they are within the bounds of realism. The main point is that an equilibrium model of taxon richness is consistent with the species–area relationship.

Simberloff concluded that Schopf's explanation was realistic and possible on the basis of diversity–area relationships seen in extant taxa.

Beyond the Area Effect

While possibly capturing the basic essence of the Permo-Triassic marine extinction, Simberloff's model probably oversimplifies the biological mechanisms involved. If area alone were the main factor, we should expect major marine extinctions in the Pleistocene as well, for sea level and shallow sea area declined dramatically. However, extinctions were low—much lower than predicted from an area effect. In fact, for many years Valentine and associates have argued that a richer array of factors was influential during major plate-tectonic events, and their understanding is important for revealing differences in extinction—say, between the Permo-Triassic and the Pleistocene (e.g., Valentine and Moores 1974, Valentine and Jablonski 1991).

In addition to considering an area effect alone, Valentine emphasized the need to relate the area effect to three scales of influence on ecological diversity: (1) the number of major provinces defined by geographic barriers and climatic variation; (2) the number of communities per province, which depends on the environmental heterogeneity in that province; and (3) the number of species per community, which may depend on the area available and, again, the heterogeneity of the area. Using these scales, Valentine noted the special features of the Permo-Triassic episode. Pangaea assembled with continuous shallow water around its margins, enabling provincial faunas to blend without major physical barriers. Climates were warmer at that time, which also reduced provinciality. A great reduction in the number of biotic provinces probably occurred. In addition, the number of communities per province probably fell as epicontinental seas declined, and zonation from shore to the continental shelf and beyond would have been compressed. Suturing of continents, with major shifts in ocean currents, would have disrupted near-shore environments and destabilized food resources, perhaps such that fewer species could have coexisted within each community. The fact that Triassic organisms— of more generalized forms and belonging to comparatively few species—became widespread adds credence to Valentine's arguments, while enriching our understanding of the kinds of organisms that survive. An interesting opportunity exists to model this greater complexity of environment–biology interactions.

11.4 MORPHOLOGICAL COMPLEXITY AND EVOLUTIONARY RATES

Simpson (1953) emphasized a general pattern: the rapid rate of morphological evolution in the mammals compared to the molluscs. "Mammals seem surely to have a much higher modal rate of evolution than molluscs" (314). Simpson recognized the pitfalls in comparisons of such different kinds of organisms, but after two chapters devoted to rates of evolution and many other discussions on such rates, his arguments were so authoritative that the pattern gained wide acceptance.

A valid question is whether the use of morphological characteristics is adequate for evaluating rates of evolutionary change. Schopf and his cohorts (1975)

addressed this question using 12 taxa that ranged from the gastropods, with the lowest rates of extinction, to the mammals, with the highest rates of extinction. Low rates of extinction indicate groups that have lived a long time and therefore have evolved less rapidly than groups whose taxa are very short-lived.

The Patterns

Schopf et al. estimated the morphological complexity of each group by counting the number of morphological terms used in that group's taxonomy. The mammalian skeleton obviously provides many more characteristics, and the need for more terms, than do the remains of gastropods or bivalves. The researchers then identified three general groupings with different ranges of complexity. The taxa with higher complexity had higher extinction rates (Figure 11-12). It seems that the fewer the characteristics available for observing change, the less likely it is that change will be revealed. This relationship influences detection of origination and extinction events at the species level as well as at higher taxonomic levels such as genera and families. Note that, in Schopf's approach, complexity applies simply to the number of discrete characteristics employed in a group's taxonomy. In the next section (11.5), we discuss another measure of the information content in structural complexity and diversity.

The Model

To test the extent to which this trend could be generated by a stochastic model, the researchers generated lineages in a way similar to that used by Raup et al. (1973), but four levels of morphological complexity were introduced: 3, 5, 10, and 20 characters per taxon. Each of the four dendrograms differed only in the number of characters used to characterize the lineage. From these lineages, extinction rates were calculated and plotted against time (Figure 11-13). The more morphological characteristics used, the higher the rate of change of taxa, i.e., the higher the extinction and evolutionary rates (cf. Section 10.1).

Thus, the pattern recognized for so long in the rates of morphological evolution may instead consist of patterns derived from differences in morphological diversity only. As Schopf et al. (1975, 69) put it, "Many previous studies of the 'rate of evolution' say more about general morphological complexity than about rate of genomic evolution." However, these arguments should not be construed as undermining genetically based evidence, such as chromosomal changes, that many mammalian taxa have evolved rapidly when compared with lower vertebrates and molluscs (see Chapter 14).

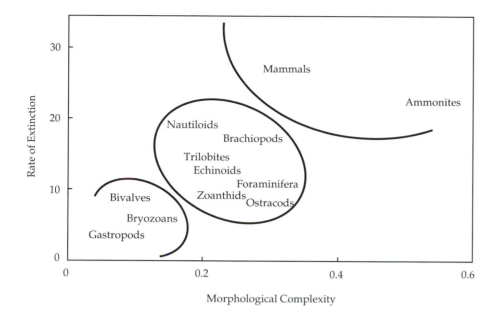

FIGURE 11-12
The positive relationship between morphological complexity and rates of extinction in 12 taxa. *(After Schopf et al. 1975)*

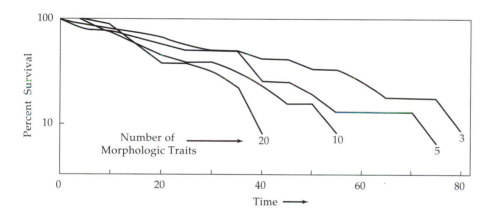

FIGURE 11-13
Lineage survival rates, which differ only according to the number of characteristics used in the hypothetical taxonomy. Taxa with the fewer morphological characters available survived the longest. *(After Schopf et al. 1975)*

11.5 SURVIVAL OF GENERALISTS AND SPECIALISTS

Another widely recognized pattern in the fossil record is that generalists tend to survive longer than specialists. In 1896, Cope proposed his **"law" of the unspecialized,** the view that specialization results in dead ends but unspecialized groups persist and advance. In spite of Simpson's (1953) argument against such a principle, many authors have used the idea in idiographic explanations of selective extinction, as Flessa et al. (1975) pointed out. A critical look at the aquatic free-living forms in the phylum Arthropoda seriously undermined the credibility of the pattern (Cisne 1974, Flessa et al. 1975).

The Ecological Niche

Schopf et al. examined the persistence of arthropod taxa in the fossil record and compared it for generalists and specialists. They used Elton's (1927) concept of the ecological niche to fuse ecological ideas about generalists and specialists with a fossil record leaving little ecological information but, in the arthropods, a wealth of morphological detail. In Elton's view, the niche of an organism is its role in the community, and the way in which a species relates to the environment is at least partly defined by the appendages it has available. For example, the cheliped of a lobster or crayfish, with its chela (pincer), is clear evidence of the animal's role as a predator. Thus, when a species' appendages show specialized use, the species can be classified as specialized, and when the appendages of an arthropod indicate little specialization, a generalist classification is warranted. With so many segments bearing appendages in aquatic arthropods, we can accurately define the continuum between generalists and specialists.

Estimating Morphological Diversity

The aquatic free-living arthropods have evolved from primitive species with little appendage specialization to more derived species with much specialization. Schopf et al. estimated the diversity of limb types per species using the Brillouin diversity index (H).

$$H = \frac{1}{N} \log_2 \frac{N!}{N_1! \, N_2! \dots N_i!}$$

N = total number of limb pairs

N_i = number of pairs of the ith limb type

The value of H ranged from about 0.3 for ancient trilobite-like fossils to 2.7 for specialized decapod crustaceans. If all limbs were alike, H would equal zero. Thus, the higher the H value, the greater the morphological specialization. Arthropod appendages may function in sensation, feeding, locomotion, communication, respiration, copulation, brooding of young, and sometimes construction of dwellings. Division of labor among appendages has increased through the evolutionary history of the aquatic arthropods.

The Pattern

A very interesting discovery is that the diversity index, H, has followed a logistic pattern through time, as if maximal diversity and maximal specialization were achieved millions of years ago (Figure 11-14). Therefore, if Cope's "law" of the unspecialized holds, taxa with high H should have persisted in the fossil record for shorter periods than those with low H. In fact, for 48 orders of these arthropods there was no correlation between the H value of an order and the duration of that order in the fossil record. We must conclude, at least on the basis of arthropod appendages, that generalized taxa do not necessarily have longer geologic ranges, and highly specialized taxa are not especially sensitive to the forces causing extinction.

11.6 TAXONOMIC SURVIVORSHIP CURVES

This study on the aquatic free-living arthropods suggests that the probability of extinction of a taxon is independent of specialized or generalized traits, and

FIGURE 11-14

The change in diversity, H, of appendages for all orders of aquatic free-living arthropods in the world through geological time, based on the mean value of about three species per order. The letters across the top of the figure are standard abbreviations for geologic periods (see Table 7-1). *(After Cisne 1974, and Flessa et al. 1975)*

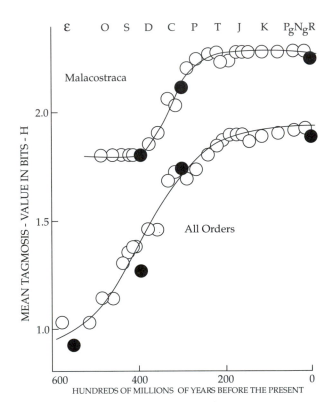

much of the modeling of paleontological records has depended on the assumption that the probability of extinction is similar for all organisms and is constant through time in any particular taxon (e.g., Raup et al. 1983). That assumption can be tested against the empirical data.

The Ecological Basis

Again, scientists needed a fusion of ecology and paleontology to understand the problem. Van Valen (1973) used **survivorship curves** that ecologists had developed to study cohorts of living species. The ecologists had borrowed the approach from actuarial tables used in the insurance business to estimate the probability of death for each age group in a human population. Figure 11-5 illustrates survivorship curves for human populations on arithmetic and semilogarithmic scales. The logarithmic scale on the

FIGURE 11-15

Survivorship curves for human populations on arithmetic and logarithmic scales. A constant rate of death yields a diagonal straight line on a semi-logarithmic scale *(After Lotka 1924, and Arms and Camp 1979)*

ARITHMETIC AGE SCALE

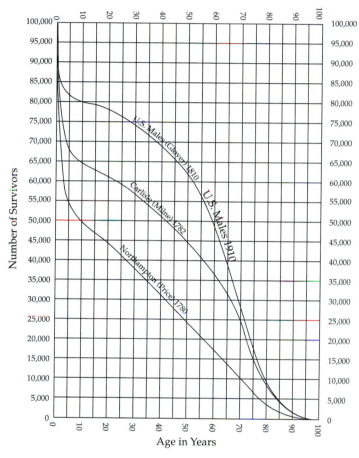

LOGARITHMIC AGE SCALE

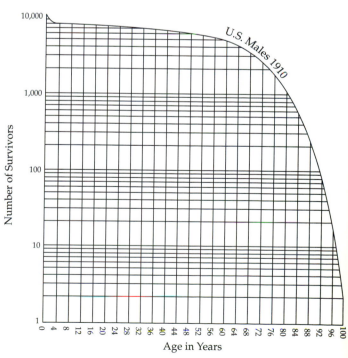

FIGURE 11-16

Taxonomic survivorship curves for extinct genera of artiodactyls and rodents and extinct pteridophyte families covering the whole fossil record. Note the constant rate of extinction in both mammalian taxa and pteridophytes *(After Van Valen 1973).*

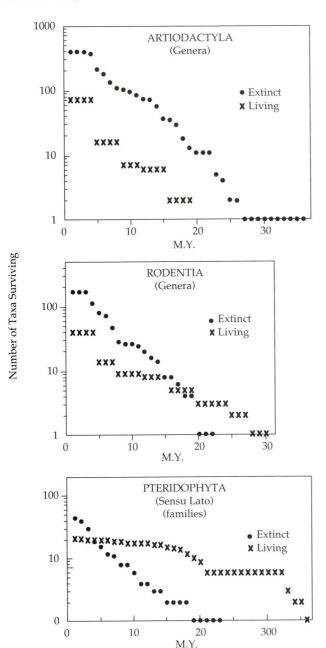

abscissa is normally used, as it indicates the force of mortality at any particular age. The slope of the line correlates with the probability of death at that age. A straight line with negative slope means constant mortality throughout life; the two extreme possibilities would be no death in early life with very high rates in the last part of life, and the converse (Figure 11-15).

Application to the Fossil Record

Van Valen (1973) applied this approach to the fossil record. He calculated the probabilities of extinction for all known subgroups in a larger group—for example, species in the dinoflagellates, genera in the foraminifers, genera in the mammals, and families in the reptiles. For the mammals he plotted the number of surviving genera on the logarithmic abscissa and the millions of years each genus survived, independent of time of origination (the age class), on the ordinate (Figure 11-16). As with the mammals, Van Valen found that extinction rates were constant for a given group, among all groups for which data were available.

Van Valen recognized some real exceptions to this general pattern, but they were very few and were related to exceptionally large extinction episodes. Raup (1975) suggested more rigorous analyses for testing linearity of the plots, but Van Valen's conclusions are essentially intact and cover some 25,000 taxa of plants and animals. Thus, the modeling assumptions made by Raup et al. (1973) and others appear to be close to reality.

The union of paleontology and evolutionary biology has been very creative. The studies described in this chapter exemplify the kinds of thinking that were developing in the early stages of paleobiology. They show how simple models can reveal general patterns and how the secrets of the fossil record are becoming integrated into modern biology.

SUMMARY

The application of ecological approaches to the fossil record has created fascinating opportunities for quantitative analysis of old debates that were originally addressed qualitatively. Five studies devoted to computer-generated phylogenies, Permo-Triassic extinctions, morphological complexity and evolutionary rates, the survival of generalists and specialists, and taxonomic survivorship curves explored the relationships between ecology and paleontology.

Computer-generated phylogenies based on the **theory of island biogeography** had many similarities to the shapes of real clades. They enabled researchers to estimate both patterns that are simulated adequately by general assumptions and those that appear to be exceptional, such as the Cretaceous–Tertiary extinctions and the **coelacanth effect.** The simple stochastic model, with minimal numbers of assumptions and constraints, enables us to probe more deeply into the nature of phylogenies in the fossil record and the mechanisms shaping clade dynamics.

The theory of island biogeography also contributed to a deeper understanding of Permo-Triassic extinctions. Here the theory was applicable directly, because the great area of shallow seas on continental shelves in the Permian and the reexpansion of the area in the Triassic provided island-like regions that sustained the rich marine invertebrate faunas. However, suturing of continents in the late Permian and lowered sea level reduced the area of epicontinental seas from 43 to 13 percent of continental area, and extinctions rose to more than 100 families of marine animals. Computer simulations indicate that a simple relationship between area and number of families may capture the essence of these extinctions, but recovery of taxa lagged behind the recovery of continental shelf area. A closer examination of area effects, however, helps to explain the magnitude of the Permo-Triassic extinctions and the qualitative changes in the surviving faunas. Low diversity of widespread, more generalized species in the survivors indicates greatly simplified ecological conditions for life in terms of reduced province diversity, reduced heterogeneity within provinces, and unstable nutrient supplies.

Examination of the relationship between morphological complexity and evolutionary rates confirmed the pattern that simple morphologies are associated with slow evolution and complex morphologies enable the detection of rapid evolution. The simple gastropods and molluscs had rates of extinction and origination much lower than those of the complex mammals. However, the detection of change is obviously correlated with the number of characteristics available for assessing change, and so the positive connection between high complexity and a high rate of evolution stems from the methods available and does not adequately address the mechanisms of evolutionary change.

Scientists examined the relative survival of generalists and specialists in the fossil record using aquatic free-living arthropods such as trilobites, crayfish, and lobsters. **Cope's "law" of the unspecialized** stated that generalists persist longer than specialists and generalists provide the basic stock from which phylogenies advance. Using the concept of the ecological niche and assuming that appendage modifications provided a view of the niche in fossil organisms, researchers estimated specialization by the diversity of appendage types in each species. It appeared that aquatic free-living arthropods have reached an upper limit of limb diversification, but there was no indication that generalized species persisted longer than specialized species in the fossil record.

Much modeling of the fossil record has assumed that extinction rates are constant. This assumption was tested with **taxonomic survivorship curves,** which revealed remarkable consistency in the fossil record with the assumption.

QUESTIONS FOR DISCUSSION

1. The null hypothesis has become very important in the modeling of paleobiological patterns. To what extent do you think the development of the null model always involves the best possible null hypothesis? Or can the null hypothesis influence the kinds of conclusions reached?

2. Can you identify taxa that each have related extant groups and a good fossil record, so that modeling on both sets of living and fossil organisms can be developed to test the validity of modeling investigations of the fossil fauna or flora?

3. If random events of origination and extinction adequately simulate real diversity of clades through time, should we acknowledge that chance plays a large role in the real world, thereby making attempts to understand the biological basis of patterns less interesting?

4. Using a modeling perspective, can you think of a taxon or taxa for study that would lend itself to an objective

test of the role of competition in the extinction–survival relationship? How would you develop the modeling approach?

5. In addition to the modeling approaches discussed in this chapter, can you think of other generalizations from extant taxa that could aid the discovery and understanding of pattern in the fossil record?

6. Do you feel that a modeling approach to the fossil record is important for separating general patterns from special cases?

7. The theory of island biogeography invokes the strong influences of population size and biotic interactions such as competition and predation. In your opinion, how realistically can these population and species characteristics be extrapolated to higher taxa, such as families, on a global scale?

8. In your opinion, which is more important in the use of models: the testing of explicit hypotheses or the development of heuristic approaches to paleobiology?

9. Extension of a model of the Permo-Triassic marine extinctions beyond the area effect alone would increase the model's complexity and specificity. Do you think such an undertaking would be valuable, and how would you deal in a general way with the additional variables?

10. The fossil records on flowering plants since Cretaceous times and on marine families of invertebrates and vertebrates since the Triassic do not provide strong support for the concept of equilibrium numbers of species and higher taxa. Both records show increasing diversification lasting over 100 million years and 200 million years, respectively, with no asymptotic maximum in evidence. Is there an approach that can resolve the conflict between the fossil record and an equilibrium theory?

REFERENCES

Arms, K., and P. S. Camp. 1979. Biology. Holt, Rinehart and Winston, New York.

Cisne, J. L. 1974. Evolution of the world fauna of aquatic free-living arthropods. Evolution 28:337–366.

Cope, E. D. 1896. The Primary Factors of Organic Evolution. Open Court, Chicago.

Diamond, J. M. 1973. Distributional ecology of New Guinea birds. Science 179:759–769.

Elton, C. 1927. Animal Ecology. Macmillan, New York.

Flessa, K. W., K. V. Powers, and J. L. Cisne. 1975. Specialization and evolutionary longevity in the Arthropoda. Paleobiology 1:71–81.

Gould, S. J. 1981. Papaeontology plus ecology as palaeobiology, pp. 295–317. In R. M. May (ed.). Theoretical Ecology: Principles and Applications. Sinauer, Sunderland, Mass.

Lotka, A. J. 1924. Elements of Physical Biology. Williams and Wilkins, Baltimore.

MacArthur, R. H., and E. O. Wilson. 1967. The Theory of Island Biogeography. Princeton Univ. Press, Princeton, N.J.

Raup, D. M. 1975. Taxonomic survivorship curves and Van Valen's law. Paleobiology 1:82–96.

———. 1977. Probabilistic models in evolutionary paleobiology. Amer. Sci. 65:50–57.

Raup, D. M., and S. J. Gould. 1974. Stochastic simulation and evolution of morphology; towards a nomothetic paleontology. Syst. Zool. 23:305–322.

Raup, D. M., S. J. Gould, T. J. M. Schopf, and D. Simberloff. 1973. Stochastic models of phylogeny and the evolution of diversity. J. Geol. 81:525–542.

Raup, D. M., and S. M. Stanley. 1978. Principles of Paleontology. 2d ed. W. H. Freeman, San Francisco.

Schopf, T. J. M. 1974. Permo-Triassic extinctions: Effects of area on biotic equilibrium. J. Geol. 82:267–274.

Schopf, T. J. M., D. M. Raup, S. J. Gould, and D. S. Simberloff. 1975. Genomic versus morphologic rates of evolution: Influence of morphologic complexity. Paleobiology 1:63–70.

Simberloff, D. S. 1974. Permo-Triassic extinctions: Effects of area on biotic equilibrium. J. Geol. 82:267–274.

Simpson, G. G. 1953. The Major Features of Evolution. Columbia Univ. Press, New York.

Valentine, J. W., and D. Jablonski. 1991. Biotic effects of sea level change: The Pleistocene test. J. Geophys. Res. 96(B4):6873–6878.

Valentine, J. W., and E. M. Moores. 1974. Plate tectonics and the history of life in the oceans. Sci. Amer. 230(4):80–89.

Van Valen, L. 1973. A new evolutionary law. Evol. Theory 1:1–30.

12

HUMAN EVOLUTION

A russet mouse-lemur (*Microcebus rufus*) from the forests of Madagascar, a representative of the order Primates, to which humans belong.

243

12.1 THE ORIGIN OF SPECIES

In the last chapter of *The Origin of Species* (1859), Darwin clearly anticipated the far-reaching impact of his theory of descent with modification through variation and natural selection.

> We can dimly foresee that there will be a considerable revolution in natural history.... Analogy would lead me one step further, namely, to the belief that all animals and plants have descended from some one prototype.... Therefore, I should infer from analogy that probably all the organic beings which have ever lived on this earth have descended from some one primordial form, into which life was first breathed. (Darwin 1859, 484)

Ultimately, he broached the most delicate point:

> In the distant future I see open fields for far more important researches. Psychology will be based on a new foundation, that of the necessary acquirement of each mental power and capacity by gradation. Light will be thrown on the origin of man and his history. (Darwin 1859, 488)

This is the only reference to the evolution of humans in *The Origin of Species*. Darwin was stating an idea that could be inferred throughout the book, that his theory applied to humans. His mention of the idea at the end of *The Origin* was meant to prime the public to deal with the subject in more detail. (Note that "man" is defined as a member of the human race in most current dictionaries and is the common name for all the species in the human family—the Hominidae. In this scientific usage, the term encompasses both male and female members of the species.)

12.2 THE DESCENT OF HUMANS

Darwin (1871) provided details on human evolution 12 years later in *The Descent of Man, and Selection in Relation to Sex*. In fact, of the seven chapters devoted to the descent of humans, Darwin spent three on mental capacity. In chapter 1 he discussed the similarity in structure between humans and other animals, common diseases of humans and other animals, similarities in the reproductive processes in all mammals, and similarities in embryonic development. He noted, for example, the close resemblance of human and dog embryos (Figure 12-1). In 1863, Huxley had pointed out in his book *Man's Place in Nature*,

> Quite in the later stages of development ... the young human being presents marked differences from the young ape, while the latter departs as much from the dog in its developments, as the human does. Startling as this last assertion may appear to be, it is demonstrably true.

FIGURE 12-1

Drawings of human **(top)** and dog **(bottom)** embryos, which appeared as figure 1 in Darwin's (1871) *The Descent of Man*. a) forebrain; b) midbrain; c) hindbrain; d) eye; e) ear; f) first visceral arch; g) second visceral arch; h) vertebral columns and muscles in process of development; i) anterior extremities; k) posterior extremities; l) tail or os coccyx.

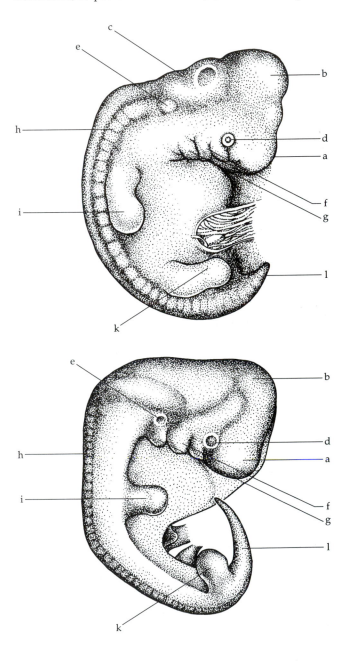

Huxley had observed the great similarity in form of ape and human embryos throughout much of their embryological development. All relevant modern techniques have since confirmed this similarity, a point we will revisit later in this chapter.

Darwin went on in chapter 1 to discuss rudimentary organs, such as "the mammae of male quadrupeds, or the incisor teeth of ruminants which never cut through the gums" (17). He discussed many vestigial muscles and the vestigial pointed ear still seen in some humans (Figure 12-2). The nictitating membrane of reptiles, amphibians, and birds remains in humans as a vestigial semilunar fold in the inside corner of the eye. Darwin concluded chapter 1 with the well-documented case he had developed (31):

> The homological construction of the whole frame in the members of the same class is intelligible, if we admit their descent from a common progenitor, together with their subsequent adaptation to diversified conditions. On any other view the similarity of pattern between the hand of a man or monkey, the foot of a horse, the flipper of a seal, the wing of a bat, is utterly inexplicable. It is no scientific explanation to assert that they have all been formed on the same ideal plan.

In chapter 2, Darwin recognized the problem of arguing similarity of form between apes and humans: the similarity does not seem to hold where mental capacity is concerned. He spent the chapter making the case "that there is no fundamental difference between man and the higher mammals in their mental faculties" (35). He argued that humans and other mammals have many instincts in common, as well as emotions, curiosity, the ability to imitate, and so on, and discussed tool use in monkeys and apes. Chapter 3 continued on the mental powers of humans and other animals. In the summary of chapters 2 and 3, Darwin concluded (105), "Nevertheless the

difference in mind between man and the higher animals, great as it is, is certainly one of degree and not of kind."

Chapter 4 was "on the manner of development of man from some lower form." Darwin recognized that monkeys living in trees developed division of labor between forelimbs and hindlimbs, with the forelimbs devoted to dexterous hands that could manipulate objects and enable facile climbing. He then recognized that if such monkeys colonized the ground rather than the trees (possibly owing to "a change in the condition of its native country" [140]), they could change in one of two directions, either becoming more quadrupedal again, like baboons, or becoming bipedal. The bipedal condition would be a great advantage, freeing the hands for self-defense with stones or clubs, for killing prey, and for gathering other food. With the evolution of bipedal gait and erect posture, many other modifications in structure became necessary, which Darwin discussed: a decline in the use of canine teeth in fighting, an increase in brain size and intelligence, nakedness of the skin ("women became divested of hair for ornamental purposes" [149]), and loss of the tail, as in the great apes.

Chapter 5 of *The Descent of Man* explored the development of intellectual and moral faculties.

In chapter 6, Darwin discussed the affinities and genealogy of humans and recognized the links among the lemurs, the monkeys, and the Old World monkeys that led to man. "Thus we have given to man a pedigree of prodigious length, but not, it may be said, of noble quality" (213).

Chapter 7 addressed the races of humans. Darwin considered the possibility that the races are distinct species but concluded that they should be regarded as subspecies. He argued that natural selection was hard put to explain the differences between human races, but sexual selection could account for many of them. Hence, he devoted the second part of *The Descent of Man* to sexual selection.

Darwin set down a very clear path for studies in human evolution and many other areas of biology.

> [The book] addresses an extraordinary number of problems that are, at this moment, on the minds of many biologists, psychologists, anthropologists, sociologists, and philosophers. It is the genius of Darwin that his ideas, clothed as they are in unhurried Victorian prose, are almost as modern now as they were when they were first published. (Bonner and May 1981, vi)

While we compare later views of human evolution with Darwin's scenario, let us keep in mind that the more recent theorists have largely built their ideas on Darwin's classic book.

FIGURE 12-2

The human ear with the vestigial pointed tip—figure 2 in Darwin's (1871) *The Descent of Man.*

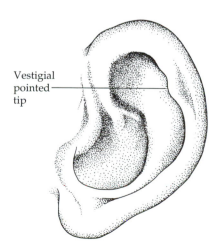

Vestigial pointed tip

12.3 REMARKABLE FEATURES OF THE HUMAN BODY AND THE ARBOREAL HYPOTHESIS ON PRIMATE EVOLUTION

Several features of human body design, although not all unique to humans, are nonetheless remarkable: binocular vision, manual dexterity and sensitivity, brain size, upright posture, and—perhaps surprisingly—extraordinarily bad design. Although many of these traits evolved synchronously, we will consider them under separate headings. As we explore explanations and justifications of the evolution of these traits, we will follow a line of argument now called the **arboreal theory of primate evolution** (Howells 1947), developed mainly in a lecture by Smith (1924) in 1912, followed by lectures in 1915 and 1916 by Jones (1917). Then Collins (1921) emphasized the adaptive function of binocular vision, completing the theory (Cartmill 1972). Scientists generally favored the hypothesis through the 1960s, although skeptics always wondered why other arboreal lineages, such as the squirrels, did not develop many of the traits that humans have ultimately inherited from the primate stock. An early critic gibed, "the only prerequisites of intellect are prehensile limbs and a convenient tree" (Hooton 1930).

After treating the arboreal theory, we will place more focus on how primates may have become so distinct from the many other lineages of arboreal mammals. The **visual predation hypothesis,** developed by Cartmill (1972, 1974a, 1974b), provides a more convincing account of the origins of features peculiar to primates.

Binocular Vision

Simons (1964) discussed binocular vision, in which both eyes face forward, enabling accurate perception of depth. Humans share this capacity with the apes and monkeys (Figure 12-3). It is not a common feature among animals.

The evolution of binocular vision occurred in the early development of arboreal (tree-dwelling) primates in the Paleocene, 63 to 58 million years ago. Species in the genus *Plesiadapis* were rodent-like and had eyes on the sides of their heads (Figure 12-4). Their claws had flattened sides, perhaps indicating the beginning of an evolution from claws to the flattened nails of apes and humans. *Plesiadapis* is not considered to be in the direct lineage of humans but is an example of an early tree-living form (cf. Chapter 7). Whether members of the genus should be regarded as primates or insectivores is open to debate (cf. Cartmill 1972, Colbert and Morales 1991). By

FIGURE 12-3

Skulls of a tree shrew (a) and some primates (b–f), with lines indicating the occipital condyle on which the skull pivots on the spine. Note the long snout and lateral eye socket of the tree shrew and the increasingly flattened profiles of the primates from b to f, coupled with frontal eye sockets and binocular vision. (b) A quadrupedal Old World monkey, *Cercopithecus*. (c) A gibbon in the genus *Hylobates*, a strongly brachiating ape. (d) *Homo erectus*, probably the immediate predecessor of modern humans. (e) A Neanderthal human, *Homo sapiens neanderthalensis*. (f) A modern human, *Homo sapiens*. Dashed lines indicate the position of the brain. The skulls are not drawn to scale. *(From Campbell 1985)*

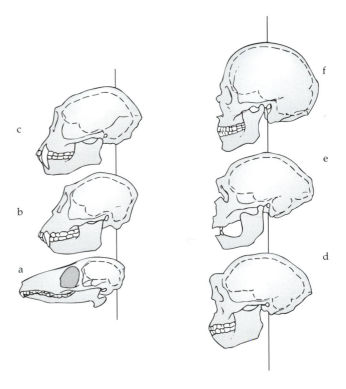

about 50 million years ago, tree-dwelling animals with eyes much more toward the front of the head had developed. The cerebral cortex of the brain, and the forehead to accommodate it, had enlarged considerably, and the snout had shortened. These animals were much more monkey-like, as in the genus *Smilodectes* (Figure 12-5), also not in the direct lineage of humans.

As primate species became specialized to life in the trees and increasingly agile, they acquired binocular vision for accurate depth perception—judgment of distances—when they jumped from branch to branch. The dexterity of their hands also increased, enabling them to grasp branches and manipulate objects. Integration of highly complex movements with

FIGURE 12-4

A reconstructed skeleton of *Plesiadapis* from the Paleocene, a detail of the skull, and an artist's conception of how the animal may have looked, based on comparison with living tree shrews. Members of this genus ranged from the size of squirrels to the size of domestic cats. *(From Simons 1964)*

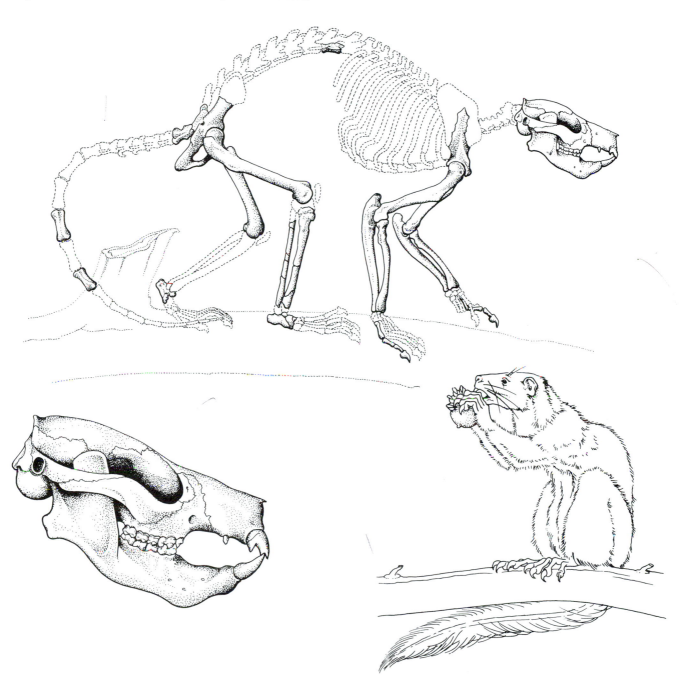

the reception of diverse stimuli from sight and touch required a larger brain. This process continued until very ape-like species with strongly frontal eyes and relatively large skulls for housing larger brains had evolved. By about 25 to 13 million years ago, genera such as *Oreopithecus*, not unlike the orangutan, were extant (Figure 12-6).

Even though the evolution of binocular vision occurred about 50 million years ago in tree-dwelling lemur-like animals, it is an essential feature in the evolution of the human species. Increased visual capacity resulted in the evolution of color sensitivity in the eye, with the development of cone cells in the retina. Color vision enabled primates to discern fine

FIGURE 12-5

The reconstructed skeleton of *Smilodectes* from the Eocene, a detail of the skull, and an artist's conception of how the animal may have looked. The lemur-like appearance is evident. Note the eye position that permits binocular vision, the forehead indicating increased brain size, and the shortened snout. *(From Simons 1964)*

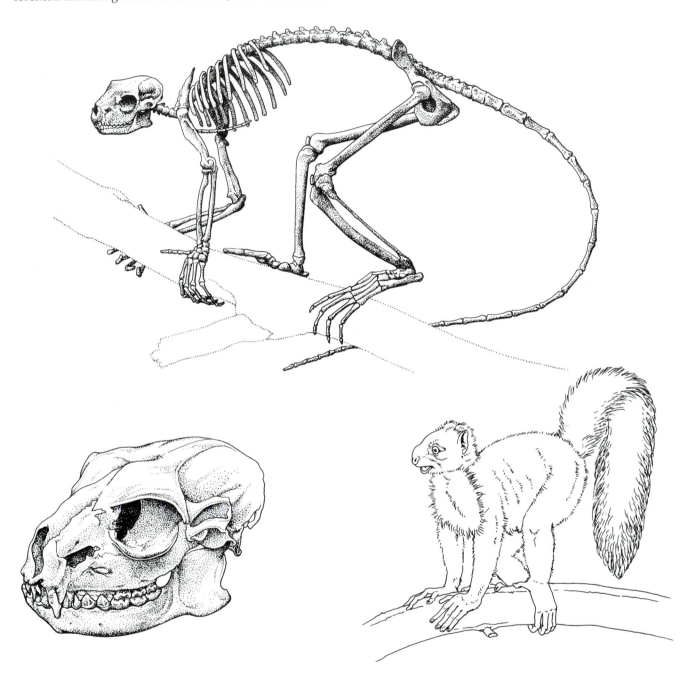

differences in shading that might indicate, for example, the precise distance to the next branch. All primates now have color vision, as do birds.

The shortening of the primate snout provided the frontal eyes with an unimpeded view (cf. Figures 12-4 and 12-5) and reduced the sense of smell, as sight became more important. The flattening of the snout also resulted in shorter jaws with fewer teeth. This

process seems to be continuing in human evolution. (We will return to the subject in Section 12.5.)

Manual Dexterity and Sensitivity

With the evolution of binocular vision came manual dexterity and sensitivity (Napier 1962). By about 14

FIGURE 12-6

A brachiating ape about 4 feet tall with a flattened facial profile, in the genus *Oreopithecus* from the Pliocene. *(From Simons 1964)*

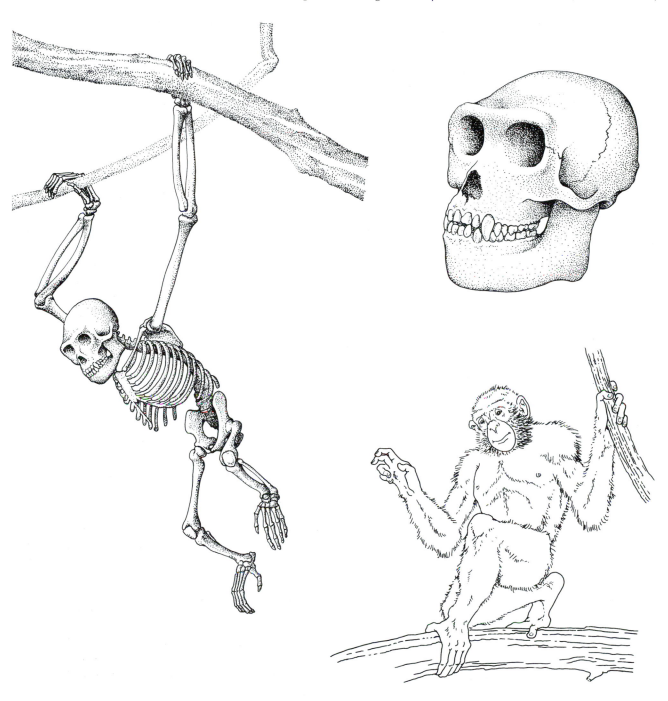

million years ago, ape-like creatures lived in the trees. Arboreal life had selected for stereoscopic color vision, manual dexterity and an opposable thumb for grasping branches, and increased brain size to meet increased needs for coordination, vision, and manual dexterity.

In addition, in the brachiating apes, the spine tended to straighten out and the legs tended to rotate into a position parallel with the spine rather than at right angles to it (Figure 12-6). These changes led to a preadaptation for upright, bipedal posture. Thus, by 14 million years ago, the monkeys and apes already had many important features that are now evident in the human hand, and they illustrate the evolutionary trends that freed hands for activities other than walking.

In fact, by about 3 million years ago, prehuman man-apes were making crude tools with hands very little different from the hands of apes and with brains only slightly larger than those of apes. This link in human evolution was assigned the genus *Australopithecus* (Figure 12-7).

The hands of *Australopithecus* were different from ours in having relatively short, much less flexible thumbs and relatively narrow fingers. As a result, the hand could exert much force in a power grip (such as when we grasp a large screwdriver in the palm of the hand for forceful driving of a heavy screw). However, the precision fingertip grip we use to manipulate small objects (such as a screwdriver when we repair a watch) was largely absent. The tools made by *Australopithecus* were therefore very crude, not so much because of lack of practice but because hands were still crudely designed with less control for delicate manipulation.

FIGURE 12-7

Australopithecus as this genus may have looked and walked.

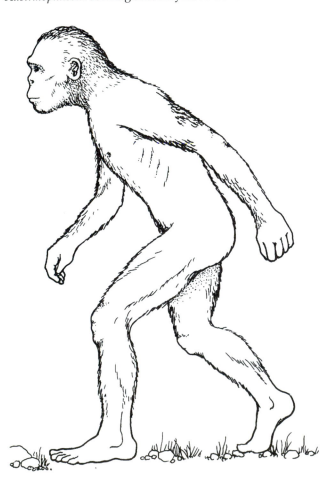

It now appears that tool use has been an important factor in human evolution and that it has both depended on and influenced the evolution of the hand. As Oakley (1958) said, "Tools makyth man." Darwin (1871) was the first to note that tool use was both the cause and the effect of bipedal locomotion. Use of one pair of limbs freed the other pair to wield tools. As tool use increased, the hind legs became specialized for walking and the hands became more specialized for making and manipulating tools. (See also Washburn 1960, Washburn and Moore 1980.)

As hands gained flexibility, the forelimbs took on additional functions, such as fighting. All human-like skeletons lack large canine teeth and look much more like female apes than male apes. Evidently, acts of aggression changed from attacks based on biting—which required a large jaw, large canines, and a very strong neck—to the use of arms, hands, and man-made weapons. With this change came reduced canine teeth, smaller jaws, and reduced neck muscles, all leading to transformation of the male form toward a form more like the female ape, which did not fight. This dramatic shift in skull design and canine tooth size is especially clear in comparisons of other primates with man (Figure 12-8) (Holloway 1974).

Brain Size

Tool-making dexterity could evolve only with increased sensitivity of the hand and increased sensory and motor control centers in the brain. Thus, both tool making and brain size increased dramatically from 3 million years ago until about 1.5 to 1.8 million years ago, when *Homo erectus* walked the earth (Figure 12-9). The thumb became longer and more maneuverable; the fingertips broadened, with more sensory cells in the skin; and the precision grip of modern humans evolved.

Brain size increased dramatically. The cortex of the brain expanded in size by a factor of two during this time, and it is clear from the organization of the cortex that the human hand played an important role in this dramatic growth. A very large part of the cortex handles sensory input from the hand and motor output to the hand (Figures 12-10 and 12-11). The area of the human cortex devoted to the hands is much larger than equivalent areas in monkeys and apes (Washburn 1960).

Brain size is obviously associated with the coordination of some very complex processes. One is the complexity and sensitivity of the human hand. Another is the need for learning to remember design, technique in tool making and hunting, and the rudiments of culture and social norms. As skills and culture increased, so the need for communication in-

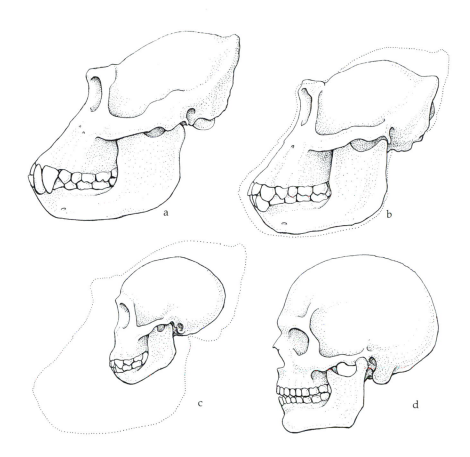

FIGURE 12-8
A comparison of gorilla (a,b,c) and human (d) skulls. In the mature human skull, sexual differences are practically lost. Note the large, robust skull of the male gorilla (a), with large canine teeth, compared with the female (b) and immature (c) skulls. Note that the human skull not only lacks large canines but is more similar to an immature ape skull than to a mature female ape skull. (*Gorilla skull from Eckhardt 1972, human skull from Holloway 1974.*)

creased, with the ultimate development of language, involving complex integration of lips, tongue, and larynx and a large commitment in the brain (Figures 12-10 and 12-11).

Nobody knows for certain when speech emerged in the human lineage. Washburn and Moore (1980) and many others have argued that speech evolved in *Homo sapiens* before the advent of the agricultural revolution about 10,000 years ago. However, given that *Homo erectus* had a large brain and a long neck to house the deep larynx, it is also likely to have used speech (Restak 1988). Indeed, if we follow the history of change in cranial shape, vocal tract, and cultural development through the lineage of all species of *Homo,* it seems that complex language capabilities began with the evolution of the genus (Deacon 1992, Schepartz 1993). Schepartz (1993) provides a fascinating synthesis of the evidence for the early evolution of language.

Let us review briefly the increases in brain size associated with skills in the human lineage. More than 4 million years ago, an ancestral hominid had a small brain (about 350 cubic centimeters) and walked on its knuckles. It may have used sticks as tools, but it did not fashion tools. Four to 3 million years ago, an advanced hominid, *Australopithecus,* emerged with about a 450-cubic-centimeter brain

(Washburn and Moore 1980), a pelvis that allowed upright posture, and the use of crudely manufactured stone tools by males. Some skulls of that time had cranial capacities of 750 cubic centimeters, but those may have belonged to another species. Leakey named skulls of 650 to 750 cubic centimeters *Homo habilis.* Some anthropologists regard them as transitional between *Australopithecus* and *Homo erectus* (cf. Day 1984).

By 1.5 million years ago, the first undisputed true human, *Homo erectus,* had a truly upright posture and a pelvis like our own. Brain size had expanded to 850 to 1,100 cubic centimeters. This human was a skillful tool maker and perhaps a speaker. *Homo sapiens,* our own species, emerged about 100,000 years ago with a brain size of about 1,400 cubic centimeters; Washburn and Moore (1980) give a range of 1,100 to 1,500 cubic centimeters. The Neanderthal human whose skull was first found in the Neander River valley in Germany (*Thal* is German for valley) had the brain size of modern humans but a slightly different cranium shape. Male neanderthals had a heavier brow ridge and a more sloping forehead than modern males; these features were not so distinct in females (Brace et al. 1979). The Neanderthal human sometimes receives subspecific status and is called *Homo sapiens neanderthalensis.*

FIGURE 12-9

Homo erectus.

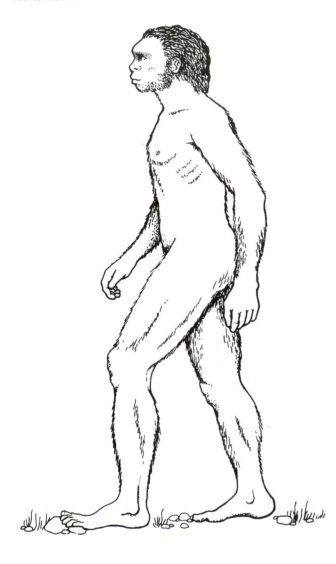

The cranium as we find it in modern humans developed by about 40,000 years ago. Precivilization humans in Europe have sometimes been referred to generally as Cro-Magnon man, but this designation is of dubious authenticity, according to Brace et al. (1979). Cro-Magnon is the type locality in the Dordogne region of France where the first fossil evidence of modern humans in Europe appeared. These Cro-Magnons were remarkable for two reasons: they possibly had contact with Neanderthals, and they left a legacy of fine art in the caves at Lascaux and Altamira (Stringer 1992), and the recently discovered Chauvet cave near Avignon.

Brain capacity more than tripled in about 3 million years, from *Australopithecus*, which used crude tools, to *Homo sapiens*. Yet humans do not have the largest brains relative to body size. In the house mouse, the ratio of brain to body volume is 1:40; in the porpoise it is 1:38; the human, 1:45; the gorilla, 1:200; and the elephant, 1:600. Of the higher and larger primates, however, the human does have the largest brain in proportion to body size. The most impressive point about the human brain is its rapid increase in size from *Australopithecus* to *Homo sapiens*.

Martin (1981) has made some interesting comparisons between brain development in humans and modern apes. He noted that the human brain is three times larger than the brain of a modern ape of the same size. Humans have infants with brains and bodies twice as large as would be predicted on the basis of length of gestation period in modern apes. And remarkably, the human brain expands fourfold from neonatal size to adult size, compared to the twofold expansion typically seen in other primates. In the face of these observations, Martin argued that humans must have acquired a predictable and rich source of food to allow for such a great energy commitment by females to the rapid growth of the fetus and the neonate brain. Consumption of plant roots, energy-rich tubers, and substantial meat, facilitated by the use of tools, would have provided the high energy required for rapid brain growth.

Upright Posture

The evolution of upright posture permitted many kinds of changes.

1. Hands were freed to specialize in tool making, weapon manufacture, and gathering and preparation of food.
2. Walking and running on two legs enabled primates to travel great distances. The striding gait of man (Napier 1967) permitted extensive travel, colonization of new localities, extensive hunting and gathering, and migration in pursuit of animals (when hunting depleted game or as the animals themselves migrated).
3. Visibility of surroundings was greater.
4. The pelvis changed from the long form of a great ape to the short, robust form of the human (Figure 12-12).
5. The vertebral column straightened out; then, for the skull to be balanced directly over the pelvis, a strong curvature of the spine developed (Figure 12-12).
6. With the strengthening of the pelvis came a narrowing of the birth canal through the pelvis.

The last-mentioned development is particularly interesting because it necessitated a profound change in reproduction between apes and humans. The reduction in the size of the birth canal occurred a little before the dramatic increase in cranial capacity.

FIGURE 12-10

Allocation of sensory centers of the human brain to the parts of the body. The size of each part of the body is drawn proportionally to the amount of the sensory cortex committed to that part. Half of a cross-section of the brain (from ear to ear) is shown. *(From Washburn 1960)*

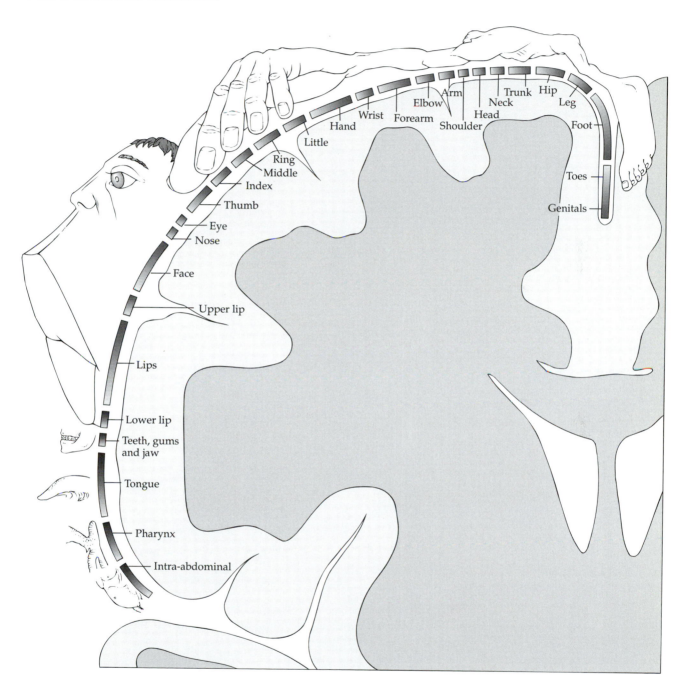

Washburn (1960) called this apparent anatomical glitch, appropriately, an **obstetrical dilemma.**

The solution was a birth much earlier in development than occurred in apes, while cranial capacity was still relatively small (cf. Table 12-1) and the brain and cranial bones were essentially in a fetal condition.

Pelvic size in present-day humans allows the brain of the neonate to be about 350 cubic centimeters at birth (Martin 1981). With the doubling of brain size during development that is typical of apes, this volume means an adult cranial capacity of 700 cubic centimeters, which fits well with large

FIGURE 12-11

Motor centers of the human brain and the parts they serve, positioned around half of an ear-to-ear cross section. *(From Washburn 1960)*

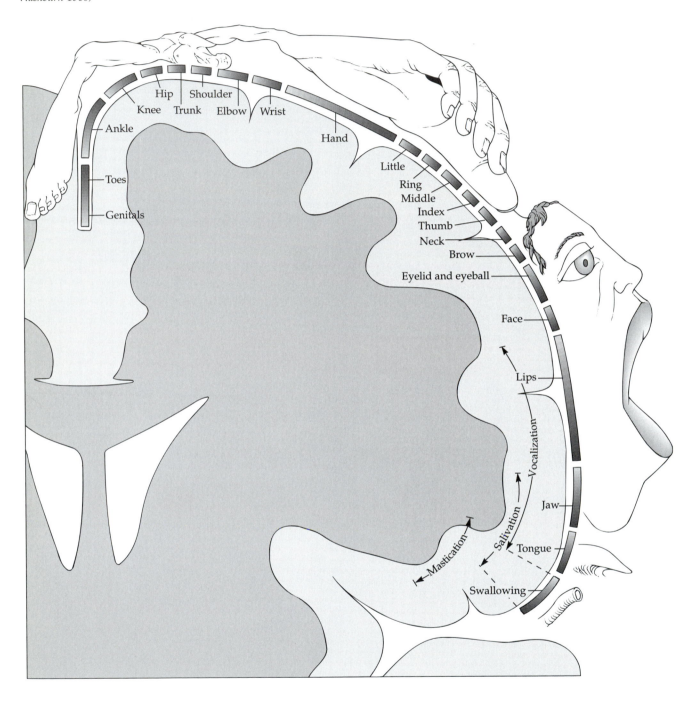

Australopithecus or small *Homo erectus.* Martin was curious about the quality of human milk that can fuel a doubling of brain growth rate compared with the apes. Does cow's milk fed to babies produce adults with more cow-like brains? Every educator has cause to ponder this question.

With a poorly developed central nervous system, the neonate was helpless. This situation could evolve only when the mother was bipedal so that she had both arms free to hold the infant. The mother–child relationship in humans necessarily became a very prolonged association, lasting several years at least.

FIGURE 12-12

Skeletons of the gorilla, which is quadrupedal, and the human, which is bipedal. Note the relatively long pelvis of the gorilla, set at about 45° off the vertical, and the squat, robust vertical pelvis of the human. *(From Napier 1967)*

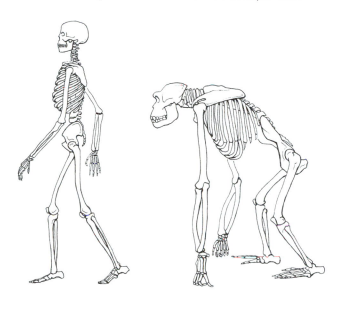

12.4 THE VISUAL PREDATION HYPOTHESIS

Cartmill (1972) noted that proponents of the arboreal theory of primate evolution rested their case on characterization of the primate order and on distinctive and complex evolutionary trends. It remained unclear, however, why not all arboreal mammals (both marsupials and placentals) had evolved the traits emphasized in the theory. What was the key to the divergence of primate design from other successful tree-dwelling mammals, such as opossums in the New World, possums and koalas in Australia, tree squirrels, raccoons, and civets?

Three key comparisons among arboreal mammals are at the center of the visual predation hypothesis. First, very active arboreal mammals can get a firm grip on the substrate with claws. Usually their hind limbs are best adapted for this secure grip, while the forelimbs test the next support, as in tree squirrels (Cartmill 1972, 1974a, 1974b). Prehensile appendages, which can grasp a limb, are not well suited for climbing large trees but are beneficial for clinging to small stems when foraging in shrubs or terminal tree branches. Hence, life in the trees is not an essential selective element in the evolution of manual dexterity.

Second, binocular vision is not a necessary feature for even acrobatic existence in the trees. Tree squirrels have very little overlap in visual fields of the two eyes, and yet their antics in the trees are equivalent to those of the most agile primates. Cartmill (1972) noted that binocular vision is most commonly observed in predators—in fact, it is practically restricted to a predatory lifestyle. Owls, hawks, and cats all depend on eyesight to detect their prey. Cats are the only mammals other than primates with strongly overlapping visual fields. Many primates

feed on exposed and easily alerted animal prey, which they carefully stalk and then strike suddenly with their prehensile hands. Stereoptic integration of the two visual fields improves the accuracy of the final strike; increase in visual-field overlap facilitates compensation for evasive movements of the prey. (Cartmill 1972, 113)

TABLE 12-1 Estimates of primate body and brain weight in neonates and adults compared with advancement factors

	Brain Weight (g)		Body Weight (g)		Advancement Factor	
	Neonatal, E_n	Adult, E_a	Neonatal, S_n	Adult, S_a	Brain, $A_e = E_n/E_a$	Body, $A_s = S_n/S_a$
Human	335	1,300	3,600	65,000	.26 (4×)	.056 (18×)
Chimpanzee	128	360	1,560	45,000	.36 (3×)	.035 (29×)
Gorilla	227	406	1,750	140,000	.56 (2×)	.013 (80×)

Data from Sacher and Staffeldt 1974.

Notes on interpretation:

1. The human brain grows about 4× in size after birth, compared to only about 3× in chimpanzee and 2× in gorilla. The A_e for the human is higher than for any other primate estimated by Sacher and Staffeldt (1974).

2. Conversely, the increase in body weight in humans, at about 18×, is less than those in chimpanzees (about 29×) and gorillas (80×).

Third, most arboreal mammals have retained large snouts and olfactory sensitivity, but the primates lost them. The distinctive feature of the primates is their complex of habits as arboreal visual predators. The one feature of the arboreal theory of primate evolution that remains intact today is that an adaptive link occurred between loss of olfactory sensitivity and gain of ophthalmic sensitivity.

The visual predation hypothesis notes the unique ecological niche the primates exploited as angiosperms radiated and as insect herbivores exploited the angiosperms. Shrubby vegetation and low limbs of trees, with many insects feeding on tender new leaves, yielded a newly rich but precarious foraging opportunity for mammals. Prehensile hands, feet, or tails enabled exploitation of a nutritious insect resource while necessitating the additional adaptations of visual orientation to prey, dexterous capture, and close and complex integration of hand and eye coordination. Thus, the diagnostic morphological features of the primates relate to ecological opportunities in Cretaceous forest communities.

12.5 PROBLEMS WITH HUMAN DESIGN

In our daily lives we feel the effects of problems with human design. Problems related to bipedal locomotion are also evident in fetal development, parturition, and neonatal development. Each of us can name many other imperfections that the evolutionary process has wrought.

Wisdom Teeth

With the shortening of the snout came a necessary reduction in jaw size, meaning that the jaw could hold fewer teeth. Evidently, this evolutionary process is continuing today. Many people do not develop wisdom teeth (third molars). Of the ones who do, many suffer and must have the teeth removed surgically because their jaws are too small to accommodate them. There is a clear evolutionary trend toward fewer teeth in the human mouth. Before the advent of surgery, there may also have been a strong selective pressure for reduced dentition. Early mammals had 66 permanent teeth, most modern mammals have 44 teeth, and humans have only 32 (Table 12-2). In Central Europe, one or more wisdom teeth are missing in 19 percent of the white human population. Agenesis of third molars is common in American Indians but rare in Africans. Brace et al. (1979) consider that reduction of wisdom teeth has been progressing for about 10,000 years and has become more common during the last few centuries.

Childbirth

Problems with childbirth clearly result from difficulty with passage of the large head of the fetus through the pelvis and from the fact that upright posture increases the probability that the fetus will become wedged in the pelvis in an awkward position: feet first, buttocks first (breech birth), or with the back of the head toward the vertebral column. Interestingly,

TABLE 12-2   Evolutionary decline in tooth number from marsupials to humans

The Dental Formula: One Side of Mouth

Number of incisors, cuspids, premolars, molars in upper jaw $\left(\dfrac{I.C.P.M.}{I.C.P.M.}\right)$
Number of incisors, cuspids, premolars, molars in lower jaw

Typical marsupial	$\dfrac{5.1.3.4}{4.1.3.4}$	Total of 50 teeth
Placental mammal	$\dfrac{3.1.4.3}{3.1.4.3}$	Total of 44 teeth
Anthropoids		
New World monkeys	$\dfrac{2.1.(2-3).(2-3)}{2.1.(2-3).(2-3)}$	
Old World monkeys	$\dfrac{2.1.2.3}{2.1.2.3}$	Total of 32 teeth
Humans	$\dfrac{2.1.2.3}{2.1.2.3}$	Total of 32 teeth with variable reductions

female beauty in many societies is associated with callipygousness—well-developed, rounded hips.

Back Problems

The strong curvature of the small of the back, in the lumbar vertebrae, clearly causes much pressure in this region—much more than in the straighter backs of monkeys. The pressure makes the muscles and vertebral column vulnerable to injury; pulled muscles, slipped discs, lumbago, and scoliosis are common maladies. Lumbago is rheumatism in the lumbar region of the vertebrae, below the thoracic region and above the sacrum; scoliosis is lateral curvature of the spine.

Hernias

Hernias, or ruptures, are splits in the muscles of the body wall that allow parts of organs, from the size of a marble to the size of a large grapefruit, to escape their normal constraints. Ruptures usually occur in the groin and are more common in men than in women because the passage of the testes into the scrotum through the body wall leaves weak spots. Upright posture puts much more pressure on the lower abdomen than exists in an animal with quadrupedal posture. Even rather slight additional pressure, as during vigorous coughing, can produce a rupture in the body wall.

In *The Scars of Human Evolution* (1951), Krogman wrote,

> It has been said that man is "fearfully and wonderfully made." I am inclined to agree with that statement—especially the "fearfully" part of it. As a piece of machinery we humans are such a hodgepodge and makeshift that the real wonder resides in the fact that we get along as well as we do. Part for part our bodies, particularly our skeletons, show many scars of Nature's operations as she tried to perfect us.

Krogman mentioned additional problems.

Varicose Veins

Upright posture created problems for blood circulation, because the heart is about 4 feet above the ground (in a 6-foot person). Thus, blood returning from the feet must overcome 4 feet of gravitational pull. The pressure developed in the veins often proves to be too much, and they become varicose.

Hemorrhoids

It is doubtful that cows, chimpanzees, or gorillas suffer frequently from hemorrhoids, but many humans do. When vertical posture upended the lower end of the large intestine, its veins became much more vulnerable to congestion. Impeded blood flow causes hemorrhoidal swellings in the lower intestine and anal sphincter.

Milk Leg

"Milk leg" during pregnancy is another problem with human design. At the division of the dorsal aorta, the femoral arteries cross over the base of the femoral veins that drain the legs. In addition, they cross where the pelvis joins the vertebral column, necessitating a prominence of bone. During pregnancy, the pressure of the fetus compresses the artery and vein against the pelvis, reducing the flow of blood from the legs and causing accumulation of fluids and consequent swelling. In quadrupeds, where the fetus presses down against the abdomen wall, this problem does not occur.

Foot Problems

The human foot is too small to bear the weight of the body with little strain. Inadequate design and size results in sore feet, calluses, and fallen arches.

Learning to Walk

Learning to walk also has its problems for human babies. When the child is born, it has a straight back like a monkey's, which then gradually bends through exercise. At about 4 months, the baby exercises its neck and holds its head erect, creating the forward curvature in the cervical vertebrae. At around 1 year, a child begins to stand up and to develop the strong curvature in the lumbar region. Therefore, part of the problem is that, while the baby is learning to walk, its skeleton has to change through exercise.

These problems with human design are particularly revealing because we can easily assess their commonness, and the marks of our long phylogenetic history become very tangible. Vestiges of quadrupedal locomotion demonstrate that the evolutionary process does not necessarily achieve perfection—far from it, in some cases. One form melds into another even though many design features could be greatly improved by a bioengineer and probably even by a plumber. There is little doubt that the human species will continue to evolve (Dobzhansky 1960, 1962), and it is interesting to speculate on the directions evolution is taking. How much do (and will) medicine and surgery modify evolutionary rates and directions? With social welfare, will postreproductive mortality still play a role in human evolution? Is there any remaining selective advantage to becoming old and wise?

12.6 FOSSIL HISTORY

While pondering these questions, let us travel back in time to examine the fossil history of modern humans and their forebears. What kind of phylogenetic trees can be constructed from the existing skeletons? In fact, there are almost as many trees for the human lineage as there are fossils, and all depend on a rather detailed understanding of osteology. Therefore, we will initially illustrate only four schemes as descriptors of the lineage, without entering into their justification (consult the sources for details). The lineages all date from the late 1970s and early 1980s, when there was considerable latitude in interpretation and, justifiably, some trepidation about too much definiteness in describing relationships.

Brace et al. (1979) favored one lineage of the five schemes they considered (Figure 12-13). Another approach is to recognize names and trends without linking all the species (e.g., Tanner 1981) (Figure 12-14). But dating of skull material clouds even this scenario, and alternative dates may result in an alternative scenario (e.g., Leakey 1981) (Figure 12-15). The phylogenies clearly depend on a rather personal interpretation of the names to use and how they should be applied to skulls and skeletons of particular fossils. A more objective approach, more like a numerical taxonomy, employs total average tooth volume (Brace et al. 1979) (Figure 12-16). Even the designation of *Australopithecus* and *Homo* as genera are not set, for Tanner (1981) assigns the species *habilis* to the former (Figure 12-14), whereas Leakey (1981) assigns it to the latter (Figure 12-15). Such uncertainty is to be expected in such a hotly debated subject, so close to the human heart (cf. Johanson and Edey 1981).

In the last decade, paleontologists have made progress in assigning fossils to genera and species; one result is a more stable interpretation of the possible relationships of fossils in the genera *Australopithecus* and *Homo*. Stability, however, should not be interpreted as an indication of correctness, as we saw in the case of the arboreal theory of primate evolution, which was widely accepted for over 50 years.

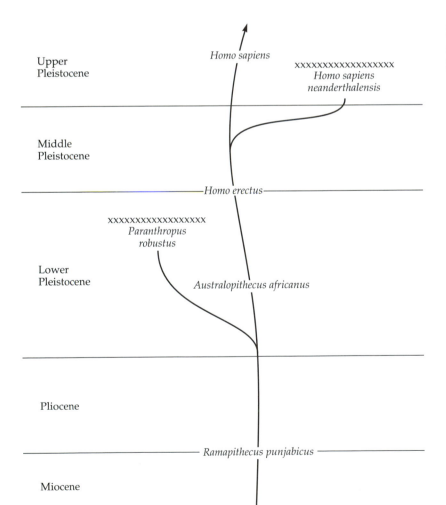

FIGURE 12-13

One possible phylogenetic tree for the human lineage. *(From Brace et al. 1979)*

FIGURE 12-14

A chronology of members of the human lineage whose skulls have been discovered and named. *(From Tanner 1981)*

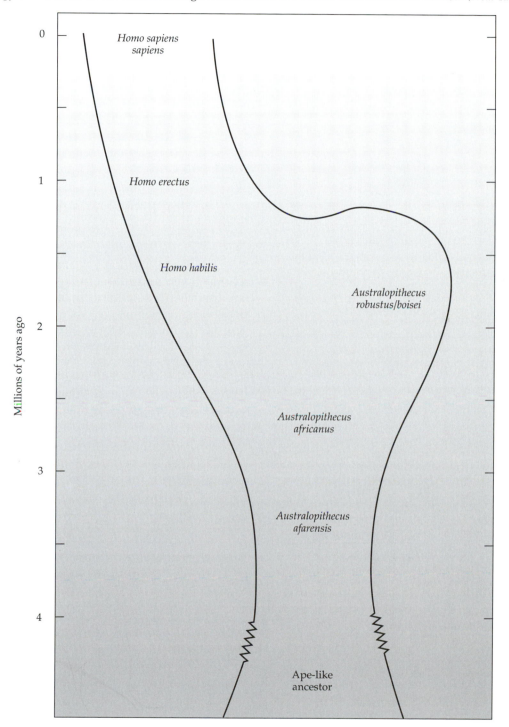

This said, there is still great uncertainty about the ancestors of the genera *Australopithecus* and *Homo*. *Ramapithecus* used to be regarded as a likely candidate, as in Figures 12-13 and 12-15. *Ramapithecus* is now thought to be a female *Sivapithecus*, which is Eurasian and more closely related to the orangutan (*Pongo*) than to the African apes, and its fossils have never been found in Africa (Simons 1992). Focus has now turned to *Aegyptopithecus* and *Proconsul* as genera in a possible lineage to *Australopithecus*. *Proconsul* shows both primitive and derived characteristics; it was a monkey-like quadruped, although, like the apes, it lacked a tail and its hands, feet, and sacrum were ape-like (Simons 1992).

FIGURE 12-15

Leakey's (1981) view of the lineage leading to *Homo sapiens*.

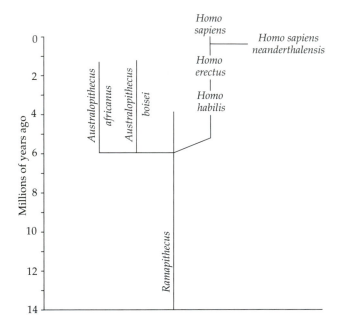

Paleontologists have usually split the genus *Australopithecus* into two groups, one with light build, or **gracile form,** and the other with the more rugged **robust form** (Wood 1992). These groups may be sufficiently different to belong to different genera—the gracile forms to *Australopithecus* and the robust forms to *Paranthropus*. Each genus would then contain two species. In *Australopithecus*, fossils from southern Africa have been called *A. africanus*, and those from East Africa *A. afarensis* (Figure 12-17). Of the robust forms, *P. robustus* is from South Africa and *P. bosei* is from East Africa. This scheme includes all the names for australopithecines, or "southern apes," used in Figures 12-13 through 12-16.

The relationships between *Australopithecus* and *Paranthropus* and their possible relationships to the genus *Homo* are under debate. Possibly no member of the genus *Australopithecus* was ancestral to humans (Figure 12-18c), or perhaps *A. afarensis* was ancestral to *Paranthropus* and *Homo*. In the latter case, *A. afarensis* could have been directly ancestral to *Homo* or ancestral to *A. africanus*, which then gave rise to *Homo*. Figure 12-18 illustrates these three hypotheses. Kimbel and Rak (1993) stress the need for uniform application of the phylogenetic species concept (Chapter 4, Section 4.11), which should enable more rigorous assignment of fossils to species. Such uniformity would be a necessary first step toward a more objective eval-

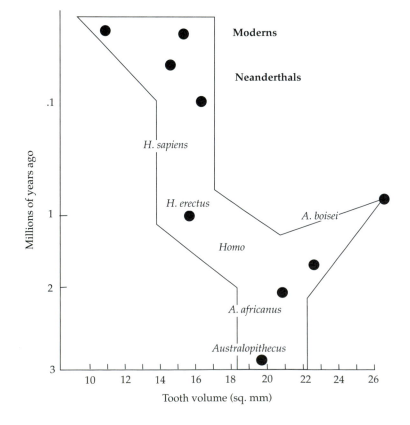

FIGURE 12-16

The relationship among total average tooth volume per individual, assigned taxonomic status of skull type, and time. *(Brace et al. 1979)*

FIGURE 12-17

The distinguishing features of australopithecines, which are regarded as including four species in two genera—the gracile *Australopithecus* and the robust *Paranthropus*. *(From Wood 1992)*

DISTINGUISHING FEATURES OF AUSTRALOPITHECINES

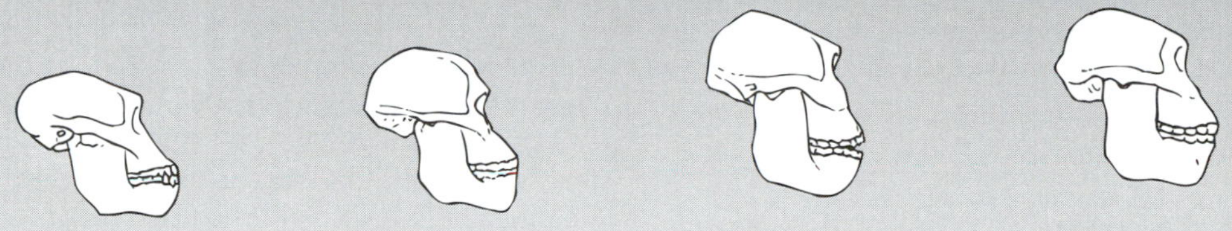

	Australopithecus afarensis	*Australopithecus africanus*	*Paranthropus boisei*	*Paranthropus robustus*
Height (m)	1–1.5	1.1–1.4	1.2–1.4	1.1–1.3
Weight (kg)	30–70	30–60	40–80	40–80
Physique	Light build; some ape-like features (e.g. shape of thorax; relatively long arms relative to legs; curved fingers and toes; marked to moderate sexual dimorphism)	Light build; probably relatively long arms; more "human" features; probably less sexual dimorphism	Very heavy build; relatively long arms; marked sexual dimorphism	Heavy build; relatively long arms; moderate sexual dimorphism
Brain size (ml)	400–500	400–500	410–530	530
Skull form	Low, flat forehead; projecting face; prominent brow ridges	Higher forehead; shorter face; brow ridges less prominent	Prominent crests on top and back of skull; very long, broad, flattish face; strong facial buttressing	Crest on top of skull; long broad, flattish face; moderate facial buttressing
Jaws/teeth	Relatively large incisors and canines; gap between upper incisors and canines; moderate-size molars	Small incisor-like canines; no gap between upper incisors and canines; larger molars	Very thick jaws; small incisors and canines; large, molar-like premolars; very large molars	Very thick jaws; small incisors and canines; large, molar-like premolars; very large molars
Distribution	Eastern Africa	Southern Africa	Eastern Africa	Southern Africa
Known date (millions of years ago)	> 4–2.5	~ 3.0–<2.5	2.6–1.2	2–1

uation of the number of species involved in human evolution and their phylogenetic relationships.

Overlapping both forms of *Australopithecus* in fossil beds in Africa, a new genus seems to have emerged, with a particularly large brain for its size, a prominent nose, and relatively thin bone in the lower jaw (Stringer 1992). This genus, *Homo*, had bipedal locomotion, and its member species used stone tools. The first true human is now generally recognized as *Homo habilis*, "handy man." Recognition of this species depended on fossils from Koobi Fora in northern Kenya, on the east side of Lake Turkana (the site designation is KNM-ER). The main forms of the genus *Homo* are thought to be related linearly—*H. habilis* to *H. erectus* to *H. sapiens*—as in Figure 12-15; Figure 12-19 shows their skulls. *Homo habilis* fossils are currently dated to 2.0 to 1.6 million years ago, *H. erectus* at 1.8 million years ago, and *H. sapiens* as early as perhaps 400,000 years ago (Figure 12-20). Humans related morphologically to the Neanderthal skull increasingly receive acceptance as a separate species,

Homo neanderthalensis, rather than a subspecies of *H. sapiens* (e.g., Rak 1993). But the weight of opinion is still on a derivation from *H. sapiens* with no living descendants. Some have recognized the need for a more objective evaluation of the variation in skeletons ascribed to a single species, such as *H. habilis* (Wood 1993). Variation in Koobi Fora remains, and the nearby Omo site, on the Omo River draining Lake Turkana, suggest two species rather than one.

To summarize the fossil history of primates, Simons (1992) created a picture with many uncertainties but that includes many of the genera discussed in this chapter (Figure 12-21). Note that *Purgatorius* is regarded as the first evidence of a primate (as mentioned in Chapter 7, Section 7.10). Simons's scheme regards *Plesiadapis* (Figure 12-4) as an archaic primate, and *Smilodectes* (Figure 12-5) is one of the earliest true primates. *Oreopithecus* floats in a realm of uncertainty. Note the possible lineage from *Aegyptopithecus* to *Proconsul* to *Australopithecus*, and *Homo*. A *Proconsul*-like ancestor to the

FIGURE 12-18

Three hypotheses on the phylogenetic relationships between the four australopithecine species and the lineage to humans. (a) *Australopithecus afarensis* gave rise to the genus *Homo* and the species *A. africanus*. (b) *A. africanus* derived from *A. afarensis* and gave rise to *Paranthropus* and *Homo*. (c) *Australopithecus* was not a precursor to the genus *Homo*. *Paranthropus aethiopicus* was a primitive form of *P. boisei*. *(From Wood 1992)*

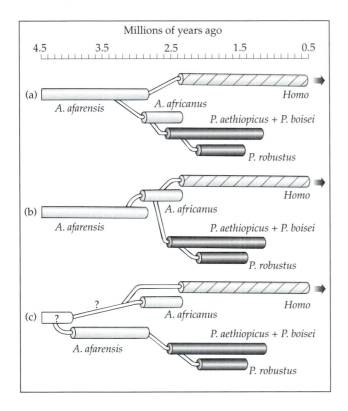

Eurasian *Sivapithecus* (and *Ramapithecus*) diverged to give rise to *Pongo*, the orangutans, perhaps.

12.7 MITOCHONDRIAL EVE

Hypotheses on the Origins of Modern Humans

Whichever phylogeny of human origins one favors, the linear arrangement of species and subspecies (whether linked by presumed lines of descent or not) conceal a fascinating set of questions. All the questions concern the origins of modern humans, *Homo sapiens sapiens*, from a more archaic stock; the nature of migrations around the Old World before or after such origins; and the possible interactions among the novel populations of modern humans and the established populations of archaic *Homo*.

FIGURE 12-19

Major differences in skull shape from *Australopithecus africanus* to members of the genus *Homo*. ("Archaic" *H. sapiens* encompasses skulls or jaws that may belong to a group distinct enough to be regarded as another species, *H. heidelbergensis* or *H. rhodesiensis*.) These fossils date back to 400,000 years ago, and so the debate on the evolution from *H. erectus* to *H. sapiens* influences the accepted date for the emergence of *H. sapiens*, in the range from 400,000 to nearer 100,000 years ago. The sites of the fossil skulls are given in parentheses. Sterkfontein is in South Africa, west of Johannesburg. KNM-ER is Koobi Fora, on the east side of Lake Turkana in northern Kenya. Skull KNM-ER 1470 was particularly important in establishing *H. habilis* as an acceptable species. KNM-ER 3733 can be regarded as the earliest *H. erectus* skull known. Broken Hill, or Kabwe, is in Zambia, on a tributary of the Zambezi River. Irhoud designates a site at Jebel Irhoud in western Morocco. *(From Stringer 1992)*

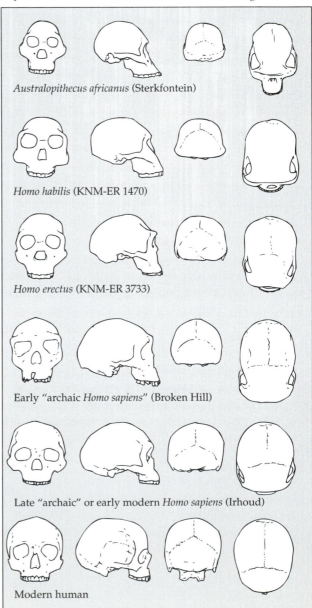

Australopithecus africanus (Sterkfontein)

Homo habilis (KNM-ER 1470)

Homo erectus (KNM-ER 3733)

Early "archaic *Homo sapiens*" (Broken Hill)

Late "archaic" or early modern *Homo sapiens* (Irhoud)

Modern human

FIGURE 12-20

Distinguishing features of species in the genus *Homo*, from *H. habilis* to modern *H. sapiens*. *(From Stringer 1992)*

DISTINGUISHING FEATURES OF EARLY HUMAN SPECIES

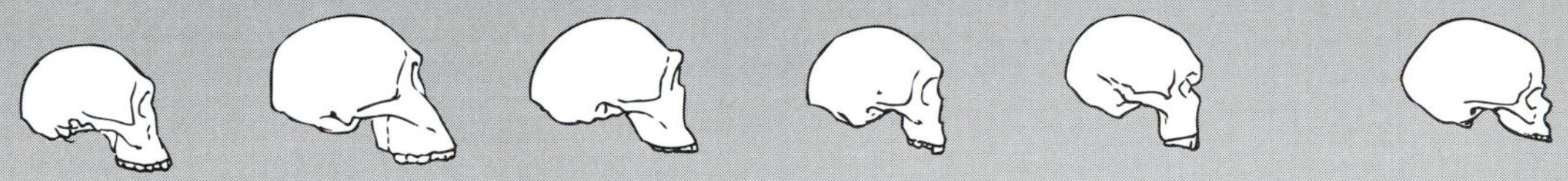

	Homo habilis (small)	Homo habilis (large)	Homo erectus	'Archaic Homo sapiens'	Neanderthals	Early modern Homo sapiens
Height (m)	1	c. 1.5	1.3–1.5	?	1.5–1.7	1.6–1.85
Physique	Relatively long arms	Robust but "human" skeleton	Robust but "human" skeleton	Robust but "human" skeleton	As "archaic H. sapiens," but adapted for cold	Modern skeleton; ? adapted for warmth
Brain size (ml)	500–650	600-800	750–1250	1100–1400	1200–1750	1200–1700
Skull form	Relatively small face; nose developed	Larger, flatter face	Flat, thick skull with large occipital and brow ridge	Higher skull; face less protruding	Reduced brow ridge; thinner skull; large nose; midface projection	Small or no brow ridge; shorter, high skull
Jaws/teeth	Thinner jaw; smaller, narrow molars	Robust jaw; large narrow molars	Robust jaw in larger individuals; smaller teeth than H. habilus	Similar to H. erectus but teeth may be smaller	Similar to "archaic H. sapiens"; teeth smaller except for incisors; chin development in some	Shorter jaws than Neanderthals; chin developed; teeth may be smaller
Distribution	Eastern (+ southern?) Africa	Eastern Africa	Africa, Asia and Indonesia (+ Europe?)	Africa, Asia and Europe	Europe and western Asia	Africa and western Asia
Known date (years ago)	2–1.6 million	2–1.6 million	1.8–0.3 million	400,000–100,000	150,000–30,000	130,000–60,000

Two hypotheses have been erected that are in need of further testing. William Howells called one the **candelabra model;** others called it the **multiregional hypothesis** (cf. Lewin 1987a, 1987b; Thomson 1992). According to this hypothesis, *Homo erectus* spread throughout the Old World and at least several populations in different locations evolved somewhat independently, but in parallel, to become archaic *Homo sapiens* and then modern humans (Thorne and Wolpoff 1992). No doubt gene flow among regions occurred, especially with selectively advantageous genes, but it was not sufficient to swamp regional characteristics and keep them from evolving through natural selection and drift (Frayer et al. 1993).

The hypothesis began as an explanation for the observations that some of the features distinguishing major human groups, such as Asians, Australian Aborigines, or Europeans, evolved over a long period in approximately the same geographic regions where these traits are found in their highest frequency. (Frayer et al. 1993, 17)

Mapped out, a candelabra-like set of populations would fan out over Africa, Europe, and Asia and evolve in parallel.

According to the alternative hypothesis, called the **Noah's Ark model** (by Howells) or the **replacement hypothesis,** modern humans evolved in one locality, spread to other parts of the Old World, and replaced all other extant populations of *Homo sapiens,* including Neanderthals. This replacement occurred with little mixing of gene pools, and so the *sapiens* subspecies remained more or less genetically distinct from archaic humans.

Both hypotheses raise questions about the origin of the migrating populations—Africa, Asia, Europe? And the Noah's Ark model conjures up scenarios of direct conflict and perhaps warfare between the spreading modern lineage and archaic populations. Fictional scenarios, understandably, have had their appeal (e.g., Auel 1980, 1982, 1985). And who was on Noah's Ark? Was it a small band or a male–female couple, the ancestors of us all? Can we trace our lineage back to an Adam and Eve in a Garden of Eden (Vigilant et al. 1991, Templeton 1993)? The dating of Eden seemed to come within our grasp

FIGURE 12-21

The major fossil forms of primates arranged in one possible phylogenetic tree, from the earliest primate, *Purgatorius*, to humans and their primate relatives. *(From Simmons 1992)*

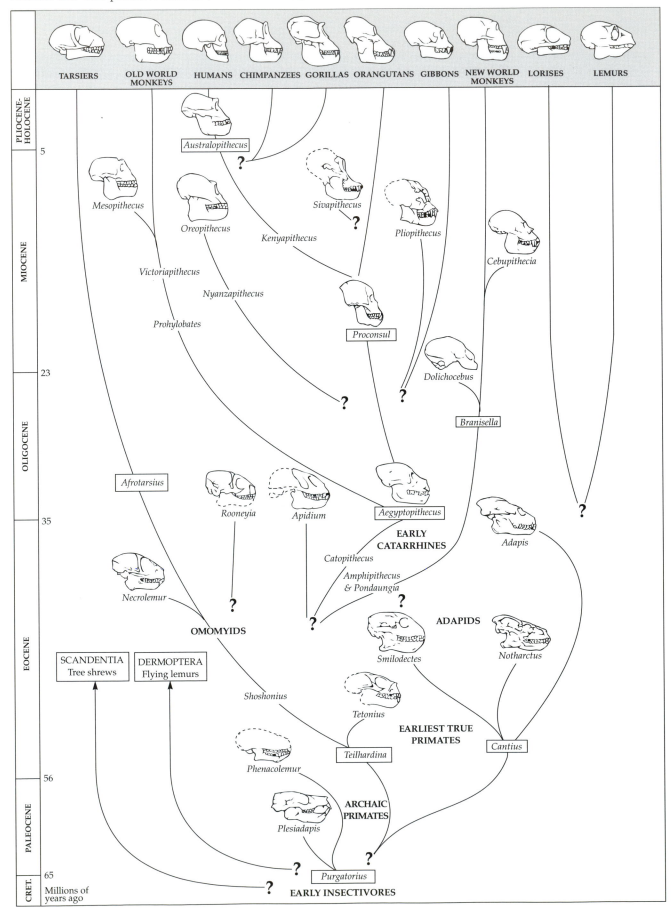

in 1987. Then again, perhaps it has slipped through our fingers.

Based on evidence from the fossils, the debate about hypotheses of human origins is doomed to linger on. Without clear definition of species and lineages, the discussion has no foundation. It revolves around population phenomena involving movement, potential genetic mixing, and large-scale geographic patterns, all demanding thousands of human fossils for adequate analysis. We can but hope that such a cornucopia of evidence will someday be accessible.

Testing Hypotheses Using Molecular Genetics

A refreshingly new tack came from molecular biologists who argued that the use of genetic analysis of living humans to investigate relationships could resolve the hypotheses. The DNA "hit the fan," so to speak, with the publication of a paper by Cann, Stoneking, and Wilson (1987) (see also Wilson and Cann 1992), which was widely interpreted as having identified and dated "Eve." The method of choice used analysis of mitochondrial DNA (mtDNA); hence the epithet "mitochondrial Eve" (e.g., Lewin 1987b).

Because mitochondria pass from generation to generation through the ova, and not the sperm, phylogenetic analyses of changes in mtDNA provide a matrilineal trace into the past. If we follow many traces back to a focal progenitor, and if the rate of change of mtDNA can be calibrated against real time, then we can distinguish the alternative hypotheses. The candelabra model predicts a date of divergence from an Eve that must have been *Homo erectus*, perhaps 1 million years ago, and relatively large accumulations of genetic differences among present human races. The Noah's Ark model predicts divergence within the time span of known fossils of *Homo sapiens*, about 100,000 years, and a relatively small amount of genetic variation among present human populations.

Researchers purified mitochondrial DNA obtained from the United States, Australia, and New Guinea, mostly from placentas. The DNA represented 147 individuals of five geographic regions and human races: Africans, Asians, Caucasians, aboriginal Australians, and aboriginal New Guineans (Cann et al. 1987). Human mtDNA contains 16,569 base pairs, and a sample of about 9 percent of this genome was used for analysis. This DNA evolves relatively rapidly compared to nuclear DNA, and so it is relatively sensitive over periods of a few tens of thousands of human generations.

The results showed remarkably little genetic divergence in mtDNA in the sample of Africans—less than in samples from other regions. This suggested a matriarch's origin in Africa with repeated colonization into Europe, Asia, Australia, and New Guinea. To calibrate the rate of change of mtDNA, Cann et al. assumed a constant rate of change in the molecule and used estimated colonization times of humans in discrete localities: 40,000 years ago in Australia, 30,000 years ago in New Guinea, and 12,000 years ago in the New World. On the basis of this calibration, Eve's existence in Eden was estimated very crudely at 200,000 years ago. The scenario revealed by this mtDNA analysis was much closer to the replacement hypothesis than to the multiregional hypothesis. Even in Asia the diversity of mtDNA types was relatively low, suggesting that populations from a relatively recent lineage had colonized Asia. This result countered the alternative prediction that a very long-term mixing of *Homo erectus* and *sapiens* mitochondrial genotypes had occurred in a population.

As we shall see in the next section and in Chapter 18, uncertainties arise when phylogenies are developed with molecular techniques and calibrated against real time. Do molecules evolve at a constant pace? Is the independent estimate of real time accurate? Is the touted phylogenetic lineage necessarily the most realistic? Is the test accurate enough to distinguish between a hypothesis of a 1-million-year-old Eve and one of a 100,000-year-old Eve? Does the estimated date relate to the origin of a lineage, a bottleneck in a much older lineage, or the single relic of an older lineage? The questions mount, and the debate heats up. Brown (1990) and others (e.g., Goldman and Barton 1992, Thomson 1992, Aiello 1993, Frayer et al. 1993, Templeton 1993) have encapsulated the developing discourse. Eve does not reveal her secrets easily—perhaps because modern techniques cannot detect her.

Templeton (1993) voiced distinct reservations about the Eve hypothesis, noting particularly major flaws in the mtDNA data. Templeton's analysis indicated that (1) the geographical location of an ancestor is ambiguous; (2) timing of the ancestor is ambiguous and likely to be much more than 200,000 years ago; (3) data indicate the presence of a small but continuous gene flow since the common mitochondrial ancestor; and (4) genetic data imply the lack of a single source population for the genetic variation present today.

12.8 RELATIONSHIPS WITH LIVING RELATIVES

Humans and Apes

A question remains: How closely related are we to our living relatives, the great apes? Despite Huxley's (1863) recognition of the great similarity between the embryologies of apes and humans, humans were shocked at the close connections.

A rush of activity to compare homologous proteins in many kinds of animals began in the early 1960s; Goodman (1961) is an early example. The proteins studied included hemoglobins, cytochromes, and serum albumins, with methods varying among amino acid sequencing, cross-reactivity of proteins, and immunological techniques. Sarich and Wilson (1966) used an immunological technique, quantitative microcomplement fixation, to study similarities of primate serum albumins. The technique provides an index of dissimilarity, or immunological distance (I.D.), between two homologous proteins. Sarich and Wilson found indexes that agreed very well with the known relationships in the primates (Table 12-3) and used them to estimate times of divergence between these groups (Sarich and Wilson 1967a, 1967b), having shown that the rate of change of serum albumin is steady and has not slowed down in the apes as suggested by Goodman (1961, 1962, 1963, 1967). Chapter 17 provides further discussion of molecular clocks and divergence times of lineages.

Sarich and Wilson (1967a) used data on fishes, amphibians, reptiles, and birds to develop the relationship, based on dehydrogenases, between immunological distance (I.D.) and time of divergence.

$$\log I.D. = kT$$

where k is a constant and T is the time of divergence in millions of years. They assumed that a similar relationship holds for serum albumins in primates and calibrated the relationship on the basis of the split between Old World monkeys and hominoids.

> Although the primate fossil record is fragmentary, it does, in combination with the available immunological evidence, provide sufficient evidence to suggest that the lineages leading to the living hominoids and Old World monkeys split about 30 million years ago. (Sarich and Wilson 1967a, 1202)

Sarich and Wilson used a mean I.D. between hominoid and Old World monkeys of 2.3 (cf. Table 12-3); thus, T became 30 million years and k could be calculated as 0.012 (log 2.3 = $k \times 30$). With a mean I.D. of 1.13 between humans and African apes and a k value of 0.012, the time of divergence is estimated as only 5 million years ago! Sarich and Wilson calculated the remainder of the phylogeny in the same way: the orangutan separated from the African apes about 8 million years ago, and the gibbon and the siamang separated about 10 million years ago (Figure 12-22).

These estimates appeared to make our species much too young for many scientists to accept them, although almost 10 years later Sarich and Cronin

TABLE 12-3 Indexes of dissimilarity (I.D.)[1] between humans and other species

Group and Species	I.D.
Hominoidea—Humans and Apes	
Human with:	
Human	1.0
Pan troglodytes, chimpanzee	1.12
Gorilla gorilla, gorilla	1.12
Pongo pygmaeus, orangutan	1.15
Hylobates lar, gibbon	1.29
Symphalangus syndactylus, Siamang	1.25
Cercopithecoidea—Old World Monkeys	
Human with:	
Macaca mulatta, macaque	2.05
Cercopithecus aethiops, guenon	2.46
Ceboidea—New World Monkeys	
Human with:	
Aotes trivirgatus, night monkey	2.5
Cebus capucinus, capuchin	6.3
Prosimii—Prosimians Such as Galagoes and Lemurs	
Human with:	
Galago crassicaudatus, galago	10.0
Lemur fulvus, lemur	14.0
Nonprimates	
Human with:	
Bos taurus, cattle	23
Sus scrofa, swine	29

[1] Developed by Sarich and Wilson (1966), using albumins. Maximum similarity is I.D. = 1.0.

(1976) maintained their validity. On the basis of the fossil record, Simons (1976) estimated that the hominid–cercopithecoid split occurred 32 to 49 million years ago, and his other estimates of splitting times are all less recent than those of Sarich and Wilson. Simons made another important estimate for *Ramapithecus*, the then-supposed link between the ancestral great apes and modern humans (Figures 12-13, 12-15). The seven main sites for *Ramapithecus* have been dated from 9 million years ago to 14 to 16 million years ago. This suggests that the split between African apes and humans must have occurred more than 9 million years ago. Simons (1976, 59) summarized,

> The fossil evidence, when considered together with the paleogeographic, summates to suggest very strongly that the branch points suggested by Sarich

FIGURE 12-22

The phylogenetic tree leading to the genera *Homo* (humans), *Pan* (chimpanzee), and *Gorilla* (gorilla), identified by the methods of Sarich and Wilson (1967a). *Pongo* is the orangutan, *Symphalangus* is the siamang, and *Hylobates* is a gibbon.

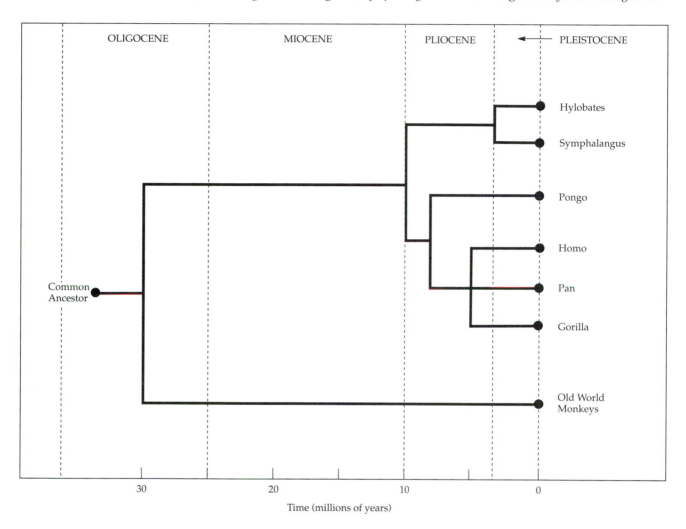

and Cronin in their contribution to this volume are too young. Either the "clock" is incorrectly calibrated or it doesn't keep proper time.

Walker (1976) agreed with Simons. He estimated the split of the orangutan from the African apes as being no more recent than 15 million years ago, compared to Sarich and Wilson's estimate of 8 million years.

Using similarity between repeated DNA sequences in primates, Gillespie (1977) estimated a rate of nucleotide change of 0.2 percent per million years. He claimed that the appearance of the baboon–macaque–mangabey group could be timed rather accurately at 10 to 15 million years ago and used this figure to estimate the rate of change of nucleotides in repeated DNA sequences. According to Gillespie's calculation, the divergence between man and chimpanzee occurred between 5 and 10 million years ago

(Figure 12-23), a less contested span of time than the estimate by Sarich and Wilson but still on the late side for those arguing more from the fossil record. Sibley and Ahlquist (1984), however, used DNA–DNA hybridization to develop a phylogeny for the hominoid primates and found a divergence time for chimp and man that was still remarkably recent, 6.3 to 7.7 million years ago. Estimates of the other branching sequences of the lineages were Cercopithecoidea (Old World monkeys), 27 to 33 million years ago; gibbons, 18 to 22 million years ago; orangutan, 13 to 16 million years ago; and gorilla, 8 to 10 million years ago.

The researchers developed the common ground of the fossil and molecular evidence for the divergence times of the apes when they united *Ramapithecus* with *Sivapithecus* after more fossils of the latter were found. The oldest date for *Sivapithecus* fossils,

FIGURE 12-23

Phylogenetic relationships of humans **(top)**, chimpanzees, and other primates, according to Gillespie (1977).

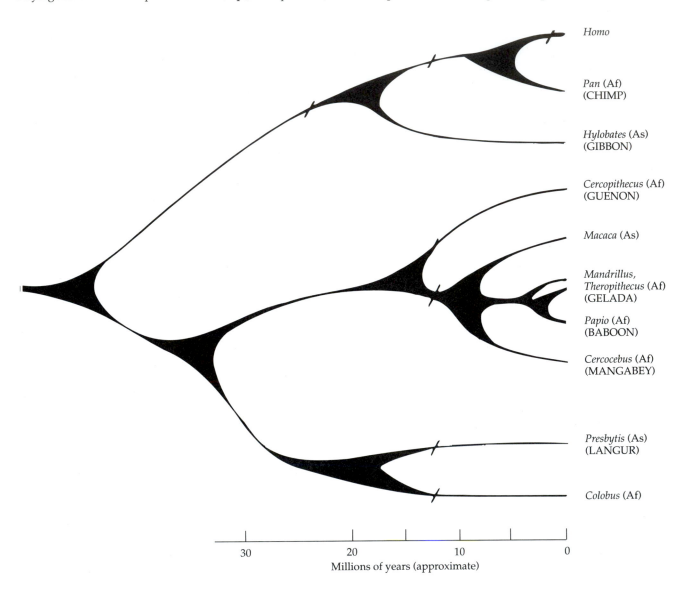

12.5 million years ago, comes from the Siwalik sequence in Pakistan (Kelley 1992). As a potential ancestor of the orangutan (*Pongo*) (cf. Figure 12-21), *Sivapithecus* permits estimated divergence times for living apes that are consistent with the molecular data.

Similarity of Humans and Chimpanzees

Whatever the actual timing of divergence of chimpanzee and human, they are surprisingly similar genetically. King and Wilson (1975) summarized the available data. In comparisons of protein sequences and immunological distances, some proteins are identical: fibrinopeptides, cytochromes *c*, and hemoglobin chains α, β, and γ. Myoglobin and the hemoglobin δ chain each differ by a single amino acid substitution. Sequences of human and chimpanzee polypeptide chains are, on average, 99 percent similar. Darwin and Huxley would have been very surprised to learn of this close linkage.

On the basis of electrophoretic comparisons of proteins, humans and chimpanzees are about 50 percent similar. In tests of polypeptide products of 44 different structural genes in human and chimpanzee, 21 were identical, 19 were totally different, and in the remaining 4 cases different frequencies of the alleles occurred. Using an index of s_i for the probability that human and chimpanzee would be identical at a particular locus i,

$$s_i = S x_{ij} y_{ij}$$

where x_{ij} is the frequency of the jth allele at the ith locus in the human population, and y_{ij} is the same for the chimpanzee population. Then the mean of these s_i values gives the estimate of the proportion of alleles at an average locus that are electrophoretically identical: $S = 0.52$. Thus, on average, 52 percent of our alleles are identical with those of the chimpanzee.

Studies on nucleic acid hybridization between chimpanzee and human compare the thermostability of human–chimp hybrid DNA *in vitro* with the stability of DNA from each species separately. Mitochondrial DNAs appear to be identical in these tests. With the use of nonrepetitive DNA sequences, nucleic acid sequence differences are about 1.1 percent. Thus, in a length of DNA 3,000 bases or 1,000 amino acids long, there would be 33 nucleotide sequence differences between the species.

King and Wilson (1975) noted a pattern in which the DNA is more different than at the protein level. For every difference in amino acid sequence, about four base differences occur in the DNA. This disproportion may well be due to the redundancy in the genetic code. Differences estimated using hybrid DNA may thus overestimate the differences in functional proteins between the two primates. This problem is not apparent in electrophoresis, sequencing of proteins, or microcomplement fixation, because those techniques study the functional proteins themselves.

Bruce and Ayala (1978) performed another study, using electrophoresis to evaluate similarities between humans and the apes. With 23-gene loci coding for blood proteins, they found the relationships given in Table 12-4. Remarkably, these estimates placed humans and the great apes closer in relationship than the sibling species of *Drosophila* studied by Ayala et al. (1974); for *Drosophila*, genetic similarity was 0.517 when 36-gene loci were evaluated.

The conclusions reached in all kinds of comparisons between chimpanzees and humans (and other apes) is that they are remarkably similar at the molecular level—so similar, in fact, as to justify, on the basis of molecular structure alone, status as sibling species. On the basis of genetic identity in *Drosophila* species, all apes would be expected to be congeneric. For example, genetic similarity in sibling species of *Drosophila* was 0.35 (Ayala et al. 1974), whereas similarity between the least-related ape, the siamang, and humans was 0.33 (Table 12-4).

TABLE 12-4   Indexes of Nei's genetic identity (I) and genetic distance (D) between humans and apes

Species	I	D
Pan troglodytes, chimpanzee	0.680	0.386
Pan paniscus, pygmy chimpanzee	0.732	0.312
Gorilla gorilla, gorilla	0.689	0.373
Pongo pygmaeus abelii, Sumatran orangutan	0.710	0.347
Pongo pygmaeus pygmaeus, Bornean orangutan	0.705	0.350
Hylobates lar, gibbon	0.489	0.716
Hylobates concolor, gibbon	0.429	0.847
Symphalangus syndactylus, siamang	0.333	1.099

Estimated by Bruce and Ayala (1978).

Genetic and Morphological Similarities

This genetic similarity is in marked contrast with the large anatomical differences between the apes and man, as noted by King and Wilson (1975). No one would call a chimpanzee a sibling species with man; no one has ever confused the two taxonomically. Nearly every bone in the body differs in a chimp–human comparison, and differences in posture, locomotion, feeding, communication, and general ecology are easily observed. Chimpanzees and humans are so different phenotypically that they are placed in different genera, *Pan* and *Homo*, and even different families, Pongidae and Hominidae.

King and Wilson (1975) regarded *Pan* as illustrating a very conservative morphological change from the common ancestor with man, whereas the human has changed dramatically in morphology from the common ancestor. In contrast, molecular divergence has occurred at about the same rate in the two species (Figure 12-24). King and Wilson thus argued that organismal evolution and molecular evolution operate independently of each other to a considerable extent. Regulatory mutations can have large effects at the organismal level with only small effects at the genetic level. King and Wilson recognized two kinds of regulatory mutations: (1) point mutations in a promoter, or operator gene, which would affect production of protein but not the amino acid sequence, and (2) chromosomal inversions, translocations, additions and deletions, and fission and fusion, which would alter the order of genes on a chromosome. Such gene rearrangements may have significant effects on gene expression, although the mechanisms are not

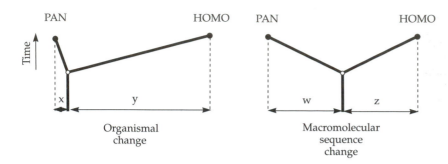

FIGURE 12-24
According to evidence developed by King and Wilson (1975), chimpanzees (*Pan*) and humans (*Homo*) illustrate similar rates of molecular evolution but very different rates of organismal evolution, based on morphological characteristics.

understood. We do know that chromosomal change correlates with evolutionary rates of organisms (Wilson, Maxson, and Sarich 1974; Wilson, Sarich, and Maxson 1974). Subsequent studies have corroborated these relationships (Wilson et al. 1975, Bush et al. 1977, Prager et al. 1976, Maxson and Wilson 1979) and are discussed in Chapter 14, "Genetic Systems."

King and Wilson (1975) also noted that the number of chromosomes in humans (46) and chimpanzees (48) are similar, but banding patterns differ on a proportion of chromosomes. At least ten large inversions and translocations and one chromosomal fusion have occurred since the two lineages diverged (see also Yunis and Prakash 1982). Certainly, we need more understanding of gene regulation in mammals. The timing of gene expression may differ greatly between humans and apes, which would alter developmental processes, especially such critical systems as the brain and vertebral column (King and Wilson 1975).

A Frog's Perspective

For an objective comparison of genetic change and morphological change, some method of quantifying morphological differences was necessary. Cherry et al. (1978) developed such a method. To avoid any bias, they employed linear traits typically used in the taxonomy of frogs to obtain a "frog's perspective" on morphological change in human and chimpanzee divergence compared with radiation in the frogs. They used nine traits in all: shank length, head length, nostril-to-lip distance, forearm length, vertebral length, eye-to-nostril distance, head width, eye-to-tympanum distance, and toe length.

Cherry et al. calculated an overall estimate of morphological similarity by using the formula

$$M = \frac{1}{n} \sum_{i=1}^{n} \frac{(|\bar{x}_i - \bar{y}_i|)}{\bar{\sigma}_i}$$

where M = morphological distance, n = number of traits (= 9), $\bar{x}_i$ = mean value of ith trait in species X, $\bar{y}_i$ = mean value of ith trait in species Y, and σ_i =

mean standard deviation of the ith trait. Therefore, M provides the mean number of standard deviations by which two species differ for each trait measured.

When frog comparisons are juxtaposed with human and chimpanzee comparisons, the results are striking (Figure 12-25). In the frogs, M values correlate well with taxonomic distance. The amazing result is that comparison of frogs in different suborders gives values of M less than the value given by the comparison of human and chimpanzee. The differences in

FIGURE 12-25
The estimated morphological distance, M between humans (*Homo*) and chimpanzees (*Pan*), in comparison with morphological distances in frogs at the taxonomic levels of subspecies (sS), species (S), genera (G), subfamilies (sF), families (F), superfamilies (SF), and suborders (sO). According to these estimates, the morphological distance of humans and chimpanzees (different genera) is about 4.5, much greater than the differences of frog genera, estimated at about 1.75. (*From Cherry et al. 1978*)

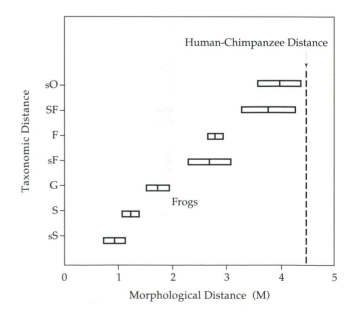

morphology between *Pan* and *Homo* (different families) are two standard deviations larger than family-level differences in the frogs. Contrasting with this pattern was the genetic pattern, in which frogs in different suborders were at least 30 times farther apart biochemically than were humans and apes. This suggests that regulatory changes in development may arise from processes involving not measurable genetic change but chromosomal change. Thus, the study of humans and their living relatives has led to the perception of major differences in evolution at the molecular and organismal levels.

Of course, all is not resolved in the evolution of the human species and its relatives. Templeton (1983) argued that human design is more conservative morphologically than the design of chimpanzees and gorillas. A debate about how to evaluate morphological distance has developed and will continue (Hafner et al. 1984, Wilson et al. 1984). Questions about the uneven rates of molecular clocks persist (e.g., Miyamoto et al. 1987) (cf. Chapter 17, Section 17.7). Tracing mitochondrial DNA back to the beginning of the human family tree has opened up new questions and opportunities (Lewin 1987a, 1987b, Frayer et al. 1993). Needless to say, the fascination with human evolution will never wane, and perhaps we will never answer all the tantalizing questions. Nevertheless, the scientific method continues to help us develop an evolutionary understanding of ourselves and our place in nature.

SUMMARY

In *The Origin of Species,* Darwin only hinted at the application of his theory to human evolution, but 12 years later in *The Descent of Man* he enunciated the extension of his theory to our own species. Using embryology, vestigial characteristics, and homology, he argued the case for close relationships among humans, apes, and monkeys, even presenting their mental powers, instincts, and emotions as being—although quantitatively different—qualitatively similar. Darwin built a sound foundation for the development of evolutionary science of the human species.

Scientists have erected two hypotheses to account for primate evolution and special traits such as binocular vision and prehensile hands with nails rather than claws. The **arboreal hypothesis** of primate evolution was the first to argue that adaptations to life in the trees preadapted a lineage for bipedal locomotion. However, many arboreal mammals, such as tree squirrels, have retained claws and almost nonoverlapping visual fields, and so the hypothesis failed to provide a clear mechanistic explanation of why primates are so different from other arboreal mammal groups. The **visual predation hypothesis** argues that the unique features of primate ecology and morphology derive from primates' habits as insectivorous predators in shrubs and low-canopy vegetation. Binocular vision enabled primates to accurately attack and capture highly mobile prey, and prehensile appendages permitted them to forage among small stems at the canopy margins.

Some remarkable features of the human body include binocular vision, manual dexterity and sensitivity, brain size, upright posture, and a skeleton and vascular system exhibiting the "scars" of human evolution. Binocular vision and manual dexterity no doubt evolved in tree-dwelling organisms and preadapted members of the human lineage for tool use and tool construction. Bipedal locomotion and upright posture freed the hands for tool manufacture, fighting, hunting and gathering with implements, and care of helpless infants. Brain size increased radically, with expanded sensory centers and motor centers integrating hand movement, facial expressions, and ultimately speech. Bipedal locomotion and upright posture in an organism designed by evolution to be quadrupedal gave humans a legacy of common body ailments that make our evolutionary history palpable to most of us.

The fossil history of the human lineage indicated a possible relationship between a *Ramapithecus* ancestor and the lineage to humans, although paleontologists currently favor fossils from Africa, such as *Aegyptopithecus* and *Proconsul*, as precursors of *Australopithecus*. The relationships among the species of *Australopithecus* and *Homo* can be grouped into three alternative hypotheses in need of further testing. But the emergence of the genus *Homo* seems to have been followed by a linear sequence from *H. habilis* to *H. erectus* and *H. sapiens*. Robust forms of *Australopithecus* and Neanderthal humans are regarded as branches off the main lineage. The robust australopithecines could be a separate genus, *Paranthropus*, and neanderthals are sometimes regarded as a full species, *H. neanderthalensis,* rather than a subspecies of *H. sapiens*.

An old debate about the origins of modern humans involves the **multiregional hypothesis** and the **replacement hypothesis.** Did *Homo erectus* spread over the Old World and evolve in different regions in parallel, with small amounts of gene flow, producing multiple origins of modern humans, or was there a single origin of *Homo sapiens sapiens,* followed by population spread and replacement of archaic *sapiens,* with little mixing of genotypes among resident and colonizing populations? The answer obtained by using mitochondrial DNA to trace phylogenetic relationships among modern humans supported the replacement hypothesis, which points to an origin in Africa about 200,000 years ago. The debate surrounding this conclusion is unresolved.

Humans and apes are remarkably similar genetically, suggesting the possibility of more recent divergence from a common ancestor than the fossil record indicates. Human and chimpanzee polypeptide chains are on average 99 percent similar, making chimpanzees our closest living relatives. When whole proteins are compared with electrophoretic techniques, we are about 50 percent similar.

Although chimpanzees and humans are closely related genetically, they look very different, suggesting that genetic change and morphological change can proceed at different rates. Chromosomal evolution may have little effect on structural proteins but have dramatic effects on development rates and, therefore, ultimately body design. Comparisons of the morphology of frogs with that of humans and chimpanzees indicate very conservative morphological change in frogs, even frogs of different suborders, but radical divergence between humans and chimpanzees, which are merely of different families.

The debate on human evolution continues on many fronts. Developing technologies and approaches promise to unravel many of the secrets of our past.

QUESTIONS FOR DISCUSSION

1. With the human body we can experience first-hand the result of phylogenetic constraints on design, and the inadequacies of adaptation in compensating for these constraints. Do you think that this empirical vantage point provides us with a unique case for understanding the relative strengths of phylogenetic constraints and adaptation?

2. In the modern human, at what age, or stage in the life history, is natural selection likely to become a weak force in human evolution?

3. If the human species survives another million years, what kinds of changes would you expect to see in design? Is it possible they would justify recognition of a new species?

4. Can you imagine alternative evolutionary routes to those taken in the human lineage that could have circumvented the obstetrical dilemma?

5. Does the evolution of the human lineage provide any clues to the presence or absence of a biological basis for long-term pair bonding between mates?

6. Given that the arboreal theory of primate evolution failed to account for radical differences in design between the primates and other arboreal mammals, why do you think the theory was largely accepted for 60 years, until the visual predation hypothesis was developed as an alternative hypothesis?

7. How would you plan a research program to improve the objectivity in defining species in the genus *Homo*? How would you erect criteria for distinguishing among species and subspecies?

8. In the multiregional hypothesis on the origin of archaic *Homo sapiens,* which additional hypotheses would you develop and test that could distinguish between it and the alternative replacement hypothesis?

9. What does the high genetic similarity between humans and chimpanzees reveal about the evolutionary process and the understanding of rates of genetic and phenotypic change?

10. If members of different species in the genus *Homo* coexisted in a region, can you imagine the kinds of scenarios in which they might have interacted? Consider conditions relating to exploitation of resources such as food and shelter, social structure, foraging patterns, and the nature of aggression within and between species.

REFERENCES

Aiello, L. C. 1993. The fossil evidence for modern human origins in Africa: A revised view. Amer. Anthropol. 95:93–96.

Auel, J. M. 1980. The Clan of the Cave Bear. Hodder and Stoughton, London.

———. 1982. The Valley of the Horses. Crown, New York.

———. 1985. The Mammoth Hunters. Crown, New York.

Ayala, F. J., M. L. Tracey, L. G. Barr, J. F. McDonald, and S. Perez-Salas. 1974. Genetic variation in natural populations of five *Drosophila* species and the hypothesis of the selective neutrality of protein polymorphisms. Genetics 77:343–384.

Bonner, J. T., and R. M. May. 1981. Introduction, pp. vii–xli. In C. Darwin. The Descent of Man and Selection in Relation to Sex. Princeton Univ. Press, Princeton, N.J.

Brace, C. L., H. Nelson, N. Korn, and M. L. Brace. 1979. Atlas of Human Evolution. 2d ed. Holt, Rinehart and Winston, New York.

Brown, M. H. 1990. The Search for Eve. Harper and Row, New York.

Bruce, E. J., and F. J. Ayala. 1978. Humans and apes are genetically very similar. Nature 276:264–265.

Bush, G. L., S. M. Case, A. C. Wilson, and J. L. Patton. 1977. Rapid speciation and chromosomal evolution in mammals. Proc. Nat. Acad. Sci. U.S.A. 74:3942–3946.

Campbell, B. 1985. Human Evolution: An Introduction to Man's Adaptations. 3rd ed. Aldine, New York.

Cann, R. L., M. Stoneking, and A. C. Wilson. 1987. Mitochondrial DNA and human evolution. Nature 325:31–36.

Cartmill, M. 1972. Arboreal adaptations and the origin of the Order Primates, pp. 97–122. In R. Tuttle (ed.). The Functional and Evolutionary Biology of Primates. Aldine–Atherton, Chicago.

———. 1974a. Pads and claws in arboreal locomotion, pp. 45–83. In F. A. Jenkins (ed.). Primate Locomotion. Academic Press, New York.

———. 1974b. Rethinking primate origins. Science 184:436–443.

Cherry, L. M., S. M. Case, and A. C. Wilson. 1978. Frog perspective on the morphological difference between humans and chimpanzees. Science 200:209–21.

Colbert, E. H., and M. Morales. 1991. Evolution of the Vertebrates. 4th ed. Wiley–Liss, New York.

Collins, E. T. 1921. Changes in the visual organs correlated with the adoption of arboreal life and with the assumption of the erect posture. Trans. Ophthalm. Soc. U.K. 41:10–90.

Deacon, T. W. 1992. The neural circuitry underlying primate calls and human language, pp. 121–162. In J. Wind, B. Chiarelli, B. Bichakjian, and A. Nocentini (eds.). Language Origin: A Multidisciplinary Approach. Kluwer Academic Publishers, Dordrecht.

Darwin, C. 1871. The Descent of Man, and Selection in Relation to Sex. Murray, London.

Day, M. H. 1984. The fossil history of man. 3d ed. Carolina Biological Supply, Burlington, N.C.

Dobzhansky, T. 1960. The present evolution of man. Sci. Amer. 203(3):206–217.

———. 1962. Mankind Evolving. Bantam, New York.

Eckhardt, R. B. 1972. Population genetics and human origins. Sci. Amer. 226(1):94–103.

Frayer, D. W., M. H. Wolpoff, A. G. Thorne, F. H. Smith, and G. G. Pope. 1993. Theories of modern human origins: The paleontological test. Amer. Anthropol. 95:14–50.

Gillespie, D. 1977. Newly evolved repeated DNA sequences in primates. Science 196:889–891.

Goldman, N., and N. H. Barton. 1992. Genetics and geography. Nature 357:440–441.

Goodman, M. 1961. The role of immunochemical differences in the phyletic development of human behavior. Human Biol. 33:131–162.

———. 1962. Evolution of the immunological species specificity of human serum proteins. Human Biol. 34:104–150.

———. 1963. Serological analysis of the systematics of recent hominoids. Human Biol. 35:377–436.

———. 1967. Deciphering primate phylogeny from macromolecular specificities. Am. J. Phys. Anthropol. 26:255–276.

Hafner, M. S., J. V. Ramsen, and S. M. Lanyon. 1984. Bird versus mammal morphological diversity. Evolution 38:1154–1156.

Holloway, R. L. 1974. The casts of fossil hominid brains, pp. 74–83. In G. Isaac and R. E. F. Leakey (eds.). Human Ancestors (readings from Scientific American). W. H. Freeman, San Francisco.

Hooton, E. A. 1930. Doubts and suspicions concerning certain functional theories of primate evolution. Human Biol. 2:223–249.

Howell, F. C. 1965. Early Man. Time, New York.

Howells, W. W. 1947. Mankind So Far. Doubleday, Garden City, N.Y.

Huxley, T. H. 1863. Evidence as to Man's Place in Nature. Williams and Norgate, London.

Johanson, D. C., and M. A. Edey. 1981. Lucy: The Beginnings of Humankind. Simon and Schuster, New York.

Jones, F. W. 1917. Arboreal Man. [Reprinted 1964.] Hafner, New York.

Kelley, J. 1992. Evolution of apes, pp. 223–229. In S. Jones, R. Martin, and D. Pilbeam (eds.). The Cambridge encyclopedia of human evolution. Cambridge University Press, Cambridge.

Kimbel, W. H., and Y. Rak. 1993. The importance of species taxa in paleoanthropology and an argument for the phylogenetic concept of the species category, pp. 461–484. In W. H. Kimbel and L. B. Martin (eds.). Species, Species Concepts, and Primate Evolution. Plenum, New York.

King, M.-C., and A. C. Wilson. 1975. Evolution at two levels in humans and chimpanzees. Science 188:107–116.

Krogman, W. M. 1951. The scars of human evolution. Sci. Amer. 185(6)54–57.

Leakey, R. E. 1981. The dawn of man. Family Weekly, Sept. 13, 6–8.

Lewis, R. 1987a. Africa: Cradle of modern humans. Science 237:1292–1295.

———. 1987b. The unmasking of mitochondrial Eve. Science 238:24–26.

Martin, R. D. 1981. Relative brain size and basal metabolic rate. Nature 293:57–60.

Maxson, L. E. R., and A. C. Wilson. 1979. Rates of molecular and chromosomal evolution in salamanders. Evolution 33:734–740.

Miyamoto, M. M., J. L. Slightom, and M. Goodman. 1987. Phytogenetic relations of humans and African apes from DNA sequences in the ch-globin region. Science 238:369–373.

Napier, J. 1962. The evolution of the hand. Sci. Amer. 207(6):56–62.

———. 1967. The antiquity of human walking. Sci. Amer. 216(4):56–66.

Oakley, K. 1958. Tools Makyth Man. Smithsonian Report. Washington, D.C.

Prager, E. M., D. P. Fowler, and A. C. Wilson. 1976. Rates of evolution in conifers (Pinaceae). Evolution 30:637–649.

Rak, Y. 1993. Morphological variation in *Homo neanderthalensis* and *Homo sapiens* in the Levant: A biogeographic model, pp. 523–536. *In* W. H. Kimbel and L. B. Martin (eds.). Species, Species Concepts, and Primate Evolution. Plenum, New York.

Restak, R. 1988. The Mind. Bantam Books, New York.

Sacher, G. A., and E. F. Staffeldt. 1974. Relation of gestation time to brain weight for placental mammals: Implications for the theory of vertebrate growth. Amer. Natur. 108:593–615.

Sarich, V. M., and J. E. Cronin. 1976. Molecular systematics of the primates, pp. 141–170. *In* M. Goodman and R. E. Tashian (eds.). Molecular Anthropology: Genes and Proteins in the Evolutionary Ascent of the Primates. Plenum, New York.

Sarich, V. M., and A. C. Wilson. 1966. Quantitative immunochemistry and the evolution of primate albumins: Micro-complement fixation. Science 154:1563–1566.

———. 1967a. Immunological time scale for hominid evolution. Science 158:1200–1203.

———. 1967b. Rates of albumin evolution in primates. Proc. Nat. Acad. Sci. 58:142–148.

Schepartz, L. A. 1993. Language and modern human origins. Yearbook of Phys. Anthropol. 36:91–126.

Sibley, C. G., and J. E. Ahlquist. 1984. The phylogeny of the hominoid primates, as indicated by DNA–DNA hybridization. J. Mol. Evol. 20:2–15.

Simons, E. L. 1964. The early relatives of man. Sci. Amer. 211(1):50–62.

———. 1976. The fossil record of primate phylogeny, pp. 35–62. *In* M. Goodman and R. E. Tashian (eds.). Molecular Anthropology: Genes and Proteins in the Evolutionary Ascent of the Primates. Plenum, New York.

———. 1992. The fossil history of primates, pp. 199–208. *In* S. Jones, R. Martin, and D. Pilbeam (eds.). The Cambridge Encyclopedia of Human Evolution. Cambridge Univ. Press, Cambridge, England.

Smith, G. E. 1924. The Evolution of Man. Oxford Univ. Press, London.

Stringer, C. B. 1992. Evolution of early humans, pp. 241–251. *In* S. Jones, R. Martin, and D. Pilbeam (eds.). The Cambridge Encyclopedia of Human Evolution. Cambridge Univ. Press, Cambridge, England.

Tanner, N. M. 1981. On Becoming Human. Cambridge Univ. Press, Cambridge.

Templeton, A. R. 1983. Phylogenetic inference from restriction endonuclease cleavage site maps with particular reference to the evolution of humans and the apes. Evolution 37:221–244.

———. 1993. The "Eve" Hypothesis: A genetic critique and reanalysis. Amer. Anthropol. 95:51–72.

Thomson, K. S. 1992. The challenge of human origins. Amer. Sci. 80:519–522.

Thorne, A. G., and M. H. Wolpoff. 1992. The multiregional evolution of humans. Sci. Amer. 266(4):76–83.

Vigilant, L., M. Stoneking, H. Harpending, K. Hawkes, and A. C. Wilson. 1991. African populations and the evolution of human mitochondrial DNA. Science 233:1303–1307.

Walker, A. 1976. Splitting times among hominoids deduced from the fossil record, pp. 63–77. *In* M. Goodman and R. E. Tashian (eds.). Molecular Anthropology: Genes and Proteins in the Evolutionary Ascent of the Primates. Plenum, New York.

Washburn, S. L. 1960. Tools and human evolution. Sci. Amer. 203(3):63–75.

Washburn, S. L., and R. Moore. 1980. Ape into Man: A Study of Human Evolution. 2d ed. Little, Brown, Boston.

Wilson, A. C., and R. L. Cann. 1992. The recent African genesis of humans. Sci. Amer. 266(4):68–73.

Wilson, A. C., L. R. Maxson, and V. M. Sarich. 1974. Two types of molecular evolution: Evidence from studies of interspecific hybridization. Proc. Nat. Acad. Sci. 71:2843–2847.

Wilson, A. C., V. M. Sarich, and L. R. Maxson. 1974. The importance of gene rearrangement in evolution: Evidence from studies on rates of chromosomal, protein, and anatomical evolution. Proc. Nat. Acad. Sci. 71:3028–3030.

Wilson, A. C., G. L. Bush, S. M. Case, and M.-C. King. 1975. Social structuring of mammalian populations and rate of chromosomal evolution. Proc. Nat. Acad. Sci. U.S.A. 72:5061–5065.

Wilson, A. C., J. G. Kunkel, and J. S. Wyles. 1984. Morphological distance: An encounter between two perspectives in evolutionary biology. Evolution 1156–1159.

Wood, B. A. 1992. Evolution of the australopithecines, pp. 231–240. *In* S. Jones, R. Martin, and D. Pilbeam (eds.). The Cambridge Encyclopedia of Human Evolution. Cambridge Univ. Press, Cambridge, England.

———. 1993. Early *Homo:* How many species? pp. 485–522. *In* W. H. Kimbel and L. B. Martin (eds.). Species, Species Concepts, and Primate Evolution. Plenum, New York.

Yunis, J. J., and O. Prakash. 1982. The origin of man: A chromosomal pictorial legacy. Science 215:1525–1530.

13

BIOLOGICAL CLASSIFICATION

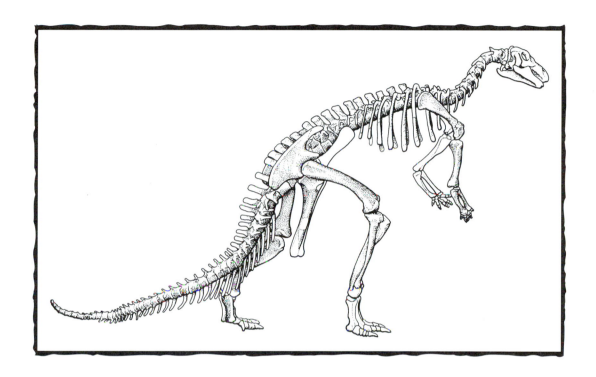

Reconstruction of the skeleton of a *Camptosaurus,* a late Jurassic dinosaur. Does it belong in the class Reptilia or in the subdivision Archosauria with crocodiles and birds?

Classification concerns the grouping of species into sets of similar organisms at any level in the taxonomic hierarchy: similar species into a family, similar families into an order, similar orders into a class, and so on. Systematics is the branch of biology concerned with classifying organisms into logical evolutionary groupings. What seems to be a very logical process is plagued with arguments on methodology. Since such arguments relate to our understanding of evolutionary relationships, classification is relevant to evolutionary theory.

One could argue that the subject of biological classification should appear early in a book on evolution because classification is so central to discussions on comparative evolutionary biology. However, the major debates on classification occurred much later than many of the debates we have addressed so far, and motivation to surmount the difficulties of classification has developed gradually.

Take, for example, the treatment of human evolution in Chapter 12. How can we recognize fossil species and genera objectively? How many species of *Homo* have actually existed? Is the phylogenetic species concept an operational, workable concept for fossil assemblages and living species? If Niles Eldredge (1993, 3) can ask the question "What, if anything, is a species?" we almost certainly need to ask, "What, if anything, is a genus, family, order, class or phylum?" And if much concern exists about species status in primates, of which we are so fond (as documented in Kimbel and Martin [1993]), then there is certainly room for equal concern about the remainder of living organisms.

The theme we develop in this chapter is a historical perspective on the classification debate over the past four decades or so. The debate is instructive because its focus is on capturing the true natures of species and higher taxa and the relationships among them. Can we reconstruct the evolutionary pathways that led to the organisms we study today as fossils or extant species?

The details of methodology in classification are crucial to an understanding of the debate. We must know the processes before we can examine their validity. Somewhat paradoxically, methodology seems peripheral to an understanding of biological evolution, and yet its results are central.

Let us start with some definitions:

Systematics: the scientific study of the kinds and diversity of organisms and of any and all relationships among them (Simpson 1961, 7)

Classification: the ordering of organisms into groups or sets on the basis of their rela-

tionships—that is, their association by contiguity, similarity, or both (Simpson 1961, 9)

Taxonomy: the theoretical study of classification, including its bases, principles, procedures, and rules (Simpson 1961, 11)

13.1 THE EVOLUTIONARY SCHOOL OF CLASSIFICATION

Mayr (1981) pointed out that until the 1950s and 1960s all taxonomists, including Darwin, belonged to what can now be called the **traditional** or **evolutionary school of classification.** Mayr summarized evolutionary classification as based "on observed similarities and differences among groups of organisms, evaluated in the light of the inferred evolutionary history" (510). The traditional approach thus considers all available characteristics, relationships, ecological roles, and patterns of distribution of the organisms under study. It "attempts to reflect both of the major evolutionary processes, branching and the subsequent diverging of the branches." This school follows Darwin in recognizing that classification must be based on genealogy (the major theme adopted by the cladistics school of classification—see later) but also agrees with Darwin that genealogy alone is not enough for a classification (a point not accepted by the cladistics school).

The evolutionary classification of a group of organisms can be summarized in a **phylogram,** which records both the branching points in a lineage and the degrees of subsequent divergence. Wiley (1981) has called the use of phylograms the "Mayr–Simpson school," for two of the most eminent systematists of their time.

Other schools of classification emerged in the 1950s and 1960s and have generated considerable debate and even acrimony. These are the phenetics school and the cladistics school.

13.2 THE PHENETICS SCHOOL OF CLASSIFICATION

The **phenetics school** is based on the appearance of organisms; the term "phenetics" comes from a Greek word meaning the outward manifestation of an object. All types of classification are based on the appearances of phenotypes; however, the phenetics school argued that classifications should be based *exclusively* on phenetic characters, and that similarity between organisms should be measured objectively and explicitly by the use of a set of important criteria.

1. Numerous characters are evaluated.
2. All characters have equal weight.
3. Each character is measured numerically.
4. Measures of similarity and clustering of species and taxonomic distances are calculated via algorithms or standard procedures, usually involving computer programs.

From this analysis emerges a diagram of relationships called a **phenogram.**

The argument of the pheneticists is that calculated phenetic distances are directly applicable to the classification of the taxon under study. Sokal and Sneath (1963) wrote the first book on phenetic classification, calling it **numerical taxonomy.** It was Mayr (1981) who pointed out that, since all taxonomic methods employ some form of numerical analysis, a much better term for this approach is **numerical phenetics.** The school of numerical phenetics emphasizes a "theory-free" classification based on the use of unweighted characters (Mayr 1981).

13.3 THE CLADISTICS SCHOOL OF CLASSIFICATION

Hennig (1950, 1966) founded the **cladistics school** (Mayr 1981), or **phylogenetics school** (Wiley 1981), in which classification is based exclusively on the branching pattern in a phylogenetically (genealogically) related group of organisms. A **cladogram** is a summary of phylogenetic relationships, showing a sequence of dichotomies, each one the splitting of a parental species into two derived species. The word "cladistics" comes from a Greek word meaning branch. To understand cladistic classification, one must understand the branching pattern in a phylogeny.

13.4 PROCEDURES IN EVOLUTIONARY CLASSIFICATION

How do the distinct schools of classification proceed in their efforts to systematize nature? Let us start with the traditional school. Mayr (1969) recognized four steps in evolutionary classification.

Overview

A. Preparatory activities
  1. Sorting individuals into phenotypic groups (phena) and these into populations
B. Genuine classification
  2. Assigning populations to species and naming species

3. Grouping species into higher taxa and naming these taxa, including:
  (a) Determination of relationships between species and higher taxa, such as genus and family
  (b) Formal delimitation of taxa
4. Ranking taxa in a hierarchy of categories

The preparatory activities (A1) require a full knowledge of the biological phenomena relating to a group: ecology, life history, existence of a complex life cycle, sexual dimorphism, polymorphism, seasonal variation, geographic variation, any caste systems, and so on. Application of evolutionary theory is needed for the recognition of reproductive isolation, genetic systems, kinds of species, and other evolved attributes of the taxon under study.

Species Identification

Researchers take a reasonable sample of individuals from a group and scrutinize them for conformity or differences in certain characters. These may be **meristic** (countable) **characters,** such as the number of setae per sclerite on an insect; **metric** (measurable) **characters,** such as length or width of parts; or **qualitative characters,** such as color or pattern. Wherever possible, statistical analyses are used to determine differences and similarities, and distributions of characters and groups of characters are mapped.

With the knowledge acquired from preparatory activities, the researchers assign populations to species while applying the appropriate species concept (B2).

Higher Taxa

The grouping of species into higher taxa (B3) involves several steps. The researchers determine the nearest relatives of each species and search for gaps between groups of species to identify clusters and groups of clusters. Then they decide which clusters to recognize formally as genera and which to designate less formally, as species groups. They arrange the genera into groups of progressively higher recognized taxa in the hierarchy, such as families, orders, classes, and so on.

Discovering nearest relatives depends on what is meant by "relationship." To a pheneticist the word means simply similarity; to a cladist it means genealogical relationship. In evolutionary classification, "relationship" means inferred genetic similarity as determined by both distance from branching points and subsequent extent of divergence.

Weighting of Characters

To the evolutionist, many relationships are obvious, and the key task is to weigh similarities as evidence of relationship (Mayr 1969). Different characters contain different amounts of information concerning ancestors. **Weighting of characters** is defined as a method for determining the relative importance of the phyletic information content of each character (Mayr 1969). This process usually emphasizes qualitative statements about weight, or importance, over quantitative criteria.

Characters that delimit a group usually emerge by trial and error. Tooth structure in mammals and wing venation and genital structure in insects receive strong weighting.

> The only reason why high weight is given to certain characters is that generations of taxonomists have found these characters reliable in permitting predictions as to association with other characters and as to the assignment of previously unknown species. (Mayr 1969, 219)

> It has been said with good reason that the trial-and-error method of improving classification is ponderous and uneconomical. To undertake a successful *a posteriori* weighting of characters requires a thorough knowledge of the history of previous classifications of a given group and an ability to make value judgments. Yet no clearly better method has so far been found. (Mayr 1969, 219)

Choice of Characters

Then we must ask, "What characters are most important and deserve the most weight?" First, **complex structures,** such as wings or genital armatures of insects, have greater weight than simple characteristics such as tibia length, because complex structures are likely to differ between even closely related species, and it is unlikely that convergent evolution would produce very similar complex structures. The mutual possession of derived characteristics also receives great weight. Taxa should be defined by **shared derived characters (synapomorphies)** and not **shared ancestral characters (symplesiomorphies).** Hennig (1950, 1966) was the first to state this principle, but all evolutionary taxonomists practice it intuitively. A symplesiomorphy simply shows that two taxa have not lost an ancestral characteristic, but they need not be closely related. However, the sharing of an evolutionary novelty, a synapomorphy, by two taxa is almost invariably a sign of close relationship and only rarely involves convergence. Of course, characters that are more or less constant and consistent have greater weight than those that are highly variable, such as skin or hair color in many vertebrates. Mayr

(1969) also noted what he called **Darwin's principle.** "The less any part of the organization is concerned with special habits, the more important it becomes for classification" (Darwin 1859, 414).

Delimitation of Higher Taxa

After the weighting of characters and the grouping of species on that basis, the next question arises: How are taxa delimited? Mayr (1969) noted five considerations:

1. The distinctness of the group, involving the sizes of the gaps among groups
2. The evolutionary role of the group and the uniqueness of its adaptive zone
3. The degree of difference among groups involving the distance between the "mean" of characters for each group and the dispersion around this mean, which defines gap size (Figure 13-1)
4. The size of the taxon—because taxa should ideally be of approximately equal size to facilitate information retrieval (information under one heading should be neither too much nor too little)
5. Ranking of related taxa at an equivalent level in the taxonomic hierarchy

Achieving the equivalence of item 5 is not easy. For example, a family should have a similar significance for flies and beetles, and the genus should have similar status in the Carnivora and Rodentia. Crowson (1958) argued for the adoption of norms for each taxonomic group. For example, the order Rodentia could be the paradigm for a mammalian order, and any other order would then use similar kinds of characteristic combinations. Scientists have never implemented Crowson's obviously important argument on a broad scale (Mayr 1969).

Delimitation of taxa above the species level seems to be extraordinarily arbitrary. An objective, uniform treatment is not available. For instance, an adequate operational definition of a **genus** is impossible. Cain (1956) defined the genus as the lowest higher category and the lowest of all categories established strictly by comparative data. Mayr (1969, 92) defined a genus as "a taxonomic category containing a single species, or a monophyletic group of species, which is separated from other taxa of the same rank [i.e., other genera] by a decided gap." He recommended that the size of the gap be in inverse ratio to the size of the taxon; for example, the larger the genus is, the more need there is to split it, and the smaller the gap deemed sufficient. Because convenience in handling taxa of equivalent size is considered more important than reflection of biologically meaningful taxon size, an evolutionary classification loses much of the phylogenetic history of a group.

FIGURE 13-1

A diagrammatic view of the gap-size criterion for de-limiting taxa. When gap sizes are small **(top)**, the taxa are regarded as closely related and are distinguished at perhaps the species or genus level. Large gap sizes **(bottom)** call for higher-ranked taxonomic distinctions—perhaps families or orders.

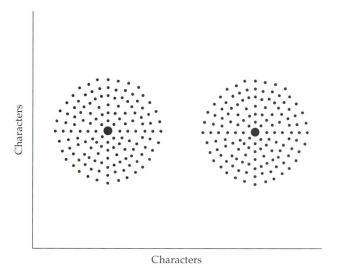

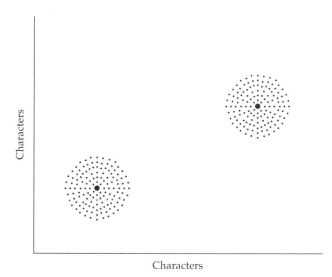

In a similar manner, Mayr (1969, 94) defined a **family** as "a taxonomic category containing a single genus or a monophyletic group of genera, which is separated from other families by a decided gap." The same kinds of definitions are used for order, class, and so on.

In conclusion, any argument that an evolutionary classification is subjective, arbitrary, and probably unstable, because new information is likely to affect gap size and taxon size, seems justified on the basis of careful scrutiny of writings by preeminent scientists in their fields. The real questions then become: Is there a better method? Do alternative approaches to classification achieve a more objective and stable classification?

13.5 PROCEDURES IN PHENETIC CLASSIFICATION

Unit Characters

Sokal and Sneath (1963) outlined the classification procedures of the phenetics school. The basic unit of information used by numerical taxonomists is the **unit character**—an attribute of an organism about which a single statement can be made (Sneath 1957). We can describe admissible unit characters as follows.

1. They must be meaningful on the basis of scientific judgment. For example, the number of leaves on a tree may be a meaningless and inadmissible character, but the number of spines per leaf may well be admissible.
2. Characters must vary, but they must not vary so much that they become meaningless.
3. Characters must be uncorrelated with other unit characters. Any property that is logically a consequence of another must be excluded.

Unit characters may take several forms. **Two-state (all-or-none) characters,** such as winged and wingless, may be used. A **quantitative multistate character** can be expressed as a single numerical value or coded into a class (1, 2, 3, and so on); for example, length of part or the number of bristles in a region of the body may be used in this way. **Qualitative multistate characters,** such as sculpting on the exoskeleton of an insect, cannot be arranged in any logical order; they need to be recoded as two-state characters. Sokal and Sneath (1963) spent three pages dealing with the problem of logically correlated characters, showing the kind of difficult decisions that must be made in the selection of admissible unit characteristics.

The characters chosen should include all those available—ideally with 40 or so as a minimum, 60 being more desirable, and 100 offering more statistical power but little improvement with use of more than 100 characters. Conventional diagnostic characters can be used, but they must be accompanied by many other characters to avoid the subjective bias of conventional taxonomy. Characters may involve any biologically meaningful kinds of information: morphological, physiological, behavioral, ecological, and distributional.

Operational Taxonomic Units

Use of the term **operational taxonomic unit** (OTU) for the group of organisms to be classified avoids any preconceptions about the group's identity as a particular taxon. Then a matrix can be developed that records the OTUs for comparison and all the quantified characters of each OTU (Table 13-1). Taxonomic resemblance between pairs of OTUs is calculated, using some kind of similarity coefficient (*S*). Sneath and Sokal (1973) list four kinds of coefficients: distance, association, correlation, and probabilistic similarity. Selection of the similarity coefficient is usually by trial and error or by personal preference and past experience.

When this process is complete, similarity coefficients are arranged in a **resemblance matrix,** or **similarity matrix,** containing every pairwise comparison (Table 13-2). Taxonomists use many different techniques, generally called clustering methods or cluster analysis, to search for patterns in this matrix. Some methods work from the bottom up, starting with separate similarity coefficient values. These **agglomerative methods** group *S* values into successively fewer groups until a single set is reached that contains all OTUs. **Divisive methods** work from the top down, starting with one group and dividing into smaller and smaller groups.

Phenograms

The resultant phenograms are displays of clusters that vary depending on the computer program used to develop them. For example, Lance and Williams (1967) showed that they could obtain very different results by varying their parameter β between +1 and

TABLE 13-2 Similarity matrix of similarity coefficients for every pairwise comparison of OTUs in the matrix

		OTU				
		1	2	3	4	5	...
OTU	1					
	2	S_{21}				
	3	S_{31}	S_{32}			
	4	S_{41}	S_{42}	S_{43}		
	5	S_{51}	S_{52}	S_{53}	S_{54}	
	⋮	⋮	⋮	⋮	⋮	

−1 (Figure 13-2). Sneath and Sokal (1973) recommended the use of both cluster analysis and an ordination analysis, the latter yielding, preferably, a three-dimensional view of the taxon under study, with more detail than the phenogram (Figures 13-3 and 13-4).

The taxonomist then draws conclusions based on an understanding of the similarities between the OTUs represented in the phenogram and based on ordination. Scientists infer that these methods yield evolutionary patterns as well as phenetic patterns: "Phylogenies are deduced necessarily from the phenetic relationships" (Sneath and Sokal 1973, 313). Very little development of methodology for formally determining higher taxonomic categories, such as genera, families, or orders, has occurred. No objective criteria seem to be available for generating higher taxonomic categories from the phenogram.

When the researcher includes fossils with present-day OTUs, a phenogram can represent a phylogenetic set of relationships, or a phylogenetic tree. Thus, phenetics can address the same questions the cladists or evolutionary taxonomists ask about evolutionary relationships among organisms.

13.6 PROCEDURES IN CLADISTIC CLASSIFICATION

Character Analysis

Classification procedures in the cladistics, or phylogenetics, school start with a **character analysis.** The researcher compares taxonomic characteristics within the taxon under study to identify homologous and nonhomologous relationships. **Homology** and **homoplasy** are differentiated. Then the homologous characters are distinguished on the basis of whether they

TABLE 13-1 Matrix of operational taxonomic units (OTUs) and the characters quantified for each unit

	OTU						
Character	1	2	3	4	5	6	...
A	—	—	—	—	—	—
B	—	—	—	—	—	—
C	—	—	—	—	—	—
D	—	—	—	—	—	—
E	—	—	—	—	—	—
⋮						

Dashes indicate the positions of actual values for each OTU for each character.

FIGURE 13-2

Phenograms illustrating variation in the degree of clustering generated by a computer program. Varying the parameter β between +0.98 and −1.00, Lance and Williams (1967) produced very different phenograms from the same set of 20 OTUs. Note that, as β declines, subgroupings in the phenogram become more apparent, and at a value of β = −0.25, groupings are discrete enough to aid the classification of the OTUs.

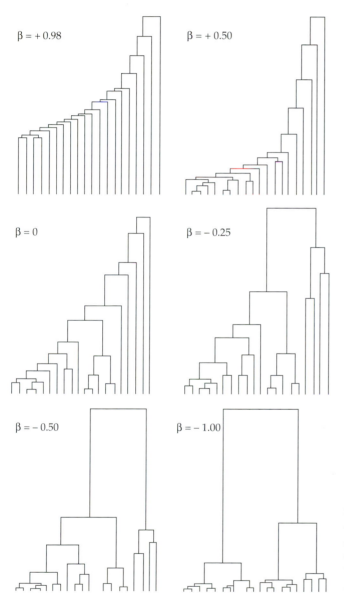

are **plesiomorphic characters** (ancient homologues) or **apomorphic characters** (recent homologues).

Wiley (1981, 122) defines a plesiomorphic character as follows: "Of a pair of homologues, the plesiomorphic character is the character that arose earlier in time and gave rise to the later, apomorphic, character." Synonyms for plesiomorphic are primitive, an-

cestral, and generalized. Wiley (122) defines an apomorphic character as follows: "Of a pair of homologous characters the apomorphic character is the character evolved directly from its preexisting homologue." Synonyms for apomorphic are derived, advanced, and specialized. The single strand of DNA in prokaryotes is a plesiomorphic character whereas multistranded DNA in the eukaryotes is an apomorphic character.

In practice, we need a clear reference point to estimate the relative positions of characters in terms of being ancestral or derived. The process called **outgroup comparison** provides that reference point. An **outgroup** is a taxon outside the group under close scrutiny, and it must have diverged before the group under study diversified, thereby providing a reference for the recognition of basic ancestral characters. For example, as we saw in Chapter 7, a lineage from fish to amphibians to reptiles to birds and mammals provides the basic stock for comparison. Amphibians are an outgroup for the analysis of evolutionary relationships among reptiles, birds, and mammals. The amphibian characters are clearly ancestral to the derived groups and therefore provide a suit of characters for comparison with the apparent emergence of new or derived characters. Characters that derived after the amphibians include the amniote egg (Figure 7-19), internal fertilization, and reproduction on land. Without reference to the outgroup of amphibians, we could not evaluate these characters as ancestral or derived.

Development of a Phylogenetic Tree

Once the characters have been assigned their statuses in relation to homology and primitive and derived characters, the taxa under study are arranged in a phylogenetic tree based on their symplesiomorphies and synapomorphies. Wiley (1981, 123) defines a synapomorphic character or synapomorphy, as "a homologous character found in two or more taxa that is hypothesized to have arisen in the ancestral species of these taxa and in no earlier ancestor." Thus, the common presence of symplesiomorphies determines the monophyletic nature of the group under study, but synapomorphies define the branching pattern in this monophyletic group.

An Example Using Mygalomorph Spiders

An example will help to clarify the process of developing a phylogeny. Actinopodidae and Migidae are families of mygalomorph spiders. They share a common feature not found in other mygalomorph spiders: the eyes are spread across the carapace and not clustered. This feature, then, is a synapomorphic character that binds the two families together while

Correlation

Corr-Skel

1 *Catharacta skua*
2 *Stercorarius pomarinus*
3 *Stercorarius parasiticus*
4 *Stercorarius longicaudus*
7 *Pagophila eburnea*
45 *Rissa tridactyla*
46 *Rissa brevirostris*
5 *Gabianus pacificus*
18 *Larus argentatus*
20 *Larus fuscus*
23 *Larus dominicanus*
25 *Larus marinus*
26 *Larus glaucescens*
27 *Larus hyperboreus*
19 *Larus thayeri*
28 *Larus leucopterus*
10 *Larus heermanni*
22 *Larus occidentalis*
16 *Larus delawarensis*
21 *Larus californicus*
29 *Larus ichthyaetus*
6 *Gabianus scoresbii*
13 *Larus belcheri*
8 *Larus fuliginosus*
12 *Larus hemprichii*
14 *Larus crassirostris*
30 *Larus atricilla*
17 *Larus canus*
36 *Larus melanocephalus*
34 *Larus pipixcan*
39 *Larus ridibundus*
44 *Rhodostethia rosea*
48 *Xema sabini*
32 *Larus cirrocephalus*
33 *Larus serranus*
35 *Larus novae-hollandiae*
38 *Larus maculipennis*
37 *Larus bulleri*
41 *Larus philadelphia*
42 *Larus minutus*
9 *Larus modestus*
47 *Creagrus furcatus*
11 *Larus leucophthalmus*
40 *Larus genei*
49 *Chlidonias hybrida*
50 *Chlidonias leucoptera*
51 *Chlidonias nigra*
77 *Sterna albifrons*
86 *Procelsterna cerulea*
53 *Gelochelidon nilotica*
52 *Phaetusa simplex*
74 *Sterna superciliaris*
91 *Rynchops nigra*
93 *Rynchops albicollis*
92 *Rynchops flavirostris*
69 *Sterna lunata*
70 *Sterna anaethetus*
88 *Anous tenuirostris*
89 *Anous minutus*
56 *Sterna hirundinacea*
57 *Sterna hirundo*
58 *Sterna paradisaea*
59 *Sterna vittata*
61 *Sterna forsteri*
62 *Sterna trudeaui*
63 *Sterna dougallii*
66 *Sterna sumatrana*
64 *Sterna striata*
80 *Thalasseus bengalensis*
83 *Thalasseus elegans*
84 *Thalasseus sandvicensis*
85 *Larosterna inca*
71 *Sterna fuscata*
90 *Gygis alba*
87 *Anous stolidus*
54 *Hydroprogne tschegrava*
78 *Thalasseus bergii*
79 *Thalasseus maximus*
68 *Sterna aleutica*
72 *Sterna nereis*
73 *Sterna albistriata*

Correlation

FIGURE 13-3

A phenogram based on cluster analysis for 81 species of gulls and terns, developed by Schnell (1970b).

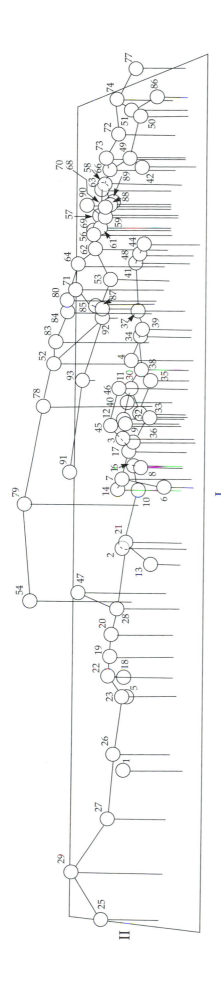

FIGURE 13-4

An ordination analysis for 81 species of gulls and terns, based on a principal-components analysis illustrated in three dimensions. *(From Schnell 1970a)*

separating them from other mygalomorph spiders (Platnick and Shadab 1976) (Figure 13-5).

The posterior sigilla of female members of the Actinopodidae appear as excavated depressions; they are simply marks on the ventral plate of the cephalothorax. This feature does not appear in the Migidae or in any other mygalomorph spiders; it thus established that the Actinopodidae are monophyletic, and it is a synapomorphous character separating the family Actinopodidae from others (Figure 13-6).

FIGURE 13-5

A phylogeny of mygalomorph spiders, showing how the synapomorphic characteristic of eye condition separates two families from the other mygalomorph families. *(From Platnick and Shadab 1976)*

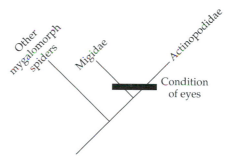

FIGURE 13-6

Recognition of excavated depressions in the posterior sigilla of females, unique to the family Actinopodidae, results in this synapomorphous characteristic establishing the monophyletic condition of the family. *(From Platnick and Shadab 1976)*

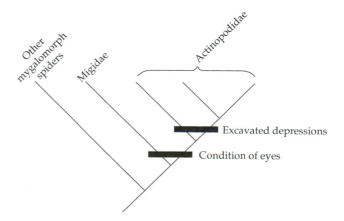

Four possible hypotheses on the phylogenetic relationships between the actinopodid spider genera (*Neocteniza* [N], *Actinopus* [A], and *Missulena* [M], considered by Platnick and Shadab (1976).

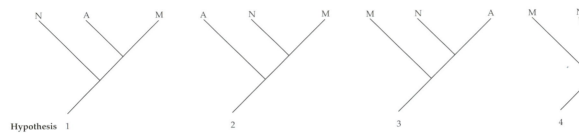

Hypothesis 1   2   3   4

Next, the phylogeneticist would examine the divergence of genera within the Actinopodidae. Inspection of three genera, *Neocteniza*, *Actinopus*, and *Missulena*, leads to four possible hypotheses on their phylogenetic relationships (Figure 13-7). Because the male *Neocteniza* had a distinct color pattern on its leg, lacking in all other mygalomorphs, Platnick and Shadab argued that this character is a synapomorphy for the genus. In addition, *Neocteniza* has a recurved thoracic groove—a deep T-shaped fissure in the dorsal surface of the carapace—whereas *Actinopus* and *Missulena* have procurved grooves. Since most mygalomorphs have recurved grooves, this observation suggests that *Neocteniza* is closer to the primitive stock than *Actinopus* and *Missulena*. The procurved grooves become a synapomorphy separating *Actinopus* and *Missulena* from *Neocteniza*. In this way the phylogenetic tree develops and all hypotheses on relationships but one are eliminated (Figure 13-8). Consideration of many other characteristics reinforces the final phylogeny (Figure 13-9).

The accepted hypothesis for the actinopodid spider genera, affirming that hypothesis 1 in Figure 13-7 was correct on the basis of leg color pattern in *Neocteniza* and procurved thoracic grooves in *Actinopus* and *Missulena*. (Based on Platnick and Shadab 1976).

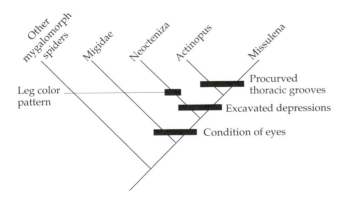

The final phylogeny for actinopodid spiders developed by Platnick and Shadab (1976), showing additional characteristics beyond those discussed in the text. Of the 22 characteristics used, only a few are labeled. Dark squares show the apomorphic states of characteristics, and white squares indicate plesiomorphic states.

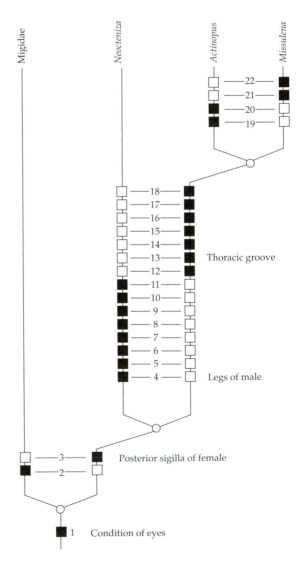

Vicariance Biogeography

Hennig (1966) proposed methods for checking a phylogenetic tree, one of which involves **vicariance biogeography.** Wiley (1981, 8) defined a vicariance event as a geographic separation of a continuous biota such that the biota becomes two or more geographic subunits. Such an event may therefore result in speciation, intraspecific geographic variation, or no apparent impact. If speciation is involved, vicariance events result in phylogenetic trees of species. If intraspecific geographic variation is involved, the phylogenetic trees are of populations within species.

Biogeography means the study of distribution of organisms in space through time (Wiley 1981, 277). Therefore, vicariance biogeography means the study of phylogenetic splitting events resulting from geographic isolation, the timing of those events, and the resulting parallel phylogenetic trees in members of the biota subjected to the same vicariance events.

Track Synthesis

The vicariance biogeographer proceeds as outlined by Wiley (1981), collecting data on the distribution of monophyletic groups, plotting them on maps, and drawing tracks around the distributions of monophyletic groups. A search for replicated patterns of distribution among different monophyletic taxa, called **track synthesis,** then occurs (Croizat 1952, 1958, 1964). Areas of endemism are then identified, and two questions are posed: (1) what are the interrelationships among the organisms inhabiting these areas of endemism, and (2) how do these relationships relate to the geographic and geologic histories of the areas themselves? To answer these questions, the biogeographer erects phylogenetic hypotheses, based on the organisms in the areas as just described, and generates an area cladogram based on geographic and geologic relationships.

Congruence

The researcher then seeks **congruence** between the phylogenetic cladogram and the area cladogram. When the phylogenetic cladogram is erected, the area inhabited by each species is substituted for the species name to generate the area cladogram. Complete congruence of the two patterns confirms the phylogenetic hypothesis. If complete congruence between phylogeny and area occupied occurs for many taxa, it indicates the commonness and generality of the factors causing vicariance.

A Geographic Cladogram

From the study of local cladograms and area cladograms, similar relationships can emerge for the global distribution of taxa. A **geographic cladogram** is relatively simple to construct for continental relationships, because detailed understanding of plate tectonics has illuminated the relationships of areas and their division and joining over the past 200 million years. A geographic cladogram of continents after the breakup of Pangaea can be developed (Rosen 1978) (Figure 13-10).

> If there is a correspondence between a geographic cladogram and one or more biological area cladograms, then we may infer that the causes associated with the geographic cladogram also caused the sequence of vicariance seen in the phylogenetic hypothesis. (Wiley 1981, 294)

One of Hennig's most important contributions to systematics was this objective and independent test of the phylogenetic cladogram.

The theme in vicariance biogeography is that most distributions of taxa result from vicariance events, and long-distance dispersal is of lesser importance. However, we can evaluate the effects of dispersal using vicariance biogeography, because a sequence of closely related species corresponds to a migration route of known continental connections. Vicariance biogeography is therefore an integral part of the approach used by the cladistics school. Ashlock (1974) reviewed the value of biogeography in classification.

13.7 CRITICISMS OF METHODOLOGY

Of course, every school of classification argues for its own validity while denigrating the other schools. The debates on methods of classification are illuminating because they are at the core of systematics and reveal much of its philosophy. Let us therefore explore some of the criticisms leveled at each school.

FIGURE 13-10

Rosen's (1978) geographic cladogram of continents.

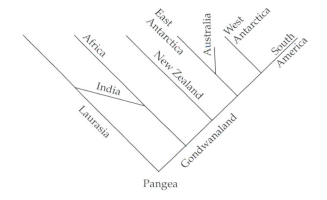

Evolutionary Classification

Criticisms of the school of evolutionary classification have been extensive, for it was dissatisfaction with this school that generated the newer subdisciplines. Sokal and Sneath (1963) pointed out that the modern synthesis contributed little to the understanding of evolution above the species level and that a reevaluation of the logical bases of taxonomy emerged much later, in the late 1950s and early 1960s. This reevaluation recognized several problems with the traditional school.

Attempts Too Much

The traditional school tries to do too much. It involves too many aspects of similarity and difference between organisms, and as a result it handles none of them well. That is, the evolutionary school is a "jack of all trades, master of none." It attempts to simultaneously classify, name, determine degrees of similarity, and map phylogenetic relationships.

Lacks Objectivity

The criteria used in developing useful characters for study are not objective. Scientists have used very personal and subjective divisions and have not applied them in a standard manner. Simpson (1961) recognized that little theoretical basis existed on which a rigorous protocol could be formed.

Early systematics was based on Aristotelian logic, which required the discovery of the real nature, or *essence*, of a group. The search for a single diagnostic character that would both bind a taxon together and separate its members from one another was an important part of early taxonomy. But identification of the *essence* of a taxon relied heavily on biological intuition and not necessarily on objective criteria.

The search for essences led to a monothetic view of taxonomy, in which groups are formed by rigid and successive logical divisions so that the possession of a unique set of features is both sufficient and necessary for membership in the group. Hence, scientists could clearly define dichotomous keys to describe affinities between species. However, serious errors in classification can result when a species or other taxon differs by a single character from the monothetically defined group. The more basic the character in the series of division, the more distant from its relatives a species will appear.

Uses Circular Reasoning

The development of phylogenies uses circular reasoning. Since *The Origin of Species*, scientists have recognized that species or other taxa have features in common because of their **common descent.** A taxon is a **monophyletic** array of related forms—that is, a group of forms with a single ancestor.

Sokal and Sneath (1963) pointed out that the difficulty with a phylogenetic approach to classification is that phylogenies are unknown in most cases. When phylogenies are constructed, they argue, the logical fallacy of circular reasoning is usually apparent. This is because a taxonomist uses three kinds of data:

1. **Resemblance.** The members of a taxon resemble one another more than they do other forms.
2. **Homologous characters.** The members of a taxon share characters of common origin.
3. **Common lineage.** A taxon is defined by its members being descended from a common stock.

The point is that item 3, common lineage, is rarely if ever known and is inferred by examination of data in items 1 and 2. In turn, conclusions in item 2 are often drawn from speculations in item 3. Basing a classification on the use of a few characters increases the problems of circular reasoning and extrapolation.

Uses Unstandardized Taxonomic Ranks

Taxonomic ranks and use of gaps among taxa are not standardized. Groupings are identified on the basis of the gaps in characters between them. As a result, evolutionists have often placed group boundaries where gaps in the fossil record exist, and these boundaries may well be defined by accidental discovery or nondiscovery of fossil forms (Sokal and Sneath 1963, Wiley 1981).

In addition, evolutionists tend to create new taxa when a taxon becomes cluttered with members. The members of the former taxon and the new taxa must then be grouped in a taxon of higher rank. Consequently, taxa become defined by the number of lower taxa present rather than by a standard level of similarity between members of a taxon and a standard level of difference between members of different taxa of the same rank.

We can see these discrepancies in another way when we consider the classification of birds and insects. For example, 8,600 species of birds make up a class, with about 27 orders, 160 families, and 2,400 genera (Townes 1969). By contrast, 60,000 species of ichneumonid wasps are lumped into a single family in the order Hymenoptera, with 25 subfamilies and only about 1,250 genera. Townes argued that vertebrate systematists would have to regard many ichneumonid genera as families: there are about 48 species per genus in ichneumonids and 54 species per family in birds.

No objective criteria are generally applicable for defining the rank of a particular taxonomic category.

Accepts Paraphyletic Groups

Acceptance of paraphyletic groups confuses the phylogenetic relationships among organisms. Evolutionary taxonomists name and rank paraphyletic groups (grades), whereas cladists (the phylogenetic school) accept only the validity of monophyletic groupings, as defined by Hennig (1966).

The cladists define a **monophyletic group,** or a **clade,** as a group of all the species descended from a single ("stem") species (Hennig 1966) (Figure 13-11). Simpson and others have used monophyly in a different sense, defining a monophyletic group as one whose most recent common ancestor is a cladistic member of that group (Ashlock 1973) (Figure 13-11).

FIGURE 13-11

Illustrations of the definitions of terms used in phylogenetic studies. *(Based on Ashlock 1973)*

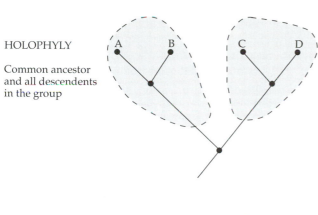

HOLOPHYLY

Common ancestor and all descendents in the group

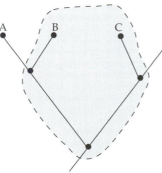

PARAPHYLY

Common ancestor in group but not all descendents in the group

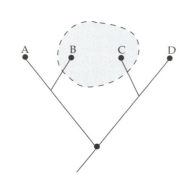

POLYPHYLY

No common ancestor in group

Therefore, to distinguish between the two viewpoints, monophyly as defined by the cladists is called **holophyly**—a useful descriptor because it signifies inclusion of all members of a group stemming from a common ancestor.

A **paraphyletic group** is defined as a group of species that includes a common ancestor and some, but not all, of its descendents (Farris 1974) (Figure 13-11). A **polyphyletic group** is a group based on convergent similarity, in which the common ancestor is not included (Hennig 1966) (Figure 13-11).

Uses Paraphyletic, Not Holophyletic, Classification

Wiley (1981) emphasized the importance of distinctions between holophyletic and paraphyletic views of monophyly. Evolutionary taxonomists tend to separate taxa according to their distinctness and the sizes of the gaps between nearest related taxa. When a group exploits a major new adaptive zone, it tends to receive a high rank in the taxonomy. For example, birds have a class to themselves, but the crocodiles, a taxon coming from the same monophyletic stock, are left as an order within the class Reptilia. Thus, the Crocodilia are paraphyletic to the class Aves, and their classification should show this relationship. The evolutionary and phylogenetic classifications come to look very different (Figure 13-12). The evolutionary classification is:

> Class Reptilia (reptiles)
>   Order Anapsida (turtles)
>   Order Lepidosauria (snakes, lizards, etc.)
>   Order Crocodilia (crocodiles, alligators, etc.)
> Class Aves (birds)
> Class Mammalia (mammals)

The phylogenetic classification is:

> Superdivision Amniota
>   Division Anapsida
>   Division Sauropsida
>   Subdivision Lepidosauria
>   Subdivision Archosaurai (dinosaurs, pterosaurs, etc.)
>   Supercohort Crocodilia
>   Supercohort Aves
>   Division Therapsida
>   Subdivision Mammalia

Wiley (1981) discussed the implications of the evolutionary classification. Formation of groups on the basis of their morphological distinctness implies the following.

1. Turtles, lepidosaurs, and crocodiles are equally distinct from each other. This is true (Figure 13-12).
2. Crocodiles are as distinct from birds as they are from mammals. This is false, because crocodiles

Perspectives of **(top)** evolutionary taxonomists and **(bottom)** phylogenetic taxonomists on the relationships among reptiles, birds, and mammals. *(Based on Wiley 1981)*

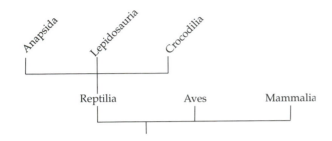

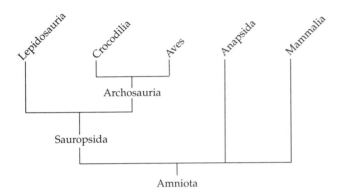

are much more like birds than they are like mammals.

3. Crocodiles are as distinct from birds as turtles and lepidosaurs are. This is also false, because birds and crocodiles are more similar than birds and the other reptiles.

The evolutionary classification is therefore misleading.

In contrast, the phylogenetic classification is informative. Wiley (1981) argued that the phylogenetic approach is realistic even in terms of adaptive zones, because crocodiles and birds build nests, provide parental care for their young, and establish and defend territories by singing songs, "although the songs of alligators are certainly less pleasing to us than the songs of meadow larks" (Wiley 1981, 265). Wiley also pointed out that we can retrieve the history of speciation and radiation from the phylogenetic classification, for it reveals, for instance, that birds evolved from a reptile stock—which the evolutionary classification does not reveal.

Mayr (1974, 112–113) has argued against this view.

A rigid application of their dogma forces cladists to break up the reptilian grade into many separate

"classes" to designate particular reptilian lineages as "sister groups" of birds and mammals. [**Sister groups** are defined as groups descended from a common ancestral group.] The fact that no one would place the crocodiles outside the reptiles if birds did not exist reveals how artificial and arbitrary this procedure is. The essential unity of the reptiles is best illustrated by the continuing argument among paleontologists as to which orders of reptiles are most closely related to which others.

Phenetic Classification

Criticisms of the phenetics school are also extensive. The phenetics school claims that groupings based on overall similarity result in stable and natural classifications, whereas phylogenetic classifications are unstable and unnatural. Wiley (1981) recognized this claim as the major bone of contention between the pheneticists and the phylogeneticists.

Generates Unnatural and Uninformative Groups

Wiley argued that phenograms are very unstable because they can differ if the same data are analyzed by different algorithms or if the same specimens have different data collected from them and are analyzed by the same algorithm.

In terms of naturalness, Farris (1977) argued that not every feature of an organism is useful in determining its natural relationships. Thus, when a phenetic analysis includes all or many features, some must be uninformative or even misleading, and the most important criteria do not receive enough attention. A phenogram therefore results in an unnatural classification.

When Mayr (1969) considered phenetics, he also found much to criticize, in addition to the abovementioned points. For example, the results of adult classifications differ dramatically from the results of classifications based on larval characters of holometabolous insects. Even when 77 adult characters and 71 larval characters of *Aedes* mosquitoes were used, the classifications turned out to be very different (Rohlf 1963).

Has Inadequate Character Choice and Weighting

It is also difficult to define a character. In a comparison of a small animal and a large one, many measurable characters relate to size alone, and so they covary and are not independent.

Many unweighted characters must be used. In such groups as arthropods, they are easy to find. But many taxa are morphologically uniform and a large number of characters would be very difficult to identify; examples include birds, frogs, and lower fungi. Mayr argues that the number of characters is not important, but their "taxonomic weight" is. Most characters have low information content or are completely redundant. Several authors have found that they can more readily obtain groupings, and the groupings are much more useful, when the number of characters used is reduced, thus eliminating the "noise" in the large data set.

Strong stabilizing selection, divergent selection, or sexual selection can yield very different amounts of difference between species, although the genetic basis for these differences may not differ so much. For example, a phenetic classification may not even recognize sibling species, whereas the birds of paradise have been under such strong sexual selection that they would appear to be very different from one another in a phenogram—despite the possibility that genetic identities of sibling species (say, of *Drosophila*) and genetic identities of birds of paradise may be quite similar.

> Only appropriate weighting can convert the phenetic distances between the genera of birds of paradise into a biologically meaningful classification. This proves the falseness of the assumption that each character is so polygenic that any random sample of characters will accurately reflect the properties of the genotype. (Mayr 1969, 208)

The converse problem is that convergent characters are not treated adequately and may result in misleading phenograms.

Last but not least, collecting data on 60 or more characters per species makes the classification of 1,000, 5,000, or 20,000 species in a group almost impossible. Yet that is frequently the taxonomist's task.

Cladistic Classification

Ernst Mayr has been a major advocate for the evolutionary school of classification and therefore a major critic of alternative schools. He made the following comments on the cladistics school in 1974 (they were reprinted in 1976). Mayr raised three major points, involving (1) arbitrary decisions, (2) misleading ranking, and (3) neglect of the facts. A brief treatment of his criticism follows.

1. **Arbitrary decisions.** "The transfer of well-known and universally understood terms to entirely new concepts cannot fail to produce confusion" (Mayr 1976, 441). For example, historically the term "phylogeny" has applied to all ancestor-descendant relationships. But Hennig (1966) restricted the word

to the branching sequences in a cladogram and the origins of taxa. "His diagrams are cladograms and not at all phylogenetic trees" (Mayr 1976, 442).

2. **A misleading conceptualization of ranking.** "The cladists, following the erroneous assumption that phylogeny is a unitary process (consisting only of branching), assert that classifying likewise is a single-step procedure and that the grouping of taxa simultaneously also supplies their rank" (Mayr 1976, 459). In contrast, Mayr saw proper classification as having two essentially separate steps. The first involved the **grouping** of lower taxa into higher taxa—usually the assemblage of species into genera, families, and so on. The second step, **ranking**, placed these higher taxa in their correct places in the taxonomic hierarchy. The rank of a taxon had usually depended on how different a group was from others, how unusual the group was relative to others, and the extent of adaptive radiation derived from the common ancestor.

3. **Operational neglect of evident facts.** The creation of branching patterns neglects important evolutionary information on how far descendants have deviated from ancestors, a critical factor in the recognition of gaps among taxa and the ranking of taxa. In addition, similarity in characteristics between taxa may derive from a common ancestor or from convergence and other factors; distinguishing between these processes is difficult, and yet it is central to classification. Mayr states that such niceties of classification are usually ignored by cladists.

Clearly, the schools of classification were polarized, but it is also apparent that each approach strengthened the field of systematics. Added to the strong tradition of the evolutionary school were the powerful quantitative approaches and analytical methods of the phenetics school. The cladists contributed a refreshing ability to test a classification with independent sets of data derived from vicariance biogeography or parasite phylogenies. Biological classification is much richer and stronger for these developments.

Evolutionary classification is still the major practice for dealing with the vast numbers of species in need of identification and systematization. However, for the depiction of evolutionary diversification and patterns in adaptive radiation in and among taxa, the cladist's view is very revealing. In the end, only nature has the truth, and features of each mode of classification will help us draw closer to that truth.

SUMMARY

Biological classification is the grouping of similar organisms, based on evolutionary relationships, into a taxonomic hierarchy ranging from species to kingdoms. Taxonomists may undertake this **systematic** ordering of the diversity of life by way of three different schools of thought, each with its own strengths and weaknesses.

The **evolutionary school of classification** is the traditional approach; it uses any information available to resolve issues on evolutionary pathways, branching of lineages, and divergence after branching and to construct a **phylogram** of relationships. The **phenetics school of classification** emphasizes the use of many **numerical characters** to construct a **phenogram** showing similarity and clustering of taxa, usually derived from computer-assisted statistical analysis of large data sets. The **cladistics (phylogenetics) school of classification** emphasizes the branching pattern in phylogenies, which it sees as sets of dichotomous events. A **cladogram** summarizes the phylogenetic relationships, showing a sequence from one parental stock to all derived taxa.

Using all available evidence, evolutionary classification proceeds in a series of groupings: populations into **species,** species into **genera,** genera into **families,** and so on. The characters used are weighted according to their perceived importance on the basis of reliability, constancy, and complexity. Gaps between sets of species are used to designate higher taxa, such as genera and families, although gap size varies according to the taxon under consideration.

Phenetic classification first identifies many **unit characters,** which are numerical in nature or numerically coded, per **operational taxonomic unit** (OTU)— commonly a species. A **similarity matrix** compares OTUs, and some kind of **cluster analysis** is employed to generate a **phenogram** of hypothesized relationships in the taxon under study.

Cladistic classification starts with **character analysis** to identify **homologous characters** and whether they are **plesiomorphic** or **apomorphic.** In comparisons of taxa, the **monophyletic** nature of the group is determined by the presence of common earlier-evolved characters, **symplesiomorphies,** and branching patterns are developed by use of common derived characters, **synapomorphies.** The example of mygalomorph spiders illustrates this process. The **cladogram** can be checked, sometimes by using **vicariance biogeography** to create an area cladogram based on the geographic and geologic relationships of the landmasses inhabited. **Congruence** between the phylogenetic cladogram and the area cladogram reinforces the validity of the cladistic analysis.

Evolutionary classification has been the target of criticism because it attempts to do everything but cannot do everything well, it lacks objectivity, and it involves circular reasoning. It uses unstandardized taxonomic ranking and accepts paraphyletic groups. The classification of reptiles, birds, and mammals illustrates the difference between **paraphyletic** and **holophyletic** classifications.

Some evolutionary biologists, identifying the problems with phenetic classification, include the development of unnatural and uninformative phenograms because of the large databases used without the potential bias of weighted characters.

Cladistic classification has its own set of detractors, who detect such problems as redefinition of well-known terms, other arbitrary decisions, and misleading ranking of taxa, as in the classification of reptiles, birds, and mammals.

Each approach to biological classification has made important contributions to the progress of systematic biology, and a synthesis of methodologies provides the strongest approach for ordering the diverse organisms on this earth.

QUESTIONS FOR DISCUSSION

1. Do you think that there is a lot of common ground between the three major schools of biological classification, and where do you think that overlap occurs?

2. In all methods of classification, do you think that it is biologically realistic to accept such apparently nebulous and variable taxa as genera and families, or should objective and standardized criteria be applied to the recognition of such taxa?

3. In the taxa you are familiar with, do you think that their classifications reflect a realistic evolutionary scenario, or is an ancestor–descendent relationship obscured?

4. Do you think that a rigorous application of holophyly will necessarily change classification at such major taxonomic levels as kingdoms, phyla, classes and orders?

5. With about 180 living species of primates, the evolution of a relatively large brain may appear to have yielded relatively little evolutionary opportunity. How would you evaluate this statement objectively, and what criteria would you use?

6. Do you think that the process of classification improves as more rules on the methodology are developed?

7. Vicariance biogeography may be used to test a phylogenetic hypothesis, but Hennig suggested other independent tests such as the use of parasites on the lineage under study. How would you test a phylogenetic hypothesis on a group of hosts using organisms that are closely associated ecologically such as symbiotic parasites or mutualists?

8. How can the situation be resolved in which current higher taxa are delineated by the existence of gaps in characters, between genera or families for example, while a cladistic approach, which emphasizes ancestor–descendent lineages, reduces the importance of gaps? For example, birds and dinosaurs have usually been regarded as distinct, but they are in the same holophyletic group.

9. To what extent do you think that the newer schools of classification have enabled a highly objective science to be developed?

10. If you were faced with the need to classify a set of sibling species, how would you proceed, and which method of classification would you use?

REFERENCES

Ashlock, P. D. 1973. Monophyly again. Syst. Zool. 21:430–438.

———. 1974. The uses of cladistics. Ann. Rev. Ecol. Syst. 5:81–89.

Cain, A. J. 1956. The genus in evolutionary taxonomy. Syst. Zool. 5:97–109.

Croizat, L. 1952. Manual of phytogeography. Dr. W. Junk, The Hague.

———. 1958. Panbiogeography. Croizat, Caracas.

———. 1964. Space, Time and Form: The Biological Synthesis. Croizat, Caracas.

Crowson, R. A. 1958. Darwin and classification, pp. 102–129. In S. A. Barnett (ed.). A Century of Darwin. Harvard Univ. Press, Cambridge, Mass.

Eldredge, N. 1993. What, if anything, is a species? pp. 3–20. In W. H. Kimbel and L. B. Martin (eds.). Species, Species Concepts, and Primate Evolution. Plenum Press, New York.

Farris, J. S. 1974. Formal definitions of paraphyly and polyphyly. Syst. Zool. 23:548–554.

———. 1977. On the phenetic approach to vertebrate classification, pp. 923–950. In M. K. Hecht, P. C. Goody, and B. M. Hecht (eds.). Major Patterns in Vertebrate Evolution. Plenum, New York.

Hennig, W. 1950. Grundzüge einer Theorie der Phylogenetischen Systematik. Deutscher Zentralverlag, Berlin.

———. 1966. Phylogenetic Systematics. Univ. of Illinois Press, Urbana.

Holmes, E. B. 1980. Reconsideration of some systematic concepts and terms. Evol. Theory 5:35–87.

Kimbel, W. H., and L. B. Martin (eds.). 1993. Species, Species Concepts, and Primate Evolution. Plenum Press, New York.

Lance, G. N., and W. T. Williams. 1967. A general theory of classificatory sorting strategies, [Part] I: Hierarchical systems. Computer J. 9:373–380.

Mayr, E. 1969. Principles of Systematic Zoology. McGraw–Hill, New York.

———. 1974. Cladistic analysis or cladistic classification? Zool. Syst. Evol.-forsch. 12:94–128.

Mayr, E. 1976. Cladistic analysis or cladistic classification? pp. 433–476. In E. Mayr (ed.). Evolution and the Diversity of Life—Selected Essays. Belknap Press of Harvard Univ. Press, Cambridge, Mass.

———. 1981. Biological classification: Toward a synthesis of opposing methodologies. Science 214:510–516.

Platnick, N. I., and M. U. Shadab. 1976. A revision of the mygalomorph spider genus Neocteniza (Araneae, Actinopodidae). Amer. Mus. Novit. 2603:1–19.

Rohlf, F. J. 1963. Classification of Aedes by numerical taxonomic methods (Diptera: Culicidae). Ann. Entomol. Soc. Amer. 56:798–804.

Rosen, D. E. 1978. Vicariant patterns and historical explanation in biogeography. Syst. Zool. 27:159–188.

Schnell, G. D. 1970a. A phenetic study of the suborder Lari (Aves), [Part] I: Methods and results of principal components analysis. Syst. Zool. 19:35–57.

———. 1970b. A phenetic study of the suborder Lari (Aves), [Part] II: Phenograms, discussion and conclusions. Syst. Zool. 19:264–302.

Simpson, G. G. 1961. Principles of Animal Taxonomy. Columbia Univ. Press, New York.

Sneath, P. H. A. 1957. The application of computers to taxonomy. J. Gen. Microbiol. 17:201–226.

Sneath, P. H. A., and R. R. Sokal. 1973. Numerical Taxonomy. W. H. Freeman, San Francisco.

Sokal. R. R., and P. H. A. Sneath. 1963. Principles of Numerical Taxonomy. W. H. Freeman, San Francisco.

Townes, H. 1969. The genera of Ichneumonidae, Part 1. Mem. Amer. Entomol. Inst. 11:1–300.

Wiley, E. O. 1981. Phylogenetics: The Theory and Practice of Phylogenetic Systematics. Wiley, New York.

III

MICROEVOLUTION

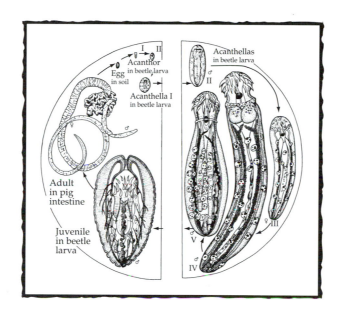

14

GENETIC SYSTEMS

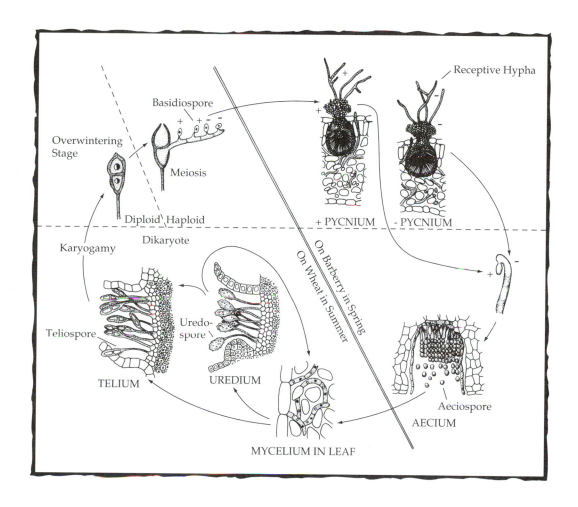

The life cycle of *Puccinia graminis*, the black-stem rust fungus pathogen of grasses and barberry, illustrating alternation of hosts and sexual and asexual phases of reproduction.

We may call all the factors that affect the hereditary behavior of a species and its evolutionary potential the **genetic system** of the species (White 1954). Systems evolve within environmental constraints, and certain systems adapt or preadapt a species for exploitation of certain types of environments. Thus, each genetic system has a strategic advantage over others under particular circumstances. It is to our advantage to understand which genetic systems result in strategies that are highly adaptive and the environmental constraints under which a strategy suffers. That is, given a certain genetic system that provides the basics of heredity in a lineage, to what further advantage can the evolutionary process employ that system?

To understand the genetic system of a species, we need to study (1) the mode of reproduction, be it bisexual, parthenogenetic, or by fission; (2) the population structure, including size, sex ratio, vagility, and extent of inbreeding; (3) the clinal variation in characteristics, or polymorphisms; and (4) the chromosome cycle and many cytological details of gamete production (White 1954, 1973). We already addressed some of these features of species in our discussion of modes of speciation.

14.1 KINDS OF SPECIES

Mayr (1963) stated that very little is known about correlations between ecological properties and specific genetic systems, and he was willing to undertake only a tentative discussion of the sets of factors that are likely to affect the evolutionary potential of species. He designated those factors as criteria for determining an organism's kind of species. Mayr's partial list includes 13 criteria:

1. System of reproduction
2. Degrees of intra- and interfertility
3. Presence or absence of hybridization
4. Variation in chromosome number or pattern
5. Difference in origin
6. Structure of species
7. Size of populations
8. Sequence of generations
9. Amount of gene flow
10. Pattern of distribution
11. Environmental tolerance
12. Rate of evolution
13. Phenotypic plasticity

All the factors in Mayr's list deserve consideration as part of the genetic system of a species, and ecological and evolutionary concerns are essential to an understanding of the biology of a species. Perhaps a surprising aspect of Mayr's emphasis that species can differ greatly depending on their own blend of traits is his adamance that only one speciation process was needed to account for their divergence (see Chapter 5).

In this chapter we emphasize the importance of the mode of reproduction of a species, and in the next chapter, "Change in Gene Frequencies," we will examine several other aspects of genetic systems. In fact, much of this book relates to the very broad topic of genetic systems, or kinds of species.

14.2 MODES OF REPRODUCTION

A classification of the modes of reproduction would include many types, according to Williams (1975), who emphasizes the adaptive significance of sexuality (Table 14-1). Table 14-2 provides some definitions of terms in Table 14-1. We have already discussed some of the terms in these tables—for example, when we considered microspecies in Chapter 4 and when we treated agamic systems of reproduction in plants in Chapter 5. Also, readers should attempt to combine with the material in this chapter their preexisting knowledge of types of reproduction in the many taxa of organisms studied in introductory courses in biology, botany, microbiology, and zoology.

Williams (1975) noted a major dichotomy in modes of reproduction. One branch includes all forms of reproduction in which meiosis, crossing over, and recombination are involved. Such **recombinational reproduction** results in progeny with genotypes different from the maternal genotype and different from each other. The other branch includes forms of reproduction without meiosis, in which mitotic cell division produces progeny that are genetic copies of the mother—hence the term **nonrecombinational reproduction.** A third case, also addressed in Table 14-1, combines both of these modes of reproduction in complex life cycles involving alternation between sexual and asexual reproduction, known as **heterogonic life cycles** or **cyclical parthenogenesis.** Such complex life cycles have evolved repeatedly in parasitic organisms where stages involving transmission from one host species to another are interspersed with repeated asexual reproduction that multiplies the maternal genotype before dispersal, with its risks (Figure 14-1).

Within the class of recombinational reproduction, Williams (1975) recognized several types as defined in Table 14-1. Type 2, which we call **typical,** includes what one normally thinks of as sexual reproduction in animals and plants. Motile male haploid gametes from one individual fuse with the more sedentary female gemetes from another individual to form a diploid zygote. Most large organisms, both plants and animals, are of this type. As organisms become progressively smaller, departures from this so-called

TABLE 14-1   Modes of reproduction recognized by Williams (1975)

A. **Recombinational reproduction.** The maternal genotype is not preserved. Usually some form of sexual reproduction is involved.
   1. **Primordial.** Offspring genotypes differ from parent because mechanisms of genotypic maintenance are poorly developed. Bacterial transduction and transformation are examples.
   2. **Typical.** Haploid gametes must unite to form a diploid zygote.
      (i) **Conservative.** Isogamy of haploid protists, with no polar bodies formed. Sexual reproduction in most protozoans and simple algae fits this category.
      (ii) **Costly.** A 50 percent genetic loss occurs during oogenesis because of polar body formation in the female, and each progeny is only 50 percent related to the parent. Sexual reproduction in most higher organisms—some algae, many plants, and many animals, including humans—fit here.
   3. **Degenerate**
      (i) **Selfing by hermaphrodites.** Examples include some snails, nematodes and parasitic helminths, and Mendel's peas and other automictic plants.
      (ii) **Parthenogenesis.** This mode applies whenever the maternal genotype is not preserved. Automictic or meiotic thelytoky is common in plants, insects, and nematodes.
B. **Nonrecombinational reproduction.** The maternal genotype is preserved because mitosis, not meiosis, is the form of cell division in reproduction. Asexual reproduction in involved.
   1. **Primitive.** This form of asexual reproduction is always fissile or vegetative. Examples are the mitotic fission of protists and vegetative reproduction in sponges.
   2. **Derived**
      (i) **Vegetative.** Clonal development from one genotype by mitosis, including production of ramets in plants, polyembryony in higher animals, and fission in some worm groups.
      (ii) **Amictogametic.** Apomictic or mitotic production of eggs or spores, such as mitotic thelytoky, arrhenotoky, deuterotoky, agamospermy, and pseudogamy.
C. **Heterogonic life cycles,** or **cyclical parthenogenesis.** A cycle of recombinational reproduction followed by nonrecombinational reproduction. Examples include a wide range of organisms: the cycles of aphids and gall wasps; complex life cycles in parasitic helminths and rust fungi; and the gametophyte-sporophyte cycle in mosses, ferns, liverworts, and lycopods.

typical mode of reproduction, including **primordial** and **conservative** recombinational reproduction, **selfing** or **automixis,** and **parthenogenesis,** become more frequent.

Within the class of nonrecombinational reproduction, Williams again noted a division between the so-called **primitive** modes of duplication through fission and the **derived** modes in higher plants and animals. Derived reproduction includes **vegetative** growth of clonal plants and multiplication of zygotes by mitosis in animals, known as polyembryony. Any form of reproduction that has degenerated from typical sexual reproduction to the use of mitosis for egg and embryo production fits into the derived category of nonrecombinational reproduction. Table 14-1 aggregates these forms as amictogametic, meaning without the mixing of gametes.

This classification of modes of reproduction may seem complex, for it attempts to encompass the range of modes seen in all the kingdoms of life on earth. It should certainly cause us to pause when we use terms such as "normal sexual reproduction" or think of ourselves as employing "typical sexual reproduction."

14.3 THE ADAPTIVE NATURE OF MODES OF REPRODUCTION

Sexual Reproduction

The advantages of typical sexual reproduction, as we have called it in Table 14-1, are that recombination constantly produces new genotypes and thus conditions in a changing environment can be tracked rather

---

TABLE 14-2 Definitions of some modes of reproduction

Parthenogenesis: Unisexual reproduction by a female from an egg, without male gametes.

Arrhenotoky: The parthenogenetic production of haploid males by a female, from unfertilized eggs. Example: male Hymenoptera.

Deuterotoky: The parthenogenetic production by a female of either male or female diploid progeny from unfertilized eggs. Example: the sexual generation of aphids.

Haplodiploidy: The parthenogenetic production of haploid males from unfertilized eggs by a female (i.e., arrhenotoky), coupled with the production of diploid females from fertilized eggs, often by the same female. Examples: ants, bees, wasps, and other members of the insect order Hymenoptera.

Thelytoky: The parthenogenetic production by a female of diploid female progeny from unfertilized eggs. Two forms are possible:

Apomictic thelytoky: Production of diploid females without meiosis and therefore without (*apo*) mixing (*mixis*) of chromosomal strands during crossing over. Also called **agamospermy** and **mitotic thelytoky**. Examples: some hawkweeds and dandelions (cf. Chapter 5).

Automictic thelytoky: Production of diploid females with meiosis involved, but two haploid cells soon fuse to form a diploid cell. Fusing haploid cells may include the egg nucleus and the second polar nucleus, two polar nuclei, or two cleavage nuclei. The genetic similarity between parent and offspring differs, depending on the types of nuclei that fuse. Example: some parasitic nematodes.

Pseudogamy: Contact with a sperm or pollen cell stimulates the production of progeny from an unfertilized egg after cleavage, without incorporation of male genetic material into the zygote. Example: some nematodes, fishes, and plants.

FIGURE 14-1

A comparison among life cycles of parasitic species that alternate between two host species and between sexual and asexual episodes of reproduction. The examples include a liver fluke, a gall aphid, a gall adelgid, and a rust fungus. The rust fungus life cycle is illustrated at the beginning of this chapter. Asterisks denote transmission phases from one host to another. The reproductive method is indicated as either asexual (asex) or sexual (sex), and the tenure on or in a host appears below each life cycle. The nomenclature for the life-cycle stages retains the typical terms for each taxonomic group. Each cycle starts with a sexually produced egg or spore and ends with the same stage. *(From Price 1992)*

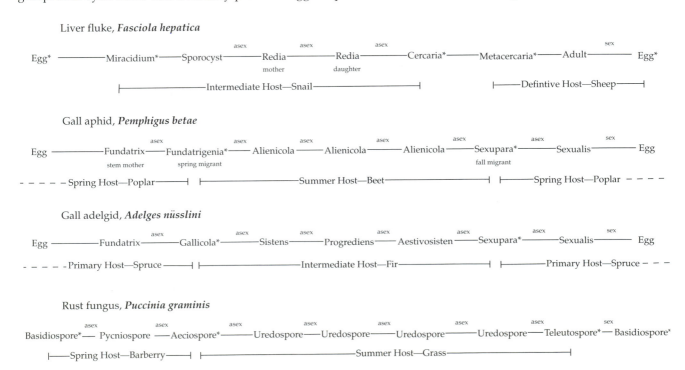

closely. With an endless series of new genotypic "experiments," exploitation of new resources may also occur. The store of new genotypes from recombination may be almost limitless in a moderately large population. There are some disadvantages to sexual reproduction, however.

1. Two individuals, a male and a female, are required for reproduction, and so population structure must evolve to maintain a high probability of meetings between opposite-sex individuals.

2. The remarkable anomaly in sexual reproduction is that r, population growth rate from a single female, is half that of an asexual female if investment in each sex is equal. A female that produces only females produces twice as many females as a sexual female that has offspring in a 50:50 sex ratio. Therefore, population growth rate in the asexual species is twice that in the sexual species. Scientists engage in a major discussion on why sex is so beneficial that it somehow compensates for loss in population growth rate (Williams 1975, Maynard Smith 1978, Margulis and Sagan 1986, Michod and Levin 1988). We will return to the evolution of sex later in this chapter.

3. The adaptiveness of new genotypes produced through typical sexual reproduction is uncertain, particularly since progeny are likely to become widely dispersed. Thus, we should expect high mortality rates due to intense selection pressure, but evolution may be rapid as a result.

4. Recombination disrupts high-fitness genotypes in every generation. Williams (1975) stressed, as the key to the adaptive advantage of sex, strong selection pressure coupled with the production (through recombination) of a small number of genotypes with very high fitness (Figure 14.2).

Degenerate Forms of Reproduction

Degenerate forms of recombinational reproduction have fewer apparent disadvantages than typical sexual reproduction. Hermaphrodites can exist at lower densities than dioecious organisms, because every member of the population is a potential mate. In addition, hermaphrodites may be self-fertile, and thus one individual can found a new colony; or, if inbreeding is tolerated, one individual can save a lineage from extinction. Such an automictic individual may even be the progenitor for a new species.

Parthogenesis

Automictic Thelytoky

The next logical step in this trend is for parthenogenesis with the maintenance of meiosis—automictic thelytoky—to evolve. In this situation a single female

FIGURE 14-2

Williams's (1975) general model, showing the advantage of sexual reproduction. Note that sexual reproduction produces higher fitness than does asexual reproduction, even though mean fitness is higher in the latter. That is because sexual reproduction generates rare genotypes with higher fitness than that achieved in asexual reproduction. (*After Williams 1975*)

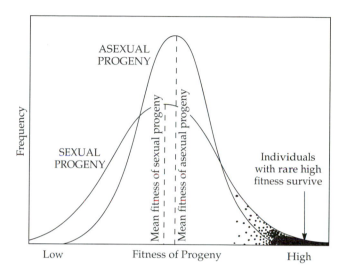

can always found a new colony, and inbreeding depression is not a problem because females do not mate. Because two similar cells may combine to form the zygote, all individuals in a population are likely to be homozygous. Progeny cannot differ substantially from the parental type, which is presumably adequately adapted to local conditions. Therefore, mortality because of maladapted phenotypes is likely to be relatively low, and natural selection locally mild. In a changing environment, however, the rate of adaptive shifts in the genotype may be too slow to prevent rapid loss of fitness.

Apomictic Thelytoky

In apomictic thelytoky no meiosis occurs, and since similar cells do not fuse to form a zygote, any heterozygosity that exists will remain in subsequent generations. Any mutations that occur with zero or positive selective advantage increase this heterozygosity, and theoretically a clone can develop in which every locus is heterozygous. Individuals may be even more genetically diverse than in sexually reproducing species. Again, evolution depends on mutation rates, which are low, but given stable conditions, a well-adapted genotype can be duplicated rapidly and available resources preempted rapidly. Agamospermy in plants results in duplication of the maternal genotype, but the use of seeds has advantages of dispersal and dormancy that other cloning methods generally lack. Agamospermy enables an escape from the sterility resulting from amphiploidy and from odd polyploids such as triploids. This reproductive method preserves particularly adaptive heterozygous genotypes and can replicate them rapidly with seed production (Grant 1971).

Haplodiploidy

In the haplodiploid arrhenotokous system, the haploid males have all genes expressed in the phenotype; thus, all come under the force of natural selection if they are not selectively neutral. Recessive genes cannot act as a store of unexpressed variability. Recombination does occur, however; and, although much less genetic polymorphism may be present in a population (but see Sheppard and Heydon 1986), evolution can be extremely rapid since all deleterious recessives are rapidly selected out of the population.

Polyploidy

Parthenogenesis, particularly thelytoky, makes the evolution of new species of organisms through polyploidy particularly likely. Polyploidy, in its simplest form, results from a diploid automictic thelytokous organism, presumably through a malfunction of the meiotic process, and so the reduction division does not occur, and two diploid cells fuse to form tetraploid progeny. In apomictic thelytoky, two diploid cells may fuse to form a tetraploid zygote. Tetraploids may then produce further polyploids. Thus, polyploid animals are usually found among parthenogenetic typological species such as weevils in the genus *Otiorrhynchus*. In bisexual organisms, new polyploid cells have an extra set of chromsomes without an equivalent set of homologous chromosomes with which to pair, and the sex determination mechanism and meiotic divisions are likely to be disrupted.

Polyploids among parthenogenetic animals and plants are particularly resistant to genetic change and thus very slow to evolve. In tetraploids, for example, a mutation at a certain locus is balanced by three other alleles, which may all be of the original type. As a result, mutations are much less likely to be expressed, and the phenotypic variability in populations is likely to be low. Nevertheless, Suomalainen (1962) found considerable genetic variation in parthenogenetic polyploid weevils in the genus *Otiorrhynchus*.

The question of why polyploidy has become so common in plants certainly demands discussion. Grant (1971) addressed several factors. Amphiploids preserve the benefits of hybrid vigor and physiological homeostasis found in hybrid stock. In addition, a highly heterozygous genotype and the potential for biparental reproduction are maintained. Understandably, the vast majority of polyploid plant species are amphiploids. Polyploids are more strongly buffered genetically than diploids, since four alleles instead of two are present in a tetraploid at any one locus. This makes expression of genetic change less likely, but if heterozygosity is truly an adaptation to cope with regulation in a heterogeneous environment, then additional alleles at a locus should be an advantage.

The consequence of these advantages is that polyploid species do seem to be more widely distributed geographically than their diploid parents (Grant 1971). Patterns in ferns studied by Manton (1950) and *Gilias* studied by Grant (1964, 1971) support this generalization.

But Ehrendorfer (1980) has criticized the long-held view that polyploids become more abundant on a latitudinal gradient from the tropics to the Arctic. Since life forms of plants have different frequencies of polyploids, researchers should compare only similar vegetation types. Perennial species in a genus tend to be polyploids, and annual species tend to be diploids, based on work by Müntzing (1936) and Stebbins (1938). Which is cause and which effect is not clear. On the one hand, perennial species more readily enable the establishment of a polyploid, as there is a longer period in which chromosome doubling can occur if a sterile hybrid is the progenitor. On the other hand, longevity seems to correlate with high DNA content in the cell, so that polyploidy could result in the perennial habit (Cavalier-Smith 1978).

Ehrendorfer (1980) makes the generalization that new polyploid species tend to originate under unstable environmental conditions, in successional communities being established by invading species. "Under these conditions the faster and more efficient mobilization of hybrid variability and the better genetic adaptability of new polyploids often made them superior to their immediate diploid ancestors" (58). Thus, disturbed habitats, such as those left by retreating glaciers, are likely to have high proportions of polyploid species. Perhaps a greater proportion of an area is disturbed in an arctic or alpine zone than at lower latitudes and elevations, permitting new colonizations, the coming together and hybridization of new species pairs, and hence many polyploid species.

In more climax-type communities, the opportunities for new polyploids are very restricted, and should they evolve, their areas and habitats are often quite narrow in comparison with those of their ancestors (Ehrendorfer 1980).

Thus, in summary, polyploidy is common in plants mainly because of three factors (Grant 1971), which set plants apart from animals.

1. Long-lived organisms, usually with means of vegetative propagation, provide much time for genomes to adjust to the new chromosome duplication and become reproductively functional after a change in ploidy.
2. Polyploidy commonly develops after chromosome repatterning during primary speciation; since this is a common form of speciation, polyploidy is common.
3. Natural interspecific hybrids are common in plants and are frequently stabilized by amphiploidy.

Cyclical Parthogenesis

The last mode of reproduction we need to discuss is the cycle of sexual and asexual reproduction, which combines the best of both worlds (Figure 14-1). Asexual reproduction accomplishes rapid reproduction in favorable habitats; then, when conditions become unfavorable, sexual reproduction takes over. Recombination ensures highly diverse progeny, and these sexually produced propagules disperse widely. Those that settle in favorable places multiply asexually. Williams (1975) called this strategy the aphid–rotifer model, as both types of organisms are good examples of such heterogenic life cycles.

It should be clear that different modes of reproduction produce species or populations with very different evolutionary possibilities. Environmental factors—the ecological theater in which a population finds itself—determine which mode is favored. Thus, whenever we think of a certain set of ecological conditions, we should try to ascertain which aspects of the genetic system are most adaptive. How do genetic systems of arctic versus tropical organisms differ? Are the genetic systems of parasites different from those of predators (see Price 1980)? Under what conditions does strong selection for parthenogenesis occur?

14.4 GENETIC SYSTEMS AMONG PARASITES—AN EXAMPLE

Parasitism will serve as an example of an exploration into these kinds of questions. Parasitism is a particularly common way of exploiting food resources; perhaps 50 percent of metazoans are parasitic. Therefore, although the following example on genetic systems does not apply to all parasites, it is relevant to a large number of species.

Parthenogenesis is common among parasitic species (Price 1980). Why is this linkage evident, and what is the evolutionary potential of this restrictive genetic system? White (1978, 317) expressed the generally held view that "thelytokous populations are blind alleys in evolution in the long term." But for parasites there are several arguments in favor of parthenogenesis in the longer term, even though parthenogenetic stocks have not resulted in major adaptive radiations.

Reduced Genetic Diversity Among Progeny

The utilization of highly stable and predictable microenvironments provided by the homeostasis of the living host places less selective pressure on the production of a great diversity of genotypes among the progeny.

Preservation of Large Gene Complexes

Particularly adaptive combinations of genes are fixed without danger of disruption (except in the case of mutation). In mitotic parthenogenesis, no recombination occurs, and in meiotic parthenogenesis, crossing over occurs between homozygous chromosomes. This locking of gene combinations may be especially important in parasites, where large banks of genes are likely to be involved with close evolutionary tracking of the host system. Disruption of such a block would generate gross maladaptations, with almost certain lethal results. Another positive feedback would reinforce the parthenogenetic mode (Figure 14-3).

Reduced Evolutionary Rates

In many cases, the host genotype changes relatively slowly because of a long generation time compared with that of the parasite. In the extreme case of a multivoltine parasite on a long-lived tree, the ratio of generations may be 1,000:1. Then a close relative is likely to replace the tree as host. When generation times differ considerably, there must be strong selection to slow down the evolutionary potential, through parthenogenesis, of the parasite. This selection reduces the recombinational load resulting from great genetic diversity of progeny after meiosis, in a uniform and stable environment to which the parent is well adapted.

Reduced Tracking of Temporary Adaptive Optima

In the nonequilibrium state of parasite populations, massive eruptions and dramatic crashes in population size are likely to dominate the scene (Price 1980). Stebbins (1958) and Harper (1977) defined the breeding system that is best adapted to circumvent this genetically disruptive situation. In an outbreeding

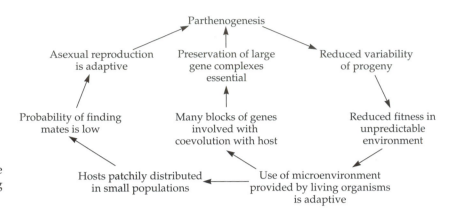

FIGURE 14-3

Positive feedback loops promote the evolution of parthenogenesis among parasitic organisms. *(From Price 1977)*

population, a population flush liberates an enormous array of genetic variability, with most genotypes being unfit for the intervening periods. Harper (1977, 773) described the situation:

> After each explosion the population returns to the "normal" condition less fit, unless in some way the display of variation has been controlled during the colonizing phase. We may perhaps interpret the predominantly inbreeding habit of colonizers and fugitives as a protection against the too rapid "tracking" of adaptive optima that are only transient. . . . Close inbreeding or apomixis are ways in which temporary forces of selection are prevented from throwing the long-term adaptation of a population off course.

This is a very apt explanation for many species with complex life cycles that include a parthenogenetic phase, such as certain parasites: aphids, parasitic helminths, and rust fungi.

Short Generations and High Fecundity

Short generation time and high fecundity result in the production of many individuals per unit of time relative to hosts. The number of mutations in a clone is similarly relatively high and is likely to provide enough genetic variation to permit the clone's longer-term viability (Price 1980). For example, the small insect herbivore *Thrips tabaci* is a common pest of agricultural crops, with many populations reproducing by thelytokous parthenogenesis. Typically, generation time from egg to adult is 14 days, mean fecundity is 80 eggs per female (the range is 12 to 109), and in warm climates at least ten generations may occur in one year. In one season in agricultural fields, populations can reach 3,600 nymphs per plant and 37 to 74 million per hectare (Price 1980).

Thrips tabaci is about the size of a *Drosophila melanogaster* individual or perhaps smaller. Presumably, it can be expected to have a similar number of genes and similar mutation rates. For *Drosophila* the esti-

mate is 10,000 pairs of genes per individual and a mean mutation rate per gene per generation of 10^{-5}. Therefore, each zygote contains 0.2 mutations on average, and (based on typical mortality and fecundity estimates) a female may leave 1.3×10^8 progeny in ten generations. These progeny are likely to contain, in the tenth generation alone, 2.6×10^7 new mutations on which natural selection can act. In addition, survivors carry mutations remaining from previous generations—about 3.8×10^6 gene mutations in all—on which natural selection will have acted. This amount of genetic variation occurs in one season, whereas the host plants, such as onions, have generation times of perhaps 2 to 3 years in nature. What is more, if mutations occurred at random, several hundred progeny in the tenth generation would have five to seven new mutations each! Numerous tenth-generation progeny would have no mutations. Possibly, this mix of some individuals with several mutations and the majority with the maternal genotype is an ideal genetic system, where numbers and generations outnumber the host plant by several to many orders of magnitude.

Replication of Maternal Environments

Perhaps small, highly specialized organisms with narrow tolerances can reach sites with conditions very similar to the maternal environment, even after a colonizing event. There may be less of a premium on major genotypic reconstructions in such specific organisms than in free-living species exposed to a more variable environment.

Progenesis

Parthenogenesis facilitates the evolution of progenesis (Gould 1977). Progenesis is the retention of formerly juvenile characteristics by adult descendants, produced by the precocious sexual maturation of an organism still in a morphologically juvenile state

(Gould 1977). Parthenogenetic species can readily adapt through progenesis because the many adult characteristics associated with mating and copulation are unnecessary, and thus juvenile forms can reproduce effectively. The adaptive feature of progenesis is that rapid maturation is possible and therefore the intrinsic rate of natural increase can be very high. For example, the cecidomyiid, *Mycophila speyeri,* undergoes one larval molt, reproduces as a true larva, and produces 38 offspring in a cycle of only five days. The sexual adults result from two larval molts and one pupal molt and require 2 weeks to develop. Populations of *Mycophila* invading mushroom beds may increase exponentially for 5 weeks and, as a result of progenetic reproduction, reach densities of 20,000 reproductive larvae per 30 centimeters square.

Gould (1977) identified the conditions under which progenesis is highly adaptive: in unstable environments, where colonization is frequent, where small size is adaptive, and in parasitic species. Of course, the first three conditions are common to the parasitic way of life.

Relative Fitness of Parthenogenetic Parasites

How much fitness is lost in parasitic species that are parthenogenetic, relative to free-living organisms that retain sexual reproduction? It seems that copying the maternal genotype is a safe strategy for parasites in many cases, and new mutations may slightly increase fitness in an already well-adapted lineage. Therefore, we may modify Williams's (1975) model explaining the advantages of sexual reproduction (Figure 14-2) for parthenogenetic parasites as in Figure 14-4. This argument for the evolution of parthenogenesis in parasites indicates the kinds of considerations we should apply to the understanding of the adaptive nature of genetic systems.

14.5 WHY DID SEX EVOLVE?

If parthenogenesis works so well for many organisms, why did sex evolve? Is it simply that larger organisms, with longer generation times, can persist only with the sexual system that reconstitutes genetic diversity during each sexually reproductive phase? This cannot be an adequate explanation, because the predominant form of reproduction in even small organisms is sexual. No extensive adaptive radiation has occurred in metazoans or plants from a parthenogenetic parental stock (except perhaps the bdelloid rotifers). So what is the overpowering selective pressure for sexual reproduction?

The most viable arguments about the evolution of sex follow. The first three emphasize the rapid genera-

FIGURE 14-4

Williams's (1975) general model, modified to compare possible relationships between the fitness range of sexual progeny in a variable environment and that of parasitic asexual progeny in the highly predictable environment of the living host. *(From Price 1980)*

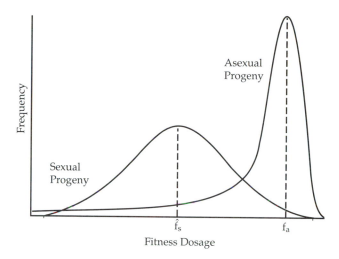

tion of new genotypes, resulting in the potential for rapid evolution. They dwell on short-term gains in fitness as a necessary buffer against extinction. The last two arguments concentrate on purging, repairing, or masking the genetic system when defects occur. They point out the long-term benefits of typical sex that would reinforce the short-term gains in sexual compared with asexual reproduction. All these hypotheses may provide parts of the explanation of why sexual reproduction is so common in nature.

The Red Queen's Hypothesis

Lewis Carroll's (1872) *Through the Looking Glass* may contain the answer. Alice and the Red Queen suddenly begin running across the chessboard, and they never seem to get anywhere.

The curious part of the thing was, that the trees and other things round them never changed their places at all: however fast they went, they never seemed to pass anything. "I wonder if all the things move along with us?" thought poor puzzled Alice. And the Queen seemed to guess her thoughts, for she cried "Faster! Don't try to talk!" . . . "Well, in *our* country," said Alice, still panting a little, "you'd generally get somewhere else—if you ran very fast for a long time as we've been doing."

"A slow sort of country!" said the Queen. "Now, *here,* you see, it takes all the running *you* can do, to keep in the same place. If you want to get somewhere else, you must run at least twice as fast as that!"

This passage is analogous to what Van Valen (1973) saw in the fossil record that prompted him to develop the **Red Queen's hypothesis.** Biotic forces, in the forms of predators, parasites, competition from other species, prey, or food plants, provide any organism with an environment that is perpetually changing because species come and go and populations evolve. Therefore, selective pressures are always multiple in space and changing through time. Thus, the genotype in a lineage must change rapidly through time—it must keep "running"—to stay in place in the community and avoid extinction. This means that selection for the reproductive mode that enables the most rapid evolutionary change will be strong. This is sexual reproduction.

Production of Rare Genotypes

Presumably, the environmental biotic components with the highest evolutionary rates and the most deleterious effects—namely, the microbial parasites and the small metazoan parasites on plants and animals—are the strongest selective agents on sexual reproduction. Levin (1975) argued that such parasites select for high rates of recombination, so that new genotypes that are resistant to the parasites are produced rapidly. Jaenike (1978) then argued that the important aspect of sex was the rapid production of novel and rare genotypes. When parasites act as frequency-dependent selective agents, the rarest genotypes are always favored, and parasitism must result in selection for much genetic polymorphism (Clarke 1976). Therefore, to stay in place in a community with rapidly changing selective regimes caused by such organisms as parasites, sex and its consequent rare genotypes produce highly fit individuals—those likely to register at the lower right in the normal distribution of fitness dosage (Figure 14-2).

Hamilton (1980) modeled the case of frequency-dependent selection by parasites in which the most common genotype is the most vulnerable. In a two-locus haploid selection model, when selection intensity is strong, a sexual species has an advantage over any asexual strain. This occurs even when fecundity is assumed to be twice as great in the asexual strains as in the sexual species. Thus, strong selection for recombination can overcome the loss in r (defined in Section 14.3) due to sex. In such a case, asexual strains die out or remain at very low frequencies.

The Pathogen Ratchet, or Rice's Ratchet

Rice (1983a, 1983b) emphasized the importance of sexual reproduction in the reduction of parent–offspring transmission of parasites. In asexual species, progeny are very similar to the parent in phenotype. Therefore, pathogens of the parent are preadapted to successfully invade the progeny. In sexual species, the progeny most different from the parent are favored, for they are least likely to be infected by the parent's parasites. Thus, in the short term the rare genotype among progeny, if also rare in the population, becomes inaccessible to many parasites tracking the lineage or present in the local population. Without sex, a lineage would accumulate well-adapted parasites, and fitness would decline. This hypothesis is known as the **pathogen ratchet** or **Rice's ratchet.**

Muller's Ratchet

Other hypotheses on the evolution of sex emphasize the importance of concealing defective DNA or purging the lineage of it. Muller (1964) advanced the hypothesis that is now known as **Muller's ratchet.** Harmful mutations occur much more frequently than beneficial ones, and they tend to accumulate in a lineage whether it is sexual or asexual. The **mutational load** thus increases. Selection acts against deleterious mutations, but mutations are generated anew and recessive mutations persist; they establish a mutation selection equilibrium.

Muller argued that the mutational load in an asexual lineage cannot decrease below that in the least-loaded clones, but the load can increase in all clones. Thus, every new deleterious mutation in a lineage clicks the ratchet one more notch toward oblivion as the mutational load increases. By contrast, in sexual species, a lineage can purge all or a large part of the mutational load, because recombination produces some progeny with a reduced load or no load (Figure 14-5). Thus, whereas in asexual lineages mutational load inevitably increases, deleterious mutations are purged from lineages using recombination (see also Heller and Maynard Smith 1979, Leslie and Vrijenhoek 1980).

Repair of DNA Damage and Masking Mutation

Bernstein et al. (1985) developed another argument. They argued that the most important problems in replication of genetic information are DNA damage and mutation. The advantages of sex are that recombination enables repair of damage and outcrossing masks deleterious mutations. The genetic variation resulting from sexual reproduction is an effect of these processes, not the primary object of selection.

Enzymes for repair of DNA damage are known to exist, and when one strand of DNA is damaged, the undamaged strand serves as a template for the replacement of the excised DNA. However, double-stranded damage is common, and no template is

FIGURE 14-5

The process in which a sexual lineage purges deleterious mutations, leaving some progeny free of mutational load. One pair of homologous chromosomes appears in each diploid cell, and an x marks a deleterious mutation.

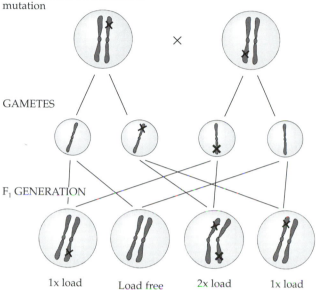

PARENTAL GENERATION
All members have deleterious mutation

GAMETES

F₁ GENERATION

1x load   Load free   2x load   1x load

available for it. Such damage can be repaired only after recombination, when undamaged DNA becomes available as a template. So long as meiosis is preserved, recombinational repair occurs. But in vegetative and apomictic lineages, double-stranded damage may well increase.

Outcrossing also has the advantage of masking deleterious mutations. Automictic and selfing organisms generate homozygosity, and so masking ability is low. In apomictic and vegetative organisms, mask-

ing is high but no recombinational repair occurs. Only in outcrossing species is masking ability generally good (intermediate) and recombinational repair available (Table 14-3). The conclusion is that the primary advantages of sex with outcrossing are that DNA repair and masking of mutations are both available.

Together, the advantages of sex emphasized by Williams (1975, Van Valen (1973), Levin (1975), Jaenike (1978), Hamilton (1980), Rice (1983a, 1983b), Muller (1964), and Bernstein et al. (1985) make up an impressive list. Developments in this field will be interesting to follow. A solution to one of the fundamental problems in biology may be near (see Margulis and Sagan 1986, Michod and Levin 1988).

14.6 CHROMOSOMAL EVOLUTION AND POPULATION STRUCTURE

Even though sex and outcrossing are so common, it may come as a surprise that evolutionary rates can differ greatly depending on population structure, another aspect of the genetic systems of organisms. Wilson et al. (1975) found that chromosomal evolution was much more rapid in placental mammals than in other vertebrates and in molluscs (Table 14-4). This finding corresponded with the knowledge that placental mammals have evolved rapidly in anatomy and lifestyle (Simpson 1953, Romer 1966). It also raised the questions of how chromosomal evolution can be rapid in placental mammals and how chromosomal rearrangements can become fixed rapidly relative to other vertebrates. As Bush (1975) argued in his review of modes of speciation in animals, it is the social structure of placental mammals that binds small groups of interbreeding animals together, thereby probably facilitating rapid fixation of chromosomal rearrangements.

TABLE 14-3   Classification of reproductive systems in diploid organisms

System	Masking Ability	Recombinational Repair	Net Fitness[1]
Automixis	Low	Yes	− +
Selfing	Low	Yes	− +
Outcrossing	Intermediate	Yes	+ +
Apomixis	High	No	+ −
Vegetative	High	No	+ −

Based on Bernstein et al. 1985.

[1] Net fitness is given as a combination of masking ability (− for low and + for high) and recombination repair (− for no and + for yes). Only in the outcrossing reproductive system do two + signs appear in the net fitness column.

T A B L E   1 4 - 4   Rates of karyotypic evolution in polytypic genera of vertebrates and molluscs

Group	Average Age of Genera in Millions of Years	Karyotypic Changes per Lineage per 10^8 Years	
		Arm No.	Chromosome No.
Placental mammals			
Rodents	4.6	9.6	8.2
Primates	4.4	7.1	7.1
Ungulates	4.3	4.3	7.2
Carnivores	11.6	3.1	1.4
Bats	10.7	2.1	1.2
Whales	6.3	1.7	0
Other vertebrates			
Marsupials	1.9	1.3	0
Snakes	12.4	2.1	0.5
Lizards	23.0	1.3	1.1
Frogs	16.7	0.8	1.0
Teleost fishes	18.8	1.5	1.1
Molluscs			
Prosobranch snails	64.7	—	0.3
Other snails	49.0	—	0.4
Bivalves	77.0	—	0.1

From Wilson et al. 1975.

Social Structuring of Populations

Breeding individuals in a social unit frequently number ten or less (Wilson 1975), and even in highly mobile placental mammals such as horses, fixation can be rapid. In the genus *Equus*, chromosome numbers range, in a stepwise cline, from $2n = 32$ in *E. zebra* in southwestern Africa to $2n = 66$ in *E. przewalskii* in Eurasia. An effective population comprises family groups, each one consisting of a stallion and several mares that live together for life (maybe 20 years or more) once they are in the group. Inbreeding in this situation could fix a chromosomal rearrangement rapidly. Related males and females might be likely to mate with each other during temporary isolation of a family group for two or three generations coupled with loss of a female from the group and replacement by a daughter or sister.

Where such social structuring is lacking, an increase in mobility, often coupled with increase in body size, reduces rates of chromosomal evolution. Whales have the lowest rates of chromosomal evolution of the placental mammals. The only flying mammals, the bats, also have relatively low rates. The smallest placental mammals, the rodents, with their very limited home ranges, show the highest rates of

chromosomal evolution in Table 14-4. Among rodents, speciation can be rapid and occur in relatively confined areas. The tight family ties of the primates no doubt more than compensate for their relatively large bodies and home ranges, leading to rapid chromosomal evolution.

Wilson et al. (1975) underestimated the number of karyotypic changes in a lineage because they dealt only with the number of karyotypes observed and did not estimate the minimum number of chromosomal mutations that must have occurred to produce the observed range of karyotypes. In 1977, Bush et al. published new estimates that took this issue into account; their numbers supported the general pattern but showed horses and primates to be speciating even more rapidly than rodents (Figure 14-6).

Population Structure in Parasitic Wasps

The next question is clear: Is tight social structuring with inbreeding likely to prevail in any categories of organisms other than the vertebrates? Good evidence exists that many parasitic wasps in the order Hymenoptera exhibit extreme forms of inbreeding (e.g., Askew 1968). One example is the chalcidoid *Nasonia*

FIGURE 14-6

The relationship between rate of chromosomal evolution and rate of speciation in major groups of vertebrates. A significant correlation ($r^2 = 0.83$) exists between the two rates. Note that, for the rapidly evolving groups, the rate of chromosomal evolution (0.4 to 1.4 events per lineage per million years) approaches the rate of speciation (1.6 to 2.8 new species per lineage per million years). *(After Bush et al. 1977)*

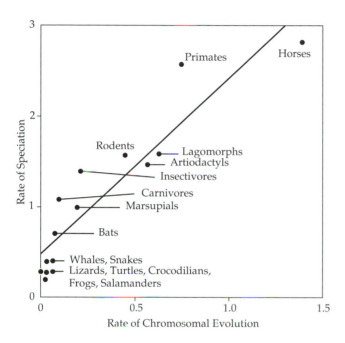

mosomal rearrangements are an important part of speciation events.

Goodpasture and Grissell (1975) studied the karyotypes of nine species of *Torymus* (Hymenoptera: Torymidae), which are parasitoids largely of gall-forming insects. The haploid chromosome number varies from five to six, five evidently being the primitive number. Each species has a unique karyotype; two species have five metacentric chromosomes, one species has four metacentrics and two telocentrics, four species have five metacentrics and one acrocentric, and so on. The pathways along which these karyological changes could occur are quite simple (Figure 14-7).

Considering that Goodpasture and Grissell (1975) calculate that only 0.001 percent of the species of chalcidoids in the world have known karyotypes, any generalizations may be premature. The fact is that the small sampling of parasitic wasps indeed supports a prediction based on patterns in mammals. This is suggestive that the pattern may be very general in species with tight social structuring. We have made the argument (Price 1980) that such population structure is very common among parasites in general, be they plants, animals, fungi, protists, or Monera, because a reproductively competent individual may frequently accomplish colonization of a new patch, and the progeny are most likely to mate with each other in the absence of nonsiblings. In the parasitic nematodes, for example, even apparently single races of a

vitripennis studied by King et al. (1969). The females lay their clutches of eggs in the puparia of flies such as house flies and carrion flies. Being arrhenotokous, these wasps have a sex ratio strongly skewed toward females. Males are brachypterous (short-winged), indicating poor dispersal abilities, and almost always a male emerges first from the host puparium and sets up a territory on its surface. He then proceeds to defend his territory against his few brothers and mates with his many sisters as they emerge from the host. Thus, in the majority of cases, inbreeding is very strong, effective population size is very small, and fixation of point or chromosomal mutations can occur very rapidly—in one or two generations if families are isolated, as is typical of highly specialized organisms utilizing very patchy environments.

Correlated with this social structure are the unquestionably rapid evolutionary rates in chalcidoid Hymenoptera (Askew 1968). Species are exceedingly numerous and seem to be in a very dynamic state. These and related parasitic wasps make up about 26 percent of the insect fauna in Britain. And the only karyological work on the group suggests that chro-

FIGURE 14-7

Some pathways of karyological change for species of *Torymus* parasitoid, postulated by Goodpasture and Grissell (1975). The haploid karyotype is given, with M = metacentric, T = telocentric, and A = acrocentric chromosomes.

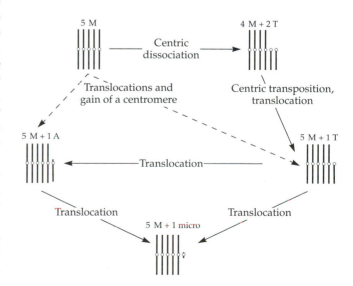

species such as *Meloidogyne hapla,* race A, has chromosome numbers $n = 14, 15, 16, 17,$ and 34 (Triantaphyllou and Hirschmann 1980).

Population Structure in Plants

Levin and Wilson (1976) have identified the same general pattern of high rates of evolutionary change in the chromosomes of seed plants (Figure 14-8). Some of the detailed comparative data are very interesting (Table 14-5). Herbs have evolved much more rapidly than other growth forms, mostly through the development of polyploids, but with aneuploidy contributing about 27 percent to chromosomal change.

Why are herbs evolving so much more rapidly than other growth forms? Levin and Wilson (1976) argued that annual and short-lived perennial herbs occupy patchy, disturbed, and transient habitats. Many species are self-fertilizing and have small populations of 20 individuals or less. This is the population structure that is conducive to chromosomal evolution. Plants living in more permanent habitats, such as long-lived perennial herbs, shrubs, and trees, generally have large populations and are outcrossing, and fixation of new karyotypes is likely to be very slow. This explanation may well account for the pattern seen in Figure 14-8 and Table 14-5.

We should consider many other factors to fully understand the implications of genetic systems and the evolutionary potential of populations and species. The next chapter will expand upon this topic.

FIGURE 14-8

The relationship between rate of chromosomal evolution and rate of speciation in major groups of seed plants. The correlation is highly significant ($r^2 = 0.98$). *(After Levin and Wilson 1976)*

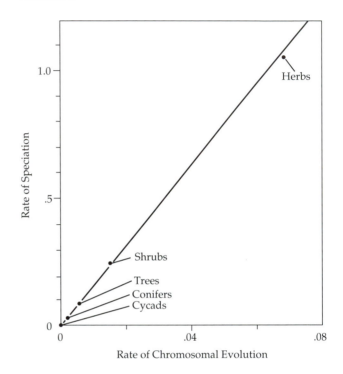

TABLE 14-5   Rates of chromosomal evolution and speciation as functions of growth form in plants

Growth Form	Mean Duration of Species in Millions of Years	True Speciation Rate	Mean Increase in Chromosomal Number Diversity per Lineage per Million Years		
			Total	Polyploid	Aneuploid
Angiosperms					
Herbs	10	1.15	0.073	0.05	0.020
Shrubs	27	0.28	0.010	0.01	0.0005
Hardwood trees	38	0.12	0.0014	0.001	0.0003
Gymnosperms					
Conifers	54	0.04	0.0001	0.00001	0.0001
Cycads	54	0.03	0.0000	0.0000	0.0000

From Levin and Wilson 1976.

SUMMARY

The **genetic system** of a species encompasses all the factors that affect the hereditary behavior of a species and its evolutionary potential. These factors include mode of reproduction, population structure, cytological characteristics, generation time, and many other concerns. This chapter emphasized modes of reproduction, the evolution of sex, and chromosomal evolution in structured populations.

Modes of reproduction are very diverse in nature. They may be divided into the modes that do not preserve the maternal genotype, because of meiosis and recombination during some form of sexual reproduction, and asexual modes in which mitosis conserves the maternal genotype. **Sexual reproduction** maximizes diversity of genotypes among progeny, which is adaptive under unpredictable conditions. **Parthenogenesis** results in many close replicates of the maternal genotype, which may well be adaptive if environments are highly predictable and stable. A parthenogenetic mode of reproduction is adaptive for some parasites—for example, those that live in or on living hosts, with homeostatic mechanisms providing relative stability of environment and many more-or-less replicated environments in the population of hosts.

The puzzle of why sex evolved has received much attention and prompted several alternative hypotheses. The **Red Queen's hypothesis** argues that rapid evolutionary potential is essential in a world of rapidly changing biotic challenges. The rapid production of rare genotypes may enable escape from parasites with high evolutionary potential. Shifts in genotypes of progeny away from the parental genotype because of sexual reproduction may prevent the accumulation of parasite lineages potentially passing from mother to offspring, with the **pathogen ratchet,** or **Rice's ratchet,** clicking off one notch of decreased fitness in a lineage for every new pathogen that colonizes. Sexual reproduction may purge deleterious mutations from a lineage in a similar manner, preventing maladaptive mutational load from increasing; this action is called **Muller's ratchet.** Another conceptual basis for the evolution of sex involves the argument that genetic diversity among progeny is a *result* of selective pressures rather than the primary cause. The primary adaptive advantage of recombination is that it enables repair of damage to DNA, and outcrossing masks deleterious mutations.

Studies on the rate of chromosomal evolution in relation to population structure have indicated in diverse taxa that, when breeding systems involve local and small populations, rapid changes occur. In fact, rates of chromosomal evolution have proved to be good predictors of speciation rates in placental mammals, plants, and perhaps parasitic wasps and other parasitic organisms.

QUESTIONS FOR DISCUSSION

1. To what extent could you argue that the mode of reproduction in a species is more important as an evolutionary factor than natural selection?

2. Given that inbreeding and other forms of reproduction resulting in reduced panmixis are so common in nature, what can this fact suggest about the mechanisms involved in the cohesion species concept?

3. Can you name most of the taxa that contain species with cyclical parthenogenesis, and why do you think this kind of complex life cycle has evolved so many times?

4. Can you discover general patterns associated with the commonness or rarity of degenerate forms of sexual reproduction that take into account perhaps size of organisms, generation times, population sizes, and the major biotic interactions involved with specific taxa?

5. If genetic systems encompass all factors that influence the evolutionary potential of a species, how should phylogenetic constraints enter into consideration of potential evolutionary rates and directions?

6. Although pseudogamy would seem to be a transient state between typical recombinational reproduction and parthenogenesis, it is remarkably well represented in some groups such as nematodes. Under what conditions do you think that pseudogamy could be maintained in a population in an equilibrium state?

7. If a parasite is parthenogenetic, will this mode of reproduction reduce significantly the impact of biotic interactions envisaged in the Red Queen's hypothesis and Rice's ratchet?

8. Has the emphasis on genetic systems such as that in the human species and other large organisms involving "typical" recombinational reproduction been misleading to some extent—for example, in generating the view that inbreeding is inevitably deleterious or that departures from the typical type are unstable over the long term?

9. The hermaphrodite condition, with selfing possible, seems to be a highly adaptive design for reproduction in plants and animals, and yet it is not observed in most terrestrial vertebrates. Can you develop a phylogenetic picture of the frequency of hermaphrodites in plants and animals, and a mechanistic explanation for the pattern?

10. Through the evolution of life in marine and terrestrial environments, can you trace any patterns in modes of reproduction by following particular phylogenetic lineages or any specific modes of life, such as lineages of carnivores, soil-dwelling heterotrophs, parasites, large primary producers, or marine phytoplankton or zooplankton?

REFERENCES

Askew, R. R. 1968. Considerations of speciation in Chalcidoidea (Hymenoptera). Evolution 22:642–645.

Bernstein, H., H. C. Byerly, F. A. Hopf, and R. E. Michod. 1985. Genetic damage, mutation, and the evolution of sex. Science 229:1277–1281.

Bush, G. L. 1975. Modes of animal speciation. Ann. Rev. Ecol. Syst. 6:339–364.

Bush, G. L., S. M. Case, A. C. Wilson, and J. L. Patton. 1977. Rapid speciation and chromosomal evolution in mammals. Proc. Nat. Acad. Sci. U.S.A. 74:3942–3946.

Carroll, L. (Dodgson, C. L.) 1872. Through the Looking Glass, and What Alice Found There. Macmillan, London.

Cavalier-Smith, T. 1978. Nuclear volume control by nucleoskeletal DNA, selection for cell volume and cell growth rate, and the solution of the DNA C-value paradox. J. Cell Sci. 34:247–248.

Clarke, B. 1976. The ecological genetics of host parasite relationships, pp. 87–103. In A. E. R. Taylor and R. Muller (eds.). Genetic Aspects of Host–Parasite Relationships. Blackwell, London.

Ehrendorfer, F. 1980. Polyploidy and distribution, pp. 45–60. In W. H. Lewis (ed.). Polyploidy: Biological Relevance. Plenum, New York.

Goodpasture, C., and E. E. Grissell. 1975. A karyological study of nine species of Torymus (Hymenoptera: Torymidae). Can. J. Genet. Cytol. 17:413–422.

Gould, S. J. 1977. Ontogeny and Phylogeny. Belknap Press of Harvard Univ. Press, Cambridge, Mass.

Grant, V. 1964. The biological composition of a taxonomic species in Gilia. Adv. Genet. 12:281–328.

———. 1971. Plant speciation. Columbia Univ. Press, New York.

Hamilton, W. D. 1980. Sex versus non-sex versus parasite. Oikos 35:282–290.

Harper, J. L. 1977. Population Biology of Plants. Academic Press, London.

Heller, R., and J. Maynard Smith. 1979. Does Muller's ratchet work with selfing? Genet. Res. 32:289–293.

Jaenike, J. 1978. An hypothesis to account for the maintenance of sex within populations. Evol. Theor. 3:191–194.

King, P. E., R. R. Askew, and C. Sanger. 1969. The detection of parasitized hosts by males of Nasonia vitripennis (Walker) (Hymenoptera:Pteromalidae) and some possible implications. Proc. Royal Entomol. Soc. London A. 44:85–90.

Leslie, J. F., and R. C. Vrijenhoek. 1980. Consideration of Muller's ratchet mechanism through studies of genetic linkage and genomic compatibilities in clonally reproducing Poeciliopsis. Evolution 34:1105–1115.

Levin, D. A. 1975. Pest pressure and recombination systems in plants. Amer. Natur. 109:437–451.

Levin, D. A., and A. C. Wilson. 1976. Rates of evolution in seed plants: Net increase in diversity of chromosome numbers and species numbers through time. Proc. Nat. Acad. Sci. U.S.A. 73:2086–2090.

Manton, I. 1950. Problems of cytology and evolution in the Pteridophyta. Cambridge Univ. Press, Cambridge, England.

Margulis, L., and D. Sagan. 1986. Origins of sex: Three billion years of genetic recombination. Yale Univ. Press, New Haven, Conn.

Maynard Smith, J. 1978. The Evolution of Sex. Cambridge Univ. Press, Cambridge, England.

Mayr, E. 1963. Animal Species and Evolution. Belknap Press of Harvard Univ. Press, Cambridge, Mass.

Michod, R. E., and B. R. Levin. 1988. The Evolution of Sex: An Examination of Current Ideas. Sinauer Associates, Sunderland, Mass.

Muller, H. J. 1964. The relation of recombination to mutational advance. Mutation Res. 1:2–9.

Müntzing, A. 1936. The evolutionary significance of autopolyploidy. Hereditas 21:263–378.

Price, P. W. 1977. General concepts on the evolutionary biology of parasites. Evolution 31:405–420.

———. 1980. Evolutionary Biology of Parasites. Princeton Univ. Press, Princeton, N.J.

———. 1992. Evolutionary perspectives on host plants and their parasites. Adv. Plant Pathol. 8:1–30.

Rice, W. R. 1983a. Sexual reproduction: An adaptation reducing parent–offspring contagion. Evolution 37:1317–1320.

———. 1983b. Parent–offspring pathogen transmission: A selective agent promoting sexual reproduction. Amer. Natur. 121:187–203.

Romer, A. S. 1966. Vertebrate paleontology. 3d ed. Univ. of Chicago Press.

Sheppard, W. S., and S. L. Heydon. 1986. High levels of genetic variability in three male-haploid species (Hymenoptera: Argidae, Tenthredinidae). Evolution 40:1350–1353.

Simpson, G. G. 1953. The Major Features of Evolution. Columbia Univ. Press, New York.

Stebbins, G. L. 1938. Cytological characteristics associated with the different growth habits in the dicotyledons. Amer. J. Bot. 25:189–198.

Stebbins, G. L. 1958. Longevity, habitat and release of genetic variability in the higher plants. Cold Spring Harbor Symp. Quant. Biol. 23:365–378.

Suomalainen, E. 1962. Significance of parthenogenesis in the evolution of insects. Ann. Rev. Entomol. 7:349–366.

Triantaphyllou, A. C., and H. Hirschmann. 1980. Cytogenetics and morphology in relation to evolution and speciation of plant-parasitic nematodes. Ann. Rev. Phytopathol. 18:333–359.

Van Valen, L. 1973. A new evolutionary law. Evol. Theor. 1:1–30.

White, M. J. D. 1954. Animal Cytology and Evolution. 2d ed. Cambridge Univ. Press, London.

———. 1973. Animal Cytology and Evolution. 3d ed. Cambridge Univ. Press, London.

———. 1978. Modes of Speciation. W. H. Freeman, San Francisco.

Williams, G. C. 1975. Sex and Evolution. Princeton Univ. Press, Princeton, N.J.

Wilson, A. C., G. L. Bush, S. M. Case, and M.-C. King. 1975. Social structuring of mammalian populations and rate of chromosomal evolution. Proc. Nat. Acad. Sci. U.S.A. 72:5061–5065.

Wilson, E. O. 1975. Sociobiology: The new synthesis. Belknap Press of Harvard Univ. Press, Cambridge, Mass.

15

CHANGE IN GENE FREQUENCIES

A naked mole rat—representative of a group with very high genetic similarity among species.

In Chapter 5, "The Origin of New Species," we discussed the ways in which gene pools become fractionated and may eventually become new species. We dealt qualitatively with long versus short periods of time, rapid versus slow evolutionary rates, little or much gene flow, and the way in which genetic drift and founder effects result in subsampling of the gene pool. Chapter 14, "Genetic Systems," explored how modes of reproduction influence the genetic systems of organisms. It would be valuable to address all these areas in quantitative terms so that we know how long a "long time" is, how rapidly differences between populations can accumulate, and how much gene flow must diminish before divergence of populations is possible. The science of **population genetics** can help us to answer some of these questions. Also, population genetics can answer such questions as how recurrent mutation affects evolutionary rates, how rapidly populations diverge under different degrees of selection, and how population size influences rates of fixation of new gene arrangements.

15.1 HOW MUCH DO SPECIES DIFFER GENETICALLY?

Before we can approach genetic questions, we need to know the degrees of difference between closely related species in nature. Unless we have this information, we cannot anticipate how much divergence between populations is necessary for populations to become new species. Studies in population genetics can give us some idea of the genetic difference between closely related species, and studies in ecology sometimes provide empirical estimates of how much two species must differ in morphology, or physiology, or other ecologically relevant variables in order to permit coexistence.

Scientists can use electrophoresis to measure genetic differences between individuals, populations, and species (see Selander 1976). In electrophoresis, proteins in extracts of organisms or in whole organisms, placed in an electric field, migrate in a supporting medium such as a starch gel. Proteins with different net electrostatic charge or molecule size migrate at different rates. After electrophoresis, the researcher places the gel in specific histochemical stains to determine the positions of enzymes and other proteins in the gel.

Genetic Variation Within Populations

The geneticist examines protein products from several loci in an organism and, by sampling representative individuals in a population, obtains population estimates of genetic variation. Early (1960s) results indicated surprisingly high levels of polymorphism in proteins in individuals and populations. Lewontin and Hubby (1966) found that an average population of *Drosophila pseudoobscura* was polymorphic for 30 percent of all loci tested and that an average individual was heterozygous for 12 percent of its observable loci. They were able to make these observations despite the fact that standard methods of electrophoresis identify only about 25 percent of protein morphs because proteins may differ even when they have the same net charge and very similar molecular sizes (Lewontin 1974). Since the first results of Lewontin and Hubby, electrophoresis has become a major tool of population geneticists and evolutionists. Selander (1976) considered that, in many plant and animal organisms, populations are polymorphic at 25 to 50 percent of their loci, and individuals are heterozygous at 5 to 15 percent of their loci (as we will discuss in more detail in Chapter 16).

Genetic Variation Between Populations

Using allele frequencies acquired by individual sampling of members of populations, it is possible to compare the genetic similarity between populations and species. Nei (1972) proposed a formula for calculating **genetic identity** of genes between two populations at the *j* locus,

$$I_j = \frac{\Sigma x_i y_i}{(\Sigma x_i^2 \, \Sigma y_i^2)^{1/2}}$$

where x_i and y_i represent the frequencies of the *i*th allele in populations X and Y, respectively (see also Avise 1976). For all loci in a sample, the overall genetic identity of X and Y is defined as

$$\text{Overall genetic identity} = \bar{I} = \frac{Jxy}{(Jx \, Jy)^{1/2}}$$

where Jx, Jy, and Jxy are the arithmetic means over all loci of Σx_i^2, Σy_i^2 and $\Sigma x_i \Sigma y_i$, respectively.

$$\text{Genetic distance } D = -\ln \bar{I}$$

Also, if these identities and distances change at a constant or roughly constant rate, we have, in effect, a protein clock in which I and D correlate with time. The clock must be calibrated from the fossil record or by inference from existing distributions (see Chapter 17).

Genetic Variation During Speciation

Ayala et al. (1974) used the electrophoretic method to calculate the genetic identities between (i) populations of the same species of *Drosophila*, (ii) subspecies, (iii) semispecies, (iv) sibling species, and (v) nonsibling species, all in the *willistoni* group of *Drosophila* in

Venezuela. Speciation in this group is by geographic isolation. In the first stage of speciation, populations in geographic isolation accumulate genetic differences and become *subspecies*. In the second stage, secondary contact hastens development of reproductive isolation through natural selection (the Wallace effect) and, until isolation is complete, populations are called *semispecies*.

Geographic populations of the same species proved very similar genetically, having a mean genetic identity of $\bar{I} = 0.970$ (Figure 15-1). Clearly, gene flow was effectively neutralizing any genetic differences that appeared in one population and not in another, or natural selection was keeping genetic identities high. After accumulation of genetic differences during geographic isolation, subspecies had $\bar{I} = 0.795$. For semispecies, $\bar{I} = 0.798$, indicating very little genetic change through natural selection during secondary contact.

Considering that some of the semispecies were regarded as practically reproductively isolated, $\bar{I} = 0.8$ would appear to be a maximum identity ($\equiv$ minimum difference) that is likely in geographically speciated gene pools. Note that very few loci (about 4 percent) have gone to fixation, meaning that at a locus alleles are completely different and $I = 0$ for

FIGURE 15-1

Genetic similarity of loci in (a) geographic populations, (b) subspecies, (c) semispecies, (d) sibling species, and (e) nonsibling species in the *Drosophila willistoni* group in Venezuela. A genetic similarity of 1 means that one locus has identical kinds and proportions of alleles in different populations. A genetic similarity of 0 means that at one locus the alleles are totally different. *(After Ayala et al. 1974)*

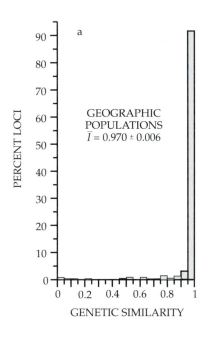

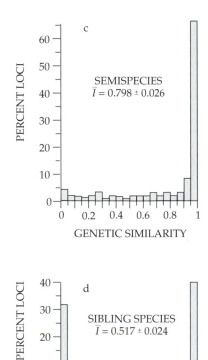

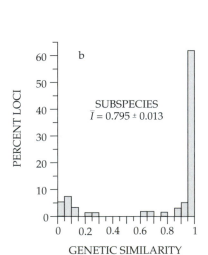

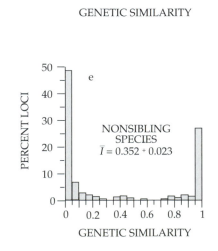

that locus (Figure 15-1). Therefore, it is rather surprising that sibling species, all of which are reproductively isolated, have much lower genetic identities than the semispecies ($\bar{I}$ = 0.517, with 32 percent of loci completely different).

Finally, closely related nonsibling species had $\bar{I}$ = 0.352, and 48 percent of loci had I = 0. Thus, we have empirical measures of the degree of difference between species when allopatric speciation occurs.

Genetic Differences Resulting from Sympatric Speciation

We should keep in mind that if sympatric speciation involving a host shift occurs, theoretically at least, we should expect perhaps only two loci to differ initially, with I = 0 for each locus and a few more modifier genes differing as adaptation to the new host progresses (cf. Chapter 5). In fact, Berlocher (1976) found genetic identities of almost 1. For example, $\bar{I}$ = 0.980 between *Rhagoletis pomonella* and *R.* new species A, which probably arose from a host shift by *R. pomonella* flies from *Crataegus* to *Cornus florida* in Mexico. Also in the *pomonella* group of *Rhagoletis*, *R. mendax* and *R.* new species B have a genetic identity of 0.989. These identities between species are even greater than the similarities Ayala et al. (1974) found in populations of the same species of *Drosophila*.

In the *suavis* group of *Rhagoletis* flies, speciation has probably been allopatric because host shifts have not been involved; all hosts are walnuts (*Juglans* spp.). Berlocher (1976) measured the following genetic identities (see also Berlocher and Bush [1982] for genetic distances): *R. suavis* and *R. completa*, 0.861; *R. juglandis* and *R. completa*, 0.718; and *R. juglandis* and *R. boycei*, 0.630. (*Suavis* and *completa* overlap in Illinois and Missouri; *juglandis* and *completa* do not overlap except in a tiny part of Arizona; *juglandis* completely overlaps *boycei*.) This sample represents the range in $\bar{I}$ values seen in the 4 × 4 matrix of species comparisons. Thus, in one case, the genetic identity for an allopatrically speciated gene pool is much higher than Ayala et al. (1974) found. All values in the *suavis* group are much higher than the value of $\bar{I}$ = 0.352 for nonsibling species (such as the *suavis* species). Remember, however, that whereas Berlocher (1976) studied variation in 15 proteins, Ayala et al. (1974) studied twice as many, thus increasing the probability of detecting differences between gene pools.

Genetic Differences When Chromosomal Mutations Result in Speciation

Through translocations, inversions, fusions, and other gene rearrangements, chromosomal evolution may change identities between parental and new species very little and result in high values of genetic identity. For example, Nevo et al. (1974) found six karyotypes of the pocket gopher, *Thomomys talpoides*, representing a complex of at least five species with genetic similarities as high as $\bar{I}$ = 0.996, with $2n$ = 44 in one species and $2n$ = 46 in the other (they differ karyotypically by only one centric fusion). Other species pairs have $\bar{I}$ = 0.961 ($2n$ = 48 and 46), $\bar{I}$ = 0.947 ($2n$ = 40 and 44), and, lowest of all, $\bar{I}$ = 0.858 ($2n$ = 60 and 48). Nevo et al. (1974) even found that populations in the same species had lower genetic identities than closely related species. Nevo and Shaw (1972) found even higher values of $\bar{I}$ among the four karyotypically distinct sibling species of the fossorial mole rats in the *Spalax ehrenbergi* complex in Israel, making these mole rats the genetically most similar species ever found at that time.

Genetic Differences Resulting from Ecological Factors

In some species, at least, large differences in allozyme frequencies (low $\bar{I}$ values) may develop after the speciation process is essentially complete, after reproductive isolation mechanisms are more or less effective. Indeed, more genetic divergence may result from the selective pressure for ecological divergence between species and adaptation to new environments than from the speciation process itself. For example, Steiner (1977) found a positive relationship between the number of host plants utilized by Hawaiian *Drosophila* species and enzyme polymorphism. The more substrates a species utilized, the more enzymes were utilized and selected for. Presumably, most or all of this genetic divergence in response to divergence of host plant utilization occurred after the speciation process was complete. Many other ecological factors may select for protein polymorphisms and divergence between populations and species; Chapter 16 will discuss these more extensively.

We are left with a wide range of $\bar{I}$ values among related species and with great similarity between reproductively isolated species, as in the *Rhagoletis suavis* group, even though it is geographically speciated and without chromosomal evolution.

Unfortunately, population genetics cannot tell us how many generations should be required for divergence of gene pools to any given level of genetic identity. Lewontin (1974, 159) said,

> It is an irony of evolutionary genetics that, although it is a fusion of Mendelism and Darwinism, it has made no direct contribution to what Darwin obviously saw as the fundamental problem: the origin of species.

Much work in population genetics has concentrated on a single locus at a time and has examined rates of

FIGURE 15-2

Genotype frequencies predicted on the basis of probabilities of union between gametes in the sperm and egg pools of the preceding generation.

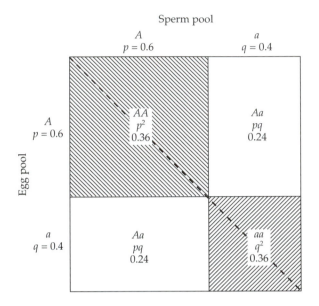

change of alleles at this locus in response to mutation pressure, selection, gene flow, and so on. This approach provides quantitative insight into the number of generations evolution might take, from the fixation of one allele through a mutation event, development of transient polymorphism at the locus, ultimate fixation of the new mutant allele, and other important processes.

15.2 HARDY–WEINBERG EQUILIBRIUM: THE NULL HYPOTHESIS

Before we can study the evolutionary process through change in gene frequencies, we need to understand the null hypothesis, which is the way in which gene frequencies are determined in the absence of evolution. A naive view, taken by Yule in 1902, is to assume that a dominant mutant eventually becomes represented in a population in the ratio of 3:1, since this is the expected ratio in the F_1 generation when two heterozygotes reproduce. But Mendelian genetics, which deals with the progeny of crosses between individuals of known genotype, is inadequate for the treatment of large randomly mating populations. In 1908, the English mathematician G. H. Hardy and the German physician W. Weinberg independently bridged the gap between Mendelian genetics and population phenomena, thereby forming the basis of modern population genetics.

Hardy (1908, 49) responded to Yule's prediction by asserting that "there is not the slightest foundation for the idea that a dominant character should show a tendency to spread over a whole population, or that a recessive should tend to die out"—assuming (1) a large population, (2) random mating, (3) sexes with equal gene frequencies, and (4) uniform fertility. Under these conditions—that is, in the absence of evolution—the proportions of genotypes in one generation depend directly on the probabilities of union between gametes derived from the preceding generation.

Thus, if p represents the frequency of allele A in a population and q represents the frequency of allele a—these two alleles being the only two in the population, so that $p + q = 1$—then we obtain the proportions of the genotypes in the next generation by multiplying the frequencies of the different gametes of the sperm pool by those in the egg pool. And if Hardy's assumptions are fulfilled, the genotype frequencies in the next generation are the squares of the gene frequencies among gametes, since males and females have equal gene frequencies. Therefore, $(p + q)^2 = p^2 + 2pq + q^2$, and these frequencies remain constant in the absence of evolutionary pressures. We call this result the **Hardy–Weinberg law** or the **Hardy–Weinberg equilibrium**.

For example, suppose the proportion of allele A is 0.6 in a population ($p = 0.6$) and a is 0.4 ($q = 0.4$, $p + q = 1$); then the genotype frequencies in the next generation are as in Figure 15-2: $AA = 0.36$, $Aa = 0.48$, and $aa = 0.16$. Other populations may have the same gene frequencies but different genotypic frequencies (Table 15-1). None of these populations will reach Hardy–Weinberg equilibrium until the next generation after random mating.

In the absence of destabilizing forces such as mutation, selection, genetic drift, gene flow, and so forth, the Hardy–Weinberg equilibrium is maintained indefinitely (gene and genotype frequencies are constant), and no evolution takes place. The Hardy–Weinberg equilibrium represents a null hypothesis against which we can estimate the forces that cause deviations from that equilibrium state—the forces of evolution.

15.3 WHAT CAUSES DEVIATIONS FROM HARDY–WEINBERG EQUILIBRIUM?

Evolution is essentially the change in gene frequencies in a population or species (with certain important exceptions, such as chromosomal evolution). We therefore need to understand the factors that cause such a change and the expected rates of change. Let

TABLE 15-1 Genotypic frequencies of populations not at Hardy–Weinberg equilibrium, each with gene frequencies of $p = 0.6$ and $q = 0.4$

Population	Genotype Frequency			Gene Frequency	
	AA	Aa	aa	p	q
1	0.60	0.00	0.40	0.6	0.4
2	0.20	0.80	0.00	0.6	0.4
3	0.50	0.20	0.30	0.6	0.4

See Figure 15-2 for a method of calculating gene frequencies.

us return to aspects of the genetic system of organisms discussed in Chapter 14 (Mettler and Gregg 1969).

Mode of Inheritance

Let us consider sexually reproducing organisms with diploid adults and haploid gametes and the factors that can cause a change in gene frequencies. We need to know whether each locus is autosomal or sex-linked, the number of alleles per locus, and the relationships of dominance and epistasis (interaction between genes at different loci) before we can calculate frequencies of genes and genotypes in the next generation.

Population Size

Sampling error from one generation to another changes gene frequencies and increases as population size decreases. Thus, when using the logic derived from the Hardy–Weinberg law, we assume populations of infinite size with zero sampling error. It is then interesting to examine how population size influences Hardy–Weinberg equilibrium and the rate of change of beneficial and deleterious alleles—that is, the rate of evolution. Genetic drift and the founder effect become important as effective population sizes depart from being very large.

Mating System

The Hardy–Weinberg law assumes random mating, or panmixis, so that a given individual has an equal probability of mating with any individual of opposite sex in the population. Thus, any tendency for inbreeding or self-fertilization, which we know to be common, changes genotype frequencies. Any other tendency for assortative mating (mating within subgroups of a population characterized by genetic similarities between mates) also changes gene frequencies. For example, small organisms with similar genetically controlled heat requirements tend to become sexually active at the same time and mate with each other more frequently than with individuals with different heat requirements. Sexual selection normally results in strongly skewed mating systems, because panmixis is lost and often a small number of individuals are responsible for much of the reproduction in a population.

Mutation

Mutation changes gene frequencies unless mutations to a new allele occur at the same rate as the reverse, mutation to the original. Commonly, a mutation pressure exists, meaning that mutation occurs in one direction only or in one direction more frequently than in the reverse direction.

Gene Flow

Movement of individuals between populations usually changes gene frequencies in both the parental and recipient populations, because the dispersing individuals are unlikely to be fully representative of the donor gene pool or the recipient gene pool. Indeed, the tendency to disperse may be under genetic control. The extent of this gene flow can be a powerful influence on evolutionary rates, as Sewell Wright (e.g., 1940) has emphasized—especially in small populations. "Migration" is a term commonly used interchangeably with "gene flow," but we prefer to reserve the term "migration" for the directed, repeated, and reciprocal movements of individuals such as migrating birds, insects, fishes, turtles, and ungulates.

Natural Selection

Natural selection results in the "differential perpetuation of genotypes" (Mayr 1963, 183), or "the unequal transmission of genes to subsequent generations by different genotypes" (Mettler and Gregg 1969, 34).

Clearly, when natural selection can confer enormous increments of fitness to the carriers of single-gene differences, it causes rapid changes in gene frequencies. For example, the mutant black morph of the peppered moth, *Biston betularia* (discussed in Chapter 1), had a 52 percent advantage over the typical morph in polluted areas in Britain (cf. Kettlewell 1956). While we must acknowledge the importance of natural selection in changing gene frequencies, many other factors, including the aforementioned five, are also effective and on occasion may be more influential than natural selection in the evolution of a species. However, it is still natural selection that provides a directional force in the evolutionary process.

Gametic Disequilibrium

We may define gametic disequilibrium as the "nonrandom association of alleles at different loci into gametes" (Hedrick 1983, 339). All the factors already discussed may contribute to gametic disequilibrium in a population. In addition, linkage of alleles on the same chromosome reduces their independent assortment into gemetes. **Genetic hitchhiking,** in which a more-or-less neutral allele travels through the generations by its association with an allele at another locus that has a selective advantage, contributes to disequilibrium. Such multilocus factors quickly become complex. Although their complexity is realistic, we will continue by considering rates of change in gene frequencies at one locus.

15.4 RATES OF CHANGE IN GENE FREQUENCIES

What rates of change in gene frequencies do these various factors cause, and how important are such factors likely to be in evolution and speciation events? Let us take the simplest approach possible by studying one factor at a time, with emphasis on the implications of a certain formulation rather than its derivation. (Derivations may be found in Falconer [1960, 1981], Mettler and Gregg [1969], Roughgarden [1979], and Hedrick [1983, 1984].)

Mutation

If a mutation occurs in one direction only and at a constant rate, it is an irreversible mutation. We can calculate the change in gene frequency caused by irreversible mutations, i.e., the recurrent mutation pressure. We express mutation rate as the probability that one allele will mutate in a generation, which may

be 1 in 10,000 alleles, or 10^{-4}. Thus, if mutation occurs from gene A to gene a,

and

$$p_t = p_0 e^{-\mu t}$$

$$q_t = 1 - p_0 e^{\mu t}$$

where p_t = frequency of gene A after t generations, q_t = frequency of gene a after t generations, p_0 = frequency of gene A after 0 generations, t = number of generations, and μ = mutation rate. The mutation rate for certain alleles in humans has been shown to be 10^{-4} mutations per generation. Therefore, 10,000 generations are required to reduce the initial gene frequency of A to about one third ($p_{1000} = p_0 e^{-1} = p_0 \frac{1}{2} \cdot 7183 \simeq p_0 \frac{1}{3}$). Without the action of natural selection, the rate of change in gene frequencies due to recurrent mutation is very low.

Mutation rates are commonly lower than 10^{-4}—more like 10^{-5} and 10^{-6} per generation. Here, if $\mu = 10^{-5}$, 69,000 generations are required to reduce p_0 by half, which gives us an estimate of the half-life of A. If $p_0 = 0.96$, 69,000 generations are required to reach $p = 0.48$, another 69,000 generations to reach $p = 0.24$, and about 350,000 generations to change A with $p = 0.96$ to near fixation of a at $p = 0.03$ and $q = 0.97$. Clearly, such enormous numbers of generations for the substitution of one allele by another suggests that mutations in themselves cannot cause rapid rates of evolutionary change among metazoans and plants.

Reversible mutation, $A \Longleftrightarrow a$, slows this process via the rate of back mutations, and an equilibrium will be reached in which

$$p = \frac{v}{\mu + v}$$

where μ = rate of mutation of allele A to allele a, and v = rate of mutation of allele a to allele A. If $\mu = 2v$ (e.g., $\mu = 2 \times 10^{-5}$ and $v = 10^{-5}$), then $p = 0.333$.

However, when populations are extremely large and generations extremely short—as in bacteria and other microorganisms, where t is in minutes—mutation can be an important force in evolutionary change. Mutation changes gene frequencies and may even lead to fixation of an allele in rare cases, but it is most important as a source of variation rather than a force for evolutionary change.

Natural Selection

If a mutation influences the fitness of its bearer, then gene frequencies change under the influence of natural selection, a process that can be much more rapid than change through mutation pressure.

If locus A has two alleles, a_1 and a_2, then the three genotypes a_1a_1, a_1a_2, and a_2a_2 will have different fitness values W_0, W_1, and W_2 if natural selection is acting. From these fitness values we can calculate the force of selection on each genotype—the coefficients of selection (s):

Genotypes a_1a_1 a_1a_2 a_2a_2

Fitness W_0 W_1 W_2

and $\dfrac{W_0}{W_0} = 1$ $\dfrac{W_1}{W_0} = 1 - s_1$ $\dfrac{W_2}{W_0} = 1 - s_2$

where a_1a_1 is the fittest genotype and the genotypes a_1a_2 and a_2a_2 suffer decrements in fitness of s_1 and s_2, respectively.

We can superimpose these relationships on the Hardy–Weinberg genotype frequencies as follows:

Genotypes	a_1a_1	a_1a_2	a_2a_2	Total
Frequency in generation 0, before selection	p^2	$2pq$	q^2	1
Fitness	W_0	W_1	W_2	—
Proportional contribution to next generation	p^2W_0	$2pqW_1$	q^2W_2	$\overline{W}(=<1)$
Frequency in generation 1, after selection	$\dfrac{p^2W_0}{\overline{W}}$	$\dfrac{2pqW_1}{\overline{W}}$	$\dfrac{q^2W_2}{\overline{W}}$	1

$\overline{W}$, the mean fitness, will always be less than 1 since selection is operating (e.g., W_2/W_0 may equal $1 - s_2$). Thus, to obtain frequencies after selection, we divide the proportional contributions to the next generation by $\overline{W}$ and restore the totals of genotype frequencies to 1.

In practice, the geneticist observes the proportions of genotypes before and after selection has occurred and calculates the selection coefficients from the empirical data (Table 15-2). In this case, a_1 is not dominant over a_2 since the heterozygote is less fit than the homozygote a_1a_1 and the selection coefficient against the a_2a_2 genotype is twice that of the heterozygote.

When we have estimated selection coefficients, we can calculate the rate of change of gene frequencies per generation if we know the mode of inheritance. For example, if the favorable allele (a_1) is dominant, then the change in gene frequency is

$$\Delta q = \frac{-sq_0^2(1 - q_0)}{1 - sq_0^2}$$

$$q_1 = \frac{q_0(1 - sq_0)}{1 - sq_0^2}$$

and

$$q_n = q_{n-1}\frac{(1 - sq_{n-1})}{1 - sq_{n-1}^2}$$

where q_0 = frequency of gene a_2 at 0 generations, q_1 = frequency of gene a_2 after 1 generation, s = selection coefficient, and n = number of generations. Table 15-3 gives other cases.

In the special and simplifying case where AA and Aa have the same fitness, equal to 1 ($s = 0$), and the fitness of aa is 0 ($s = 1$), then

$$q_n = \frac{q_0}{1 + nq_0} \quad \text{and} \quad n = \frac{q_0 - q_n}{q_0q_n} = \frac{1}{q_n} - \frac{1}{q_0}$$

Selection coefficients may be large, causing rapid Δq. For example, Kettlewell (1956) measured the survival of three genotypes of the peppered moth, *Biston betularia*, in polluted woods near Birmingham, England, where the darker morphs had an adaptive advantage in relation to the selective agents, birds that hunt visually (see Chapter 1). Kettlewell released

TABLE 15-2   Genotype frequencies in a population before and after selection

	a_1a_1	a_1a_2	a_2a_2
Frequency in generation 0, before selection	0.25	0.50	0.25
Frequency in generation 1, after selection	0.35	0.48	0.17
Relative survival value	$\dfrac{0.35}{0.25} = 1.4$	$\dfrac{0.48}{0.50} = 0.96$	$\dfrac{0.17}{0.25} = 0.68$
Relative fitness	$\dfrac{1.4}{1.4} = 1.0$	$\dfrac{0.96}{1.4} = 0.70$	$\dfrac{0.68}{1.4} = 0.40$
Selection coefficient	$1 - 1 = 0$	$1 - 0.7 = 0.3$	$1 - 0.4 = 0.6$

Results are from a hypothetical experiment and illustrate the calculation of selection coefficients

TABLE 15-3 Effect of inheritance mode on change in gene frequency of allele a_2 under selection pressure

Mode of Inheritance	Mean Fitness After Selection $(\overline{W})$	Change in Frequency of Allele a_2 (Δq)
1. Complete dominance of a_1 Selection against a_2	$\overline{W} = 1 - sq^2$	$\Delta q = \dfrac{-sq^2(1 - q)}{1 - sq^2}$
2. Complete dominance of a_1 Selection against a_1	$\overline{W} = 1 - sp^2 - 2pqs$	$\Delta q = \dfrac{sq^2(1 - q)}{1 - s(1 - q)^2}$
3. No dominance Selection against a_2	$\overline{W} = 1 - spq - sq^2$	$\Delta q = \dfrac{-\frac{1}{2}sq(1 - q)}{1 - sq}$
4. Heterozygote advantage (overdominance) Selection against homozygotes	$\overline{W} = 1 - s_1p^2 - s_2q^2$	$\Delta q = \dfrac{(1 - q)q[s_1 - q(s_2 + s_1)]}{1 - s_1(1 - q)^2 - s_2q^2}$

After Mettler and Gregg 1969.

known numbers of marked moths in a wood—exposing them to natural bird predation—and recaptured the survivors at light traps or traps baited with virgin females. The gene for melanism in *Biston betularia* is completely dominant (Kettlewell 1959).

In one experiment in a polluted area, Kettlewell released 154 carbonaria (melanic) moths and 65 typical moths (white with black spots) and recaptured 82 carbonaria and 16 typical. These figures were the basis for an estimate of the selection coefficients, assuming that CC and Cc were equally represented in the population (Table 15-4).

With this selection coefficient of 0.54 against the homozygous recessive, we can calculate how many generations it would take to change the frequency of the black mutant from about 2 percent (assume CC = .01, Cc = .01, cc = .98, p = .015, q = .985), as in unpolluted areas, to the 87 percent (assume CC = .44, Cc = .44, cc = .12, p = .66, q = .34) observed in populations near Birmingham. Since this species has one generation per year, such a calculation can provide an estimate of how rapidly a population would become largely melanic if an area were to change suddenly from being unpolluted to having the extent

TABLE 15-4 Calculation of selection coefficients for *Biston betularia* genotypes

	CC, carbonaria	Cc, carbonaria	cc, typical	Total
Number before selection	77	77	65	218
Number after selection	41	41	16	98
Frequency before selection	.35 (77/218)	.35	.29	1
Frequency after selection	.42 (41/98)	.42	.16	1
Relative survival value	1.20 (.42/.35)	1.20	.55	
Relative fitness	1.00 (1.2/1.2)	1.00	.46 (.55/1.2)	
Selection coefficient	0 (1 − 1)	0	.54 (1 − .46)	

Data are from Kettlewell's release-and-recapture studies.

of pollution that existed near Birmingham in 1953 and 1955, when Kettlewell performed his experiments. Using the selection coefficient of 0.54 calculated from Kettelwell's data and $p = .02$, we find that it would take only nine generations to reach $q = 0.32$.

Haldane (1956) estimated that, from the first mutant in a population in a polluted woodland, it would take only 37 years (= generations) for the frequency of the melanic allele (C) to change from $p = 10^{-5}$ to $p = 0.8$. Haldane assumed that the fitness of c was .5 that of C, although he also assumed that the heterozygote had a selective disadvantage.

Perhaps a selection coefficient of 0.54 is unusually large, but we have so few good estimates in nature that it is hard to judge. A number of studies have found the selective action of predation to be severe (Price 1984). We should remember, too, that gene frequencies change least rapidly when the favorable allele is completely dominant, as in the *B. betularia* case, because the heterozygote protects recessive traits that, when expressed, are being selected against. Thus, gene frequencies could change as rapidly with lower selection coefficients if there were no dominance, if selection were against the dominant allele, or if the heterozygote were favored (Table 15-3).

There is no doubt, however, that natural selection can change gene frequencies rapidly and is the great molding force of the evolutionary process. This molding force can take three basic forms (Figure 15-3). Selection may favor individuals in the center of the normal distribution of phenotypes, in which case a

population becomes more strongly represented by these phenotypes, and the tails of the distribution contract (Figure 15-3a). The effect is to maintain the mean of the distribution and keep it in a stable condition through episodes of selection; hence, we apply the term **stabilizing selection.** Should selection favor one end of a normal distribution of phenotypes, then such **directional selection** shifts the normal distribution away from that direction—for example, to the right in Figure 15-3b. Under certain conditions, perhaps involving a population that is diversifying into two different microhabitats, what was the most frequent phenotype might no longer be well adapted to the new conditions, and **disruptive selection** would result (Figure 15-3c). The frequency of phenotypes around the mean of the distribution would decline and two peaks of high frequency would develop, together with a broadening of the distribution of phenotypes. Thus, the typically normal distribution of phenotypic characters would be disrupted. Should such selection continue, perhaps the new peaks would become more distinct with time.

Gene Flow

Gene flow can influence evolutionary rates of populations in two ways, with opposite effects. Relatively high gene flow into a population can swamp the effects of gene change resulting from local conditions, natural selection or random events, and can slow or prevent divergence of populations. As Wright (1940,

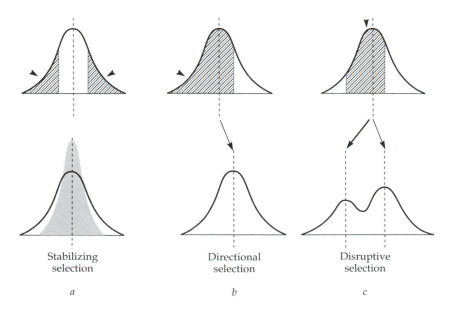

Stabilizing   Directional   Disruptive
selection   selection   selection

a   *b*   *c*

FIGURE 15-3

The three kinds of natural selection that work on normal distributions before selection of characters in a population. Countable or measurable characters such as number of seeds per pod, height per individual, and shoot length per stem, commonly vary over a certain range on the *x* axis, with frequency of individuals in any part of the range (*y* axis) showing a normal distribution. The top row of figures represents populations before selection, the means of the distributions (dashed lines), and the portions of the populations that will be selected against (shaded areas). Solid pointers indicate the foci of selection. The bottom figures show the results of selection. (a) With stabilizing selection, the mean remains the same but the variation around the mean diminishes. (b,c) Means of distributions may change (arrows).

1943, 1949) has emphasized, if gene flow is low and occurs between populations with rather different gene pools, it provides new alleles to the population, provides new sources of variation, and functions as mutation in accelerating change in gene frequencies and the evolution of the population.

The general relationships between gene flow and change in gene frequencies are given by the equations

$$q_1 = mq_m + (1 - m)q_0 = q_0 - m(q_0 - q_m)$$

and

$$\Delta q = -m(q_0 - q_m)$$

where q_0 = frequency of allele a in generation 0 in recipient population, q_1 = frequency of allele a in generation 1 in recipient population, m = proportion of individuals in recipient population composed of dispersers from a donor population = rate of gene flow (a constant), and q_m = frequency of allele a in donor population (a constant). These formulas invoke all conditions of the Hardy–Weinberg equilibrium except that gene flow is present and unidirectional. They assume that gene flow is constant; that the donor population is infinitely large so that q_m does not change, although loss from the gene pool occurs through dispersal; and that the recipient population is necessarily smaller so that new arrivals change gene frequencies to some degree.

Thus, if gene flow is a trickle—say, 20 reproductively competent individuals into a population of 20,000 adult individuals (m = .001)—then change in the recipient population is very small. If the donor population has a frequency of gene a of 0.30 (q_m = 0.3) and the recipient population a frequency of a = 0.10 (q_0 = 0.10), then:

$$q_1 = .001 \times 0.3 + (1 - .001)0.1 = 0.1002$$

$$q_2 = .001 \times 0.3 + (1 - .001)0.1002 = 0.1004$$

This change is likely to be significantly less than the rates of divergence between donor and recipient populations due to natural selection and other factors, and so it slows divergence very little. In addition, the carriers of gene a may introduce to the recipient populations new alleles that are highly adaptive in the new environment, thus accelerating the process of natural selection.

Gene flow may be high, however. In an extreme case, 10,000 adult individuals might migrate into a population of 20,000 adult individuals in each generation. If the original gene frequencies are as in the former example, then

$$q_1 = 0.5 \times 0.3 + (1 - 0.5)0.1 = 0.20$$
$$q_2 = 0.5 \times 0.3 + (1 - 0.5)0.20 = 0.25$$

The maximum, 0.30, is almost reached, and input from the donor population effectively swamps any divergence in frequencies of this gene.

Thus, for small recipient populations, even a few immigrant individuals may represent a significant m, and if m is constant for each generation, divergence may be impossible. Clearly, then, population structure becomes an important part of understanding the quantitative aspects of evolution. Population size and distance between populations are important influences on rates of gene fixation and the probabilities of gene flow between populations. These probabilities may be so low that gene flow is intermittent or even very rare. Then its contribution to the genetic diversity of a small gene pool probably outweighs the swamping effect of high rates of gene flow. Therefore, we will examine population structure in more detail next.

Genetic Drift

The most unrealistic assumptions in the Hardy–Weinberg law are that populations are large (ideally, of infinite size) and mating is random. Panmixia in large populations ensures no loss of genes from the gene pool of one generation and no change in gene frequency if natural selection, mutation, and gene flow do not operate, because sampling error is zero or close to it. In a very large sample from the preceding generation, all genes are likely to be represented, and in the same proportions. As the sample becomes smaller, however, the sampling error increases and gene changes from one generation to another become more pronounced. The effects of this genetic drift can be very influential in the evolutionary process, particularly since we know that many species are composed of small, relatively isolated populations (see Chapter 14).

Genetic drift is the alteration of gene frequencies through sampling error. It can be important in evolution under three conditions: continuous drift, intermittent drift, and the founder effect (Wilson and Bossert 1971).

Continuous Drift

When a population remains small, sampling error is effective in each generation. For example, a population of 10 individuals may produce 100 progeny, of which only 10 survive to reproduce. If allele frequencies are equal at a locus (p = q = .50), the probability that this frequency will be represented in the 10 survivors is relatively low:

$$\frac{10!}{5!5!}\left(\frac{1}{2}\right)^{10} = 0.246$$

This binomial probability estimates that in only a quarter of the cases will the gene frequency of $p = q = .50$ be preserved. There is also a small probability—2 in 1,000—that the ten survivors will all have the same allele, it will become fixed, and the other will be lost to the population.

$$2\left(\frac{1}{2}\right)^{10} = 0.002$$

Intermittent Drift

A large population may shrink to a small size so that sampling error is intermittently large, either in a single breeding season or over several seasons.

The Founder Effect

A few individuals or only one reproductively competent individual may found a new population. In either case, the founders represent only a small fraction of the parental gene pool. Repeated founder effects can cause very rapid divergence of new populations from the parental population. Carson's (1968, 1975) concept of the flush-and-crash cycle makes the combination of intermittent drift and the founder effect very important in the process of evolution and speciation (see Chapter 5).

Sampling error is a probabilistic phenomenon; the distribution of q is binomial in a very large sample of populations; and, although q may change in any one population, changes in q in all populations are zero (i.e., mean $\Delta q = 0$)—the gains cancel out the losses. In any one population, however, the amount of variation in allele frequencies from one generation to another depends, in the absence of evolution, on population size and allele frequency at the start of a generation. The expected variation in allele frequencies among progeny populations is

$$\text{Variance of } \Delta q = \frac{pq}{2N}$$

where N = population size and $2N$ is the number of gametes that produce N progeny. As population size decreases, sampling error and variance in allele frequencies increase; genetic drift in each generation is considerable. As allele frequency deviates from equality ($p = q = 0.5$), however, the variance decreases (Table 15-5).

This random walk of gene frequencies inevitably leads to the loss of some alleles, fixation of others, and in general a loss of heterozygosity in a population—an important consequence of genetic drift (Figure 15-4).

Another important result is that, although genetic drift reduces variation within a population, it increases variation *among* populations. Thus, where sampling error is significant, genetic drift can play a major role in divergence of small populations. Wright (1940) calculated that, in the absence of other evolutionary forces, fixation and loss occur at a rate of about $1/(4N)$ per locus per generation. Therefore, in a panmictic population, the time for an allele to become fixed or lost is, on average, $4N$ generations, although any one allele may become fixed or lost much more rapidly or slowly than this mean value.

Clearly, population size has a dramatic effect on the influence of genetic drift. On average, a

---

TABLE 15-5   Relationship between starting allele frequency (p) and expected variance of change in allele frequency of progeny as population size increases and p declines

	Initial Values of p					
$2N$	0.5	0.4	0.3	0.2	0.1	
10	0.025	0.024	0.021	0.016	0.009	variance is
20	0.0125	0.012	0.0105	0.008	0.0045	reduced by
50	0.005	0.0048	0.0042	0.0032	0.0018	an order of
100	0.0025	0.0024	0.0021	0.0016	0.0009	magnitude

variance is reduced by 2.8 fold

After Giesel 1974.
Note: As population size increases, p declines.

FIGURE 15-4

Several possibilities for the random walk of allele frequencies through genetic drift, showing that the random walk may result in fixation or loss of an allele, or the preservation of an allele over many generations. The bell-shaped curves illustrate the variance of the expected change in gene frequency. Note that the variance decreases as p departs from a value of 0.5. *(After Giesel 1974)*

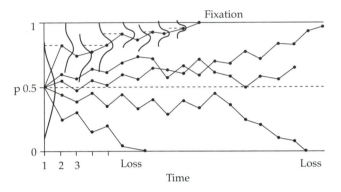

population of ten individuals has alleles fixed in 40 generations, with a wide range of probabilities of fixation around this mean. Thus, some alleles may become fixed in a few generations during rather short periods of isolation between populations. In contrast, in a large population of, say, 10,000 to 100,000 individuals, alleles hardly ever become fixed in a few generations, and on average it takes from 40,000 to 400,000 generations for fixation or loss. Fixation of chromosomal rearrangements operate under the same probabilities.

To understand how important genetic drift is in real populations, we therefore need to know the sizes of effective breeding populations and how commonly these are small. We also need to keep in mind that natural selection greatly accelerates the rates of change of favorable alleles and gene rearrangements. We saw in Chapter 14 that population sizes may frequently be small, at least temporarily, making intermittent drift important. Carson (1968, 1975) considered this situation to be common in Hawaiian *Drosophila* species. We have also pointed out that inbreeding is common among parasites, panmixia is prevented, and fixation can proceed much more rapidly because the effective population size is drastically reduced (Price 1980). Founder effects are common in parasitic organisms, as well. All these factors reduce Wright's estimate of fixation rates so that, in small populations of 10 to 100 individuals, fixation can easily occur in two to ten generations during only brief periods of isolation. For small, rather specialized organisms, such as parasites and many *Drosophila* species, genetic drift must be a very important factor. Also, whenever social structuring is strong, even large organisms may have very small effective population sizes (Chapter 14).

Other chapters of this book refer to the effects discussed in this chapter. Repeatedly, small populations at the edges of their species ranges have entered into scenarios of evolution; we have discussed rapid evolutionary rates and seen that evolution in the apparent absence of natural selection plays an important role. Thus, we need to keep this chapter's topics in mind through much of the other parts of this book (especially Chapters 5, 8, 14, 16, 17, and 18).

SUMMARY

The behavior of a population's genes through generations of sexually reproducing organisms is a crucial aspect of evolution. The extent of difference between species, the causes of changes in gene frequencies, and the rates of those changes are key aspects of the evolutionary process.

Genetic variation within populations can be surprisingly high—30 percent of loci in a population of *Drosophila*, for example, and 12 percent heterozygosity in an average individual. We can estimate similarity between populations of the same species as the proportion of loci that are the same and have the same frequency of alleles, ranging between $\bar{I} = 1$ for complete identity and 0 for complete dissimilarity. During speciation events, populations diverge in their identities, with one set of estimates for geographically speciating *Drosophila* providing values of

$\bar{I} = 0.97$ for populations of the same species, $\bar{I}$ of about 0.80 for subspecies and semispecies, $\bar{I} = 0.50$ for sibling species, and $\bar{I} = 0.35$ for closely related species that are morphologically distinct and allopatrically speciated. Sympatric speciation, however, such as in *Rhagoletis* fruit flies, may result in sibling species with genetic identities as high as $\bar{I} = 0.98$. In addition, speciation events involving chromosomal mutations may conserve high genetic similarities, as in pocket gophers ($\bar{I}$, 0.99 to 0.86) and mole rats. Ecological factors may well select for much more genetic divergence between species than is involved in the speciation process itself.

In the absence of evolution, the **Hardy–Weinberg equilibrium** provides the null hypothesis for gene frequencies in populations. Assuming that populations are very large, mating is random, there is an equal sex ratio, and fertility is uniform, then genotype frequencies remain constant. Departure from Hardy–Weinberg equilibrium results in evolutionary change that may be caused by deviations from the assumptions made. Modes of inheritance, population size, mating system, mutation, gene flow, natural selection, and gametic disequilibrium are examples.

Rates of change in gene frequencies and rates of evolution differ significantly according to the factor that influences deviations from Hardy–Weinberg equilibrium. **Recurrent mutation** can produce only very slow change in gene frequencies. **Natural selection** may result in rapid evolution, depending on the strengths of the selection coefficients and the dominance status of the gene under selection. **Gene flow** has variable effects on evolutionary rates, according to the relative sizes of immigrating and recipient populations. **Genetic drift** can have strong effects in small populations where sampling error is high, becoming strongest where founder effects minimize the proportion of genes from the donor population carried by the founders. **Population size** therefore influences fixation rates of alleles, and where effective population sizes are small, fixation may occur in a few generations.

QUESTIONS FOR DISCUSSION

1. If sibling species of *Drosophila*, which look alike, can have a genetic identity of about $I = 0.52$ but humans and chimpanzees, which appear distinctly different and are placed in different genera, have an identity of about 0.70, do you think that electrophoretic studies can actually aid in understanding phenotypic change and the process of genetic divergence of lineages?

2. Sibling species of *Drosophila* have a genetic identity of about 0.52 and sibling species of *Rhagoletis* may have a genetic identity as high as 0.98. Does this difference suggest that very different speciation processes have been involved or that the ecological relationships in the genera are very different?

3. Do you think that it is still true that population genetics has not helped in understanding the origin of new species?

4. If the conditions for Hardy–Weinberg equilibrium as a null hypothesis are almost never met in natural populations, do you think it is fruitless to use it in population genetics?

5. There is a large variety of genetic systems even if we consider modes of reproduction alone, many of which cause departures from Hardy–Weinberg conditions. In each example you consider, how will this affect the use of the Hardy–Weinberg equilibrium in calculating gene frequencies in subsequent generations?

6. Much population genetics has dealt with one-locus models. Do you think that this approach focuses unrealistic attention on genes with strong effects and with high probability for change in frequency, which may overemphasize the possibility of rapid evolutionary change in a population?

7. Unidirectional gene flow can cause rapid evolutionary change in a small recipient population according to the formulae provided in this chapter. What kinds of species and in what situations are such conditions likely to be seen in plants and animals?

8. Genetic drift is a concept that seems to be difficult for some to understand. Can you devise a demonstration that illustrates the role of genetic drift clearly?

9. In a demonstration that illustrates genetic drift, how would you show that sampling error can be greatest when allele frequencies are equal in a population?

10. Among taxa that you are familiar with, can you suggest ways in which the effective population size, or the size of the breeding population, is much lower than the total population size?

REFERENCES

Avise, J. C. 1976. Genetic differentiation during speciation, pp. 106–122. *In* Ayala, F. J. (ed.) Molecular Evolution. Sinauer, Sunderland, Mass.

Ayala, F. J., M. L. Tracey, L. G. Barr, J. F. McDonald, and S. Perez-Salas. 1974. Genetic variation in natural populations of five *Drosophila* species and the hypothesis of the selective neutrality of protein polymorphisms. Genetics 77:343–384.

Berlocher, S. H. 1976. The genetics of speciation in *Rhagoletis* (Diptera: Tephritidae). Ph.D. dissertation, Univ. of Texas.

Berlocher, S. H., and G. L. Bush. 1982. An electrophoretic analysis of *Rhagoletis* (Diptera: Tephritidae) phylogeny. Syst. Zool. 31:136–155.

Carson, H. L. 1968. The population flush and its genetic consequences, pp. 123–137. *In* R. C. Lewontin (ed.). Population Biology and Evolution. Syracuse Univ. Press, Syracuse, N.Y.

———. 1975. The genetics of speciation at the diploid level. Amer. Natur. 109:83–92.

Falconer, D. S. 1960. Introduction to Quantitative Genetics. Ronald, New York.

———. 1981. Introduction to Quantitative Genetics. 2d ed. Longman, Harlow, England.

Giesel, J. T. 1974. The Biology and Adaptability of Natural Populations. Mosby, St. Louis.

Haldane, J. B. S. 1956. The theory of selection for melanism in Lepidoptera. Proc. Royal Soc., London. Ser. B. 145:303–306.

Hardy, G. H. 1908. Mendelian proportions in a mixed population. Science 28:49–50.

Hedrick, P. W. 1983. Genetics of Populations. Science Books International, Boston.

———. 1984. Population Biology: The Evolution and Ecology of Populations. Jones and Bartlett Publishers, Boston.

Kettlewell, H. B. D. 1956. Further selection experiments on industrial melanism in the Lepidoptera. Heredity 10:287–301.

———. 1959. Darwin's missing evidence. Sci. Amer. 200(3):48–53.

Lewontin, R. C. 1974. The Genetic Basis of Evolutionary Change. Columbia Univ. Press, New York.

Lewontin, R. C., and J. L. Hubby. 1966. A molecular approach to the study of genic heterozygosity in natural populations, [Part] II: Amount of variation and degree of heterozygosity in natural populations of Drosophila pseudoobscura. Genetics 54:595–609.

Mayr, E. 1963. Animal Species and Evolution. Belknap Press of Harvard Univ. Press, Cambridge, Mass.

Mettler, L. E., and T. G. Gregg. 1969. Population Genetics and Evolution. Prentice–Hall, Englewood Cliffs, N.J.

Nei, M. 1972. Genetic distance between populations. Amer. Natur. 106:283–292.

Nevo, E., Y. J. Kim, C. R. Shaw, and C. S. Thaeler, Jr. 1974. Genetic variation, selection, and speciation in Thomomys talpoides pocket gophers. Evolution 28:1–23.

Nevo, E., and C. R. Shaw. 1972. Genetic variation in the subterranean mammal, Spalax ehrenbergi. Biochem. Genet. 7:235–241.

Price, P. W. 1980. Evolutionary Biology of Parasites. Princeton Univ. Press, Princeton, N.J.

———. 1984. Insect Ecology. 2d ed. Wiley, New York.

Roughgarden, J. 1979. Theory of Population Genetics and Evolutionary Ecology: An Introduction. Macmillan, New York.

Selander, R. K. 1976. Genic variation in natural populations, pp. 21–45. In F. J. Ayala (ed.). Molecular Evolution, Sinauer, Sunderland, Mass.

Steiner, W. W. M. 1977. Niche width and genetic variation in Hawaiian Drosophila. Amer. Natur. 111:1037–1045.

Weinberg, W. 1908. Über den Nachweis der Vererbung beim Menschen: Jahreshefte Vereins Vaterland. Naturkunde Würtemberg 64:368–382.

Wilson, E. O., and W. H. Bossert. 1971. A Primer in Population Biology. Sinauer, Sunderland, Mass.

Wright, S. 1940. Breeding structure of populations in relation to speciation. Amer. Natur. 74:232–248.

———. 1943. Isolation by distance. Genetics 28:114–138.

———. 1949. Population structure in evolution. Proc. Amer. Phil. Soc. 93:471–478.

Yule, G. U. 1902. Mendel's laws and their probable relation to intra-racial heredity. New Phytol. 1:193–207, 222–238.

16

GENETIC AND PROTEIN POLYMORPHISM

Young female cone

Winter

Mature female cone

The European larch (*Larix decidua*), a deciduous conifer, represents the gymnosperms, a group of usually long-lived woody perennials. Many have wide geographic ranges and display outcrossing by wind dispersal of pollen.

16.1 THE EVOLUTION OF A GENE SUPERFAMILY

The Early Role of Cytochrome P450

About 2 billion years ago, when photosynthesis was producing free oxygen in the atmosphere, many bacteria had their first exposures to that gas. It was very toxic. Any protein with the capacity to reduce the oxygen became highly adaptive in the new environment. Probably some early form of a cytochrome P450 molecule was the novel detoxification enzyme that proceeded to evolve in the bacteria (Gonzalez and Nebert 1990). Carriers of the ancestral gene for cytochrome P450 thus were preadapted for an atmosphere increasingly contaminated with oxygen, and lineages of both the bacteria and the gene thrived and spread.

Decontamination and detoxification involving metabolism of chemicals foreign to the body have been persistent biochemical challenges throughout the evolution of life on earth, from the bacteria to humans. Therefore, we should not be surprised that, although cytochrome P450 molecules vary widely among species, all P450 proteins contain highly conserved regions that indicate a common ancestry, and we can glimpse another direct evolutionary linkage between bacteria and humans (Nebert et al. 1989).

Polymorphism in Genes and Proteins

Indeed, by 1989, 78 unique P450 genes had been sequenced, and by 1991 scientists knew the structures of 154 of these P450 genes. The rate of discovery of new genes suggests that mammalian species may each have at least 60, and as many as 200, individual P450 genes (Nebert et al. 1989, Nebert and Nelson 1991). Such extreme polymorphism in genes and proteins requires some explanation, both in terms of the mechanisms causing such a rich genetic stock of protein morphs and in terms of the adaptive functions and ecological necessities for such variety. Both proximate and ultimate explanations help us to understand the evolution of protein polymorphism and the reasons it occurs so commonly in populations and species.

Phylogenetic Development of the Cytochrome P450 Superfamily

Nebert and his associates have traced the phylogenetic development from one ancestral P450 gene in bacteria about 2 billion years ago to present-day organisms such as rats and humans (Nebert et al. 1989, Gonzalez and Nebert 1990, Nebert and Nelson 1991). According to estimates by Nebert et al. (1989), the cytochrome P450 superfamily of genes radiated into 14 families by about 600 million years ago (Figure 16-1). Although the picture is rapidly growing more complex—for example, by 1990, 20 gene families had been identified (Gonzalez and Nebert 1990)—Figure 16-1 provides a good visual summary of the sequence of events. Subsequent radiations in certain families increased genetic diversity enormously. Notably, Family II diverged into eight subfamilies by 300 million years ago, and its members contained fully 50 percent of the unique genes known in 1989 (Figure 16-1).

Adaptive Significance of Gene and Protein Polymorphism

The Role of Ecological Diversity

Is this extreme protein polymorphism necessary and adaptive, especially in such homeostatically regulated organisms as mammals? With perhaps 60 to 200 individual P450 genes per mammalian species, the answer must lie in the extremely high diversity of ecological challenges to organisms in the forms of chemicals foreign to the body: toxins, drugs, pollutants, and plant metabolites. Initially, "pollutants" such as oxygen and antibiotics, both produced by bacteria, would have challenged the lives of other bacterial species. Subsequently, ingestion of toxic foods by early eukaryotic protists would have posed metabolic problems. And ultimately, as chemical defenses in organisms increased in diversity, so the challenges to digestive systems and metabolism would have increased dramatically. As terrestrial life developed and food in the form of plants became abundant for the new herbivores, an "arms race" of toxic defenses in plants and detoxification enzymes in animals may well have escalated innumerable times, creating the basis for a great variety of cytochrome P450 genes and proteins (Gonzalez and Nebert 1990). The decontamination systems in our bodies, which create challenges for drug therapies today, were founded in a genetic stock that can be traced back some 2 billion years (Figure 16-1).

Escalation of Interactions Between Plants and Animals

The adaptive radiation of Family II of the cytochromes P450 is particularly interesting because of the success of this lineage and the evident ecological forces selecting for the family's expansion. Gonzalez and Nebert (1990, 183) noted an "explosion" of new genes during the last 400 million years, especially in Family II. The Siluro-Devonian explosion of terrestrial plants from about 430 to 350 million years ago (see Chapter 7) no doubt accelerated selective pressure for detoxification enzymes. Colonization of the land by arthropods and vertebrates, many of which

FIGURE 16-1

A phylogenetic tree of the P450 gene superfamily calculated by Nebert et al. (1989). We can roughly calibrate the evolutionary distance (*E.D.*) by setting the divergence time between bacterial and eukaryotic P450 genes at 1.4 billion years, based on the earliest eukaryotic fossils. This divergence occurred at an *E.D.* of 2.5 on the scale. Therefore, each evolutionary-distance unit of 1 converts to approximately 560 million years. Diamonds at divergences of the tree symbolize gene duplication events; the circled Roman numerals and letters between *E.D.*'s of 1 and 2 indicate P450 gene families, and those between *E.D.*'s of 0 and 1 indicate subfamilies.

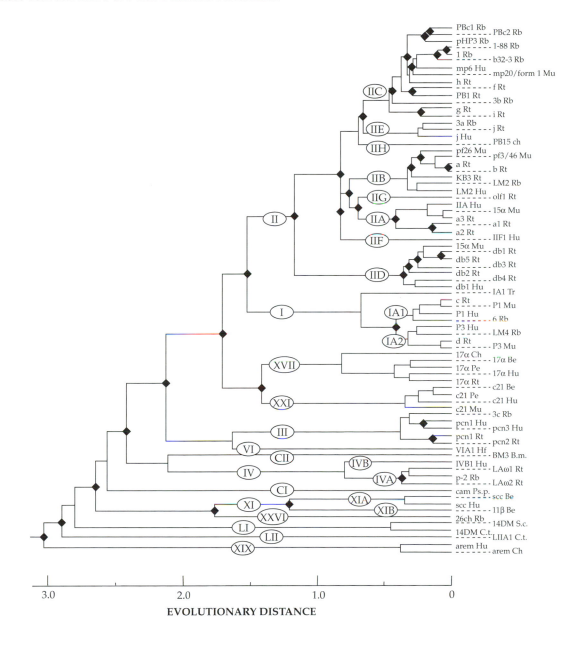

became herbivores, placed selective pressures on the rapid diversification of P450 genes.

> Animals began to ingest plants. As a means of defense, plants countered by developing new stress metabolites (phytoalexins) to make them less palatable and/or digestible; animals then responded with new P450 genes to detoxify these phytoalexins. (Gonzalez and Nebert 1990, 183)

The Way in Which Cytochromes P450 Work

Metabolism of Foreign Chemicals

How do cytochromes P450 function as agents of detoxification, and why does a single species need so many? P450 proteins are involved with the breaking down, or metabolizing, of chemicals, especially drugs and similar kinds of organic molecules that are foreign to the body. Hydroxylation of carbon, nitrogen, and sulfur compounds; dehalogenation; alkylation; deamination; and reduction are some of the common functions (Figure 16-2). A series of different P450 enzymes may metabolize any one drug until the products can be excreted safely. However, because of the general form of the enzymes' biochemical activity, any one P450 enzyme may be involved in the processing of many drugs and other toxins, but no single enzyme is capable of detoxifying all toxic compounds. Together these representatives of P450 families can tackle the breakdown of a very wide variety of organic chemicals. Such a large array of P450 enzymes correlates with the array of environmental chemicals that challenge organisms, with each family or subfamily of P450s characteristically involved with certain kinds of toxins (Table 16-1).

Induction Resulting in Increased Activity

When a drug-like compound is ingested—by a mammal, for example—a reaction in the body increases the production of certain P450 enzymes. Such foreign compounds are therefore called **inducers** of P450 enzymes. In many cases, rates of transcription from the gene to mRNA increase to accelerate synthesis of enzyme molecules (Table 16-1). In fewer cases, increases in P450 enzymes occur without increases in transcription.

TABLE 16-1 Chemicals that induce the production of cytochromes P450 and the specific gene family affected

Inducing Chemicals	P450 Gene Family or Subfamily Affected[1]
Polycyclic hydrocarbons, 2,3,7,8-tetrachlorobenzo-*p*-dioxin, benzoflavones	I, IIA
Phenobarbital, terpenoids	IIB, IIC, III, VI
Ethanol, acetone, pyrazole, imidazole	IIE[2,3]
Pregnenolone 16α-carbonitrile, dexamethasone	III
Rifampicin, triacetyloleandromycin, griseofulvin	III[3]
Clofibrate, other peroxisome proliferators	IV
ACTH (cyclic AMP)	XI, XVII, XIX, XXI
Tetradecane	LII
Camphor	CI
Phenobarbital	CII

Family names are those in Figure 16-1.

From Nebert et al. 1989.

[1] Unless otherwise indicated, an increase in P450 protein has been shown to reflect increases in the transcriptional rate of the gene.

[2] Post-transcriptional.

[3] Post-translational.

FIGURE 16-2

A schematic view of the detoxification of chemicals such as plant defensive compounds (allelochemicals) and other chemicals foreign to the organism (xenobiotics). These commonly lipophilic compounds may be ingested, treated with cytochromes P450 in various parts of the body, and either excreted as hydrophilic compounds or broken down further in a secondary phase of metabolism. (*From Brattsten 1992, based on Williams 1974*)

LIPOPHILIC ⟶ HYDROPHILIC

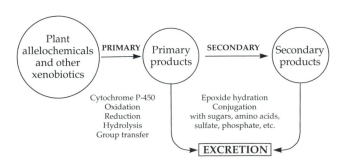

Polymorphism in Drug Responses and Ethnic Variation

Polymorphisms in P450 genes and proteins are important in drug therapies because some morphs effect strong induction in response to a drug and break it down rapidly, whereas other morphs metabolize the drug poorly and are thus hypersensitive. Unless drug testing is carried out on a large population with the relevant morphs represented, the results may be unexpected and serious. For example, the chemical debrisoquine was used in the treatment of hypertension in humans until the discovery of a polymorphic reaction in Caucasian populations in North America and northern Europe (Gonzalez and Nebert 1990). About 90 percent of the population are extensive metabolizers (EM phenotypes), and the remainder are poor metabolizers (PM phenotypes) and liable to experience undesirable reactions to the drug (Figure 16-3). PM is a recessive allele transmitted in Mende-

FIGURE 16-3

The frequency distribution in a British Caucasian population showing how the polymorphism in cytochrome P450 influences a population's ability to metabolize the drug debrisoquine, used in the treatment of hypertension. Extensive metabolizers (EMs) contained a P450 that could hydroxylate the drug, leading to excretion of low levels of debrisoquine and high levels of hydroxydebrisoquine, which resulted in low values on the $\log_{10}$ metabolic ratio scale. Poor metabolizers (PMs) with inefficient P450 enzymes excreted relatively large amounts of the drug and low levels of the metabolic product, resulting in high values on the metabolic ratio scale. *(From Gonzalez and Nebert 1990, Idle and Smith 1979)*

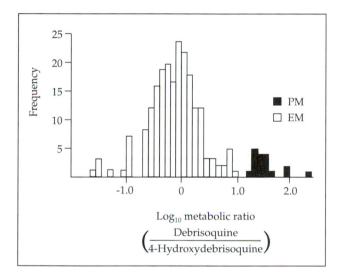

lian fashion. Susceptibility to various forms of cancer may be influenced at the EM–PM locus. In the case of lung carcinoma, for example, the EM phenotypes are more susceptible. Hence, the polymorphism may have been maintained in populations (before drug therapy and smoking) by the balancing selective effects of different kinds of environmental chemicals. Large ethnic differences in drug metabolism support this theory. The PM phenotypes can occur in as much as 50 percent of a population. Occurrence is 16 percent in Nigeria and 5 to 10 percent in northern Caucasians; the phenotype may be absent in the Cuna Amerindians of Panama and in the Japanese. Dietary differences may well be involved, because frequent use of foods with toxic ingredients may result in natural selection against the PM genotype. In small doses, plant toxins can produce pharmacological effects that are important in herbal medicines. For example, some sulfur-containing compounds that are common in the onions and related plant species or that are produced

when plants are crushed have strong bacteriocidal activity. They are the natural equivalents of the sulfur drugs of modern medicine.

Some Mechanisms Resulting in Gene Multiplication

Geneticists have also addressed the question of how one gene has proliferated into such an extensive superfamily of genes. What are the proximate genetic mechanisms that have resulted in very different P450 genes but conserved enough small sequences to allow the development of a "family portrait"? Many molecular mechanisms have been involved, no doubt, but two seem especially important: **gene duplication** and **gene conversion.**

Gene Duplication

Genes that pass down lineages are called **orthologous genes.** They may be tracked to a common ancestor, such as α-hemoglobin genes in vertebrates. But such genes may undergo gene duplication in a lineage—resulting, for example, in α- and β-hemoglobin genes after divergence of the genes. Further gene duplication increases the ultimate polymorphism in a lineage, as in the hemoglobins now found in humans, α, β, γ, and δ. The descendants of the duplicated ancestral gene are known as **paralogous genes.** Paralogous genes occur in the P450 families—such as four members in the cytochrome P450 IID subfamily (CYP2D) (Figure 16-4). Thus, gene duplication can produce multiple copies of an ancestral gene in a lineage, and every orthologous gene can evolve to perform a different function. Many of the bifurcations in the P450 lineage have resulted from gene duplications, as indicated in Figure 16-1, where the 78 genes derive from at least 39 gene duplications.

Gene Conversion

The second important process in the evolution of the P450 superfamily is gene conversion, which results from nonreciprocal recombination following crossing over between homologous chromosomes during meiosis (Figure 16-5). Thus, sequences of genes or parts of genes are rearranged, a process that may have repeated itself many times in the radiation of the P450 superfamily (Gonzalez and Nebert 1990).

This example of adaptive radiation of a gene into a large superfamily illuminates one reason why so much protein polymorphism occurs in populations and species. It is a particularly fascinating case because we can trace the P450 lineage back so far and because the family is so important in determining susceptibility to the large variety of environmental chemical hazards—and therefore in the medical

FIGURE 16-4

Four rat genes in the cytochrome P450 subfamily CYP2D, showing their similarities in structure. The full nucleotide sequence for part of the CYP2D5 gene is provided, and only differences appear for equivalent parts of the other genes, 2, 3, and 4. The solid lines indicate regions in the genes that are thought to have been involved with gene conversion events. Vertical rows of dots indicate boundaries between introns and exons *(From Gonzalez and Nebert 1990)*

```
        Intron 7                                          Exon 8
CYP2D5  CTCGGCAGTGCCACCCCTGTCCATCTTCATGCCCTGTGCCTTC TGCCAG  GGGACGACCCTCATCATCAACCTGTCGTCCGTGCTGAAGGATGAGACTGTCTGGGAGAAGCCCCTCCGCT  4955
CYP2D2  A    CT A A     T  CA A A   C T A               A      CC       C
CYP2D3  A A  CA  -       GC AA A                         A      CC       C A                                          A
CYP2D4  A A  CA  -       GC AAT A  T C                   A         C      C A                 C

                                                                     Intron 8
TCCACCCTGAACACTTCCTGGATGCCCAGGGCAACTTTGTGAAGCATGAGGCCTTCATGCCATTCTCAGCAG  GTACCTCTGGGGGGCCTGACTCCCTGTCTATTCCCAGGGTGGCTGGGA  5075
   T A                                                                          G                              A
                                                                                G

                                                Exon 9
GGGTGGAGCTCCAATCAGGCCCCAGACTGACTGAGCCCCTGTCCCACCACAG  GCCGCAGAGCATGCCTTGGGGAGCCCCTGGCCCGCATGGAGCTCTTCCTCTTCTTCACCTGCCTCCTG  5195
          G                          T                              G
              A                      T                              G
                                                                    G

CAGCACTTCAGCTTCTCCGTGCCGGCCGGACAGCCCCGGCCCAGG ACCCTTGGCAACTTTGCTATTTCAGTTGCCCCCTTGCCCTACCAGCTCTGTGCTGCGGTACGGGAGCAAGGACACTA  5317
   G        G T  C G                          A   GT AC AT GC   AT ACA T   A      AT CT CT   G G GGG
A G         CA T                             GA TA TGT   CT CTCC AG T CC          ATTCAA TT      A GA
   G        CA T                             GA TA  T   G GC CTGACCA  GCGC        T ACCC CT A-GG  GG AC

A  TTCCAGTCCAGC-TC
G C    -T TT -
     T      C
GCATCTCA TCACTG
```

FIGURE 16-5

An example of reciprocal and nonreciprocal recombination following crossing over between homologous chromosomes during meiosis. *(From Watson 1970)*

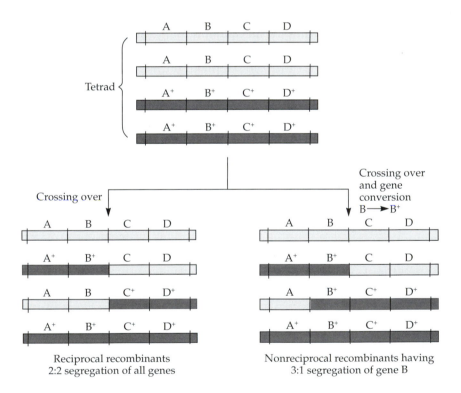

treatment of humans. The P450 enzymes also pose vexing problems for humans, because the pests we would like to control have their own banks of inducible P450 enzymes ready to detoxify our pesticides (Brattsten 1992). It is perhaps a little unpleasant to realize that our P450s have a common heritage!

16.2 THE MAINTENANCE OF ALLELE POLYMORPHISM AT ONE LOCUS

Another aspect of protein polymorphism is the common occurrence of more than one allele at a single locus, resulting in some proportion of a population being heterozygous. How are multiple variations of a gene, whose enzyme products perform very similar functions, maintained in a population? Are the variants equally effective and without selective differences, or does each contribute uniquely to the organism's ability to survive the vagaries of the environment? The EM–PM locus in the P450s suggested a clue to such polymorphisms, with one enzyme promoting drug metabolism and the other providing greater resistance to chemically induced cancers. But the mechanism of mutation in a hemoglobin gene is better understood.

A Mutation in a Human Hemoglobin Gene

Ancient in terms of human history, but recent relative to the origin of cytochrome P450, is a novel mutation in a gene that codes for part of the human hemoglobin molecule. Each normal red blood cell carries hemoglobin molecules in the millions. A normal hemoglobin molecule (HbA) in an adult human is made of four polypeptide chains: two α chains with 141 amino acid residues each and two β chains with 146 amino acid residues each. Each chain winds around a heme, containing iron, that is responsible for picking up oxygen in the lungs and transporting it to cells in the remainder of the body. The four polypeptide chains then interact with one another to form a complex three-dimensional structure—the hemoglobin molecule (Figure 16-6). The empirical formula for normal hemoglobin is $(C_{738}H_{1166}O_{208}N_{203}S_2Fe)_n$. When stretched out into a two-dimensional chain, the sequence of amino acids can be identified as follows in the β chain:

Sequence in DNA   CAGaGTAaAACa
                    TGCaGGTaCTC . . . . . .
Sequence in RNA   GUCaCAUaUUGa
                    ACGaCCAaGAG . . . . . .
Amino acid   Valine Histidine Leucine
               Threonine Proline Glutamine . . . . . .

A small mutation in the sixth codon from CTC to CAC caused translation of the mRNA to valine instead of glutamine. This changed the hemoglobin structure and its resultant properties, with extensive repercussions in molecular activity, cellular structure, human physiology, and the evolution of the human species. Scientists gave the mutant form the symbol HbS.

Change in Molecular Activity and Cellular Structure

The glutamine in normal hemoglobin is electrically charged and hydrophilic, or water-attracting. But valine is uncharged and hydrophobic, or water-repelling. This change in electrical charge causes molecules of HbS to aggregate into chains of polymerized hemoglobin molecules in the capillaries after releasing oxy-

FIGURE 16-6

A hemoglobin molecule, showing two α and two β polypeptide chains, each wrapped around a heme and each bonded to the other polypeptide chains in a three-dimensional complex. Note that the mutation producing HbS hemoglobin is positioned near the ends of the β polypeptide chains. *(From Cerami and Peterson 1975)*

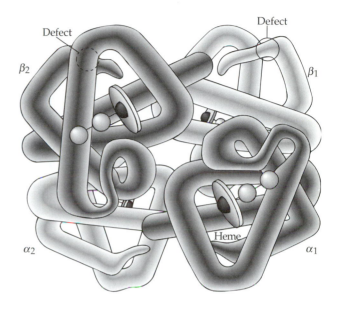

gen (Cerami and Peterson 1975). Under low oxygen tension, the formation of many long fibrils pushes the red blood cells out of shape, and they commonly elongate, developing pointed ends and becoming stiff with fibrils. A crescent or sickle-like shape is common, hence the designation HbS ("S" for *sickle*).

Physiological Problems

Sickled cells tend to get caught in the narrow capillaries and block blood flow (Figure 16-7). If all red blood cells, or erythrocytes, in a human individual contain HbS hemoglobin, then blockages become so common that the resultant poor blood circulation causes severe symptoms: tiredness, headache, muscle cramps, severe pain, and (understandably) irritability. The body destroys damaged cells, creating a shortage of red blood cells, and anemia develops. In fact, the physiological effects of **sickle-cell anemia** (sickle-cell disease), the disease caused by HbS, are frequently so devastating that death occurs early in life.

Inheritance of HbA and HbS

The two alleles of HbA and HbS at the same locus behave as Mendelian codominant traits. Hence, the three possible phenotypes have different properties.

FIGURE 16-7

A schematic representation of the human circulatory system, showing red blood cells, or erythrocytes, flowing down from the lungs (**left**) to the heart, which pumps them through arteries to the rest of the body, depicted as a network of capillaries (**right**). In the capillaries, oxygen leaves the hemoglobin and enters the cells, and the deoxygenated hemoglobin tends to aggregate and distort the erythrocytes into crescent-like and other shapes, in which condition they pass in veins back to the heart and lungs. Note that in the capillaries sickled cells tend to block free passage of blood. Arrows indicate direction of blood flow. *(From Cerami and Peterson 1975)*

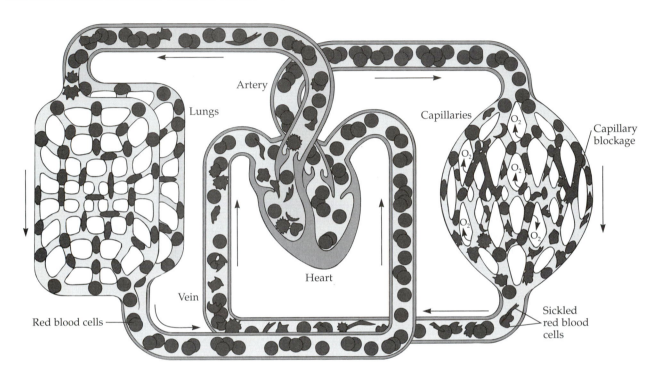

The homozygote HbAHbA produces no sickle-type hemoglobin. The heterozygote, HbAHbS, produces about 60 percent normal and 40 percent sickle-type hemoglobin, and the homozygote HbSHbS produces 100 percent of the mutant hemoglobin. Bearers of the mutant homozygote therefore suffer the effects of sickle-cell anemia. Heterozygotes have a preponderance of normal hemoglobin and suffer no obvious deleterious effects unless they are physically stressed—for example, at high altitudes. They are **carriers** of the **sickle-cell trait**.

Frequency of Alleles in Natural Populations

Under most circumstances, the frequency of the mutant allele HbS stays very low in a population because of the allele's lethal effect on the homozygotes—selection against the mutant is stringent. Yet in large areas of Africa the HbS mutant became very common. Some native African populations have 40 percent of individuals heterozygous with the HbS trait, and frequencies of 10 to 20 percent are widespread. In matings between two carriers, the probability is one in four

that a child will be HbSHbS and suffer the consequences of the disease; in about 80 percent of cases, this means early death. Hence, to maintain high frequencies of HbS in a population, the HbAHbS genotype must confer a strong selective advantage to the individual—so strong that it more than compensates for the deleterious effects in the mutant homozygote.

Adaptive Advantage of the Heterozygote

Anthony C. Allison (1954, 1956) solved the puzzle of the sickle-cell trait's commonness in some African populations. He noted that the incidence of HbS coincided with the distribution of malignant tertian malaria, a commonly lethal disease caused by the protist parasite *Plasmodium fulciparum* (Figure 16-8). In a broad belt across Africa, children were exposed to malaria, which is transmitted by blood-sucking mosquitoes, almost all year long. The probability of contracting the disease was very high. Children were particularly vulnerable until they were old enough to develop immunity.

FIGURE 16-8

A map of the African continent, showing the frequency of the sickle-cell gene, HbS, in the native African population. Note that the broad belt of HbS distribution from west to east Africa is within the tropics, where the incidence of the frequently lethal malignant tertian malaria is also high. High frequencies of HbS in the east are associated with Lake Victoria on the equator (part of the Rift Valley system) and coastal Mozambique. *(From Allison 1956)*

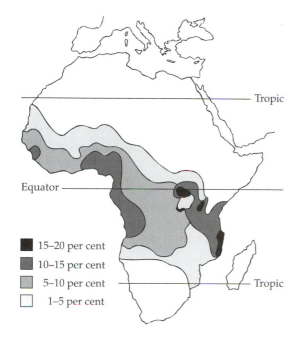

- ■ 15–20 per cent
- ■ 10–15 per cent
- ■ 5–10 per cent
- □ 1–5 per cent

Allison (1954) found that children with the sickle-cell trait (HbAHbS) had a considerably higher probability of surviving malaria than those who were homozygous for normal hemoglobin (HbAHbA); the heterozygote had perhaps a 25 percent advantage where malaria was most severe. Since the 1950s, researchers have discovered that sickling kills the malaria-causing *Plasmodium* parasite, which develops inside red blood cells. In addition, the parasite produces change in pH within the cells, which increases the probability of cell sickling. Thus, in areas with malignant tertian malaria, the selective advantage of HbS in the heterozygous form balanced the selective disadvantage of HbS in the homozygous form, and the genetic polymorphism in populations was a **balanced polymorphism** maintained by **heterozygote advantage.**

Consequences of a Lost Heterozygote Advantage

As native Africans migrated to other parts of the world, outside the zones of tertian malaria, the advantage of the HbAHbS genotype was lost. With only

negative selective pressure on HbS, its frequency should decline. This has been the case. Assuming that slave ships from West Africa carried to the United States a Negro population with 20 to 25 percent carriers of the sickle-cell trait, selection against the trait over the past 300 years has reduced the frequency of HbS. Matings with other ethnic groups alone may have reduced the HbS frequency to 15 percent. Selection over about 12 generations would have caused the loss of an additional 6 percent HbS, resulting in the presently observed frequency of about 9 percent in the black population. This means that 3 in 1,000 children born to black parents have sickle-cell anemia. Whether medical treatment will halt the continued evolution of HbS frequency in North America remains to be seen. Almost certainly, as tertian malaria is controlled in Africa, gene frequencies will continue to evolve and HbS frequencies will decline.

The decline in frequency of the HbS allele is comparable to the decline in the melanic mutant of the peppered moth, *Biston betularia*, after pollution abatement in the British Isles, discussed in Chapter 1. Both results were predictable from evolutionary theory.

The Importance of the Sickle-Cell Case

Sickle-cell anemia was the first case of a disease being understood at the molecular level. A single amino acid substitution changed the function of the hemoglobin molecule, which in turn changed the cell form and human physiology, resulting in the molecular disease sickle-cell anemia. The relationships between humans and *Plasmodium* parasites also changed, with a selective advantage developing for carriers of the HbS allele, and populations in Africa evolved as HbS frequencies increased. This evolutionary trend has been reversed in populations unexposed to malaria.

The sickle-cell case demonstrates that we can trace and explain the effects of a single mutation all the way to the evolution of human populations. We can understand the persistence of a balanced polymorphism maintained by heterozygote advantage in Africa. Not only do we comprehend the proximate mechanisms resulting in a human disease, but we have the ultimate explanation for the evolution of this disease.

16.3 PATTERNS IN GENETIC AND PROTEIN POLYMORPHISM

The cases of the cytochromes P450 and the hemoglobin HbA–HbS system provide powerful evidence to account for the existence of large amounts of genetic and protein polymorphism and heterozygosity in natural populations. Of course, human diseases have given humans a strong impetus to understand these phenomena; the P450 and HbS cases are of particular interest in human epidemiology and population genetics. Given the amount of research required to understand these cases and the humanitarian pressure to develop knowledge of them, we can begin to understand why so much genetic polymorphism in nonhuman species remains to be explained in detail. Nevertheless, the search for patterns in natural genetic polymorphisms is a first step toward understanding the mechanisms that cause such patterns.

Polymorphism in Plants

From 1968 to 1988, about 650 studies on the electrophoretically detectable genetic diversity in plant species and populations were published, covering about 450 species and 165 genera. Starting from these sources, Hamrick and Godt (1990) searched for patterns in genetic variation on the basis of eight criteria: taxonomic status, life form, geographic range, regional distribution, breeding system, seed dispersal, mode of reproduction, and successional status (Table 16-2). This kind of broad-ranging survey adds a dimension to our understanding of genetic systems of species (discussed in Chapter 14). It helps us group species in terms of their genetic systems.

Overall, in plant species, 50.5 percent of loci were polymorphic, with a mean of 1.96 alleles per locus. In populations within these species, 34.2 percent of loci were polymorphic. Certainly these estimates demonstrate an impressive amount of genetic and protein polymorphism in natural populations. Variations around these mean figures at the species level were generally significant. Even so, explanatory power was relatively weak; the criteria account for only 24 percent of the variation in genetic diversity. Of this 24 percent, geographic range accounted for the largest part, 32 percent; life form added 25 percent; breeding system, another 17 percent; and seed dispersal mechanism, 17 percent.

Taxonomic Status and Correlated Traits

Under the first of the eight criteria, taxonomic status, gymnosperms had a higher level of polymorphism than monocotyledonous angiosperms, and both had greater polymorphism than dicotyledonous angiosperms (Table 16-2). The explanations for such differences are complex, because so many of the eight traits studied are correlated to some extent. Correlations existed among taxonomic status, life form, geographic range, breeding system, seed dispersal, and successional stage. The gymnosperms studied were

long-lived coniferous woody perennials, many with wide geographic ranges, outcrossing by wind, sexual reproduction, and largely occurring in late succession—all traits that tend to produce high genetic diversity. As we should expect for late successional species that frequently occur in dense monocultures, with extensive ranges over much environmental heterogeneity and with wind pollination and obligate sexual reproduction, genetic polymorphism was high.

Geographic Range

Geographic ranges of species exhibited a strong pattern, with relatively low polymorphism in endemic species, at 40 percent of loci, increasing to 59 percent in widespread species. Which is cause and which effect is hard to determine, but again correlated traits are probably important because trees with wide ranges (for example) are commonly obligately sexual, wind-pollinated outbreeders. The high polymorphism in boreal–temperate regions also correlates with the large number of conifers in those cold climates.

We can see other interesting patterns in breeding systems, with wind pollination, seed dispersal via animal transport, and late succession correlating with high polymorphism and with sexual reproducers having relatively high polymorphism (Table 16-2).

Variation in Populations

Hamrick and Godt (1990) discovered a strong correlation between genetic variation at the species level and at the population level, with genetic diversity values accounting for 79 percent of the variance. Within populations, genetic diversity correlated most strongly with the traits associated with breeding system, geographic range, taxonomic status, and life form.

All of these patterns help us to understand the complex interplay of factors discussed in other chapters: mode of reproduction, population size and structure, the potential for movement of gametes or propagules between populations, generation time, habitat heterogeneity, the role of mutualism (in pollination and seed dispersal), and other factors. Whether we will ever attain a comprehensive understanding of the extent to which each factor contributes to genetic polymorphism is an open question. Clearly, we need much more information on the biochemical role of each allele at polymorphic loci and why such polymorphisms are maintained. This information will be a long time coming. We also need a deeper knowledge of how alleles at different loci interact and the extent to which natural selection works on gene complexes rather than at single loci. Is it only when selection is strongest, and genetic consequences dire, as

TABLE 16-2   Allozyme variation within species of plants

Criterion	Number of Species	Mean Number of Populations	Mean Number of Loci	Percentage of Loci Polymorphic	Genetic Diversity[1]
Taxonomic status					
Gymnosperms	55	8.5	16.1	70.9a[2]	0.173a
Monocots	111	18.7	15.5	59.2b	0.181a
Dicots	329	11.9	16.8	44.8c	0.136b
Life form					
Annual	190	18.5	14.9	50.7b	0.161a
Short-lived perennial					
Herbaceous	152	8.8	17.2	41.3c	0.116b
Woody	17	18.1	25.2	41.8−[3]	0.097−
Long-lived perennial					
Herbaceous	4	6.0	13.8	39.6−	0.205−
Woody	110	9.3	17.0	64.7a	0.177a
Geographic range					
Endemic	81	6.5	17.8	40.0c	0.096c
Narrow	101	8.8	16.9	45.1bc	0.137b
Regional	193	10.4	16.7	52.9ab	0.150b
Widespread	105	25.5	14.6	58.9a	0.202a
Regional distribution					
Boreal–temperate	19	7.8	16.2	79.7a	0.186a
Temperate	348	11.6	15.6	48.5b	0.146a
Temperate–tropical	30	37.3	15.1	58.8b	0.170a
Tropical	76	10.6	21.3	49.2b	0.148a
Breeding system					
Selfing	123	20.3	16.2	41.8b	0.124b
Mixed—animal	64	8.9	14.4	40.0b	0.120b
Mixed—wind	9	10.0	12.5	73.5a	0.194a
Outcrossing—animal	172	10.7	17.7	50.1b	0.167ab
Outcrossing—wind	105	10.7	16.7	66.1a	0.162ab
Seed dispersal					
Gravity	198	10.1	16.9	45.7b	0.136bc
Gravity-attached	15	29.2	18.6	69.3a	0.166ab
Attached	55	20.8	16.5	68.8a	0.204a
Explosive	27	12.4	18.6	30.4c	0.092c
Ingestive	67	17.6	13.2	45.7b	0.176ab
Wind	111	8.7	16.6	55.4b	0.144bc
Mode of reproduction					
Sexual	407	13.1	16.7	51.6a	0.151a
Sexual and asexual	66	9.2	15.0	43.8b	0.138a
Successional status					
Early	226	18.5	16.6	49.0b	0.149a
Middle	152	7.6	16.4	47.6b	0.141a
Late	95	9.4	16.3	58.9a	0.161a

Based on Hamrick and Godt 1990.

[1] Genetic diversity (H_{es}) was calculated for each locus using the formula $H_{es} = 1 - \Sigma p_i^2$, where p_i is the mean frequency of the ith allele. The mean of all H_{es} values over all loci was then used in the table.

[2] Within each major criterion, means followed by the same letter in a column are not significantly different at the 5 percent probability level.

[3] The sign − indicates data excluded from analysis because of small sample size.

in the cases of industrial melanism and sickle-cell anemia, that we can thoroughly understand the evolution of gene frequencies at a single locus?

Polymorphism in Animals

Taxonomic Comparisons

Animals are not as easily classified as plants into the many kinds of categories examined by Hamrick and Godt (1990), and the main comparisons among animals have emphasized simple taxonomic similarities and differences. There are some additional interesting contrasts, as shown in Table 16-3. One striking difference is that animals tend to have lower genetic polymorphism than plants. Within the animals, vertebrates generally exhibit lower variation than invertebrates. The group "other marine invertebrates" shows high genetic diversity, probably because the release of gametes into water is common and currents determine the settling of immatures, resulting in large panmictic populations. In this sense, these invertebrates are similar to any plants with outbreeding mating systems and wind dispersal of pollen and seeds. Why the small sample of marine snails exhibits such low polymorphism and heterozygosity is not clear. Haplodiploid wasps and bees show reduced genetic variation relative to other insects, possibly because with one sex haploid, balancing selection and heterozygote advantage are less likely to persist.

The relatively low genetic polymorphism and heterozygosity in fish and homeothermic vertebrates is interesting. It possibly reflects the maintenance of stable bodily conditions by various means, which would reduce the importance of genetic diversity. Marine fish, especially, live in stable environments in terms of physical conditions and the kinds of foods consumed. In contrast, rodents and other mammals may well depend on microbial enzymes and genetic polymorphism to handle their diverse foods and the corresponding environmental diversity within their ranges.

Does Substrate Dictate Heterozygosity?

Each gene codes for a polypeptide chain, and the chains form into the enzymes and other proteins that are measured in electrophoretic studies. We may therefore predict that enzymes with invariable substrates will be monomorphic, with individuals homozygous, and that enzymes with variable substrates are likely to be more polymorphic, with many individuals heterozygous. A heterozygous individual produces two enzymes at a locus, instead of the single

TABLE 16-3 Genetic polymorphism based on electrophoretic analysis of allozymes in animal taxa

Taxon	Number of Species	Percent of Loci Polymorphic	Mean Heterozygosity
Invertebrates			
Marine snails	5	17.5	8.3
Land snails	5	43.7	15.0
Other marine invertebrates	9	58.7	14.7
Haplodiploid wasps and bees	6	24.3	6.2
Drosophila	43	43.1	14.0
Other insects	23	32.9	7.4
All invertebrates	93	39.7	11.2
Vertebrates			
Fishes	51	15.2	5.1
Amphibians	13	26.9	7.9
Reptiles	17	21.9	4.7
Birds	7	15.0	4.7
Rodents	26	20.2	5.4
Mammals	46	14.7	3.6
All vertebrates	135	17.3	4.9
All plants (from Hamrick and Godt 1990)	473	50.5	—

Based on Hartl 1980 and Hedrick 1983.

TABLE 16-4   Mean heterozygosity in animal taxa for enzymes with different functions

	Enzyme Category		
Organism	Variable Substrate	Regulatory Enzymes	Nonregulatory Enzymes
Drosophila	0.205	0.210	0.086
Other insects	0.289	0.281	0.094
Noninsect invertebrates	0.169	0.100	0.122
Fishes	0.063	0.110	0.066
Amphibians	0.118	0.227	0.062
Reptiles	0.079	0.039	0.039
Birds	0.088	0.096	0.151
Mammals	0.048	0.056	0.032
Average	0.175	0.161	0.073

From Hedrick 1983, after Powell 1975

enzyme from a homozygous locus, and is likely to benefit from the greater diversity of enzymes if substrates are variable. Regulatory enzymes may also be polymorphic if their functions are affected by temperature (for example), which could be more important in poikilotherms than in homeotherms. Nonregulatory enzymes include structural and ribosomal proteins.

Powell (1975) conducted a survey that uncovered some interesting patterns in mean heterozygosity in animal taxa. He categorized enzymes as working on variable substrates, regulatory enzymes, or nonregulatory enzymes. Variable-substrate enzymes and regulatory enzymes both generally showed higher levels of heterozygosity than nonregulatory enzymes (Table 16-4), as predicted. Terrestrial poikilotherms showed higher levels of heterozygosity for the variable-substrate and regulatory enzymes than did homeotherms, except for lizards—which may achieve

greater metabolic stability behaviorally when compared with amphibians, for example. Fishes exhibited levels of heterozygosity comparable to those of homeotherms.

16.4 IS GENETIC AND PROTEIN POLYMORPHISM ALWAYS ADAPTIVE?

This chapter has dwelt upon a few approaches to the search for explanations of polymorphism in species and populations. The search will continue. But could it be in vain in many cases? Could much variation be selectively neutral?

We need to place the study of genetic polymorphism against the background of history so that we can better appreciate the development of such questions. In the next chapter, we will do just that.

SUMMARY

Perhaps 2 billion years ago, a gene became the progenitor of a large superfamily of **cytochrome P450 genes.** Able to break down toxic substances in the environment, the lineage became important to bacteria; insects; mammals, including humans; and many other groups. The chemicals **(xenobiotics)** foreign to the bodies of these organisms were so varied that the gene families proliferated through time, with adaptive radiation probably accelerating as animals

started to consume terrestrial plants. When a xenobiotic challenges the body of an organism, P450 enzymes are induced and increase their capacity to detoxify, usually by accelerating transcription from the gene to mRNA.

Many polymorphisms in P450 genes occur in humans. The existence of both extensive metabolizers (EMs) and poor metabolizers (PMs) in a population complicates the development of drug therapies. The

P450 polymorphisms vary ethnically, probably reflecting long-standing dietary differences, particularly in plant foods and their modes of preparation.

The radiation of the P450 superfamily involved extensive gene duplication and gene conversion. This example of adaptive radiation is important to our understanding of the origins of extensive genetic variation and to the use of drug therapies in medical treatments of humans.

Normal hemoglobin (HbA) and **sickle-type hemoglobin** (HbS), exemplifying **blood polymorphism** in humans, illustrate how genetic heterogeneity can be maintained in a population when **heterozygote advantage** results in a balanced polymorphism. A mutation in a single codon resulted in HbS, a trait that, in the homozygous condition, results in polymerization of molecules, distortion of erythrocytes, reduced circulation of blood, and the many symptoms associated with **sickle-cell anemia, or sickle-cell disease.** Until the advent of modern medical practices, this disease was commonly fatal before reproductive age, forcing the question of how the trait was maintained at high frequencies in some African populations. The answer lay in the discovery that high frequencies of HbS coincided with frequent infection by the protist parasite *Plasmodium fulciparum,* which causes the often lethal disease malignant tertian malaria. Heterozygotes (HbAHbS) gained an advantage over both homozygotes (HbAHbA and HbSHbS) in the presence of malaria because of greater resistance to the disease, without experiencing the symptoms of sickle-cell anemia. In regions to which African Negroes have emigrated, out of malarial zones, a predictable decline in HbS frequency has been evident. The story of research on sickle-cell anemia demonstrates the development of a detailed understanding of a molecularly based disease, and it shows how a single mutation can change the function of a molecule, with a series of ramifications for cell function, parasite survival, human physiology, and the evolution of gene frequencies in human populations.

Scientists have observed a great amount of genetic and protein polymorphism in many species and populations of plants and animals. Investigated via electrophoretic techniques, plant species show a mean of 50 percent of loci polymorphic; the range is from 45 percent in dicotyledonous angiosperms to 71 percent in gymnosperms. The range in animals is from 15 percent polymorphic loci in mammals to 59 percent in some groups of marine invertebrates, with levels of polymorphism generally somewhat lower than in plants. Potential explanations for differences in genetic diversity among taxa and among species categorized according to criteria such as geographic range, breeding system, and mode of reproduction fail to account for a large part of the variation. We do know that widespread species with large populations and ranges and with outbreeding systems that include extensive dispersal of gametes or propagules generally exhibit relatively high genetic diversity. The different functions of gene products may also explain some genetic diversity, because enzymes working on variable substrates and regulatory enzymes, especially in some poikilotherms, tend to be more polymorphic than nonregulatory enzymes. Nevertheless, so much genetic polymorphism commonly occurs in natural populations that scientists have wondered whether all of it is under the influence of natural selection.

QUESTIONS FOR DISCUSSION

1. Concerning the argument on the selective pressures on the diversification of cytochrome P450 genes, could you develop objective tests to examine, for example, the role of toxic plants as selective agents in proliferation of these genes?

2. Do you think that the study of polymorphisms in drug responses in humans and animals will help to elucidate the reasons for the high numbers of P450 genes in individuals and populations?

3. Does the scenario developed for the cytochrome P450 gene family suggest to you that molecular evolution is under the strong influence of natural selection, or would you consider chance events such as gene duplication and more or less random mutation to be of similar importance?

4. Do you think that the sickle-cell trait will vanish in areas of the world where malaria no longer exists?

5. For the case of the mutation in the β-hemoglobin chain that resulted in the sickle-cell trait, do you think that a mutation at another point in the chain would have a similarly strong effect, or is the effect of a mutation likely to vary greatly depending upon the position of the mutation in the molecule?

6. If the mutation resulting in the sickle-cell trait had occurred on the α-hemoglobin chain, would the effects be identical?

7. Can you imagine the kinds of effects a mutation might have if it occurred on either hemoglobin chain, but close to the heme group?

8. There are various interesting comparisons to be made concerning the amount of genetic variation in plants adapted to different ways of life. In pairwise comparisons can you detect the likely reason for differences—for example, which percentage of loci that are polymorphic is higher in wind-pollinated outcrossing plants than in animal-pollinating plants, and higher in late successional plants than in early successional plants?

9. If variable substrates result in selection for genetic polymorphism, why do you think that plants tend to have such high levels of polymorphism compared with many animal groups?

10. Why do you think that vertebrates in general have lower genetic polymorphism than many invertebrate groups and plants?

REFERENCES

Allison, A. C. 1954. Protection afforded by sickle-cell traits against subtertian malarial infection. Brit. Med. J. 1:290–301.

Allison, A. C. 1956. Sickle cells and evolution. Sci. Amer. 195(2):87–94.

Brattsten, L. B. 1992. Metabolic defenses against plant allelochemicals, pp. 175–242. *In* G. A. Rosenthal and M. R. Berenbaum (eds.). Herbivores: Their Interactions with Secondary Plant Metabolites. 2d ed. Vol. 2, Ecological and Evolutionary Processes. Academic Press, San Diego.

Cerami, A., and C. M. Peterson. 1975. Cyanate and sickle-cell disease. Sci. Amer. 232(4):44–50.

Gonzalez, F. J., and D. W. Nebert. 1990. Evolution of the P450 gene superfamily: Animal–plant "warfare," molecular drive and human genetic differences in drug oxidation. Trends Genet. 6:182–186.

Hamrick, J. L., and M. J. W. Godt. 1990. Allozyme diversity in plant species, pp. 43–63. *In* A. D. H. Brown, M. T. Clegg, A. L. Kahler, and B. S. Weir (eds.). Plant Population Genetics, Breeding, and Genetic Resources. Sinauer, Sunderland, Mass.

Hartl, D. L. 1980. Principles of Population Genetics. Sinauer, Sunderland, Mass.

Hedrick, P. W. 1983. Genetics of Populations. Science Books International, Boston.

Idle, J. R., and R. L. Smith. 1979. Polymorphisms of oxidation at carbon centers of drugs and their clinical significance. Drug Metab. Rev. 9:301–317.

Nebert, D. W., and D. R. Nelson. 1991. P450 gene nomenclature based on evolution. Methods Enzymol. 206:3–11.

Nebert, D. W., D. R. Nelson, and R. Feyereisen. 1989. Evolution of the cytochrome P450 genes. Xenobiotica 19:1149–1160.

Powell, J. R. 1975. Protein variation in natural populations of animals. Evol. Biol. 8:79–119.

Watson, J. D. 1970. Molecular Biology of the Gene. 2d ed. W. A. Benjamin, Menlo Park, Calif.

Williams, R. T. 1974. Inter-species variations in the metabolism of xenobiotics. Biochem. Soc. Trans. 2:359–377.

17

NON-DARWINIAN EVOLUTION

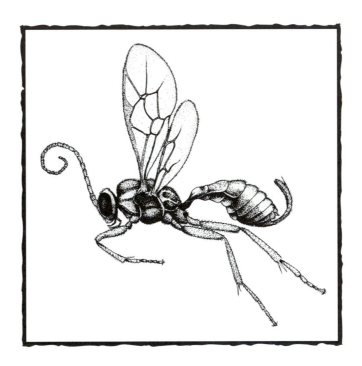

A nonsocial haplodiploid wasp, *Lathrostizus lugens*, probably has relatively high levels of genetic variation compared with its social relatives in the order Hymenoptera.

17.1 GENETIC VARIATION IN POPULATIONS

In 1963, Mayr noted that, up to that time, estimates of genic polymorphism in natural populations had ranged from 50 percent of loci down to less than 1 percent of loci. Such estimates were more like guesses, based on what *should be* present rather than what *was* present. Even so, a good deal of cytological evidence about genetic variation had accumulated. That evidence included polymorphisms for chromosomal inversions and translocations, frequency of rare visible mutations at many loci, frequencies of chromosomes that were deleterious when homozygous, and the amounts of the deleterious effects. In addition, scientists could observe some examples of single-locus polymorphisms, such as sickle-cell anemia. Dobzhansky (1951) and Mayr (1963) published books that provided authoritative descriptions of the state of the art in genetic variation.

But the approaches used in population genetics up to the early 1960s did not allow researchers to address one of the most fundamental questions: How much genetic variation exists in a population, or at what proportion of loci is a diploid individual likely to be heterozygous, and what proportion of loci in a population is likely to be polymorphic? And yet, as Hubby and Lewontin (1966, 577) stressed,

> A cornerstone of the theory of evolution by gradual change is that the rate of evolution is absolutely limited by the amount of genetic variation in the evolving population. Fisher's "Fundamental Theorem of Natural Selection" (1930) is a mathematical statement of this generalization, but even without mathematics it is clear that genetic change caused by natural selection presupposes genetic differences already existing, on which natural selection can operate. In a sense, a description of the genetic variation in a population is the fundamental datum of evolutionary studies; and it is necessary to explain the origin and maintenance of this variation and to predict its evolutionary consequences.

Thus, a major effort in genetics during the previous 50 years (from early in the 20th century) had been directed at estimating the amounts and kinds of genetic variation that existed in natural populations, as shown by Dobzhansky (1951) and Mayr (1963).

17.2 THE ADVENT OF ELECTROPHORETIC STUDIES

Together in 1966, Lewontin and Hubby produced a radical shift in emphasis in the study of natural genetic variation with the presentation of two papers (Hubby and Lewontin 1966, Lewontin and Hubby

1966). They realized that most mutations in a structural gene should result in a substitution, a deletion, or an addition of at least one amino acid in the polypeptide produced by the gene. In a certain proportion of cases, this caused changes in the net electrical charge on the polypeptide and the charge on the enzyme or protein of which it was a part. Moreover, since enzymes and proteins were made up of polypeptides from one or two structural genes, such differences in charge were likely to segregate in a Mendelian way. Harris (1966) had begun using the same approach for the study of human blood enzymes. Lewontin, Hubby, and Harris started a revolution in the study of genetic variation in populations, for it was clear that their approach was applicable to the study of almost any multicellular organism.

Lewontin and Hubby studied the electrophoretic mobility of enzymes and other proteins, using the method described by Hubby (1963; see also Selander 1976). They reported a large amount of genetic variation in five populations of *Drosophila pseudoobscura* at 18 loci (Lewontin and Hubby 1966). The average population was polymorphic at 30 percent of all loci studied, and 12 percent of the loci of an average individual were heterozygous, in a range of 8 to 15 percent. The authors also knew that these were underestimates of the real variation in the population, because not all amino acid changes would alter the net charge of the molecule.

Since that time, scientists have used electrophoresis to energetically study genetic variation in natural populations (Lewontin 1974, Selander 1976). In general, the results have supported Lewontin and Hubby's (1966) findings that a great amount of variation occurs in natural populations (Table 17-1). There is impressive variation even at the lowest end of the genetic diversity scale.

17.3 THE EMERGENCE OF THE PROBLEM OF GENETIC DIVERSITY

Lewontin and Hubby (1966) immediately recognized the dilemma they had unveiled. How could this quantity of variation persist in natural populations? What were the mechanisms influencing the genetic diversity of natural populations? They offered three possible explanations.

(1) Alleles Are Irrelevant in Natural Selection

According to this hypothesis, alleles have no relevance to natural selection but are adaptively equivalent isoalleles. Then genetic drift would drive populations to homozygosity, except that recurrent mutation and migration counteract such a trend.

TABLE 17-1 General estimates on genetic variation in natural populations

Taxon	Mean Percent of Loci Polymorphic per Population	Mean Percent of Loci Heterozygous per Individual
Relatively low levels of genetic variation		
Birds	14.5	4.2
Haplodiploid wasps (social)	24.3	6.2
Marine snails	17.5	8.3
Rodents	20.2	5.4
Large mammals	23.3	3.7
Relatively high levels of genetic variation		
Drosophila	52.9	15.0
Other insects	53.1	15.1
Haplodiploid wasps (nonsocial)[1]	44.0	16.3
Marine invertebrates	58.7	14.7
Land snails	43.7	15.0
Plants	46.4	17.0

Based on Selander (1976), only 10 years after Lewontin and Hubby's (1966) classic paper.

[1] Data from Sheppard and Heydon 1986.

Lewontin and Hubby noted that, if this were the case, local populations could be found with either one allele or the other because the alleles would be functionally equivalent. They did not observe this pattern but found a high frequency of heterozygosity per individual. Also, they noted that laboratory strains maintained a lot of genic polymorphism, which also suggested that the variation was necessary and important. They concluded that complete selective neutrality was not a satisfactory explanation.

(2) Mutation Rates Are High

In this view, selection tends to eliminate alternative alleles, but mutation restores them. The suggestion is that mutation can keep pace with natural selection, meaning either that mutation rates must be unusually high (higher than 10^{-3} per locus per generation)—which is unlikely—or that natural selection must be very weak. This argument suggests, as did hypothesis 1, that natural selection is a minor element in the evolutionary process and that alleles are close to selective neutrality.

(3) Selection Favors Heterozygotes

This hypothesis avoids the problems with hypotheses 1 and 2, since heterosis can maintain genic variation in a population of any size in spite of mutation and migration; however, it comes with two problems of its own.

Very Weak Selection

We must assume that the two homozygotes are very weakly selected against; otherwise, the total amount of differential selection in a population would become enormous. Thus, we must return to options 1 and 2. Lewontin and Hubby (1966, 606) provide an example:

Suppose two alleles are maintained by selecting against both homozygotes to the extent of 10% each. Since half of all individuals are homozygotes at such a locus, there is a loss of 5% of the population's reproductive potential because of that locus alone. If our estimate is correct that one third of all loci are polymorphic, then something like 2,000 loci are being maintained polymorphic by heterosis. If the selection at each locus were reducing population fitness to 95% of maximum, the population's reproductive potential would be only $(0.95)^{2000}$ of its maximum or about 10^{-46}. If each homozygote were 98 percent as fit as the heterozygote, the population's reproductive potential would be cut to 10^{-9}. In either case, the value is unbelievably low. While we cannot assign an exact reproductive value to the most fit multiple heterozygous genotype, it seems quite impossible that only one billionth of the reproductive capacity of a Drosophila population is being realized. No Drosophila female could conceivably lay two billion eggs in her lifetime. . . . We then have a dilemma. If we postulate weak selective forces, we cannot explain the observed variation in natural populations unless we invoke much larger mutation and migration rates than are now considered reasonable. If we postulate strong selection, we must assume an intolerable load of differential selection in the population. (Lewontin and Hubby 1966, 606–607)

Adaptive Superiority of Heterozygotes

The second problem is that if heterosis maintains genic polymorphism, then it must be demonstrated that the heterozygotes are adaptively superior. Each heterozygotic individual does have two forms of a particular protein. And variation in the physiochemical characteristics of the same functional protein might very well enhance the flexibility of an organism living in a variable environment. Instead of polymorphism being simply a population advantage, every individual has the advantage.

Thus, Lewontin and Hubby set the stage for the grand debate on **non-Darwinian evolution,** or the **neutral theory of molecular evolution.** Is genic variation in natural populations adaptive or neutral? That was the central question.

17.4 THE NEUTRAL THEORY OF MOLECULAR EVOLUTION

King and Jukes (1969) and Kimura and Ohta (1971) championed the neutral theory of molecular evolution. The theory does not deny the existence of natural selection and Darwinian evolution, but it asserts that at the molecular level most polymorphisms and most mutations that become common in populations are adaptively neutral. For example, King and Jukes (1969) concluded that of the spontaneous changes in DNA fed into the gene pool, the number of neutral mutations likely to occur is far greater than the number of adaptive ones. Protein molecules are constantly "probed" as a result of point mutations, and the genome becomes virtually saturated with changes that are not weeded out through natural selection. King and Jukes concluded that most proteins contain regions where substitutions of many amino acids can take place without producing appreciable changes in protein function. They noted as evidence the great variability in primary structure in homologous proteins from various species and the rapid rate at which molecular changes accumulate during evolution. Let us explore examples of these two aspects of the neutral theory.

Variation in Homologous Proteins

Addressing the great variability in protein structure, King and Jukes (1969) developed a table of proteins with the calculated number of amino acid differences and the number of substitutions per codon per year (Table 17-2). Significantly, these kinds of changes appear to occur at random in relation to the site of the relevant codon (Table 17-3). King and Jukes (1969) provided an example for cytochrome c, noting that this protein appears to have identical and clearly defined functions in the cells of all eukaryotes. Also, the cytochromes c of several organisms are fully interchangeable in *in vitro* studies of intact mitochrondria. Cytochromes with different amino acid combinations can perform equally well. For human diabetics, pig and bovine insulins are effective substitutes for the human hormone. Therefore, we have essentially a null hypothesis that evolutionary change at the molecular level occurs by random mutation and random drift.

TABLE 17-2 Rates of amino acid substitutions in mammalian evolution

Protein	Total Number of Comparisons of Amino Acids	Observed Number of Amino Acid Differences	Observed Number of Differences per Codon	10^{-10} Substitutions per Codon per Year
Insulin A and B	510	24	0.047	3.3
Cytochrome c	1,040	63	0.061	4.2
Hemoglobin α chain	432	58	0.137	9.9
Hemoglobin β chain	438	63	0.144	10.3
Fibrinopeptide A	160	76	0.475	42.9
Bovine hemoglobin fetal chain (bovine line of descent only)	438	97	0.221	22.9
Guinea pig insulin (guinea pig line of descent only)	255	86	0.337	53.1

Based on examples provided by King and Jukes (1969).

TABLE 17-3   The distribution of numbers of amino acid changes compared for 110 sites in cytochrome c chains

Number of Changes per Site	Number of Sites Having the Specified Number of Changes	The 29 Invariable Sites Removed in the 0 Change Row	Poisson Distribution Describing the Pattern If Changes Are Randomly Distributed
0	35	6	6
1	17	17	16
2	18	18	20
3	19	19	18
4	10	10	12
5	6	6	6
6	3	3	3
7	1	1	1
8	1	1	0.3
9	0	0	0.1
	110	81	

Based on comparisons of such animals as human, horse, rabbit, pig, and gray whale. Note that the observed pattern of changes is close to the expected pattern of random change predicted by the Poisson distribution. However, the 29 invariable sites must be removed before the fit is adequate.

Rapid Change in Protein Structure

Concerning the rapid change in protein structure through evolutionary time, King and Jukes (1969) estimated that the average rate of evolutionary change is 16×10^{-10} substitution per codon per species per year. Kimura (1968) and Jukes (1965) estimated that the total molecular evolution in vertebrate species proceeds at about one amino acid substitution every 2 years. This is an impressive rate of change. If such changes were under the influence of natural selection, the selection would have to be extremely strong. Perhaps a more reasonable explanation is the one that Lewontin and Hubby (1966) discounted: most amino acid changes must be due to the passive fixation of selectively neutral mutations.

Constancy of Rates of Change

A third property of molecular evolution is that amino acid substitutions seem to proceed at a constant rate through evolutionary time. For example, scientists can observe amino acid differences between the proteins of pairs of mammal species, and the fossil record is well enough known to enable us to estimate how long ago the lineages of the two species in the comparison diverged. We can therefore plot the number of changes against time, and the slope of the line provides an estimate of the rate of change. Kimura (1979)

gave one example by comparing the total number of amino acid differences observed in seven proteins, in 16 pairwise comparisons of mammals (Figure 17-1).

Molecular Clocks

The rates of change of proteins seem to be so constant that, in the absence of a good fossil record, use of the rates as "clocks" has become an important method for establishing phylogenies and the timing of lineage separations. In the literature one can find phylogenetic trees plotted against time, as though a fossil record existed, but in fact time is calculated on the basis of protein differences. For example, Maxson and Wilson (1974) plotted a lineage of *Hyla* frogs over 80 million years. They based their time scale on the assumption that differences between the albumins of the two lineages used in the analysis have accumulated at an average rate of 100 immunological distance units per 60 million years. This estimate would seem to be valid, because the clock could be set not by knowledge of the fossil record but by known times of continental separation due to plate tectonics. In a comparison between marsupials and hyline frogs in the New World and Australia, Maxson et al. (1975) showed that the known dates of continental separation of Gondwanaland enabled an estimate of time since the divergence of phylogenies, and the rate of change of albumin agreed with that used for the *Hyla*

FIGURE 17-1

The straight-line relationship between time since divergence of two lineages and the number of nucleotide substitutions since divergence, based on the number of amino acid differences observed in seven proteins. Solid circles are for 11 pairs of mammals, and open circles are for five pairs of primates. The line, drawn using only the solid circles, illustrates the slower rates of substitution in primate lineages. *(From Kimura 1979)*

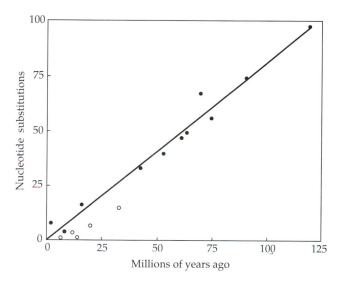

frogs. Fitch (1976) evaluated the quality of various molecular clocks and concluded that several exist, although they are rather sloppy.

Kimura (1979) aptly summarized the neutral theory: "At the molecular level most evolutionary change and most of the variability within a species are not caused by selection but by random drift of mutant genes that are selectively equivalent."

17.5 THE DARWIN–MULLER HYPOTHESIS

Although the neutral theory has not gone unchallenged by Darwinian evolutionists, the gap between the two theories is not as great as one might think. As Lewontin (1974, 198) asserted,

> The so-called neutral mutation theory is, in reality, the classical Darwin–Muller hypothesis about population structure and evolution brought up to date. It asserts that when natural selection occurs it is almost always purifying, but that there is a class of subliminal mutations which are irrelevant to adaptation and natural selection. This latter class, predictable from molecular genetics and enzymology, is what is observed, they claim, when the tools of electrophoresis and immunology are applied to individual and species differences.

Lewontin called this the Darwin–Muller hypothesis because Muller had argued in the 1940s (e.g., Muller 1949) that evolution occurred largely through cryptic genetic change rather than natural selection. Remember also that Darwin did not deny the existence of neutral characters in his *Origin of Species*. Thus, Muller's view countered Mayr's (1949) classical Darwinian selectionist position.

17.6 WHERE DOES NATURAL SELECTION OPERATE?

What, then, can we say about the importance of natural selection in the evolutionary process? We can examine this question on three different scales, all relevant to the debate: adaptation in whole organisms, three-dimensional structure of proteins, and two-dimensional structure of proteins.

Whole Organisms

No evolutionary biologist denies that whole-organism morphology and anatomy are under the influence of natural selection and that the majority of characters are adaptive. The elephant's trunk, the camel's hump, the fish's gills, and the human brain have all evolved under the strong influence of natural selection (Kipling 1902 notwithstanding).

Three-Dimensional Structure of Proteins

No evolutionary biologist denies that the three-dimensional structure of a protein is an essential feature of its function and that purifying natural selection must maintain this structure (Figures 16-6, 17-2, and 17-3). Selection will most likely be against any mutation that changes an amino acid and thereby changes tertiary structure. For example, King and Jukes (1969) noted that there are about 29 amino acid sites in cytochrome *c* molecules that are invariable across many species. These sites are needed for combining with the heme group, for interacting with cytochrome *c* oxidase, and possibly for other functions. It is clear that the active sites on a molecule cannot accommodate new amino acid sequences, and natural selection acts constantly to purify the codons for these sites of new mutations. Natural selection on the active sites in protein molecules is indeed very strong. As King and Jukes (1969) pointed out, proteins and sites within proteins differ with regard to the stringency of their requirements; some can change and others cannot.

Another example of how molecules can vary comes from the α- and β-hemoglobin chains in humans. Harris (1970) listed 59 variants in humans. Of

FIGURE 17-2

One β polypeptide chain of hemoglobin, showing its three-dimensional form. The 26 numbered sites are invariant in all animals tested, suggesting that they are important in defining the structure and function of the molecule and are under stringent natural selection. Figure 16-6 illustrates the hemoglobin molecule. *(From Zuckerkandl 1965)*

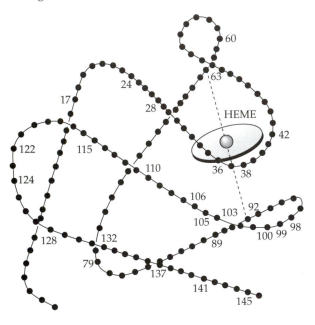

those, 43 have no known pathological effects; 5 cause methemoglobinemias because they are near the active site of the heme iron and are therefore mildly pathological; and 11 cause instability of the hemoglobin molecule, which results in various degrees of hemolytic anemia. Thus, about 75 percent of the amino acid substitutions have no known effect on hemoglobin function and are probably selectively neutral, but others do influence hemoglobin function and are strongly or weakly selected against on the basis of the severity of their pathological effects.

Two-Dimensional Structure of Proteins

Much of the random change in amino acids can occur in the secondary structure of proteins, so long as it does not alter the tertiary structure and is not at the active site or sites of the molecule. For large molecules such as proteins, it seems quite reasonable that even the majority of amino acid sites can be quite variable and can change in a seemingly random manner.

> Natural selection is indirectly operative in the patterns of neutral evolutionary change in that only functionally equivalent isoalleles are allowed the small possibility of fixation through random genetic drift.

Those alleles which do become fixed through drift are not a random selection of all substitutional mutations, but alleles which have been "selected" for innocuousness. (King and Jukes 1969, 795)

17.7 EVIDENCE OF NATURAL SELECTION ON MOLECULES

Serum Esterase Polymorphism

If variation in protein structure is not random, then it is necessary to demonstrate the nonrandom distribution of protein morphs and the adaptive significance of this variation. Then the action of natural selection can be demonstrated to maintain the variation. A good early example of this approach was Koehn's 1969 study. He showed that a polymorphism occurred at the serum esterase locus (ES-I^a and ES-I^b) in the freshwater fish *Catostomus clarkii*, a sucker or catfish. The frequency of alleles varied with latitude (Figure 17-4). The activity of the enzyme that was more frequent in southern populations (ES-I^a) increased as temperature rose from 0°C to 37°C, whereas the activity for the allele that was more frequent in northern populations (ES-I^b) increased as temperature fell. The heterozygote was best at intermediate temperatures. Koehn could explain the distribution of enzymes on the basis of their differential efficiencies at different temperatures. He identified the selective force—temperature—and the physiological mechanism through which it selected for different enzyme morphs at different parts of the fish's range.

Multilocus Systems

Clegg et al. (1972) provided a rather different kind of example of a place where natural selection must be at work in maintaining enzyme polymorphisms. They argued that multilocus systems are likely to behave differently from single-locus systems, and therefore a theory developed largely around single-locus systems cannot be used to predict multilocus dynamics.

Clegg et al. studied populations of barley, *Hordeum vulgare*, scoring individual plants for their genotypes at four enzyme loci. They then predicted the genotypes of the gametes over several generations, assuming no selection on any of the alleles. They showed that alleles tended to be selected together even when they were at nonlinked loci, and certain alleles became much more abundant than others. Also, the same alleles became abundant in two independently evolving populations. Those results showed that natural selection structured the populations into sets of highly interacting, coadapted gene complexes and acted on correlated multilocus units.

FIGURE 17-3

The two-dimensional structure of human hemoglobin when it is stretched out in a linear display of polypeptide chains, and the amino acid residues from which the chains are built. Circles and triangles appear above the 26 invariable sites noted in Figure 17-2; circles indicate residues that are the same in all hemoglobin and myoglobin chains tested, and triangles mark residues that are the same for all known hemoglobin chains. Note that additional sites are invariant in the four human hemoglobins illustrated, but these sites show variation in other animals studied. *(From Zuckerkandl 1965)*

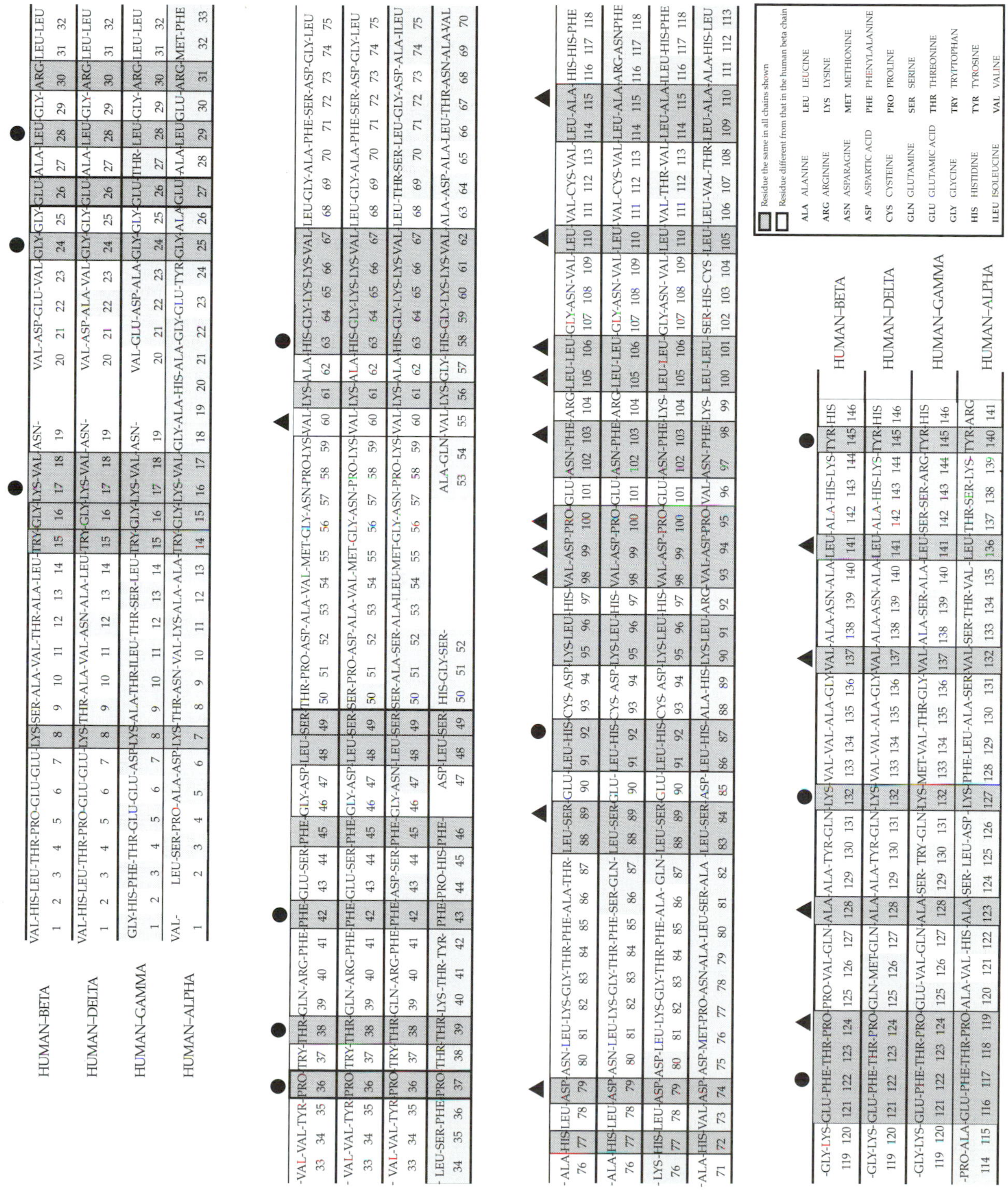

FIGURE 17-4

The distribution of fitness of genotypes on a latitudinal temperature gradient estimated by Koehn (1969) for the freshwater fish *Catostomus clarkii*. The genotypes at the serum esterase locus (ES-I) are given on the right. Note that the allele ES-I[b] has higher activity at cooler temperatures, and ES-I[a] has higher activity at warmer temperatures.

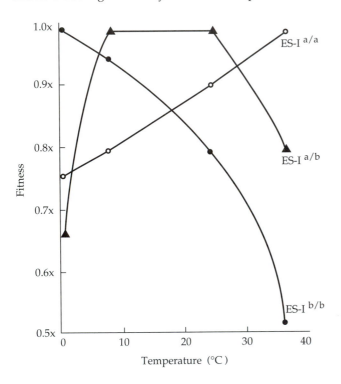

Hence, study of single-locus systems may not be able to explain the adaptive function of a protein polymorphism. The set of experiments on barley certainly helps us to understand why such a great amount of variation exists—if, indeed, alleles are commonly selected as, and function as, multilocus units.

Nonrandom Patterns in Electromorph Frequencies

Selectionists also used nonrandom patterns in electromorph frequency to counter the neutralists' arguments. The patterns simply showed that polymorphisms do not change at random but are presumably acted upon by stabilizing selection, at least in some populations. For example, if we examine polymorphisms in several different populations, we should see very different frequencies if each morph is on a random walk, and very similar frequencies if the morphs are under stabilizing selection. Ayala (1974) presented data supporting the latter possibility. He provided several tables with frequencies of different alleles at a locus—for example, allelic variation at the esterase-7 locus in eight natural populations of *Drosophila tropicalis* (Table 17-4). The frequencies of the numerous alleles were too similar for the neutral theory to be acceptable.

Neutralists may have argued both that not enough time had elapsed since isolation of populations for frequencies to diverge toward randomness sufficiently and that gene flow was sufficient to swamp random changes in frequency. Ayala coun-

TABLE 17-4 Allelic variation at the esterase-7 locus in eight natural populations of *Drosophila tropicalis*

| | Alleles[1] | | | | |
Locality	96	98	100	102	105
1. Catatumbo, Venezuela	0.04	0.08	0.48	0.32	0.04
2. Barinitas, Venezuela	0.02	0.14	0.64	0.18	0.02
3. Caripito, Venezuela	0.00	0.11	0.47	0.35	0.05
4. Tucupita, Venezuela	0.05	0.09	0.77	0.09	0.00
5. Santiago, Dominican Republic	0.00	0.15	0.43	0.35	0.05
6. Santo Domingo, Dominican Republic	0.01	0.11	0.46	0.30	0.09
7. Mayagüez, Puerto Rico	0.02	0.16	0.51	0.28	0.04
8. Barranquitas, Puerto Rico	0.00	0.06	0.34	0.41	0.16

The populations are those studied by Ayala (1974). Note the general similarity in frequencies for each allele across all sites, even though the sites are widely separated.

[1] Proportions do not sum to 1.00 because some rare alleles are not recorded.

tered their arguments by showing that at some loci the frequencies differed greatly between localities, showing that there was time enough for populations to diverge and that dispersal of individuals was not great enough to make gene flow an important factor. Natural selection must have been maintaining frequencies in independently evolving populations through stabilizing selection.

Inconstant Rates of Molecular Evolution

Another selectionist approach has been to argue that proteins evolve at inconstant rates, with directional selection early in their evolution followed by more stabilizing selection when a molecule is close to perfection in function. Researchers have determined amino acid compositions and sequences in many animals, in protein chains of globins, fibrinopeptides, cytochrome *c*, and others. In phylogenetic studies, they have aligned sequences of the same protein type in different species against one another and estimated the minimum number of mutations that could produce the differences. The researchers have then used those distances to construct a divergence dendrogram and a phylogenetic tree with ancestral sequences that produce the fewest possible mutations over the tree. Figure 17-5 illustrates some interesting features of such a phylogenetic tree constructed for 12 mammalian species, using myoglobin and α- and β-hemoglobin.

Globin Evolution

The differences between the rates of change of proteins in the lineages are significant. From the common eutherian ancestor to modern organisms, there have been 45 nucleotide replacements in the human lineage (8 + 6 + 12 + 8 + 11 + 0 in Figure 17-5), 83 in the lemurine lineage (8 + 6 + 13 + 56), and 75 in the bovine lineage (8 + 11 + 13 + 19 + 24). These differences suggest that rates of change in some lineages may be almost double those in others.

Another significant point is that humans and chimpanzees are so close. Only one nucleotide replacement separates them, whereas 43 replacements separate the ox and the sheep, two bovids (19 + 24). According to Sarich and Wilson (1967), this indicates that the separation between ancestral humans and African apes is much more recent than is generally believed on the basis of morphological evidence (see Chapter 12). However, Goodman (1976) made the argument that molecular evolution decelerated in the hominids. Indeed, Goodman presented evidence that the globins evolved at very different rates. On the basis of calculations from larger lineages than in Figure 17-5, he showed that early evolution of globin sequences in vertebrates proceeded at faster rates than later evolution (Figure 17-6). The fast evolution was apparently due to directional selection for improved function of hemoglobin, and the slow evolution was due to stabilizing selection after the improvements were more or less fixed.

Different Rates Over 680 Million Years

During the first 380 million years of the phylogenetic tree (680 to 300 million years ago), globin genes evolved at an average rate of 46 nucleotide replacements per 100 codons per 10^8 years. Note the rates of change from the branch that leads to insect–annelid globins to the myoglobin branch (52, 52, 50). The highest rate observed was during the period roughly between 450 million and 350 million years ago, when changes separated vertebrate myoglobin and hemoglobin branches and then subdivided the hemoglobin branch into α- and β-hemoglobin. At that time, the rate was about 109 nucleotide replacements per 100 codons per 10^8 years (58 + 51 for α-hemoglobin,

FIGURE 17-5

A phylogenetic tree using myoglobin and α and β chains of hemoglobin for 12 mammalian species. The numbers are estimated amino acid residue replacements in the lineage since the point of divergence. (Roman numerals are minimum estimates, and the higher italicized values include estimates for superimposed replacements.) Note that rates of change from the common eutherian ancestor differ. In the human lineage, 45 nucleotide replacements (8 + 6 + 12 + 8 + 11 + 0) have occurred, but in the lemurine lineage, 83 replacements have been documented (8 + 6 + 13 + 56). Units are nucleotide replacements per 100 codons per 100 million years. Asterisks show estimates based on incomplete data on myoglobin and may be slightly less than those for the complete molecule. *(From Goodman 1976)*

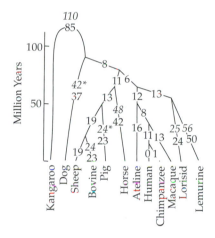

FIGURE 17-6

A phylogenetic tree using 55 globins with well-studied amino acid sequences. Note the very rapid rate of substitutions during the development of hemoglobin from about 500 million years ago to 400 million years ago, averaging about 109 nucleotide replacements per 100 codons per 10^8 years (e.g., from the myoglobin fork to α- and β-hemoglobin: α chain = 58 + 51 = 109; β chain = 58 + 47 = 105). (Roman and italicized numerals are as in Figure 17-5, but only estimates for superimposed replacements are given when available.) *(From Goodman 1976)*

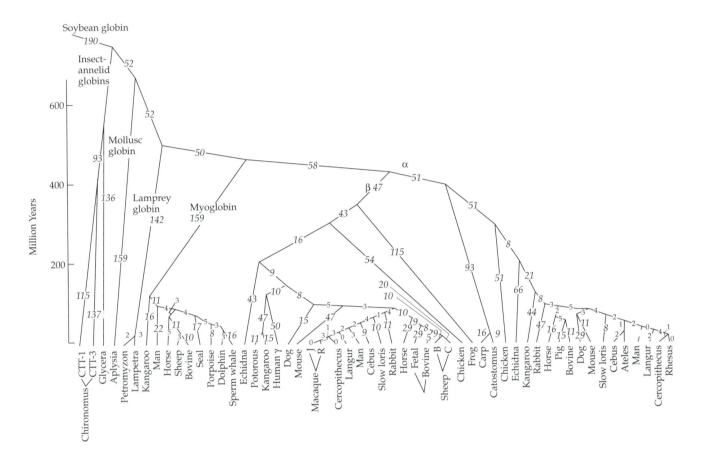

58 + 47 for β-hemoglobin). In the last 300 million years, from the amniote ancestor to the present, the hemoglobins evolved at an average rate of only 15 nucleotide replacements per 100 codons per 10^8 years, as shown by the many short arms in the last 100 million years.

Thus, the proteins may be evolving at very different rates at different times, and only when we evaluate change over extended periods of time do crude approximations of constant rates of change appear, in support of Fitch's (1976) assessment that we have some "sloppy clocks" to use. As we should expect, organisms with different body sizes also differ in generation time, with consequences for evolutionary rates. In addition, higher body temperatures of homeotherms may correlate with increased rates of DNA synthesis and nucleotide replacement. Such variables help to explain why rates of protein evolution differ in different kinds of organisms (Martin and Palumbi 1993).

In summary, it is fairly easy to discover the middle ground between neo-Darwinians and non-Darwinians, or selectionists and neutralists. Many differences in protein structure may well be due to random change in amino acids, but natural selection acts very forcefully on changes at or near active sites of proteins, and much protein structure has evolved under the influence of natural selection.

SUMMARY

Scientists' estimates of genetic variation in populations were inadequate until the advent of electrophoresis, which provided much better impressions of the extents of heterozygosity and polymorphism. The great amount of genetic variation—12 percent of loci heterozygous per individual, and 30 percent of loci polymorphic per population—posed the problem of how such genetic diversity could be maintained in populations. Could alleles be free from natural selection and neutral, could mutation rates balance rates of selection, or was selection favoring heterozygotes? If evolution occurred in the absence of natural selection, its mode was an alternative to Darwin's theory, aptly called **non-Darwinian evolution.** Because such evolution occurred in selectively neutral molecules, it was also called the **neutral theory of molecular evolution.**

Evidence for the neutrality of molecular variation to natural selection came from the great variability in homologous proteins and the randomness of amino acid substitutions. The rapid and more-or-less constant rate of change in proteins also suggested a clock with its steady pace and so was known as a **molecular clock.**

The contribution of the neutral theory of evolution reinstated and formalized earlier recognitions of differences between Darwin and Muller, among others. Anticipating the theory of non-Darwinian evolution, Muller had argued that hidden genetic change, rather than natural selection, caused evolution. The neutral theory applies particularly to changes in the two-dimensional structures of protein molecules.

The **selectionists** countered the **neutralists** by showing that (as in serum esterase polymorphisms) each protein form functioned best under different environmental conditions so that a polymorphism was maintained throughout the geographic range of a species. They also demonstrated that selection worked (at times) on genes at several loci together and that such multilocus systems might account for much protein variation. Another tactic demonstrated the nonrandom pattern of electromorph frequencies, suggesting that stabilizing selection was acting to keep frequencies more similar than would be expected from random substitutions. In addition, the selectionists developed arguments against a constant molecular clock, suggesting that evolution was rapid in the formative stages of a molecule such as hemoglobin, under the influence of natural selection, but the pace of change diminished as molecular structure neared the optimum. Rates of change also were very different in different lineages, even within the same molecular type, suggesting nonrandom forces in action.

Resolution of the issues dividing the selectionists and the neutralists can come from the recognition that many differences in homologous protein structure can result from random amino acid substitutions, but natural selection works forcefully on changes at or near the active site of the molecule and on changes affecting the three-dimensional structure and activity of the molecule.

QUESTIONS FOR DISCUSSION

1. Will it be necessary to understand the biochemical function and activity of most proteins before we fully understand high genetic polymorphism and heterozygosity?

2. In your opinion, has the neutral theory of evolution retarded the development of evolutionary theory, or has it advanced and broadened the theory?

3. Does the neutral theory weaken Darwin's stature as the father of evolutionary theory or increase it, in your opinion?

4. Given a knowledge of the biochemistry of protein molecules and their three-dimensional structure, how would you estimate quantitatively what "most" actually means for any protein in the statement "at the molecular level most polymorphisms and most mutations that become common in populations are adaptively neutral"?

5. At a rate of one amino acid substitution every 2 years in vertebrate species, what kinds of changes in protein function could result in relatively rapid evolutionary change in a species, in your opinion?

6. How many ways can you think of that could be used to calibrate a molecular clock? Which particular organisms could you associate with accurately estimated dates for calibrating such a clock?

7. One possible explanation for high polymorphism in populations is that multilocus systems are important and frequently have adaptive value. How would you evaluate this possibility?

8. The globin proteins provide a remarkable long-term picture of evolutionary change and divergence of lineages. At which points in the lineages would you argue that major adaptive breakthroughs in globin structure and function occurred?

9. In the history of the neutralist–selectionist debate, do you think it is likely that scientists knowingly overstated their case in order to create a stronger impact? Was a more balanced view available, or do you think that in the sociology of science strong and biased cases need to be presented in order to gain attention?

10. Which term is more accurate in your opinion, *neutral evolution* or *non-Darwinian evolution*, or is there a better term that you would prefer?

REFERENCES

Ayala, F. J. 1974. Biological evolution: Natural selection or random walk? Amer. Sci. 62:692–701.

Clegg, M. T., R. W. Allard, and A. L. Kahler. 1972. Is the gene the unit of selection? Evidence from two experimental plant populations. Proc. Nat. Acad. Sci. USA 69:2474–2478.

Dobzhansky, T. 1951. Genetics and the Origin of Species. 3d ed., rev. Columbia Univ. Press, New York.

Fisher, R. A. 1930. The Genetical Theory of Natural Selection. Oxford Univ. Press, London.

Fitch, W. M. 1976. Molecular evolutionary clocks, pp. 160–178. *In* F. J. Ayala (ed.). Molecular Evolution. Sinauer, Sunderland, Mass.

Goodman, M. 1976. Protein sequences in phylogeny, pp. 141–159. *In* F. J. Ayala (ed.). Molecular Evolution. Sinauer, Sunderland, Mass.

Harris, H. 1966. Enzyme polymorphisms in man. Proc. Royal Soc. Ser. B. 164:298–310.

———. 1970. The Principals of Human Biochemical Genetics. North Holland, London.

Hubby, J. L. 1963. Protein differences in *Drosophila*, [Part] I: *Drosophila melanogaster*. Genetics 48:871–879.

Hubby, J. L., and R. C. Lewontin. 1966. A molecular approach to the study of genic heterozygosity in natural populations, [Part] I: The number of alleles at different loci in *Drosophila pseudoobscura*. Genetics 54:577–594.

Jukes, T. H. 1965. The genetic code II. Amer. Sci. 53:477–487.

Kimura, M. 1968. Evolutionary rate at the molecular level. Nature 217:624–626.

———. 1979. The neutral theory of molecular evolution. Sci. Amer. 241(5):98–126.

Kimura, M., and T. Ohta. 1971. Protein polymorphism as a phase of molecular evolution. Nature 229:467–469.

King, J. L., and T. H. Jukes. 1969. Non-Darwinian evolution. Science 164:788–798.

Kipling, R. 1902. Just So Stories for Little Children. Macmillan, London.

Koehn, R. K. 1969. Esterase heterogeneity: Dynamics of a polymorphism. Science 163:943–944.

Lewontin, R. C. 1974. The Genetic Basis of Evolutionary Change. Columbia Univ. Press, New York.

Lewontin, R. C., and J. L. Hubby. 1966. A molecular approach to the study of genic heterozygosity in natural populations, [Part] II: Amount of variation and degree of heterozygosity in natural populations of *Drosophila pseudoobscura*. Genetics 54:595–609.

Martin, A. P., and S. R. Palumbi. 1993. Body size, metabolic rate, generation time, and the molecular clock. Proc. Natl. Acad. Sci. USA 90:4087–4091.

Maxson, L. R., and A. C. Wilson. 1974. Convergent morphological evolution detected by studying proteins of tree frogs in the *Hyla eximia* species group. Science 185:66–68.

Maxson, L. R., V. M. Sarich, and A. C. Wilson. 1975. Continental drift and the use of albumin as an evolutionary clock. Nature 255:397–400.

Mayr, E. 1949. Speciation and systematics, pp. 281–298. *In* G. L. Jepsen, E. Mayr, and G. G. Simpson (eds.). Genetics, Paleontology, and Evolution. Princeton Univ. Press, Princeton, N.J.

———. 1963. Animal Species and Evolution. Belknap Press of Harvard Univ. Press, Cambridge, Mass.

Muller, H. J. 1949. The Darwinian and modern conceptions of natural selection. Proc. Amer. Phil. Soc. 93:459–470.

Richmond, R. C. 1970. Non-Darwinian evolution: A critique. Nature 225:1025–1028.

Sarich, V. M., and A. C. Wilson. 1967. Immunological time scale for hominid evolution. Science 158:1200–1203.

Selander, R. K. 1976. Genic variation in natural populations, pp. 21–45. *In* F. J. Ayala (ed.). Molecular Evolution. Sinauer, Sunderland, Mass.

Sheppard, W. S., and S. L. Heydon. 1986. High levels of genetic variability in three male-haploid species (Hymenoptera: Argidae, Tenthredinidae). Evolution 40:1350–1353.

Zuckerkandl, E. 1965. The evolution of hemoglobin. Sci. Amer. 212(5):110–118.

18

RATES OF EVOLUTION

The zebra, a representative of the horse family, Equidae, illustrates the concept of quantum evolution. *(Illustration by Stephen Price)*

18.1 HISTORICAL PERSPECTIVE

Modern biologists usually perceive Darwin's view of evolutionary change as a series of gradual transitions from one type to another over long periods of time, under the influence of natural selection. Mayr (1963) called this scenario **gradual speciation,** and Eldredge and Gould (1972) used the name **phyletic gradualism.** Alternative views hold that speciation may be relatively rapid or even very abrupt. Mayr (1963) and Dobzhansky (1970) labeled such events **instantaneous speciation, speciation by macrogenesis,** and **saltational speciation,** whereas Simpson (1953) had used **quantum evolution** to denote a rapid transition from one adaptive peak to another in a diverging lineage. Although these concepts of rapid evolutionary change have proliferated since the modern synthesis, their history is long.

Saltational Evolution

The Greek philosophers, from Aristotle to Theophrastus and Pliny, espoused a belief in saltational kinds of evolutionary change (Zirkle 1959). As noted in Chapter 4, they believed in rampant hybridization that created new species of very different natures from the originals: camel × panther = giraffe; dog × goat = wild boar; camel × sparrow = ostrich. These must have been saltational events indeed!

It may come as more of a surprise that Charles Darwin subscribed to saltational evolution, not so much in *The Origin of Species* (1859) as earlier (but see the discussion of Darwin's ideas on rates of evolutionary change in Section 18.9). In 1836, presumably after seeing 13 species of finches on a set of islands formed in the fairly recent geological past, he was impressed with how rapidly species must change. He wrote, "If one species does change into another it must be *per saltum,* or species may perish" (Eiseley 1979, 47; see also Barlow 1946, 263). Eiseley (1979) argued that Edward Blyth (1835, 1836, 1837) had influenced Darwin profoundly by noting that species tend to adjoin in their distributions, so that at the edge of its range one species changes abruptly—saltationally—into another. Darwin, in an unpublished essay of 1844, noted that "sports" may develop rapidly and be perpetuated without gradual change (Stanley 1981, 136).

> So in the state of Nature some small modifications, apparently adapted to certain ends, may perhaps be produced from the accidents of the reproductive system, and be at once propagated without long-continued selection of small deviations towards that structure.

Lyell (1834) also argued that species must adapt and change rapidly or they would die out before their alterations could be of any use. Eiseley (1979) contended that both Blyth and Lyell had caused Darwin to consider that macromutative change might be the mechanism enabling adaptation as the environment changed or as a species changed its geographical distribution.

By 1859, Darwin revised his opinion. Already in October 1838, he had read Malthus's essay on population and realized its application to the selection that occurs in nature. His theory of transmutation through natural selection emphasized the gradual, often imperceptible, change of species over Lyell's almost endless time. In the modern synthesis, many scientists espoused this process, and it is the one we tend to take for granted.

Gradualism Questioned

As Mayr (1963) pointed out, the rediscovery of Mendelian genetics introduced some serious doubts about gradualism being sufficient to explain the speciation process. The doubts arose partly because many—including Bateson, Zirkle, Goldschmidt, and Morgan—regarded natural selection as an insufficiently strong creative process (Mayr 1980). They saw natural selection as a purely negative process, and not a constructive one. Morgan (1932), for example, thought that natural selection acted merely as a sieve for the materials that presented themselves as variation, and therefore it could not play a creative role in evolution.

Mayr (1980) identified other areas of contention in the early 1900s.

1. **The species argument.** Existing species are separated by discrete gaps in characteristics. Intermediates are not observed. Only minor characters, not the major characters, vary in species, and therefore the sterility barrier, or reproductive isolation, could not possibly evolve gradually.

2. **The origin of evolutionary novelties.** Selection cannot explain the origins of new structures, because incipient new organs, such as rudimentary wings, can have no selective value unless they are developed enough to be fully functional. Hence, major adaptive changes—nongradual changes of characters—must occur. For example, Punnett (1915) argued that mimicry in butterflies had to evolve by saltation, because all types intermediate between the cryptic progenitor and the mimetic descendent would be maladaptive, and single Mendelian characteristics often dictated the morph differences. Fisher (1958) addressed this case in his discussion of the theory of saltations, and Punnett's argument still has credibility (Sheppard 1962).

3. **The origin of higher taxonomic categories.** Even if it could be demonstrated that related species occasionally derive from each other by gradual natural

selection, higher taxa are far too distinct and too different to permit a belief in their origination on the basis of gradual evolution by natural selection.

4. **The integrated nature of evolutionary change.** In most phyletic lines, changes in many morphological and physiological characters occur simultaneously, showing that the whole genotype changes simultaneously. Surely, if the mechanism for this kind of change were natural selection on individual mutations, too many positive mutations would have to occur at the same time.

These were real, fundamental problems. Scientists debated them hotly between 1900 and 1930, and no clear synthesis was available. In fact, Mendelian genetics alone seemed to provide a sufficient mechanism for speciation. For example, Bateson (1894) recognized the distinctness of continuous and discontinuous variation. His seemingly very logical main conclusion was that the discontinuity of species resulted from the discontinuity of variation. De Vries (1906) also subscribed to this kind of saltational evolutionary change. According to his mutation theory, ordinary variation was distinct from mutational variation. **Ordinary variation** could not result in changes beyond the variation in a species, whereas **mutational variation** completely separated the carrier from its parents as a new species. The counterarguments of prominent neo-Darwinians such as Julian Huxley (1942), Mayr (1963), and Dobzhansky (1970) are familiar. For example, Mayr (1963, 438) explained:

> The occurrence of genetic monstrosities by mutation, . . . is well substantiated, but they are such evident freaks that these monsters can be designated only as "hopeless." They are so utterly unbalanced that they would not have the slightest chance of escaping stabilizing selection.

Modern Views

The situation is by no means black and white, even today. **Saltational change,** or change by leaps, seems to be a reality in certain phylogenetic lines, while gradual change of some characteristics seems to be a reality in others. Different rates of gradual change, rather than different processes, may explain some apparently saltational changes. Does "rapid gradual" change qualify as saltational? Can we distinguish real saltation from rapid gradualism in the fossil record? As usual, the middle ground between opposing views seems to be most realistic, and the questions really are: Under what conditions, and in what kinds of species, are we likely to see **phyletic gradualism**? What kinds of conditions and species are predisposed to accelerated rates of change, or saltational change, and when do species not change at all?

Let us first discuss the cases supported by solid evidence of saltational speciation. Dobzhansky (1970) accepted without question that saltational speciation occurs through **allopolyploidy.** Mayr (1963) considered evolution by saltation well established by good evidence from cases of **polyploidy, hybridization, chromosomal rearrangements,** and **shift in sexuality.** We discussed the first three in Chapters 4 and 5. "Shift in sexuality" refers to the commonness of an association between polyploidy in animals and parthenogenetic reproduction that overcomes the problems associated with imbalance of the sex determination system. Grant (1971) also accepted as valid the process of saltational speciation. What he called **quantum speciation** is based largely on the work and arguments of Lewis (1962, 1966; Lewis and Raven 1958), derived from studies on *Clarkia* species (see Chapter 5).

18.2 RAPID EVOLUTION IN *CLARKIA*

Lewis identified certain mechanisms in *Clarkia* during speciation. His 1962 paper on **catastrophic selection** argued that speciation is rapid, involves chromosomal rearrangements, and may be characteristic of many plant groups. Lewis noted that several species pairs of *Clarkia* exist in which characters suggest that one derived from the other, rather than both having diverged from a parental stock. In addition, the derived species was *invariably* in more xeric conditions and had chromosomal differences from the parental species. He noted that drought was the most important determinant in *Clarkia* distribution. On the basis of experimental studies, he found that whole populations went extinct relatively frequently at the edges of the species range.

Lewis noted that, although catastrophic selection has produced insecticide resistance in insects and pathogen resistance in plants during agricultural breeding trials, this resistance has not caused speciation events. Catastrophic selection in peripheral populations of *Clarkia* at times of drought was very common. Either no individuals resistant to drought survived and the population went extinct, or perhaps one or two resistant individuals survived. In the latter case, with the parental population dead, the survivors and their progeny were spatially isolated. The main reason for the survival of these rare new genotypes was that they set seed early, before the severe drought could prevent seed production, probably implicating a regulatory change in plant development. Therefore, even genotypes with reduced fertility could persist, and, with no competition from parental species, natural selection would have tended to improve fertility. Thus, a large population would have developed rapidly in a more-or-less vacant niche, and a new species would have become established. Catastrophic selection therefore may be an important mechanism in speciation and may play a significant role in dramatic changes in evolutionary rates.

18.3 RAPID EVOLUTION IN THE MARSH FRITILLARY BUTTERFLY

Studies by E. B. Ford and his father on the marsh fritillary butterfly in England highlight an example of very obvious changes in rates of evolution and illustrate the consequences of the **flush–crash cycle** that Carson (1968, 1975) would identify later. The Fords (1930) studied an isolated population of the marsh fritillary butterfly, *Melitaea (Euphydryas) aurinia*, in which changes in wing pattern could be observed relatively easily because of the pattern's com-

plexity (Figure 18-1). They carried out their study for 19 years. In addition, collectors had left records of population condition during the previous 36 years. Thus, the total preserved specimens covered a period of 55 years (Figure 18-2)!

It is remarkable that in four to five butterfly generations, from 1920 to 1924, a significant evolution in wing pattern took place, probably accompanied by pleiotropic gene effects that shifted the adaptive core of the population in response to the new environmental conditions prevailing in 1924. In contrast, little evolution had occurred between 1896 and 1920, although the environment had probably changed considerably. Around 1920, a sudden flush permitted an increase in genetic diversity during relaxed selection and subsequent rapid adaptation to the environment. Ford (1964, 14–15) described the flush phase:

> An extraordinary outburst of variability took place. Hardly two specimens were alike, while marked de-

FIGURE 18-1

The marsh fritillary butterfly, *Melitea (Euphydryas) aurinia*, studied by Ford and Ford (1930). North American members of the genus are known as checkerspot butterflies, for good reason. All specimens are males. (**Top left**) Dorsal view of a typical form. (**Top right**) Side view of a typical form feeding on a devil's-bit scabious inflorescence. (**Bottom left**) An aberrant form showing observable kinds of differences between it and typicals. (**Bottom right**) A very bizarre form with almost no markings and few scales. (*Illustration by Stephen Price*)

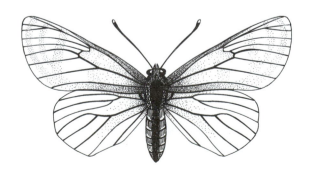

FIGURE 18-2

General trends in population size of the marsh fritillary butterfly, based on accounts in Ford and Ford (1930) and Ford (1957, 1964). Note that high variation in the population occurs under relaxed selection during flushes, and as soon as strong selection resumes, variation is low. The flush from 1920 to 1924 resulted in a substantially different typical phenotype after 1924, suggesting rapid evolution after a long period of stabilizing selection that started in about 1895. The arrow on the time axis shows the start of the Fords' field studies.

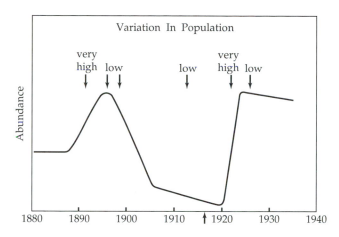

partures from the normal form of the species in color pattern, size and shape were common. A considerable proportion of these were deformed in various ways; the amount of deformity being closely correlated with the degree of variation, so that the more extreme departures from normality were clumsy upon the wing or even unable to fly.

The shift in wing characters was the result of the extreme diversity of genotypes subjected to selection in 1923 and 1924, when the population flush ended and the population stabilized so that mortality increased to about 99 percent (assuming a female lays 200 fertile eggs on average).

Why should these kinds of differences in evolutionary rates over 50 years not be mirrored by similar differences in geological time? That they are is supported by the very different evolutionary rates of hemoglobin in the periods from 500 to 400 million years ago (109 nucleotide replacements per 100 codons per 10^8 years) and from 300 million years ago to the present (15 replacements) (Goodman 1976; see also Chapter 16). And these differences are only means, suggesting that rates of change in some pairwise comparisons were orders of magnitude different.

18.4 SALTATIONAL EVOLUTION IN BOLYERINE SNAKES

Scientists have engaged in more speculative discussions on saltational evolution. On a very different theme from the foregoing, Frazzetta (1970, 1975) found it difficult to accept gradualism as a mechanism for the evolution of an intramaxillary joint in bolyerine snakes (a subfamily of the Boidae, the pythons and boas) (Figure 18-3). The joint is adaptive because it allows increased rotation of the anterior maxillary tooth tips, which improves the strike without loss of the posterior part of the maxilla—as in viperid snakes, whose teeth are adapted for gripping prey until they are finished swallowing it.

FIGURE 18-3

(**Top**) A skull of *Casarea dussumieri*, a snake in the subfamily Bolyerinae in the family Boidae, which includes pythons and boas. (**Bottom**) A skull of *Candoia bibroni*, which may be closer to the original boid stock from which *Casarea* evolved. Note the divided maxilla (mx) in *Casarea* and the single bone in *Candoia*. *(From Frazzetta 1970)*

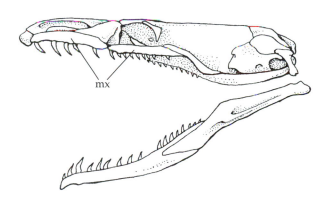

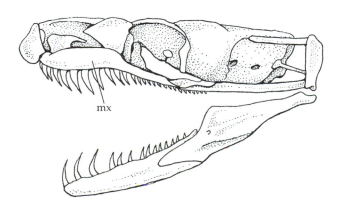

Alternative Hypotheses

The details of this example are submerged in complex osteology, but Frazzetta identified three possible routes of origin of the intramaxillary joint:

1. **A gradual process.** The maxilla would gradually become thinner and ultimately separate at the intermaxillary joint. However, in intermediate phases the thin bone would be susceptible to breaks during use, and so the process is unlikely.
2. **Pseudoarthrosis.** The first process would lead to a high probability of maxillary breakage. "When an 'injured' individual develops a pseudoarthrosis [a false joint] connecting the separated maxillary pieces, it is selectively favored" (Frazzetta 1970, 64). Continued selection would result in a population of snakes in which thinning had reached the ultimate extreme, with complete division of the maxilla.
3. **Saltation.** An individual with an intramaxillary joint would suddenly arise, manifesting a saltational change from the ancestral condition. Frazzetta favored this possibility.

Goldschmidt's Hopeful Monsters

Frazzetta realized that, to support the third possibility, he had to counter the arguments leveled against Goldschmidt's (1940) concepts of "systemic mutations" leading to **"hopeful monsters"** that rarely became selectively favored. Let us briefly review Goldschmidt's ideas. Living between 1878 and 1958, he was much influenced by the arguments from 1900 to 1930 pointing out difficulties with the concept that natural selection results in major changes in phenotypes (discussed earlier in this chapter). Thus, he did not believe that important new adaptations arose gradually.

Goldschmidt argued that each major evolutionary change occurred via a sudden and large genetic alteration, which he called a **macromutation** or **systemic mutation.** The macromutation produced a phenotype very different from the parental phenotype. Such new phenotypes he called "hopeful monsters" because the probability of their having a selective advantage was so low. In turn, Mayr (1963) called them "hopeless monsters" because he believed the probability was zero.

Goldschmidt's arguments have generally not been accepted, and Frazzetta recognized three basic factors that undermined their credibility.

1. There are no well-established cases known to show that gradualism could not have produced a new adaptive type (Mayr 1963).
2. Reproduction between the hopeful monster, with its major genetic reconstitution, and the parental type would probably be quite unsuccessful because the progeny would have low fitness. If selection cannot produce intermediate forms with selective advantage, then an intermediate-form hybrid must fail also.
3. A very unusual phenotype in a population would probably be selectively inferior to the phenotypes that have been refined by natural selection for the optimal adaptive adjustment to the environment.

A Snake Not So Hopeless

For the bolyerine snakes, then, what would be the counterarguments? Let us treat them in the same order.

1. In the bolyerine snakes, it is very difficult to visualize a smooth transition in structural change.
2. Causes of drastic phenotypic modifications may not be limited to major genetic changes. Small genetic changes that alter the timing of ontogenetic programs can have very significant effects on the mature phenotype. Goldschmidt recognized this relationship in 1938, and Matsuda (1979) also emphasized the importance of alterations in developmental processes causing major phenotypic changes (see also McKinney and McNamara 1991). Thus, genetic changes of the same type as those invoked by gradualists can have saltational effects on the phenotype. In addition, genetic incompatibility between the parental type and the new mutant would be unlikely. There is no reason to suppose that the primordial bolyerine snake could not mate with its normal boid peers.
3. Frazzetta (1970, 66) argued that structures in organisms "display a remarkable capacity for functional adjustment." For instance, the cranium of the new bolyerine snake could have accommodated phenotypically to the revised maxillary plan.

Persistence of an Inferior Phenotype

Frazzetta admitted that a mutated animal may be somewhat clumsy, and so it is necessary to imagine how a new trait can persist while selection perfects it. First, where few competitors exist, some imprecision may be tolerable. The bolyerine snakes evolved in the archipelago of islands (including Mauritius) 500 miles east of Madagascar in the Indian Ocean, and so they evolved in a depauperate fauna (cf. Lewis's views on new genotypes surviving in competition-free space). Second, where a species is relatively isolated from others, it tends to have a broad niche exploitation pattern, which perhaps leads to more flexibility of phenotype, greater potential for change, and the opportunity for individuals with the most dramatic changes in genotype to exploit a new ecological niche and so avoid intraspecific competition. Even though in the parental population the new joint may be adaptively inferior, the new phenotype may be perfectly viable in its new role. This scenario has been termed the **Ludwig theorem.** "A genotype utilizing a novel subniche could be added to the population even if it were of inferior viability in the normal niche of the species" (Mayr 1963, 245).

Some of the elements of this argument are similar to those made by Lewis (1962, 1966) on saltational evolution. The new genotype survives in a location adjacent, on some environmental gradient, to that of the parental genotype. Freedom from competition in this new environment is an important aspect of survival.

Bush (1975) also mentioned that Goldschmidt's (1940) view of the hopeful monster was perhaps forgivable. In both sympatric and parapatric models of speciation, a new genotype colonizes and exploits a new niche, just as it does in the Lewis and Frazzetta scenarios. Bush quoted Spurway (1953): "The population size N of Wright should be so small [for fixation of chromosomal rearrangements] that it may be profitable to think of some species originating from a single pair in a new Eden. Being an Adam and Eve gives a monster a chance to hope."

18.5 BEYOND THE REACTION NORM

Matsuda (1979) articulated themes that crop up again and again in this literature: Small changes in developmental processes can have large effects on phenotypes, and new phenotypes exploit new environments. Recall the brief discussion on **heterochrony** in Chapter 5. Matsuda took the view that the environment influences developmental processes, and in arthropods, abnormal development—particularly abnormal metamorphosis—always occurs in changing or unusual environments. He recognized the following general scheme of interaction:

External factors → hormones ↔ genes →
development → structure

Matsuda emphasized that environmental influences on the endocrine system are the driving forces that lead to phenotypic change, and in this position he sounds rather Lamarckian (cf. Chapter 2, Section 2.3). He invokes the concept of **genetic assimilation,** or the **Baldwin effect,** which argues that the genetic material may gradually come to code for an adaptive phenotypic change through selection that ensures expression of that phenotype.

Matsuda, like Goldschmidt (1938, 1940), argued that the normally observed phenotypes of a species constitute a very small range of what the genotype is potentially capable of producing. That is, the genotype's **reaction norm** is narrow compared with its potential range. Both scientists recognized that hormonal action could greatly influence the reaction norm so that phenotypes beyond the norm would be expressed.

Bark Louse to Bird Louse

One of Matsuda's examples may clarify the idea of reaction norms. A young bark louse (Order Psocoptera) is wandering over a tree trunk, grazing on unicellular algae, and happens to crawl onto a bird. Although the environment on the bird is very different in practically every way imaginable, the louse stays there. New food and different temperatures and humidies could greatly influence developmental processes and hormonal balance. Such changes as higher juvenile hormone production would influence development to such an extent that wings might not appear in the adult. Essentially, a sudden shift would occur from a bark louse in one order to a chewing louse in another (Order Mallophaga). This kind of initial change could take place without any genetic change, and one wonders if genetic assimilation would be important. So long as the new phenotypes stayed in the new environment, the modifications would be stable and permanent. No doubt drift and selection would play their roles in the evolution of mallophagans in response to small population sizes and to the new environment on birds, once the saltational event had occurred.

Heterochrony: From Mushroom to False Truffle

McKinney and McNamara (1991) entered into the details of why **heterochrony** is so important in evolution. This "change in timing or rate of developmental events, relative to the same events in the ancestor" (387), was noted as an important **intrinsic evolutionary factor** that caused small to large changes in the design of members of a lineage. Such factors have received much less emphasis than **extrinsic evolutionary factors,** especially natural selection.

In recent times there has sometimes been a tendency to see heterochrony merely as an instigator of rapid evolutionary change, internally steered and leaping like an evolutionary Tarzan from adaptive peak to adaptive peak. The neo-Darwinian (or modern synthetic) view, on the other hand, sees evolution as gradual and externally steered: the slow, laborious conquest of successive adaptive peaks driven by the continuous grind of natural selection. But does heterochrony only produce morphological saltations? . . . We argue that it is a more fundamental aspect of evolution, underpinning most morphological change, be it gradual or rapid (McKinney and McNamara 1991, 168).

Because small genomic changes may cause small to large morphological change, alterations in developmental timing are likely to result in a continuum of effects. Small changes in the genes regulating development could have large effects. An example concerns the evolution of false truffles, those fungal species with underground fruiting bodies that animals such as squirrels eat so avidly (e.g., States 1990). No change in the mitochondrial genome accompanied a suggested shift from an aboveground (epigeous) mushroom such as *Suillus* to the belowground (hypogeous) false truffle, *Rhizopogon*, a break from one genus to another (Bruns et al. 1989). A likely scenario is the paedomorphic retention of the belowground primordium of the fruiting body, accompanied by accelerated spore production underground (Figure 18-4). Under conditions of water stress, the prolonged development from primordium to mature fruiting body with a closed cap, as in the stomach fungi (Gasteromycetes), followed by completely hypogeous development in the false truffles, would be highly adaptive. The shift would also favor a change in spore dispersal from forcible spore discharge and wind dispersal to dispersal by animals.

Remote Genera of Plants

Van Steenis (1969) also considered abnormal development as a possible mechanism in the evolution of species and higher taxa of plants in the tropics. He noted that the tropical rain forest has many taxonomically isolated species in genera that seem to be remote from their nearest phylogenetic neighbors. These isolated species also have abnormal numbers of strange structures. He regarded them as taxonomic parallels of terata: marvels, monsters, or anomalies of form. Either saltational genetic changes or association with pathological or mutualistic organisms could produce such marvels. Genera such as *Rafflesiana,* the giant-flowered parasitic angiosperms of the tropics, and *Nepenthes,* with its pitcher-like leaves (Figure 18-5), are so different and distant from other taxa that saltation may have been involved in their origins. Still, the mechanisms in all these cases may remain elusive.

18.6 QUANTUM EVOLUTION

Simpson's (1944) classic contribution to the modern synthesis, *Tempo and Mode in Evolution,* had evolutionary rates as a theme and, in the last chapter, discussed "quantum evolution." In his 1953 book, Simpson defined **quantum evolution** as evolutionary changes of adaptive zone such that transitional forms between the old zone and the new do not persist, i.e., events that involve all-or-none reactions. He emphasized that such a change is not different in kind from phyletic evolution but is an extreme case of it. Simpson outlined many examples and asserted that even the origin of classes may result from quantum evolution. Some of his examples follow.

Brachydonty to Hypsodonty in the Equidae

Although it was a good browser, *Miohippus* in the late Oligocene had low-crowned teeth **(brachydonty).** *Merychippus* in the late Miocene, and about 15 million years later, had high-crowned teeth **(hypsodonty)** covered with cement, with complex crests (Figure 18-6). The enamel was folded so that, as the teeth wore, the complex bands of enamel continued to project slightly above the softer dentine and cement. This tooth, which lasted a long time in spite of hard wear, was adapted for grinding tough forage and seeds well—which enabled easier digestion—and for utilizing a diversification of food types from grass to shrubs and trees. Also, as horses became larger, they lived longer and needed better, longer-lasting teeth (Simpson 1953, MacFadden 1985). The evolution of hypsodonty coincided with the development in the Miocene of grasslands, which provided plentiful forage, although the grass was highly abrasive because of the silica bodies in its stems and leaves.

Thus, a dramatic shift in design occurred over 15 million years or less, whereas brachydonty had

---

FIGURE 18-4

The possible paedomorphic origin of the genus *Rhizopogon* from a *Suillus*-like ancestor and evolution through an intermediary stomach fungus such as *Secotium* or *Endoptychum.* *(After Bruns et al. 1989)*

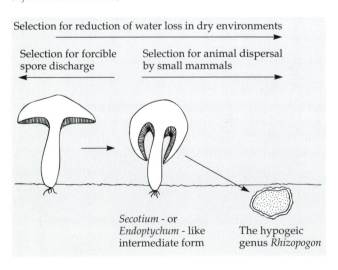

Selection for reduction of water loss in dry environments →

← Selection for forcible spore discharge   Selection for animal dispersal by small mammals →

Secotium - or *Endoptychum* - like intermediate form   The hypogeic genus *Rhizopogon*

FIGURE 18-5

A member of the genus *Nepenthes*, a tropical pitcher plant from the Old World. The genus was named in 1689, and J.D. Hooker described digestion of insects in the pitchers in 1874. Note the two forms of pitcher on a single plant, each at the tip of a leaf and each with a lid. Nectar secreted on the inner lid and the inner lip attracts insects, which slip and fall into the liquid in the pitcher. Digestion of the insects by enzymes secreted in the pitcher and by bacteria gleans added nutrients for absorbtion by the plant. Some species' pitchers can reach 50 cm long and 25 cm in diameter and are said to catch small vertebrates. This single genus, with about 70 species, has its own family, the Nepenthaceae. *(From Pietropaolo and Pietropaolo 1985)*

FIGURE 18-6

Teeth of *Miohippus* and *Merychippus*, demonstrating the change in upper molars from the early browsing horse (with low-crowned teeth) to a later grazing horse (with high-crowned teeth). Whereas the brachydont tooth of the earlier *Miohippus* was set in place when it erupted, as are human teeth, the hypsodont tooth of *Merychippus* continued to move up above the gum as the tooth surface wore down. *(From Simpson 1951)*

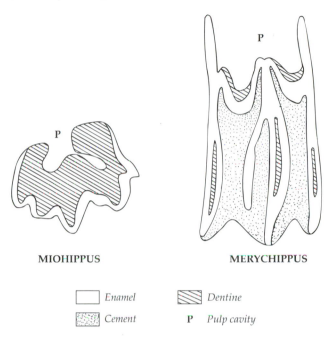

MIOHIPPUS   MERYCHIPPUS

☐ *Enamel*   ▨ *Dentine*

▦ *Cement*   **P** *Pulp cavity*

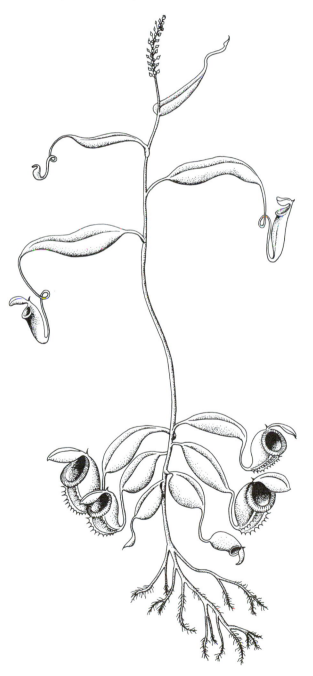

existed for more than 100 million years, in the earlier horses through the Eocene and Oligocene and presumably in their mammal ancestors from the Cretaceous.

Fifteen million years is brief in terms of the fossil record, and such a time span is most recognizable in the Cenozoic Era—the last 65 million years—in which vertebrate fossils are relatively abundant.

Gaps in the Fossil Record

Remember, however, that the Cenozoic Era has yielded no fossils of some vertebrates that lived all through that time (Simpson 1953). Coelacanth fishes left abundant fossils in some formations, but none has been found since the late Cretaceous, 75 million years ago, even though coelacanths are still living today. Another example is the sphenodonts found in the Triassic, Jurassic, and early Cretaceous. No fossils have been found from the intervening period, although sphenodonts exist today in the form of the sphenodon, or tuatara (Figure 11-7), a reptile that

FIGURE 18-7

The single-pulley astragalus of the burro, a perrisodactyl, and the double-pulley astragalus of a mule deer, an artiodactyl. Note the robustness of the tarsus of the burro, *Equus asinus*, and the thinner structure in the mule deer, *Odocoileus hemionus*. In the burro the joint is flat, providing a single-pulley system between the astragalus and the metatarsals. In the mule deer the same joint is hinged in two places with a ridge between, forming a double-pulley system. (Based on Vaughan 1986) *(Illustration by Tad Theimer)*

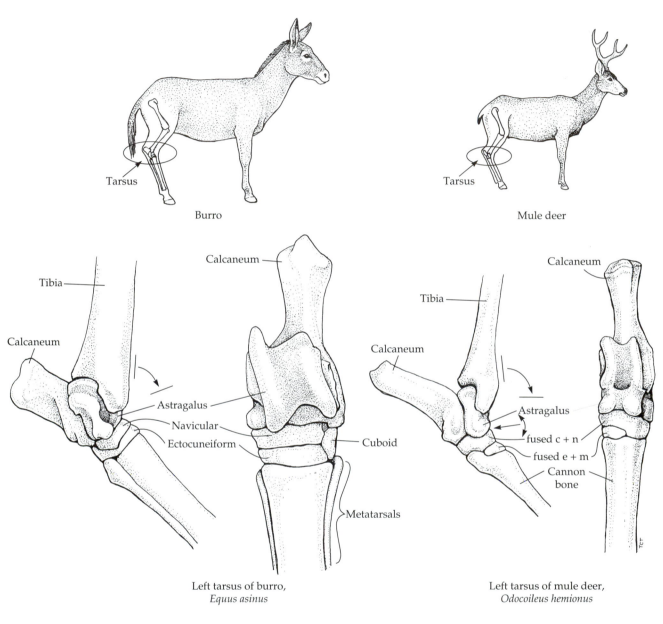

Burro

Mule deer

Left tarsus of burro,
Equus asinus

Left tarsus of mule deer,
Odocoileus hemionus

c + n = cuboid + navicular
e + m = ectocuneiform + middle
    cuneiform
Cannonbone = fused third and fourth
    metatarsals

FIGURE 18-8

The zebra, a member of the family of horses (Equidae) in the order of odd-toed ungulates (Perissodactyla). *(Illustration by Stephen Price)*

now lives only on a few islands near the coast of New Zealand (Colbert 1980). The sphenodont's "disappearance" creates a gap in the fossil record of something like 130 million years. Thus, real, well-established gaps do occur in the fossil record, even in the last 65 million years.

Such gaps make the fossil record much harder to interpret in terms of estimating rates of evolution. Perhaps it is fortuitous that the fossil record of the horses is sufficient to enable scientists to distinguish rates of evolution through the history of the lineage.

The Emergence of the Artiodactyla

The shift in ankle design from the perrisodactyls to the artiodactyls is another case in which a rather dramatic change enabled entry into a new adaptive zone. The perrisodactyls had a single pulley on the astragalus, whereas the artiodactyls have a double pulley (Figure 18-7). The double astragalus enables a great amount of flexion and extension of the hind limbs so that the animals can leap remarkably well. Compare the zebra (Perrisodactyla) in Figure 18-8 and the gerenuk (Artiodactyla) in Figure 18-9; note the differences

FIGURE 18-9
The gerenuk, a member of the family that includes cattle, sheep, goats, and antelopes (Bovidae) in the order of even-toed ungulates (Artiodactyla). *(Illustration by Stephen Price)*

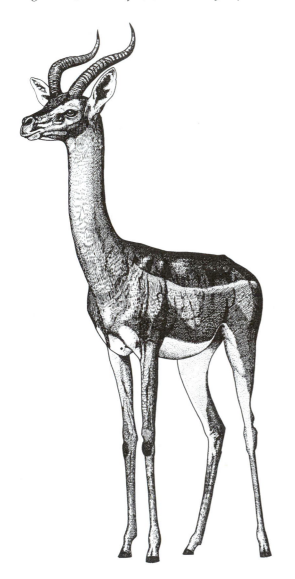

in form of the strong runner and the agile leaper. The evolution of the complex gut, which enabled rumination—the rapid consumption of large amounts of food and subsequent digestion in a protected place—was probably equally important in the impressive adaptive radiation of the Artiodactyla.

The most primitive ungulates, the condylarths, appear in the early Paleocene, about 60+ million years ago, and the first artiodactyls (with the double-pulley astragalus) appear in the early Eocene, no longer than 10 million years later. Thus, according to

the best estimate available in the fossil record, this major taxon, an order of mammals, evolved within about 10 million years.

The Rise of the Stylinodontinae from the Conoryctinae

The Taeniodonta is an order of mammals that is now extinct. Within the Taeniodonta, the more primitive Conoryctinae rapidly gave rise to a lineage with enlarged claws and powerful limbs—the Stylinodontinae—at the end of the Cretaceous (Figure 18-10). From the rather generalized form of the Conoryctinae, a dramatic shift occurred to larger animals with skulls 30 or more centimeters long, robust jaws with long canines, heavy limbs, and strong claws for digging (Figure 18-10). The canines grew persistently and, with a strong enamel layer on the anterior, remained sharp for cutting while the posterior wore into a grinding surface. The nature of the new adaptive zone can be surmised only from the fossils. The limbs equipped for digging or tearing and the teeth for cutting and grinding suggest a diet of subterranean tubers (Colbert and Morales 1991). Patterson (1949) speculated that stylinodonts adapted to eat hard-shelled fruits and nuts, using their appendages to tear off tough husks and their sharp teeth and strong jaws to crack shells. The canines may have worked like a parrot's beak; one genus was called *Psittacotherium*, *psittacus* being Greek for parrot.

"At high levels [of taxonomy] some element of quantum evolution is usually involved. That is the most important point about this mode of evolution" (Simpson 1953, 389). Simpson's concept of quantum evolution includes the existence of discontinuities between adaptive zones, so that an all-or-none evolutionary leap must be made across each discontinuity, or nonadaptive zone (Figure 18-11). Characteristics must accumulate in a population until they reach a threshold so that a rapid transition can occur from one adaptive zone to the other. Whether these characteristics develop under the influence of natural selection is controversial, Simpson said, but they are preadaptive for the transition. Once the adaptive shift occurs under the influence of strong directional selection, stabilizing selection predominates in the new adaptive zone. "The quantum change is a breakthrough from one position of stabilizing selection to another" (Simpson 1953, 391). Finally, Simpson argued that quantum evolution is usually involved in the opening, explosive phase of adaptive radiation of a group.

Simpson's writings have a persistent theme: that change occurs from a parental zone, niche, or range

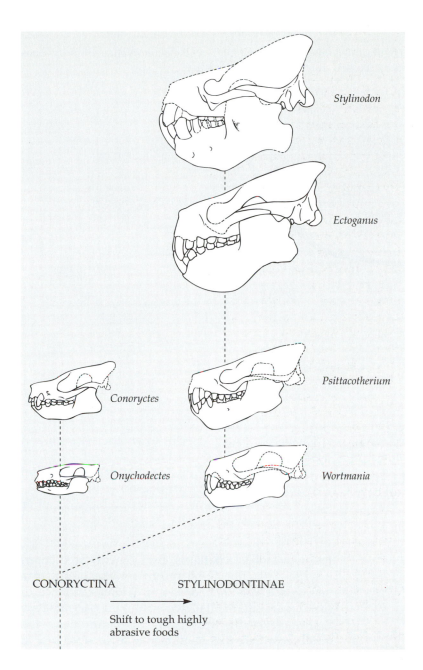

Stylinodon

Ectoganus

Conoryctes

Psittacotherium

Onychodectes

Wortmania

CONORYCTINA   STYLINODONTINAE

→

Shift to tough highly
abrasive foods

FIGURE 18-10

Patterson's (1949) reconstruction, slightly modified, of the phylogeny of the taeniodonts from the generalized Conoryctinae to the more specialized Stylinodontinae, with the development of much larger bodies and skulls, massive mandibles, and large canines for cutting and grinding.

into an unoccupied zone and ecological niche. He recognized that species, as well as higher taxa, may often arise in this way.

18.7 PUNCTUATED EQUILIBRIA

Although nowhere in their paper do Eldredge and Gould (1972) refer to Simpson's concept of quantum evolution, Simpson's model would produce a pattern they describe at length: rapid evolutionary change followed by long periods of stabilizing selection. The result is a fossil record of **punctuated equilibria:** long periods of **stasis** (no change), punctuated by brief episodes of rapid change. These scenarios are not predicted by the allopatric speciation model, which assumes gradual divergence of populations over long periods of time.

Contrasting Phyletic Gradualism and Punctuated Equilibria

Eldredge and Gould (1972) contrasted the classic **phyletic gradualism**—a term they coined—with

FIGURE 18-11

Simpson's (1953) argument for quantum evolution in the Taeniodonta from a parental stock of Conoryctinae rapidly aross a nonadaptive zone and into a new zone occupied by the new subfamily Stylinodontinae.

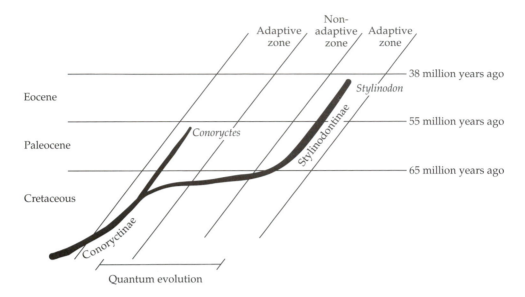

FIGURE 18-12

Possible patterns in the fossil record, showing phyletic gradualism as a clade develops (cladogenesis), evolution of a lineage through time without branching (anagenesis), and punctuated equilibrium with a preponderance of stasis and rapid episodic speciation events. The "morphological characters" axis for each alternative indicates a gradient of characters over which a lineage changes through time. Gould (1985) noted that anagenesis could be gradual or punctuated, as is illustated also in cladogenesis and punctuated equilibrium.

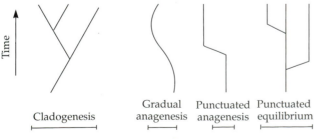

predictions from the allopatric model of speciation. They summarized phyletic gradualism as follows:

1. An ancestral population changes into modified descendants, giving rise to new species (Figure 18-12).

2. Change from one species to another is even and slow.

3. The change usually involves the whole ancestral population or a large proportion of it.

4. Transformation to new species occurs over all or most of the geographic range of the ancestral species.

From these basic patterns, two general predictions emerge. First, the fossil record for the origin of a new species should consist of a long sequence of continuous intermediate types from the parental type to the new species. Second, any discontinuities in the fossil record are due to an inadequate record, not to saltational evolutionary changes.

Eldredge and Gould contrasted the scenario of phyletic gradualism with one of punctuated equilibria. They envisaged a phyletic line that was usually in equilibrium, unchanging, disturbed only rarely by rapid and episodic events of speciation (Figure 18-12). They asserted that an important aspect of allopatric speciation is that a new species can arise when a small local population becomes isolated at the margin of the geographic range of its parent species. That assertion crops up over and over again in discussions of saltational evolution and is a valid and widely accepted theme. The consequences for paleontology are apparent enough. Since new fossil species do not arise in the place where ancestors lived, it is unlikely that the paleontologist can trace the gradual divergence of lineages by following species up a rock column.

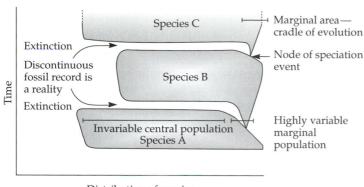

Species C

Marginal area—
cradle of evolution

Extinction

Node of speciation
event

Discontinuous
fossil record is
a reality

Species B

Extinction

Invariable central population
Species A

Highly variable
marginal
population

Time

Distribution of species

FIGURE 18-13

A general scenario for the punctuated equilibrum concept of evolutionary change. Visualize a species-change pattern that appears when time is measured vertically through a stratigraphic section of rock, and ponder how many such rock sections would be needed to reveal at least the distribution of species in the central population.

We can summarize the tenets of the allopatric model, contrasted with the phyletic gradualism model, as follows:

1. New species evolve by the splitting of lineages.
2. Species emerge rapidly.
3. New species emerge from small subpopulations of the parental type.
4. Each of these small populations has a narrow geographic range on the margin of the parental species' range.

The resultant predictions are quite different from those derived from the concept of phyletic gradualism. First, in any local part of the ancestral species range, there should be a sharp break between the ancestral and derived forms. During dispersal from the point of origin, the new species and the parental species diverged significantly in morphology. Discovery of the actual event of speciation is most un-

likely. Second, a fossil record with many gaps represents a complete local history of the evolution of a lineage.

Thus, Eldredge and Gould recognized that a saltational fossil record in any locality has associated with it a saltational speciation event. They brought paleontology up to date with modern evolutionary theory, emphasizing stasis and the relatively rapid evolutionary change in a lineage of fossil taxa. Figure 18-13 illustrates their general scenario.

An Example Using Little Frog-like Trilobites

Eldredge's (1971, 1972) studies of the trilobite *Phacops rana* (Figure 18-14) illustrate the pattern Eldredge and Gould recognized. *Phacops rana* lived in a large area of sea with its eastern margin in what are now New York State, Pennsylvania, and West Virginia. The only really variable characteristic in *Phacops* is the

FIGURE 18-14

Phacops rana, an appealing little trilobite, about 6 cm (2.3 inches) long, studied by Eldredge (1971, 1972). Imaginitively beheld, the creature is reminiscent of a small frog. Specimens were very abundant around 370 million years ago. *(Illustration by Stephen Price)*

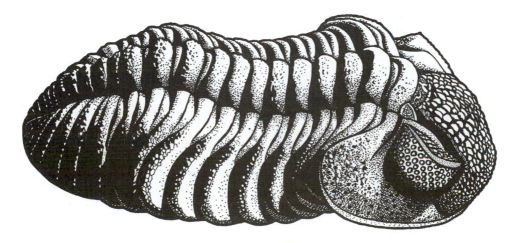

number of vertical columns, or files, of ocelli in the eye (Figure 18-15); the ancestral number is 18 dorso-ventral files. In the epeiric (shallow) sea covering a large part of a continent but in contact with ocean, the deposited strata show no variation in file number in the *Phacops* eyes within each stratum. In the marginal sea, however, there were two episodes of variation in which 18-file and 17-file *Phacops* coexisted and, much later, 17-file, 16-file, and 15-file individuals co-existed. Eldredge argued that it was from these variable marginal populations that new taxa spread into the epeiric sea. Their fossil record, in what are now the cow pastures of New York State, shows a sudden and discontinuous change over large areas of the geographic range of the lineage. A 17-file *P. rana rana* replaced the 18-file *P. rana* and was, in turn, replaced by the 15-file *P. rana norwoodensis* (Figure 18-16). It appears that two episodes of punctuated equilibrium with a long period of stasis between them occurred in this phylogeny.

To what extent such subspeciation events reflect macroevolutionary patterns and whether the epeiric sea was itself marginal to the large Tethys Sea remain debatable points. Nevertheless, the *Phacops* example effectively illustrates the concept of stasis and punctuated equilibrium in the fossil record (see also Gould and Eldredge 1977).

FIGURE 18-15

The right eye of *Phacops rana milleri* with the front facing right, showing the 18 dorso-ventral files of ocelli that were typical of the original *P. rana* stock. *(Illustration by Stephen Price)*

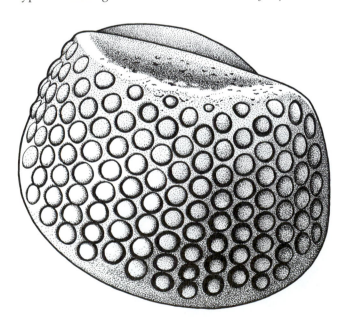

FIGURE 18-16

The phylogeny of subspecies within the species *Phacops rana*, according to Eldredge and Gould (1972). Note that evolutionary divergence from the parental stock occurred repeatedly in the marginal sea, and the new subspecies spread much more extensively into the epeiric sea.

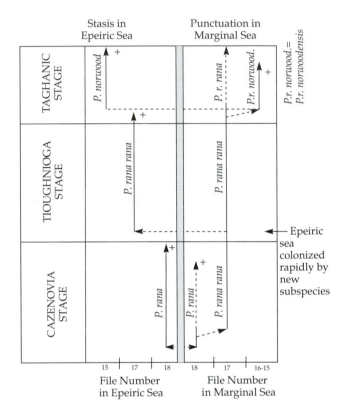

Generality of Punctuated Equilibrium

Hallam (1978) reanalyzed the example of *Gryphaea*, a Jurassic oyster discussed by Eldredge and Gould, to see how common punctuated equilibrium was, relative to phyletic gradualism, in this and other bivalves. He carried out a comprehensive biometric analysis on numerous specimens, and all but one characteristic supported the punctuated equilibrium model. The character that seemed to support the phyletic gradualism model was size, which increased gradually through time but only in the minority of species.

This increase in size of phylogenies through the fossil record has been recognized for a long time and is known as **Cope's rule** or Cope's law (Cope 1885, 1896; Newell 1949). Paleontologists have observed the pattern so frequently—Newell found no inverse trends among invertebrate fossils—that it may have set the stage very early for the general acceptance of phyletic gradualism. Hallam (1978, 23), however, generalized by saying that "phyletic size increase may be the only major gradualistic exception to the punctuated equilibria speciation model" (see also Williamson 1981).

The increasingly precise analysis of diverse fossil groups, such as the horses, yields a more complex picture (MacFadden 1985, 1986). Generally, horses have been regarded as a good example of Cope's rule, because the largest members in the group increased in size, over about 57 million years, from 50-kilogram animals such as *Hyracotherium* to the large genus *Equus* with species reaching 500 kilograms. Hidden in this broad picture are varying trends: (1) some lineages remained small; (2) some lineages showed decreased body sizes; and (3) lineages showed prolonged relative stasis (32 million years) followed by a shorter period of increasing body size (25 million years) (MacFadden 1986).

A Genetic Model

Kirkpatrick (1982) developed a genetic model showing that a continuous polygenic character can change under the influence of natural selection in the way envisaged for quantum evolution or punctuated equilibrium. He concluded that quantum evolution can occur without environmental change, genetic drift, or macromutation. It can occur when phenotypic variance of the population increases (Figure 18-17), as we saw in the marsh fritillary butterfly (Figure 18-2). The shift from one adaptive peak to another can proceed even though intermediate phenotypes have lower fitness. These shifts can be large and rapid, with a character changing by several standard deviations in tens or hundreds of generations. Such rapid change appears in the fossil record as a saltational event. Kirkpatrick (1982, 846) ended his paper by saying,

> The model described in this paper is consistent with the neo-Darwinian view and furthermore shows how subtle and gradual changes in the environment or in the internal genetics of an organism can give rise to a new fossil record with apparent discontinuities. This leads to the conclusion that rather conventional assumptions—natural selection operating in large populations—will suffice to produce the patterns observed.

Cronin (1985) provided one actual example of rapid evolutionary change coupled with environmental change. He documented episodic speciation events, associated with climatic change, and long periods of stasis in ostracods.

FIGURE 18-17

Kirkpatrick's (1982) model of quantum evolution, based on conventional population genetics.

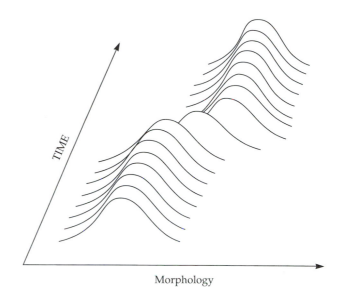

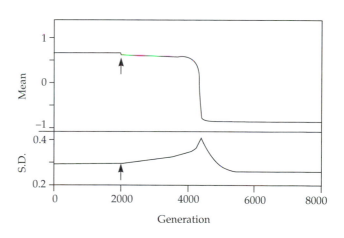

18.8 HOW FAST IS RAPID EVOLUTION?

The question of what is "rapid" evolution and what is "slow" in geologic time is clearly an open one. Gould (1982) defined geologically *instantaneous* change as 1 percent or less of the duration of subsequent stasis. Thus, a species that persisted for 10 million years would have speciated in 100,000 years. Stanley (1981) reckoned that for speciation to be *rapid*, it must occur in less than a few thousand to tens of thousands of years. Stanley gave some examples. Lake Nabugabo, at the margin of Lake Victoria in Uganda, has been isolated from Lake Victoria for only about 4,000 years, and yet five indigenous species of cichlid fish live in it and are absent from Lake Victoria and other localities. Therefore, five new species have evolved in 4,000 years. Zimmerman (1960) suggested that five or more species of the moth genus *Hedylepta* must have evolved within 1,000 years in Hawaii. They are all endemic species and specific to banana, and the plant was introduced to the islands only 1,000 years before his study. Among the pupfishes in Death Valley, California, a major shift in morphology from the parental *Cyprinodon salinus* to *C. milleri* occurred in a few thousand years (see also Humphries 1984).

Gould (1982) acknowledged the claim made by others that many scientists from Darwin to Simpson had already described punctuated equilibrium, although Gould argued that those authors were gradualists. Eldredge and Gould (1972) and Gould (1982) certainly emphasized the two critical aspects of punctuated equilibria: the geologically rapid origins of taxa and the long subsequent periods of stasis. At that time, other authors had not highlighted the sig-

nificance of the latter point. Since the 1980s, interest has been lively (e.g., Berggren and Casey 1993). Still, the issue is a matter of degree. The hypsodont tooth in horses evolved rapidly and has stayed hypsodont since its origin. All artiodactyls still have the double-pulley astragalus that evolved about 55 million years ago. In a general way, stasis has been apparent for this period of time.

18.9 DARWIN ON RATES OF EVOLUTIONARY CHANGE

Templeton (1982) argued that Darwin indeed saw evolutionary change occurring as punctuated equilibria but that, throughout *The Origin*, he simply emphasized microevolutionary changes and seldom dealt with macroevolution. When Darwin did discuss macroevolution, he made it plain that rates of evolution

FIGURE 18-18

Darwin's (1859, 1872) diagram of lineages passing through time, a few speciating frequently and diverging gradually (A and I) and many others showing stasis (e.g., B–H) and eventual extinction (e.g., B–D, K, L). The diagram fails to represent all the scenarios Darwin discussed, notably the view that "it is far more probable that each form remains for long periods unaltered, and then again undergoes modification."

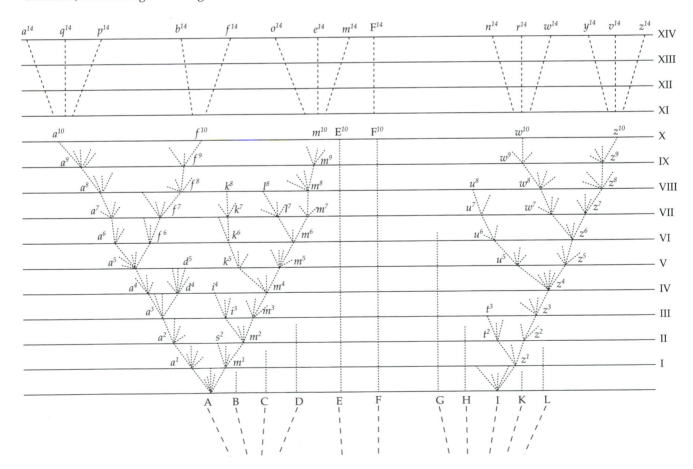

could be very different and adaptive shifts could be rapid in relation to long periods of stasis. Templeton and Giddings (1981) quoted some passages from *The Origin of Species*. On page 357 (1936 Modern Library, Random House, edition), Darwin argues that sudden changes can occur in a local population and then spread rapidly through the geographic range of the species so that "they appear as if suddenly created there, and will be simply classed as a new species." Still on that page, he adds:

> Many species when once formed never undergo any further change but become extinct without leaving modified descendants: and the periods, during which species have undergone modification, though long as measured by years, have probably been short in comparison with the periods during which they remain the same form.

On page 373, Darwin asserts:

> A number of species, however, keeping in a body might remain for a long period unchanged, whilst within the same period several of these species by migrating into new countries and coming into competition with foreign associates, might become modified; so that we must not overrate the accuracy of organic change as a measure of time.

Eiseley (1979) noted Darwin's comment on a diagram in Chapter 4 of *The Origin* (Figure 18-18):

> But I must here remark that I do not suppose that the process [of production of new varieties] ever goes on so regularly as is represented in the diagram, though in itself made somewhat irregular, nor that it goes on continuously; it is far more probable that each form remains for long periods unaltered, and then again undergoes modification. (Darwin 1872, 90)

Templeton and Giddings (1981) took the view that the modern synthesis of the 1940s had never yielded an espousal of a single mechanism in macroevolution, and the assertion that by 1950 all evolutionary biologists subscribed to phyletic gradualism was an oversimplification. Statements by three of the architects of the modern synthesis illustrate their point. Fisher (1958, 153) discussed why many speciation events and morphological transitions follow a pattern of stasis punctuated with "sudden spurts of change." Haldane (1937, 337) thought gradual continuous changes in population genetics would occur, "on a geological time scale, almost explosively." According to Wright (1949), periods of stasis were interspersed with periods of rapid adaptive transition.

Although rates of change have long been a theme in evolutionary studies, they will probably continue to be the subject of debate for decades to come. Certainly there is room for additional mechanistic empirical and theoretical studies on rates of change. Until we know much more about gene regulation of development, however, a good proportion of the literature will remain rather speculative.

SUMMARY

Rates of evolutionary change are by no means constant, and the degree of inconstancy has been debated through the ages. We can view end points in the continuum of rates of change as **stasis,** or absence of change, and **saltation,** or a rapid jump in evolution, with gradual change filling in the middle ground. Even Darwin espoused ideas of saltational change, although they were less evident in *The Origin* than in earlier notes.

The rediscovery of Mendelian genetics, with its particulate inheritance, raised questions on how gradual evolution could be, particularly when favorable mutations provided discrete steps in genotypic change.

Accepted cases of saltational change include shifts in ploidy level, hybridization, chromosomal rearrangements, shifts from outcrossing to parthenogenesis, and quantum speciation in plants. An example of **quantum speciation** involves the plant genus *Clarkia*, in which **catastrophic selection** in marginal dry habitats results in the establishment of new genotypes beyond the parental range. The marsh fritillary butterfly experienced rapid shifts in evolutionary rates, including flushes in abundance that released large amounts of genetic variation. When stringent selection resumed, the population had shifted noticeably in general phenotypic characters.

Controversial evidence of saltational evolution comes from the evolution of the intramaxillary joint in the bolyerine snakes. The example brings into focus recent conjectures equivalent to those by Goldschmidt, with his **hopeful monsters.** Whether the "monsters" were hopeful or hopeless is hard to discern, but inferior genotypes may well survive better when isolated from parental types in a new ecological setting. Were the progenitors of the human race hopeful monsters?

Saltational evolution may also occur when individuals enter radically new environments that push phenotypes beyond the reaction norm of the genotype. Remote plant genera are hard to explain without invoking some kind of evolutionary jump outside the normal range of variation.

Quantum evolution is a rapid evolutionary shift from one adaptive zone to another, frequently creating a higher taxon. Examples include the shift from brachydonty to hypsodonty in the horses, the emergence of the artiodactyls from the perrisodactyls, and taeniodont evolution into the stylinodont adaptive zone.

The concept of **punctuated equilibria** recognizes long periods of **stasis**, or **equilibrium**, punctuated with brief episodes of rapid change in a fossil lineage. This contrasts with **phyletic gradualism**, a concept of continuous change. The fossil record frequently supports the punctuational model, as illustrated by the trilobite *Phacops rana*; the oyster genus, *Gryphaea*;

and other lineages. A genetic model accounts for such rapid shifts in morphology during environmental changes and amplified genetic variance.

"Rapid evolution" is, of course, a relative term, suggesting a major break from the parental stock within a short period after a long period of stasis. Punctuational events may cover a few thousand years, 100,000 years, or 10 million years, depending on the author, the taxon under study, and the period of stasis before or after the event.

Returning to Darwin and *The Origin of Species*, it is clear that he was by no means restricted to a gradualistic view of evolution; rather, he acknowledged periods of stasis and eruptive episodes of change when he considered macroevolutionary events. Darwin indeed anticipated the avidly debated topic of evolutionary rates.

QUESTIONS FOR DISCUSSION

1. Can you imagine the dilemma that Darwin faced when sifting through all the different points of view on rates of evolution in the literature of his time? If Darwin had adopted a different position in *The Origin*, such as saltational change in the speciation process, do you think such differences would have changed the course of the debate on Darwin's book?

2. How many different processes can you envision that can result in rapid evolutionary change, and which examples would you use to support your case?

3. Rate of evolutionary change is usually unquantified in a discussion in relation to background rates and supposed rapid rates. Is it important to treat this subject in a more quantitative way that would enable, for example, direct comparisons of rates of evolution, or an appreciation of a continuum in rates?

4. To what extent do you think that the discussion on rates of evolution is largely the slight refinement of a Darwinian view of evolution or a major set of breakthroughs in understanding?

5. Are there some generalizations to be recognized that would help with a better synthesis on rates of evolution, or do you think that the field is constituted mainly of special-case scenarios?

6. Most cases made for rapid evolution emphasize pattern of change and more or less guess at the mechanisms and even the rates of change. Would more careful mechanistic studies on the processes involved help to clarify understanding? Can you identify any taxon and locality in which such studies could be undertaken?

7. In the fossil record much biological information is lost that may otherwise help in understanding rapid evolutionary rates. How would you develop a research program that would use living organisms to gain an enhanced understanding of the fossil record?

8. Considering your knowledge of evolution in the bacteria, protozoa, fungi, plants, and animals, would you agree with the statement that probably the only character that changes gradually in a lineage is size?

9. For small poikilothermic animals like insects, mites and nematodes, and the fungi, can you envisage conditions in which phenotypes may occur beyond the reaction norm, but remain stable in the new environment?

10. What is your response to the quotation, "Being an Adam and Eve gives a monster a chance to hope"? Does consideration of the human animal enlighten the discussion on rates of evolution?

REFERENCES

Barlow, N. (ed.). 1946. Charles Darwin and the Voyage of the Beagle. Philosophical Library, New York.

Bateson, W. 1894. Materials for the Study of Variation. Macmillan, New York.

Berggren, W. A., and R. E. Casey (eds.). 1993. Tempo and mode of evolution from micropaleontological data. Paleobiology 9:326–428.

Blyth, E. 1835. An attempt to classify the "varieties" of animals. . . . Mag. Natur. Hist. 8:40–53.

———. 1836. Observations on the various seasonal and other external changes which regularly take place in birds. . . . Mag. Natur. Hist. 9:399.

———. 1837. On the psychological distinctions between man and all other animals. . . . Mag. Natur. Hist. n.s. 1:1–9, 77–85, 131–141.

Bruns, T. D., R. Fogel, T. J. White, and J. D. Palmer. 1989. Accelerated evolution of a false-truffle from a mushroom ancestor. Nature 339:140–142.

Bush, G. L. 1975. Modes of animal speciation. Ann. Rev. Ecol. Syst. 6:339–364.

Carson, H. L. 1968. The population flush and its genetic consequences, pp. 123–137. In R. C. Lewontin (ed.). Population Biology and Evolution. Syracuse Univ. Press, Syracuse, N.Y.

———. 1975. The genetics of speciation at the diploid level. Amer. Natur. 109:83–92.

Colbert, E. H. 1980. Evolution of the Vertebrates. 3d ed. Wiley, New York.

Colbert, E. H., and M. Morales. 1991. Evolution of the Vertebrates. 4th ed. Wiley–Liss, New York.

Cope, E. D. 1885. On the evolution of the Vertebrata, progressive and retrogressive. Amer. Natur. 19:140–148, 234–247, 341–353.

———. 1896. The Primary Factors of Organic Evolution. Open Court, Chicago.

Cronin, T. M. 1985. Speciation and stasis in marine Ostracoda: Climatic modulation of evolution. Science 227:60–63.

Darwin, C. 1859. The Origin of Species by Means of Natural Selection. Murray, London.

———. 1872. The Origin of Species by Means of Natural Selection. 6th ed. Vol. 1. Murray, London.

de Vries, H. 1906. Species and Varieties, Their Origins by Mutation. Open Court, Chicago.

Dobzhansky, T. 1970. Genetics of the Evolutionary Process. Columbia Univ. Press, New York.

Eiseley, L. 1979. Darwin and the Mysterious Mr. X. Harcourt Brace Jovanovich, New York.

Eldredge, N. 1971. The allopatric model and phylogeny in Paleozoic vertebrates. Evolution 25:156–167.

———. 1972. Systematics and evolution of Phacops rana (Green, 1832) and Phacops iowensis Delo, 1935 (Trilobita) in the middle Devonian of North America. Bull. Amer. Mus. Natur. Hist. 147:45–114.

Eldredge, N., and S. J. Gould. 1972. Punctuated equilibria: An alternative to phyletic gradualism, pp. 82–115. In T. J. M. Schopf (ed.). Models in Paleobiology. Freeman, Cooper, San Francisco.

Fisher, R. A. 1958. The Genetical Theory of Natural Selection. 2d ed. Dover, New York.

Ford, E. B. 1957. Butterflies. 3d ed. Collins, London.

———. 1964. Ecological Genetics. 2d ed. Methuen, London.

Ford, H. D., and E. B. Ford. 1930. Fluctuation in numbers and its influence on variation in Melitaea aurinia, Rott. (Lepidoptera). Trans. Royal Entomol. Soc. London 78:345–351.

Frazzetta, T. H. 1970. From hopeful monsters to bolyerine snakes? Amer. Natur. 104:55–72.

———. 1975. Complex Adaptations in Evolving Populations. Sinauer, Sunderland, Mass.

Goldschmidt, R. 1938. Physiological Genetics. McGraw–Hill, New York.

———. 1940. The Material Basis of Evolution. Yale Univ. Press, New Haven, Conn.

Goodman, M. 1976. Protein sequences in phylogeny, pp. 141–159. In A. J. Ayala (ed.). Molecular Evolution. Sinauer, Sunderland, Mass.

Gould, S. J. 1982. The meaning of punctuated equilibrium and its role in validating a hierarchical approach to macroevolution, pp. 83–104. In R. Milkman (ed.). Perspectives in Evolution. Sinauer, Sunderland, Mass.

———. 1985. The paradox of the first tier: An agenda for paleobiology. Paleobiology 11:2–12.

Gould, S. J., and N. Eldredge. 1977. Punctuated equilibria: The tempo and mode of evolution reconsidered. Paleobiology 3:115–151.

Grant, V. 1971. Plant Speciation. Columbia Univ. Press, New York.

Haldane, J. B. S. 1937. The effect of variation on fitness. Amer. Natur. 71:337–349.

Hallam, A. 1978. How rare is phyletic gradualism and what is its evolutionary significance? Evidence from Jurassic bivalves. Paleobiology 4:16–25.

Humphries, J. M. 1984. Genetics of speciation in pupfishes from Laguna Chichancanab, Mexico, pp. 129–139. In A. A. Echelle and I. Kornfield (eds.). Evolution of Fish Species Flocks. Univ. of Maine, Orono.

Huxley, J. S. 1942. Evolution, the Modern Synthesis. Allen and Unwin, London.

Kirkpatrick, M. 1982. Quantum evolution and punctuated equilibria in continuous genetic characters. Amer. Natur. 119:833–848.

Levinton, J. S., and C. M. Simon. 1980. A critique of the punctuated equilibria model and implications for the detection of speciation in the fossil record. Syst. Zool. 29:130–142.

Lewis, H. 1962. Catastrophic selection as a factor in speciation. Evolution 16:257–271.

———. 1966. Speciation in flowering plants. Science 152:167–172.

Lewis, H., and P. H. Raven. 1958. Rapid evolution in Clarkia. Evolution 12:319–336.

Lyell, C. 1834. Principles of Geology. Vol. 2. 3d ed. Murray, London.

MacFadden, B. J. 1985. Patterns of phylogeny and rates of evolution in fossil horses: Hipparions from the

Miocene and Pliocene of North America. Paleobiology 11:245–257.

———. 1986. Fossil horses from "Eohippus" (*Hyracotherium*) to *Equus*: Scaling, Cope's Law, and the evolution of body size. Paleobiology 12:355–369.

Matsuda, R. 1979. Abnormal metamorphosis and arthropod evolution, pp. 137–256. *In* A. P. Gupta (ed.). Arthropod Phylogeny. Van Nostrand Reinhold, New York.

Mayr, E. 1963. Animal Species and Evolution. Belknap Press of Harvard Univ. Press, Cambridge, Mass.

———. 1980. Prologue: Some thoughts on the history of the evolutionary synthesis, pp. 1–48. *In* E. Mayr and W. B. Provine (eds.). The Evolutionary Synthesis: Perspectives on the Unification of Biology. Harvard Univ. Press, Cambridge, Mass.

McKinney, M. L., and K. J. McNamara. 1991. Heterochrony: The Evolution of Ontogeny. Plenum Press, New York.

Morgan, T. H. 1932. The Specific Basis of Evolution. Norton, New York.

Newell, N. D. 1949. Phyletic size increase, an important trend illustrated by fossil invertebrates. Evolution. 3:103–124.

Patterson, B. 1949. Rates of evolution in taeniodonts, pp. 243–278. *In* G. L. Jepsen, E. Mayr, and G. G. Simpson (eds.). Genetics, Paleontology and Evolution. Princeton Univ. Press, Princeton, N.J.

Pietropaolo, J., and P. Pietropaolo. 1985. Carnivorous Plants of the World. Timber Press, Portland, Oreg.

Punnett, R. C. 1915. Mimicry in Butterflies. Cambridge Univ. Press, Cambridge, England.

Sheppard, P. M. 1962. Some aspects of the geography, genetics and taxonomy of a butterfly, pp. 135–152. *In* D. Nichols (ed.). Taxonomy and Geography. Syst. Assoc. Publ. No. 4. London.

Simpson, G. G. 1944. Tempo and Mode in Evolution. Columbia Univ. Press, New York.

———. 1953. The Major Features of Evolution. Columbia Univ. Press, New York.

———. 1951. Horses: The Story of the Horse Family in the Modern World and Through Sixty Million Years of History. Oxford University Press, Oxford.

Spurway, H. 1953. Genetics of specific and subspecific differences in European newts. Symp. Soc. Exp. Biol. 7:200–237.

Stanley, S. M. 1981. The New Evolutionary Timetable: Fossils, Genes, and the Origin of Species. Basic Books, New York.

States, J. S. 1990. Mushrooms and Truffles of the Southwest. University of Arizona Press, Tucson.

Templeton, A. R. 1982. Adaptation and the integration of evolutionary forces, pp. 15–31. *In* R. Milkman (ed.). Perspectives on Evolution. Sinauer, Sunderland, Mass.

Templeton, A. R., and L. V. Giddings. 1981. Letter to the editor: Macro-evolution conference. Science 211: 770–771.

van Steenis, C. G. G. J. 1969. Plant speciation in Malesia, with special reference to the theory of non-adaptive saltatory evolution, pp. 97–133. *In* R. H. Lowe-McConnel (ed.). Speciation in Tropical Environments. Academic Press, London. [Reprinted in Biol. J. Linnaean Soc. 1:97–133.]

Vaughan, T. A. 1986. Mammalogy. 3d ed. CBS College Publishing, New York.

Williamson, P. 1981. Palaeontological documentation of speciation in Cenozoic molluscs from Turkana Basin. Nature 293:437–443.

Wright, S. 1949. Population structure in evolution. Proc. Amer. Phil. Soc. 93:471–478.

Zimmerman, E. C. 1960. Possible evidence of rapid evolution in Hawaiian moths. Evolution 14:137–138.

Zirkle, C. 1959. Species before Darwin. Proc. Amer. Phil. Soc. 103:636–644.

19

THE GENETIC BASIS OF COEVOLUTION

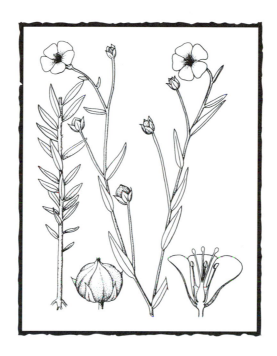

The flax plant, *Linum usitatissimum*, also called linseed, host to the fungal pathogen *Melamposa lini.* The close association between host and parasite was the original stimulant for coining the term "coevolution."

19.1 RECIPROCALLY INDUCED EVOLUTIONARY CHANGE

Coevolution may be defined as *reciprocally induced evolutionary change between two or more species or populations.* Although Mode (1958) coined the term, Ehrlich and Raven were the first authors to make it conspicuous. Their 1964 paper engendered a concept and a theme for much work on evolution since that time. They emphasized "the importance of reciprocal selective responses between ecologically closely linked organisms" (606), suggesting that interaction between plants and insects may have contributed significantly to the diversity of organisms on this earth.

The basic scenario is a plant population subjected to herbivore attack in which a new mutation or recombination causes the production of a new compound, which acts as a chemical defense. The mutated plant may then enter a new adaptive zone because of its relative freedom from herbivore attack, and

adaptive radiation from this parental stock may follow. Ultimately, new mutations or recombinations in a population of herbivores would circumvent the problems associated with the new chemical defense, and the plant's defense against the herbivore's new genotype would be reduced. The herbivores would then radiate onto the new plant species. This stepwise escalation of defense and attack might continue indefinitely, with close genetic tracking between plant and herbivore.

A Case Study of Butterflies on Plants

Ehrlich and Raven (1964, 1967) based their arguments on patterns in butterfly taxa associated with plant taxa, which demonstrate that, once a peculiar chemical defense in a plant taxon becomes ineffective against a line of butterflies, that line radiates. Figure 19-1 presents examples in which the patterns provide

FIGURE 19-1

Some relationships between butterfly taxa and plant taxa discussed by Ehrlich and Raven (1964). Note that some butterfly subfamilies are associated with plant families that commonly contain toxic compounds: Pierinae with the mustard oil–containing families Cruciferae and Capparidaceae; Ithomiinae with the potato family, Solanaceae, with members rich in alkaloids such as solanine, nicotine, and tomatine; Danainae with the dogbanes and milkweeds (Apocynaceae and Asclepiadaceae), which have latex canals and very toxic cardiac glycosides.

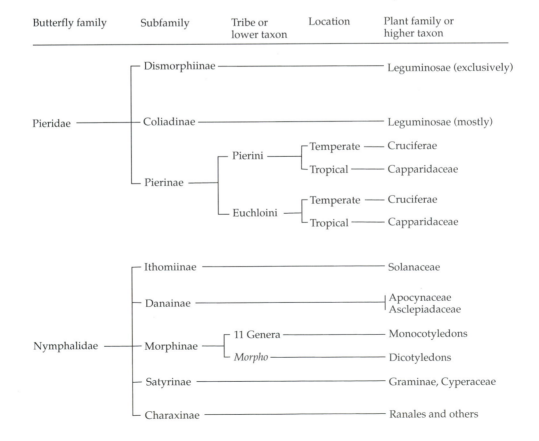

compelling support for Ehrlich and Raven's argument. Pierid butterflies have radiated extensively on two groups of plants: (1) the Leguminosae and (2) the Cruciferae–Capparidaceae families, which are closely related phylogenetically and chemically. A major source of defense in both families is mustard oils, which are effective against the majority of herbivores but no longer against pierid butterflies. In the Nymphalidae, the close ties between highly toxic plants and subfamilies are evident: Ithomiinae with the alkaloid-containing Solanaceae, and Danainae with the closely related families Apocynaceae and Asclepiadaceae, which are usually rich in alkaloids and cardiac glycosides.

Do Insect Herbivores Select for Plant Defenses?

Whereas it is quite probable that plants have selected for herbivore specialists in particular chemical constituents, it is not always as clear that herbivores have selected for new, apparently defensive chemicals in plants. Many chemicals in plants that play important defensive roles are also very active in the plants, apparently closely associated with primary metabolic pathways (Seigler and Price 1976). They exist in a dynamic equilibrium with half-lives of only a few hours. For instance, labeled nicotine injected into tobacco plants subsequently appears in amino acids, other organic acids, and sugars. Nicotine biosynthesis in a tobacco plant uses about 12 percent of fixed carbon per day, but 40 percent of this is degraded in a 10-hour photoperiod. Chemicals in plants may have been selected for many constituents of essential chemical pathways, independent of a defensive function. The point is that, using patterns such as the feeding relationships seen in the butterflies, we cannot establish with confidence that there has been *reciprocally induced* evolutionary change between plant defense and insect attack. Janzen (1980) reprimanded Ehrlich and Raven and subsequent authors who have used equivalent approaches in discussing coevolution for failing to demonstrate cause-and-effect relationships of characteristics in *both directions* between interacting species. He stressed the importance of reserving the term "coevolution" for interactive relationships that have actually produced the selective pressures for reciprocally induced traits.

If we ought to be this rigorous in our use of the term "coevolution," are there any undoubted cases of mutually induced interactive systems? Ideally, to claim such a case, we should demonstrate (1) that the traits involved in two interacting populations or species are genetically determined; (2) that, where variation in these traits occurs, the two taxa vary in tandem; and (3) that the traits and the variation are

mutually induced—that is, these genetic traits do not occur in areas where the other interactive species is absent or rare, and each trait is specific to that species or population and no other.

Very few known cases fulfill the requirements of this idealized test for real coevolution. As we shall see, however, some cases begin to meet the needs of a rigorous test. Interestingly, Mode's (1958) discussion, in which he used the word "coevolution" for the first time, included some of these cases.

19.2 THE GENE-FOR-GENE HYPOTHESIS

Flax and Its Fungal Rust Parasite

The known relationships between parasitic fungi and their host plants at the genetic level became bewilderingly complex during the 1930s. It was not until Flor studied the genetics of flax (*Linum usitatissimum*) and its rust fungus, *Melampsora lini*, in 1942 that a relatively simple picture emerged. Using two flax varieties and two strains of the rust, Flor investigated the inheritance of resistance in flax and of virulence in the rust. He found that each rust strain had a single recessive gene for virulence and that these genes segregated independently. Two tightly linked Mendelian systems existed in the host and in the parasite. In this early work, Flor (1942, 668) summarized his findings by saying "the pathogenic range of each physiologic race of the pathogen is conditioned by pairs of factors that are specific for each different resistant or immune factor possessed by the host variety."

Further studies by Flor (1946, 1947) provided additional details on the tight genetic ties between host and parasite. First he studied the inheritance of virulence of two rust races, 22 and 24, on two flax varieties, Ottawa and Bombay (Flor 1946). Table 19-1 lists the symbols used to denote the relevant alleles, and Table 19-2 illustrates the patterns of virulence and avirulence in combinations of host and parasite genotypes. In Ottawa flax, Flor found that L determined resistance to race 24 but not to race 22, and in Bombay flax, N determined resistance to race 22 but not to race 24. In race 22, vL determined virulence on Ottawa, and in race 24, vN determined virulence on Bombay flax. These results demonstrate that Mendelian genetics works out very predictably in studies of the inheritance of virulence and avirulence in the rust on the two flax varieties (Figure 19-2).

In a similar manner, Flor (1947) studied the inheritance in the two varieties of flax in relation to their resistance and susceptibility to the two races of *Melampsora lini* (Figure 19-3). Not only do the factors for virulence and avirulence in the parasite and

T A B L E   19-1   Complementary alleles in flax and rust

Host Resistance and Susceptibility	Parasite Virulence and Avirulence
L dominant allele for resistance in Ottawa to race 24	VL dominant allele for avirulence in race on Ottawa flax
N dominant allele for resistance in Bombay to race 22	VN dominant allele for avirulence in race 24 on Bombay flax
l recessive allele resulting in susceptibility to race 24 on Bombay	vL recessive allele for virulence in race 22 on Ottawa flax
n recessive allele resulting in susceptibility to race 22 on Ottawa	vN recessive allele for virulence in race 24 on Bombay flax

T A B L E   19-2   Pattern of virulence and avirulence of rust races on flax varieties

	Race 22		Race 24	
	vL	VLVL	vNvN	VNVN
Ottawa, LLnn	Virulent	Avirulent	Avirulent	Avirulent
Bombay, llNN	Avirulent	Avirulent	Virulent	Avirulent

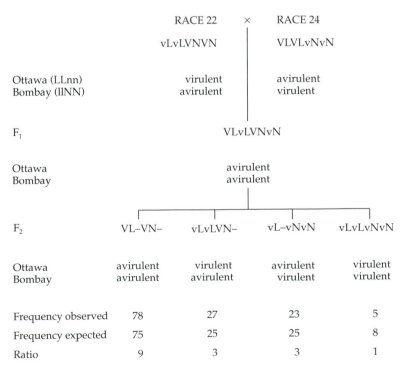

	RACE 22	×	RACE 24
	vLvLVNVN		VLVLvNvN
Ottawa (LLnn)	virulent		avirulent
Bombay (llNN)	avirulent		virulent

F₁   VLvLVNvN

Ottawa   avirulent
Bombay   avirulent

F₂	VL–VN–	vLvLVN–	vL–vNvN	vLvLvNvN
Ottawa	avirulent	virulent	avirulent	virulent
Bombay	avirulent	avirulent	virulent	virulent
Frequency observed	78	27	23	5
Frequency expected	75	25	25	8
Ratio	9	3	3	1

F I G U R E   19-2
Mendelian genetics of rust races 22 and 24 on two flax varieties, Ottawa and Bombay. The genotypes and pathogenicity on each flax variety are provided for the parental F₁ and F₂ generations.

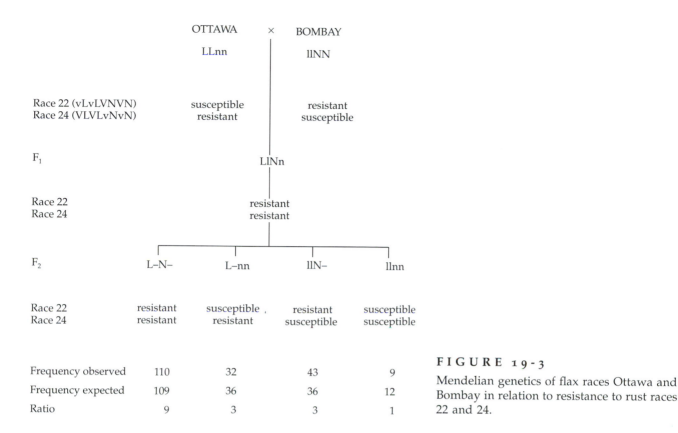

FIGURE 19-3

Mendelian genetics of flax races Ottawa and Bombay in relation to resistance to rust races 22 and 24.

resistance and susceptibility in the host perform in a classical Mendelian fashion, but the systems of inheritance are perfectly complementary. From these data, Flor (1955, 1956) developed the **gene-for-gene hypothesis:** During their evolution, a host and its parasite develop complementary genic systems, with each host gene that affects defense matched by a parasite gene that affects attack.

This gene-for-gene relationship between host and parasite precisely illustrates the "stepwise evolutionary responses" that Ehrlich and Raven (1964, 606) concluded must exist, from their study of butterflies on plants. A gene conferring resistance to a parasite would spread rapidly in a plant population in the presence of abundant parasites, but eventually a gene to overcome the plants' resistance would arise in the parasite population. Then another gene in the host population would arise, conferring additional resistance, and again a counteracting gene in the parasite population would arise. Close genetic tracking between host and parasite populations would lead to many matched pairs of genes in coevolving populations.

Commonness of the Relationship

The simplicity of the agricultural system in which the gene-for-gene relationship was first observed clearly enabled the detection of an extremely complex rela-

tionship in nature. Inbreeding and mass culture of homozygous stock simplified the genetics of plant resistance to the point where researchers could observe simple relationships, which in turn are elements of complex natural systems. Day (1974) suggested that such gene-for-gene relationships were detectable in nature and selected the *Melandrium–Ustilago violacea* system as one such relationship worthy of further study. Goldschmidt (1928) found that simple Mendelian genetics regulated the inheritance of virulence in six races of the smut fungus on different host species. Person (1967) regarded the development of gene-for-gene relationships between hosts and parasites to be the rule rather than the exception.

Indeed, the literature has by now documented or suggested this close genetic linkage for many systems. Flor (1971) listed 18 examples, all from host–fungus relationships. Day (1974) listed 27 cases: 19 from plant–fungus relationships, 2 bacteria–plant, 3 virus–plant, 1 nematode–plant, 1 insect–plant, and 1 parasitic angiosperm–plant.

Leguminous Plants and Nitrogen-Fixing Bacteria

One example of a bacterium–host gene-for-gene relationship occurs within the mutualistic association of leguminous plants and nitrogen-fixing bacteria in the genus *Rhizobium.* Nutman (1969) provided evidence suggesting a gene-for-gene relationship, as we should expect for any close-knit interaction, be it parasitic or mutualistic. Nutman's work on red clover "suggests that some major host and bacterial genes are complementary, the expression of one being wholly dependent on the other" (432). Bohlool and Schmidt (1974) suggested that legume lectins may provide a site on the legume root surface that interacts with specific polysaccharide coats on the *Rhizobium* cell, thus enabling the initiation of nodulation. Such a mechanism would certainly explain the high specificity of species and strains of *Rhizobium* for certain species and varieties of legumes. The important message is that mutualistic relationships are likely to exhibit tightly linked genetic systems, just as in parasites, but their description will be a while in coming because they have undergone less scrutiny by the plant breeder.

Wheat and the Hessian Fly

The example of the insect–plant gene-for-gene relationship is worth some study, since current discussions on coevolution so often involve plant–insect relationships. The Hessian fly, *Mayetiola destructor*, became a serious pest of wheat in North America in the 1920s. Both the wheat and the fly are indigenous to the Old World. Although they probably evolved together in North America, we cannot be certain that the resistance factors found in North American wheat strains were relics of a truly coevolved system in the Middle East.

Predicting the results of crosses between strains of the Hessian fly on the basis of Mendelian inheritance requires the knowledge that the paternal haploid genome is eliminated during spermatogenesis, as in the other cecidomyiids, and so males breed as though homozygous for the maternal traits. In one study, Hatchett and Gallun (1970) investigated heritability of virulence of the Hessian fly strains GP (an avirulent strain) and E (a virulent strain) on Monon wheat. Segregation in the F_2 generation and the backcrosses indicated that a single gene differentiated the races and that avirulence was dominant over the virulence gene. Where no segregation occurred, the cause was the elimination of the paternal genome. Figure 19-4 provides some examples of crosses made by Hatchett and Gallun. From such crosses, Hatchett and Gallun (1970) concluded that the genic systems of insect and host were complementary in a gene-for-gene relationship: "For a given race of flies to survive on a wheat variety, the insect must possess the complementary recessive gene to counter the dominant gene for resistance in the host plant" (1406).

By 1972, researchers had identified eight races of the Hessian fly, defined by three different varieties of wheat. Day (1974) pointed out that that example

was very similar to the one defined by Flor and to other studies on fungal pathogens of plants. "Almost certainly, many of the conclusions drawn from plant pathology concerning the future of genes for race-specific resistance and the part they play in epidemiology will apply to this and other insect pests" (105).

The opportunity seems great for ecologists, pathologists, geneticists, and entomologists to bring this kind of study on coevolution into the mainstream of ecological and evolutionary concepts and theory. Parallel studies of the enemies of cultivated varieties and wild-type plants, such as cultivated and wild solanums, would surely be rewarding. Mycologists, having a rather tractable host–parasite system, could make marvelous contributions by carrying their well-developed methods into the study of natural populations. Entomologists could busy themselves with small specialized parasitic species—such as gall makers, leaf miners, and bud moths—and other sessile species, such as scale insects and white flies, where the gene-for-gene relationship is most likely to exist.

Genetic Heterogeneity in Host–Parasite Systems

An interesting consequence of the gene-for-gene relationship is the enormous diversity of genotypes required in a parasite population to effectively exploit a host population that has a diverse array of resistance genes segregating among genomes. For example, 19 resistance genes in apple, each with an allele for susceptibility, related to virulence and avirulence in the parasitic fungus *Venturia inaequalis* would define 2^{19}, or 524,288, races of the parasite (Day 1974)! And that number is in a simplified agricultural system. This example should indicate the level of detail with which ecologists will ultimately have to work if we wish to understand any aspect of the host–parasite relationship: population dynamics, population genetics, niche breadth, dominance–diversity relationships, or community structure in general. When we see such genetic diversity dictated in simple systems, is there really such a problem accounting for high levels of polymorphism and heterozygosity in natural populations?

Animal Hosts and Their Parasites

Animal hosts and their parasites also contain evidence, although it is not yet well documented, that gene-for-gene relationships may exist. Bradley (1974) again greatly simplified the genetics of the host by studying susceptibility to *Leishmania donovani* among inbred mouse strains. In humans in tropical regions, this parasite causes visceral leishmaniasis, or kala-azar. Bradley found one strain of mice to be resistant

Assume A^a is a dominant gene conferring avirulence in the Hessian fly race.

Race GP is A^aA^a.

a^v is a recessive gene conferring virulence in the Hessian fly race.

Race E is a^va^v.

Example 1.   A cross between strains GP and E.

Race   GP ♀   ×   E ♂

Genotype A^aA^a a^va^v (paternal genome eliminated)

F_1 A^aa^v

Predicted   100% avirulent
Observed   100% avirulent

Example 2.   Segregation in the F_2 generation when the male has derived its virulence gene from the paternal side.

F_1 (GP♀ × E♂)♀   ×   (GP♀ × E♂)♂

Genotype A^aa^v A^aa^v (paternal genome eliminated)

F_2 A^aA^a A^aa^v

Predicted   50% avirulent   50% avirulent
Observed   100% avirulent

Example 3.   Segregation in the F_2 generation when the male has derived its virulence gene from the maternal side.

F_1 (GP♀ × E♂)♀   ×   (GP♀ × E♂)♂

Genotype A^aa^v a^vA^a (paternal genome eliminated)

F_2 A^aa^v a^va^v

Predicted   50% avirulent   50% virulent
Observed   51% avirulent   49% virulent

Example 4.   Segregation in the F_1 backcross.

F_1 and P   (GP♀ × E♂)♀   ×   E ♂

Genotype A^aa^v a^va^v (paternal genome eliminated)

                  A^aa^v a^va^v

Predicted   50% avirulent   50% virulent
Observed   40% avirulent   60% avirulent

The similar backcross (E♀ × GP♂)♀ × E♂ gave a better ratio of
47% avirulent and 53% virulent.

FIGURE 19-4

Some examples of the Mendelian genetics of Hessian fly races GP and E on Monon wheat, studied by Hatchett and Gallun (1970).

and another susceptible. By an array of crosses similar to those used by Flor and Hatchett and Gallun, he demonstrated the existence of a single major autosomal dominant gene that controlled resistance. When Trischmann et al. (1978) used the same strains of mice to study resistance to *Trypanosoma cruzi*, which produces Chagas' disease in Latin America, the pattern was reversed, with the strain resistant to *L. donovani* being susceptible to *T. cruzi* and the strain susceptible to *L. donovani* being resistant to *T. cruzi*.

Although the genetics of these systems are incompletely known, so far the patterns appear to match those of plant–parasite systems. Resistant genes are specific to individual parasite species; the dominant allele confers resistance, and the recessive allele confers susceptibility. We can predict with some confidence that the virulence genes in the parasites will be recessive. Why this pattern is so common is a subject worthy of more study and consideration.

19.3 PHYLOGENETIC TRACKING BY PARASITES ON HOSTS

If we accept, then, that the patterns observed by Ehrlich and Raven (1964) really involve coevolutionary tracking at the genetic level, there should be a significant body of literature suggesting such cases of parasites phylogenetically tracking their hosts. The literature is indeed considerable, addressing animal–parasite, plant–parasite, and plant–mutualist systems (Table 19-3). Parasitologists are sufficiently familiar with these patterns to have incorporated them into parasitological rules:

Fahrenholz's rule. In groups of permanent parasites, the classification of the parasites usually compares directly with the natural relationships of the hosts (Eichler 1948).

Fuhrmann's rule. Each order of birds has its particular cestode fauna (Noble and Noble 1976).

For 50 years before ecologists joined the discussion, the parasitologists had been examining the type of pattern Ehrlich and Raven demonstrated, and it is significant that the first use of the word "coevolution" was by a parasitologist. In fact, discussion of phylogenetic tracking of hosts by parasites has been a tradition among parasitologists, as illustrated by the significant literature cited in Table 19-3. And the more specific the parasites to their hosts, the greater the literature and the more compelling the evidence for phylogenetic tracking. Lice, which spend their entire lives on their hosts, are very specific (Price 1980), and the literature on them is extensive. Plant pathologists and entomologists have not shared the tradition of discussing phylogenetic tracking, as indicated by the relatively small literatures in those areas. Opportunities for more extensive investigation are available.

TABLE 19-3   Literature suggesting phylogenetic tracking of hosts by parasites

Parasite Group	Host Group	Source
Animal–parasite systems		
Fleas (Siphonaptera)	Mammals	Jordan 1942, Holland 1958, 1963, Traub 1985
Lice (Phthiraptera)	Mammals and birds	Harrison 1914, Metcalf 1929, Hopkins 1942, 1949, Clay 1950, 1970, Rothschild and Clay 1957, Timm 1983, Timm and Price 1980
Flukes (Trematoda)	Fishes, reptiles	Manter 1966, Brooks 1979, Brooks et al. 1981
Tapeworms (Cestoda)	Birds and rodents	Fuhrmann 1932, Rausch 1957
Mites (Acarina)	Mammals and bark beetles	Fain and Hyland 1985, Lindquist 1969
Bat flies (Streblidae) and mites	Bats	Wenzel et al. 1966, Machado-Allison 1967
Plant–parasite systems		
Butterflies (Papilionoidea)	Angiosperms	Ehrlich and Raven 1964
Rust fungi (Uredinales)	Angiosperms	Saville 1975, Anikster and Wahl 1979
Fig wasps[1]	Figs	Wiebes 1979

[1] Fig wasps should be regarded as mutualists, although gall formation in the ovary closely resembles the parasitic habit.

19.4 COEVOLUTION OF HOST IMMUNE SYSTEMS AND PARASITES

Let us switch gears considerably and discuss another case in which the genetic basis of interaction between host and parasite is partially understood. This case also illustrates aspects of coevolution. We cannot be sure that present-day parasites have had an evolutionary impact on the immune system in animals. There is no doubt, however, that parasite species living today have properties similar to those of parasites that lived while the immune system was evolving, and the immune system is so versatile that it would be ridiculous to argue that any one parasite–host interaction resulted in its evolution. Thus, we maintain that the immune system and the genetics to counter it in parasites qualify as a legitimate case of coevolution and comes close to the ideals set out early in this chapter. As we shall also see, an individual host and its parasites exhibit tight coevolutionary responses, leaving no doubt that truly reciprocal genetic interaction is involved at this level.

Evolution of the Immune System

So far, there is no evidence that invertebrates produce molecules that function as the vertebrate immunoglobulins, although cellular responses to foreign proteins are well developed. No invertebrates produce inducible specific antibodies, although we can recognize the beginnings of an immune system in the forms of immunologic specificity of cell rejection in sponges (e.g., Van de Vyver et al. 1990) and graft rejection in coelenterates, annelids, and echinoderms (Stites and Caldwell 1978). However, the most primitive vertebrates, the jawless fishes (Agnatha) (including the hagfish and lamprey), synthesize specific antibodies. Vertebrates have a two-component immune system: **cellular immunity** and antibody production, or **humoral immunity.**

Stites and Caldwell (1978) make three generalizations about the phylogenetic evolution of immunity:

1. Primitive and quasi-immunologic phenomena are present in the simplest forms of present-day animal species.
2. Cellular immunity evolved before humoral immunity.
3. A truly bifunctional immune system with dual lymphoid organs in a highly differentiated form is the most recent immunologic evolutionary development.

Thus, cellular immunity was well developed in the primitive cartilaginous fishes, and a crude version of this system occurs in the jawless fishes, whereas immunoglobulins such as those now found in humans first appeared in amphibians.

The selective pressure exerted by parasites on the defensive systems of their hosts became stronger and stronger as host organisms tended to become larger and less fecund, because escape from infection was no longer a viable strategy. The presence on terrestrial animals of blood-feeding insects, which acted as effective vectors of pathogenetic microorganisms, also increased the probability of infection.

The remarkable feature of the immunoglobulins is their incredible array of antigen-binding specificities and biological activities, with just one part of the variable region of the molecule being represented by thousands of variants (for example, in humans). The major question becomes: How do genes code for this incredible variation, and how do parasites circumvent this potent defense? Given that an organism can generate so much diversity at the variable end of the antibody molecule, an amazing variety of antigens can be identified and inactivated, and an amazing variety of parasites can be identified as nonself.

Hypotheses on the Generation of Immunoglobulin Diversity

Two hypotheses accounted for genetic diversity capable of synthesizing an enormous variety of variable ends on immunoglobulin molecules (e.g., Jerne 1973):

1. **The germ-line hypothesis.** All genes coding for the variable-region sequences of amino acids arose during evolution, through gene duplication, and all are present in the germ-cell DNA. They are inherited from the parents. Only a part of this genetic information is expressed.
2. **The somatic mutation hypothesis.** A small pool of genes is passed on in the germ line, and the numerous variants arise by a process of random somatic mutation during ontogeny of the individual. Natural selection among mutations is based on the internal antigenic experience of the individual, no doubt influenced by an individual's exposure to parasites. That is, during development of an individual, natural selection occurs for those variable-site immunoglobulin genes that are adaptive for countering parasitic challenges experienced by the individual. An organism may be said to evolve under the influence of natural selection by antigens of parasites. If we can establish that parasites have a complementary evolutionary system, we have in this example a true coevolutionary system of an incredibly dynamic kind.

Studies on the number of copies of a particular DNA sequence in the genome have suggested that the number of copies in much less than the variation in the variable region of immunoglobulins, thus lending support to the somatic mutation hypothesis. Individual mice of an inbred strain produce entirely different sets of antibody molecules, although they have the same germ-line genes. (Interestingly, Whitham and Slobodchikoff [1981] emphasized the importance of somatic mutation in plants for defense against herbivores.)

The immune response is thought to have originated from a single immune-response gene. The proximity of multiple genes with products that have similar or related activities indicates that they have probably arisen from this original immune-response gene by reduplication (Perkins 1978). Thus, the amount of genetic material devoted to immunoglobulin codes derived from the germ line and derived during somatic mutation indicates the diversity of antigenic challenges an organism or a population normally encounters. Perhaps it also indicates the diversity of countermeasures to antibody diversity maintained by some parasite species.

Clearly, somatic mutation is the mechanism responsible for the great diversity of antibodies. Indeed, French et al. (1989) used the term **hypermutation** to emphasize the abundance of mutational changes. It now appears that the germ-line genes produce antibodies of relatively low affinity in the absence of antigens from parasites or toxins. Once antigens challenge the body, activation of a somatic mutational system follows, which produces antibodies of higher affinity, and the most efficacious antibodies are the objects of selection. The result is a very versatile but highly specific immune system.

Antibody genes assemble from several DNA segments that are not contiguous on a chromosome. Because different segments may combine, a given number of codons can produce much more genetic diversity than a linear array would allow. There is additional flexibility in the ways in which gene segments can join, which increases the diversity of three-dimensional structures of the variable region of the antibody molecule, at the binding site active against antigens (Marx 1981).

This process of hypermutation is not a haphazard or accidental change in genes but is as programmed as the rest of the organism's development (Borst and Greaves 1987). Not only is assembly of genes from gene segments involved, but so are amplification and deletion of genes, and DNA rearrangements alter gene expression. "The cell rewrites its genetic code" (Weissman and Cooper 1993, 68).

How can parasites escape detection by such a versatile defensive system? Tiny organisms cannot overcome a host by force, as a predator does, but they have evolved to deceive their hosts in a multitude of ways (Bloom 1979, Goodenough 1991).

The Variable Character of Parasite Coats

Some parasites present to the host such a variable character that even a small number of parasite species during the evolution of the vertebrates could have been enough to select for the complex immune system of the higher vertebrates. The protozoans in the genus *Trypanosoma* occur in fish and amphibians (transmitted by leeches), reptiles, birds, and mammals (transmitted mostly by insects) (Molyneux 1977). Their distribution in hosts closely parallels the distribution of antibody production, or humoral immunity. Members of the genus *Leishmania* in the same family of protozoans (Trypanosomatidae) have been found in mammals and reptiles.

When a host becomes infected, enormous parasite population fluctuations follow, reminiscent of a prey population interacting with a predator (Figure 19-5). The analogy is close except that, at each peak in numbers, natural selection by antibodies has caused change in trypanosome type, and at each trough the antibodies produced by the host are different. Essentially, we have coevolution in miniature, with selection for different trypanosome types and different lymphocyte genotypes that produce the antibodies. This system mimics in a period of days what may develop over hundreds or thousands of years in plant–herbivore systems.

The fluctuation in numbers results from frequent changes in surface antigens in trypanosome populations. As antibodies in the host increase in response to a prevalent antigen surface type in trypanosomes, selection is against the commonest antigen-type individuals, and the bearers of new and rare antigen-containing surfaces have a temporary advantage and proliferate until the immune system responds and stimulates appropriate lymphocytes, which secrete antibodies and start multiplying.

It appears that the major mechanism for genetic control of changing surface antigens is that numerous genes code for surface antigens, and these genes are switched on and off in a regular sequence (Bloom 1979, Hoeijmakers et al. 1980). Each gene codes for a different antigen type, and each individual can produce a large but finite series of variable antigen types. Researchers have purified and sequenced these variant glycoproteins for amino acids at their terminal ends and have found no similarities in sequence. Each had a unique sequence, indicating that the differences do not result from mutational changes in amino acids. In effect, then, trypanosomes have evolved a system of antigen change that is remarkably similar to the system of antibody production. There is a striking complementarity in the mechanisms; programmed gene rearrangements that alter gene expression occur in both trypanosomes and humans (Borst and Greaves

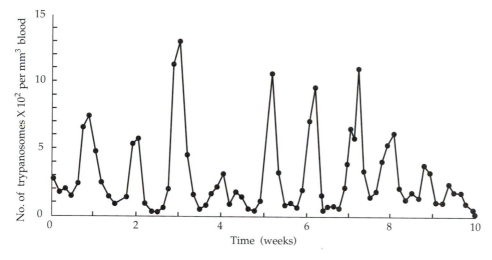

FIGURE 19-5
The number of trypanosome individuals per cubic millimeter of blood in a human with African trypanosomiasis. *(After Bloom 1979.)*

1987). Equivalent mechanisms also occur in the bacterium that causes gonorrhea, *Neisseria gonorrhoeae.*

The systems are even more similar in mechanism than one might have expected, for Doyle and coworkers (see Vickerman 1978, Bloom 1979) isolated 19 clones of a variable antigen type of *Trypanosoma brucei,* and in the absence of antibody they found new variants arising spontaneously in nine of the clones. The frequency was remarkably high at approximately 10^{-3} to 10^{-4}, suggesting that mutation may be important in generating new variants during the "ontogeny" of a population, just as it is in the ontogeny of a population of cells that develops into a vertebrate animal. Variant types were detectable because some new types grew more rapidly than others. Essentially, natural selection enabled some new mutants to become dominant. So Bloom (1979) suggested that there are probably two mechanisms of selection for variants in glycoproteins: (1) change in growth rates enabling one type to dominate, and (2) immunoselection favoring the very rare and new variant type.

At all levels of organization—species, populations, individuals, cells, and molecules—the relationship of host immune system to parasite is a very dynamic coevolutionary system. Whether the parasites use camouflage, distraction, or mimicry (Goodenough 1991), the immune system in the vertebrates strongly influences their evolution.

19.5 WHERE DOES COEVOLUTION COMMONLY OCCUR?

Can we generalize about where coevolution is likely and where it is likely to be strongest in parasite–host systems? Wherever parasites are highly specific and hosts have few or only one parasite, coevolution is likely to be finely tuned. Pimentel and Bellotti (1976) illustrated one side of this relationship. They tracked the rate of evolution of house fly populations exposed to resistance factors in their nutrient medium that simulated the resistance factors in plants. When a fly population was exposed to one resistance factor at a time, the population evolved tolerance for or resistance to the factor (e.g., citric acid, copper sulfate, or magnesium nitrate) in only seven to ten generations. However, when a fly population was exposed to six factors at once, it did not evolve resistance to those factors in 30 generations. These data suggest that, when a host is only one resistance gene ahead of the parasite population, a matching virulence gene is likely to appear quite rapidly in the parasite population; but if a potential host population is several resistance genes in advance of the parasite population, then such an evolutionary jump by the parasite is much less likely to occur.

Parasites show dramatic differences in specificity to their hosts. The percentage of species in a taxon that are specific to only one host species provides an index of specificity (Table 19-4). Generally, the degree of specificity increases as the intimacy of the relationship between parasite and host increases. For example, the Philopteridae, or bird lice, spend all their lives on the host and reproduce on the host. In contrast, hippoboscids (louse flies) are parasitic only in the adult stage, and the mature larva drops from the parent fly, which incubated it, to the ground, where it pupates. A strongly flying fly emerges to hunt for a host, which is usually one of several kinds of bird. The tight link between the host and parasite is lost, and coevolutionary ties between species are weak or absent.

In the systems of many parasites on both animals and plants, very specific coevolution is likely to be quite common.

T A B L E   1 9 - 4   Percent of species in a taxon known to be parasitic on only one host species

Insects Parasitic on Plants		Animals Parasitic on Animals	
Agromyzidae	57%	Philopteridae	87%
Gelechiidae	53%	Monogenea	85%
Miridae	22%	Braconidae	60%
Heliconius butterflies	24%	Nematoda	52%
North American butterflies	22%	Digenea	25%
British butterflies	21%	Hippoboscidae	17%

See Price 1980 for sources.

19.6 CLASSIFICATION OF COEVOLUTIONARY SYSTEMS

We should ask ourselves if this chapter's concentrated discussions of specific parasites are an adequate representation of the genetic basis of coevolutionary systems. After all, the several books on coevolution cover a much wider array of topics, including pollination systems, seed dispersal, paleontology of the vertebrates, predators and prey, competition, and others (e.g., Gilbert and Raven 1975, Futuyma and Slatkin 1983, Nitecki 1983, Spencer 1988). In those systems, however, genetically based, reciprocally induced relationships are seldom established. Indeed, the process of coevolution may be much narrower in scope than typical coverage of the subject would lead us to expect. Feinsinger (1983, 307) noted that "most plants and pollinators move independently over the landscape, not in matched pairs." Craig et al. (1988, 75) asserted that "evidence for plant–herbivore coevolution in natural systems is not conclusive."

Thompson (1989; see also 1994) reviewed the present scope of the field of coevolution and noted that, for most modes of coevolution, host and parasite systems were the most likely candidates. His classification recognized five mechanisms of coevolution and the separate process of cospeciation.

Gene-for-Gene Coevolution

Hosts and parasites have complementary gene loci for resistance and virulence, susceptibility and avirulence, with each gene in the host matched by a gene in the parasite. (Section 19-2 discussed this ideal evidence for coevolution.)

Specific Coevolution

Specific coevolution involves coadaptation of two interacting species without the tightly reciprocal genetic systems seen in gene-for-gene relationships. Two regularly competing species may diverge in characteristics under the coevolutionary influence. Mutualistic species may amplify commitments to the mutualism, as we shall see in Chapter 20. Mimicry between toxic species—for example, butterflies—may result in convergence of patterns and behaviors, with each species on a mutually induced track toward common traits.

Guild Coevolution

Coevolution involves interacting groups of similar species, such as groups of herbivores or groups of plants. This more diffuse coevolution may indeed be represented by the evolution of the immune system in vertebrates and evolution of deception in the vertebrates' parasites (discussed in Section 19-4).

Diversifying Coevolution

Thompson noted that diversifying coevolution commonly involves parasites that reduce panmixis in populations, reduce gene flow, and ultimately result in enough reproductive isolation to cause speciation events. Sexually transmitted symbionts may isolate the populations that are adapted to the parasite from those that are vulnerable. Parasites may evolve into mutualists that are so effective as to develop tight specificity—such as the gall wasps that became specific pollinators on figs (discussed in Chapter 20) and the yucca moths that pollinate yuccas.

Escape-and-Radiation Coevolution

Ehrlich and Raven (1964) initially described escape-and-radiation coevolution (discussed in Section 19-1). It involves guild coevolution that is intermittent through evolutionary history, with the escapes from interaction perhaps outlasting the interactive phases. The process differs from diversifying coevolution in that radiation of plants occurs in the absence of major parasitic herbivores. However, it is possible that many diverse host and parasite systems may have diversified in ways similar to the scenario envisaged by Ehrlich and Raven.

Parallel Cladogenesis, or Cospeciation, Need Not Involve Coevolution

Parallel radiation of interacting taxa, such as hosts and their parasites, has frequently been called coevolution, but this cospeciation need not involve reciprocally induced evolutionary changes. Host speciation during geographic isolation would isolate less mobile parasite populations even more than it would the host populations. The result would be cospeciation without hosts causing genetic change in parasites and vice versa. Parallel cladogenesis, or phylogenetic tracking, is commonly suggested for host and parasite systems (as discussed in Section 19-3, with examples in Table 19-3), and it is the result anticipated for escape-and-radiation coevolution. However, cospeciation and coevolution do not necessarily combine and certainly are almost never demonstrated.

SUMMARY

Coevolution is reciprocally induced evolutionary change in two or more species or populations. Used with reference to the hypothesized interaction between plants and butterflies, the term focused the attention of scientists. New genotypes in a plant lineage escaped attack because of increased defenses, and their progeny radiated. Eventually, a butterfly lineage circumvented the defense and colonized the radiated plants. A stepwise escalation of defense and attack would result in phylogenetic tracking of butterflies on plant lineages. However, it is usually difficult to establish that an insect herbivore or any other herbivore selected for plant traits.

The **Gene-for-gene hypothesis** provides a sound example of coevolution in which genes for resistance and susceptibility in a host are matched by genes for virulence and avirulence in a parasite. The hypothesis was based on an interaction between flax plants and a rust fungus, which also appeared later in many agricultural systems, including that of wheat and the Hessian fly. The gene-for-gene hypothesis may apply in systems of animal hosts and parasites, such as protozoan diseases of mammals.

Coevolution of host and parasite lineages may result in phylogenetic tracking of host clades by parasite clades. Many examples illustrate the linkage between radiations of host taxa and parasite taxa, but whether mutually induced evolutionary change is necessarily involved remains a debatable point.

Host immune systems became more complex with the evolution of the vertebrates; without a doubt,

parasites strongly selected for these systems. In fact, the complementary systems in hosts and parasites— generating diversity of antibodies and antigens, respectively—provide many examples of coevolution, not only between lineages but within individual hosts infected by pathogen populations. Hypotheses on the generation of immunoglobulin diversity included claims that the **germ line** contained enough genetic diversity and, alternatively, that **rapid somatic mutation** must be involved. The **Somatic mutation hypothesis** has had support since; it maintains that antibody genes assemble from several DNA segments and provide diversity in their modes of assembly and in the flexibility with which gene segments join together. In some parasites, the great array of antibodies rapidly produced by hosts is met with an equivalent programmed assembly of genes from gene segments, so that parasite coats exposed to host antibodies change antigens in an organized manner.

Coevolution is likely to be common in tightly linked, highly specific interactions such as those between hosts and parasites. Indeed, in a classification of coevolutionary systems, host–parasite examples frequently emerge as examples of **gene-for-gene coevolution, diversifying coevolution,** and **escapeand-radiation coevolution. Specific coevolution** and **guild coevolution** are probably relevant to a wider range of organisms. Escape-and-radiation coevolution is likely to cause **parallel cladogenesis** or **cospeciation** in host and parasite lineages, but coevolution is not necessarily involved in such events.

QUESTIONS FOR DISCUSSION

1. Do you consider the evidence from the phylogenetic tracking by parasites on hosts helpful in discussions on the coevolutionary process?

2. Does change in immunoglobulins within individuals due to somatic mutation actually fit the definition of evolution? Are most humans evolving rapidly?

3. Does the phenomenon of hypermutation in immuno-globulin genes reinforce the Red Queen's Hypothesis and Rice's Ratchet, or will it have minimized the selective action of parasites on the evolution and mainte-nance of sex?

4. Do you think that parasite and host interactions are particularly likely to result in coevolution, or are any other kinds of interaction equally likely to result in co-evolution?

5. In any interactions involving predators and prey with which you are familiar, can you think of traits that have been selected for in a reciprocally induced manner?

6. How would you plan a research program to investigate the possibility that guild or diffuse coevolution was a more likely process in some cases than specific coevolu-tion? What kinds of species and relationships would you select for study?

7. In the case involving convergence of pattern in coexist-ing toxic butterflies under the influence of visually hunting predators, most likely birds, is there actually direct reciprocity of induced selection between the two butterfly species, or does evolutionary change result from selection by the predators? If the latter case is the most viable in your opinion, would you agree that coevolution is taking place in this kind of system?

8. In your view, has the term "coevolution" been overap-plied, without clear documentation of reciprocally in-duced selection, equivalent to the Panglossian para-digm in which adaptation has been invoked in an unrigorous manner?

9. Concerning the endosymbiotic theory of eukaryotic cell evolution, can you envisage coevolutionary interac-tions between the prokaryotic entities as of central im-portance in the refinement of processes in the eukaryo-tic cell?

10. How would you employ the scientific method to test the hypothesis on escape-and-radiation coevolution?

REFERENCES

Anikster, Y., and I. Wahl. 1979. Coevolution of the rust fungi on Graminae and Liliaceae and their hosts. Ann. Rev. Phytopathol. 17:367–403.

Bloom, B. R. 1979. Games parasites play: How parasites evade immune surveillance. Nature 179:21–26.

Bohlool, B. B., and E. L. Schmidt. 1974. Lectins: A possible basis for specificity in the *Rhizobium*–legume root nod-ule system. Science 185:269–271.

Borst, P., and D. R. Greaves. 1987. Programmed gene re-arrangements altering gene expression. Science 235:658–667.

Bradley, D. J. 1974. Genetic control of natural resistance to *Leishmania donovani*. Nature 250:353–354.

Brooks, D. R. 1979. Testing hypotheses of evolutionary rela-tionships among parasites: The digeneans of crocodil-ians. Amer. Zool. 19:1225–1238.

Brooks, D. R., T. B. Thorson, and M. A. Mayes. 1981. Fresh-water stingrays (Potamotrygonidae) and their hel-minth parasites: Testing hypotheses of evolution and coevolution, pp. 147–175. *In* V. A. Funk and D. R. Brooks (eds.). Advances in Cladistics: Proceedings of the First Meeting of the Willi Hennig Society. New York Botanical Garden, New York.

Clay, T. 1950. The Mallophaga as an aid to the classification of birds with special reference to the structure of feath-ers, pp. 207–215. Proc. 10th Int. Ornith. Congr.

———. 1970. The Amblycera (Phthiraptera: Insecta). Bull. British Mus. (Natur. Hist.) Ent. 25:75–98.

Craig, T. P., P. W. Price, K. M. Clancy, G. L. Waring, and C. F. Sacchi. 1988. Forces preventing coevolution in the three-trophic-level system: Willow, a gall-forming herbivore, and parasitoid, pp. 57–80. *In* K. C. Spencer

(ed.). Chemical Mediation of Coevolution. Academic Press, San Diego.

Day, P. R. 1974. Genetics of Host–Parasite Interaction. Free-man, San Francisco.

Ehrlich, P. R. and P. H. Raven. 1964. Butterflies and plants: A study in coevolution. Evolution 18:586–608.

———. 1967. Butterflies and plants. Sci. Amer. 216(6):104–113.

Eichler, W. 1948. Some rules in ectoparasitism. Ann. Mag. Natur. Hist. Ser. 12, 1:588–598.

Fain, A., and K. E. Hyland. 1985. Evolution of astigmatid mites on mammals, pp. 641–658. *In* K. C. Kim (ed.). Coevolution of Parasitic Arthropods and Mammals. Wiley, New York.

Feinsinger, P. 1983. Coevolution and pollination, pp. 282–310. *In* D. J. Futuyma and M. Slatkin (eds.). Coevolu-tion. Sinauer, Sunderland, Mass.

Flor, H. H. 1942. Inheritance of pathogenicity in *Melampsora lini*. Phytopathology 32:653–669.

———. 1946. Genetics of pathogenicity in *Melampsora lini*. J. Agr. Res. 73:335–357.

———. 1947. Inheritance of reaction to rust in flax. J. Agr. Res. 74:241–262.

———. 1955. Host–parasite interaction in flax rust: Its genetics and other implications. Phytopathology 45:680–685.

———. 1956. The complementary genic systems of flax and flax rust. Adv. Genet. 8:29–54.

———. 1971. Current status of the gene-for-gene concept. Ann. Rev. Phytopathol. 9:275–296.

French, D. L., R. Laskov, and M. D. Scharff. 1989. The role

of somatic hypermutation in the generation of antibody diversity. Science 244:1152–1157.

Fuhrmann, O. 1932. Les Tenias des oiseaux. Mém. Univ. Neuchâtel 8:1–383.

Futuyma, D. J., and M. Slatkin (eds.). 1983. Coevolution. Sinauer, Sunderland, Mass.

Gilbert, L. E., and P. H. Raven (eds.). 1975. Coevolution of Animals and Plants. Univ. of Texas Press, Austin.

Goldschmidt, V. 1928. Verebungsversuche mit den biologischen Arten des Antherenbrands (Ustilago violacea): Ein Beitrag zur Frage der parasitären Spezialisierung. Z. Bot. 21:1–90.

Goodenough, U. W. 1991. Deception by pathogens. Amer. Sci. 79:344–355.

Harrison, L. 1914. The Mallophaga as a possible clue to bird phylogeny. Austr. Zool. 1:7–11.

Hatchett, J. H., and R. L. Gallun. 1970. Genetics of the ability of the Hessian fly, Mayetiola destructor, to survive on wheats having different genes for resistance. Ann. Entomol. Soc. Amer. 63:1400–1407.

Hoeijmakers, J. H. J., A. C. C. Frash, A. Bernards, P. Borst, and G. A. M. Cross. 1980. Novel expression-linked copies of the genes for variant surface antigens in trypanosomes. Nature 284:78–80.

Holland, G. P. 1958. Distribution patterns of northern fleas (Siphonaptera). Proc. 10th Int. Cong. Entomol. 1:645–658.

———. 1963. Faunal affinities of the fleas (Siphonaptera) of Alaska with an annotated list of species, pp. 45–63. In J. L. Gressitt (ed.). Pacific Basin Biogeography. 10th Pacific Sci. Cong. Bishop Museum, Honolulu.

Hopkins, G. H. E. 1942. The Mallophaga as an aid to the classification of birds. Ibis 14:94–106.

———. 1949. The host-associations of lice of mammals. Proc. Zool. Soc. London 119:387–604.

Janzen, D. H. 1980. When is it coevolution? Evolution 34:611–612.

Jerne, N. J. 1973. The immune system. Sci. Amer. 229(1):52–60.

Jordan, K. 1942. On Parapsyllus and some closely related genera of Siphonaptera. Rev. Esp. Entomol. 18:7–29.

Lindquist, E. E. 1969. Review of holarctic tarsonemid mites (Acarina: Prostigmata) parasitizing eggs of ipine bark beetles. Mem. Entomol. Soc. Can. 60:1–111.

Machado-Allison, C. E. 1967. The systematic position of the bats Desmodus and Chilonycteris, based on host–parasite relationships (Mammalia: Chiroptera). Proc. Biol. Soc. Washington 80:223–226.

Manter, H. W. 1966. Parasites of fishes as biological indicators of recent and ancient conditions, pp. 59–71. In J. E. McCauley (ed.). Host–Parasite Relationships. Oregon State Univ. Press, Corvallis.

Marx, J. L. 1981. Antibodies: Getting their genes together. Science 212:1015–1017.

Metcalf, M. M. 1929. Parasites and the aid they give in problems of taxonomy, geographical distribution, and paleogeography. Smithsonian Misc. Coll. 81(8):1–36.

Mode, C. J. 1958. A mathematical model for the co-evolution of obligate parasites and their hosts. Evolution 12:158–165.

Molyneux, D. H. 1977. Vector relationships in the Trypanosomatidae. Adv. Parasitol. 15:1–82.

Nitecki, M. H. (ed.). 1983. Coevolution. Univ. of Chicago Press, Chicago.

Noble, E. R., and G. A. Noble. 1976. Parasitology: The Biology of Animal Parasites. 4th ed. Lea and Febiger, Philadelphia.

Nutman, P. S. 1969. Genetics of symbiosis and nitrogen fixation in legumes. Proc. Roy. Soc. B 172:417–437.

Perkins, H. A. 1978. The major human histocompatibility complex (MHC), pp. 165–174. In H. H. Fudenberg, D. Stites, J. L. Campbell, and J. V. Wells (eds.). Basic and Clinical Immunology. 2d ed. Lange, Los Altos, Calif.

Person, C. 1967. Genetic aspects of parasitism. Can. J. Bot. 45:1193–1204.

Pimentel, D., and A. C. Bellotti. 1976. Parasite–host systems and genetic stability. Amer. Natur. 110:877–888.

Price, P. W. 1980. Evolutionary Biology of Parasites. Princeton Univ. Press, Princeton, N.J.

Rausch, R. 1957. Distribution and specificity of helminths in mictrotine rodents: Evolutionary implications. Evolution 112:361–368.

Rothschild, M., and T. Clay. 1957. Fleas, Flukes and Cuckoos: A Study of Bird Parasites. Macmillan, New York.

Saville, D. B. O. 1975. Evolution and biogeography of Saxifragaceae with guidance from their rust parasites. Ann. Missouri Bot. Gard. 62:354–361.

Seigler, D., and P. W. Price. 1976. Secondary compounds in plants: Primary functions. Amer. Natur. 110:101–105.

Spencer, K. C. (ed.). 1988. Chemical Mediation of Coevolution. Academic Press, San Diego.

Stites, D. P., and J. L. Caldwell. 1978. Phylogeny and ontogeny of the immune response, pp. 141–154. In H. H. Fudenberg, D. P. Stites, J. L. Campbell, and J. V. Wells. Basic and Clinical Immunology. 2d ed. Lange, Los Altos, Calif.

Thompson, J. N. 1989. Concepts of coevolution. Trends Ecol. Evol. 4:179–183.

———. 1994. The Coevolutionary Process. Univ. Chicago Press.

Timm, R. M. 1983. Fahrenholz's rule and resource tracking: A study of host–parasite coevolution, pp. 225–265. In M. H. Nitecki (ed.). Coevolution. Univ. of Chicago Press.

Timm, R. M., and R. D. Price. 1980. The taxonomy of Geomydoecus (Mallophaga: Trichodectidae) from the Geomys bursarius complex (Rodentia:Geomyidae). J. Med. Entomol. 17:126–145.

Traub, R. 1985. Coevolution of fleas and mammals, pp. 295–437. In K. C. Kim (ed.). Coevolution of Parasitic Arthropods and Mammals. Wiley, New York.

Trischmann, T., H. Tanowitz, M. Wittner, and B. R. Bloom. 1978. Trypanosoma cruzi: Role of the immune response in the natural resistance of inbred strains of mice. Expl. Parasit. 45:160–168.

Van de Vyver, G., S. Holvoet, and P. Dewint. 1990. Variability of the immune response in freshwater sponges. J. Exper. Zool. 254:215–227.

Vickerman, K. 1978. Antigenic variation in trypanosomes. Nature 273:613–617.

Weissman, I. L., and M. D. Cooper. 1993. How the immune system develops. Sci. Amer. 269:64–71.

Wenzel, R. L., V. J. Tipton, and A. Kiewlick. 1966. The streblid batflies of Panama (Diptera Calypterae:Streblidae), pp. 405–675. *In* R. L. Wenzel and V. J. Tipton (eds.). Ectoparasites of Panama. Field Mus. Nat. Hist., Chicago.

Whitham, T. G., and C. N. Slobodchikoff. 1981. Evolution by individuals, plant–herbivore interactions, and mosaics of genetic variability: The adaptive significance of somatic mutations in plants. Oecologia 49:287–292.

Wiebes, J. T. 1979. Co-evolution of figs and their insect pollinators. Ann. Rev. Ecol. Syst. 10:1–12.

20

MUTUALISTIC SYSTEMS
OF COEVOLUTION

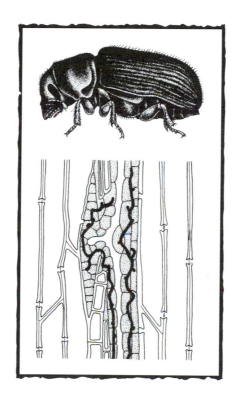

A bark beetle in the genus *Dendroctonus* and the blue-stain fungus *Ceratocystis*. The beetle carries the fungus to a new host tree. The infection kills the tree, and as the tree dies, beetles breed in it.

Mutualist systems are likely to coevolve. Close association almost inevitably requires adaptive adjustments in each species, and if the relationship is truly mutualistic—meaning that two species survive better in the presence of each other than alone—the relationship is likely to be long-term in evolutionary time.

20.1 CONVENTIONAL WISDOM ON MUTUALISM

Scientists have never adopted mutualism as a general ecologal and evolutionary theme with the same enthusiasm as they have the pairwise relationships involved with competition and predation (cf. Boucher 1985). And conventional wisdom has it that mutualism is common in the tropics only because of the mellow, stable environment; elsewhere, the stability properties of mutualistic systems render them inviable in the long term. Williamson (1972), for example, while devoting seven chapters to two-species interactions, dismisses mutualism in a single sentence, noting that "its importance in populations in general is small" (95).

Orians and his team (1974) have nurtured this conventional wisdom, noting that the importance of mutualisms seems to increase from temperate to tropical climates. They provided the following examples:

1. There are no obligate ant–plant mutualisms north of 14°.
2. There are no nectarivorous or frugivorous bats north of 32–33°.
3. There are no orchid bees north of 24°.
4. Numbers of plant species with extrafloral nectaries decline rapidly north of 24°.

They noted that even within the tropics mutualism is most prevalent where conditions are most favorable—in warm, wet evergreen forests.

Futuyma (1973) also argued that the high diversity in tropical systems results in finely adjusted accumulation of coevolved species, perhaps a commonness of mutualistic relationships, and therefore a fragile balance in the community.

20.2 MODELS OF MUTUALISM

Instability Results in Extinction

Theoretical studies have given more credibility to the aforementioned buttresses of conventional wisdom. May (1976) concluded that stable and predictable environments, as in the humid tropics, favor the long-term persistence of mutualistic relationships. His argument related to obligate mutualists, such as a pollinator needing pollen from a plant and an outcrossing plant needing the pollinator. A pollinator population, Y, obeying the logistic equation would have a carrying capacity proportional to the plant population, X. Consequently, equilibrium numbers of Y must lie along the line $dY/dt = 0$ (Figure 20-1).

The equilibrium values for the plant population are concave because, if plants become too rare, pollinators become inefficient and the plant population declines, and at high densities the population is self-limiting and reaches an asymptote independent of pollinator abundance. According to May's model, all points outside the hatched area in Figure 20-1 move toward the stable equilibrium, and all points in the hatched area move toward extinction.

If this system existed in a highly unpredictable environment, populations would eventually fall into the hatched area and to extinction. Hence, the conclusion is that prolonged persistence is unlikely. May's predictions were not supported by Gilbert's (1977) studies on the mutualism between *Heliconius* butterflies and the cucurbits they pollinate. The adaptive traits in this system cause greater constancy in adult populations through the increased probability of successful reproduction. Thus, time lags are a necessary component in models of this mutualistic system.

FIGURE 20-1

May's model of two mutualistically interacting populations, X and Y, such as plant and pollinator populations. The threshold density of pollinators is the minimum number necessary to sustain the plant population at the unstable point. Arrows indicate the directions in which populations move. (*After May 1976*)

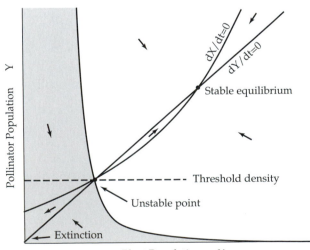

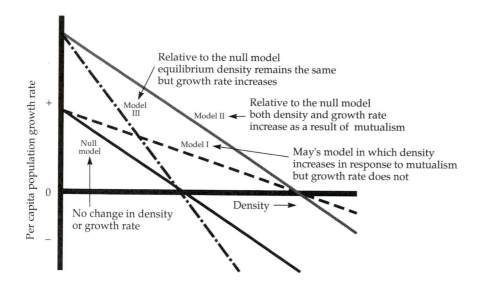

Per capita population growth rate

+

0

−

Null model

Model III

Model II

Model I

Density →

Relative to the null model equilibrium density remains the same but growth rate increases

Relative to the null model both density and growth rate increase as a result of mutualism

May's model in which density increases in response to mutualism but growth rate does not

No change in density or growth rate

FIGURE 20-2

Addicott's models of per capita population growth rate of a species as a function of its own density, given a constant density of a second species, for the three basic models of mutualism and the null model. (*After Addicott 1981*)

Stabilizing Effects of Mutualism

Addicott (1981) also challenged May's conclusion. He argued that many models of mutualism are realistic, and May's model is simply one possibility in which mutualism increases the equilibrium density of the species but does not affect the maximum possible growth rate (Addicott's model I) (Figure 20-2). Addicott modeled two other probable relationships: model III, in which the two species affect each other's population growth but not the equilibrium density, and model II, in which mutualism increases both growth rate and equilibrium density. In the null model, neither growth rate nor equilibrium density is influenced.

Then Addicott examined, by computer simulation, the time required to return to equilibrium after a perturbation away from equilibrium. He plotted a group of such times on a polar coordinate system where the angular axis is the initial angle of displacement from equilibrium and the radius represents the return time to equilibrium. Addicott's plots include simulations for several values of α, indicating the strength of the mutualistic interaction (just as α is an estimate of competitive effect in the Lotka–Volterra equations).

In model I, the stronger α becomes, the more destabilizing the interaction, and return times are usually longer than in the null model. In model II the reverse occurs: return times are the same as or less than in the null model, and higher α's have shorter return times for many displacement values. In model III, return times are invariably shorter than in the null model, and the stronger the mutualistic effect, the shorter the return time for all displacement values. When all models are compared using $\alpha = 0.5$, the relationships appear as in Figure 20-3.

Addicott's (1981) simulations indicate that both models II and III produce more stable mutualistic relationships as the mutualism becomes increasingly

FIGURE 20-3

The results of computer simulations of Addicott's models, plotted on a polar coordinate system. For each model, the distance to the center of the figure represents the return time to equilibrium after a perturbation, and the angle of displacement gives the extent of the perturbation. Note that model I is destabilizing relative to the null model, and models II an III are more stable, principally because an increased per-capita population growth rate results in a rapid return to equilibrium density (After Addicott 1981)

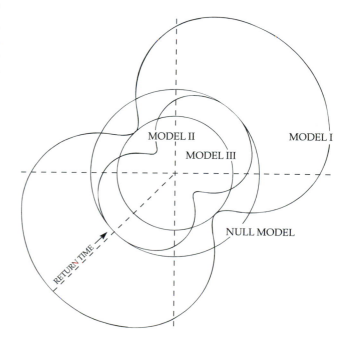

MODEL II

MODEL III

MODEL I

NULL MODEL

RETURN TIME

beneficial. That is, there is likely to be coevolution toward more and more obligatory relationships until the two species become inseparable. This is the condition May (1976) explicitly avoided, because he was interested in the dynamics of two-species systems "each of which has . . . a life of its own" (69), such as the plant–pollinator example he used in his modeling. It is only in Addicott's model I, which is the model that May used, that the mutualistic relationship is destabilizing.

Thus, in the mutualistic systems in which the growth rate of populations increases (models II and III)—whether or not the equilibrium density of populations increases—we should expect the evolution of increasing interdependency between mutualistic species. The stronger this interdependency has become, the more persistent the relationship should be. We might predict, then, that the more intimate the coevolved system, the greater the adaptive radiation(s) resulting from the initial beneficial interaction between species.

20.3 CLASSIFICATION OF TYPES OF MUTUALISM

What does the empirical evidence tell us about the validity of this armchair biology? A classification of types of mutualism into (1) **intracellular mutualisms,** where one organism lives within the cells of another; (2) **intraorganismal mutualisms,** where one organism lives intercellularly or in special organs (see also Douglas 1992); and (3) **mutualisms between organisms,** where whole organisms that are essentially free-living associate in mutualistic relationships (Table 20-1) helps to order the large number of examples available. We could add another category at a higher level—essentially as (4) **mutualisms between populations** of both involved species rather than between individuals. Pollination systems, for example, depend on populations of plants and populations of pollinators rather than on close relationships between individuals of each species. In fact, many pollination systems have explicit adaptations that prevent too close a relationship and thereby maximize outcrossing between plants.

20.4 PATTERNS IN RADIATIONS OF MUTUALISTS

A general pattern of radiation becomes evident when the scientist couples the classification with an estimate of the number of species involved with each type of mutualism and makes a guess about the number and size of taxa that may have radiated as a result of acquisition of a mutualist.

TABLE 20-1 Types of mutualism, with examples

Intracellular

Mitochondria in prokaryote cells to form eukaryotes (Margulis 1970, 1975, 1976, 1981)

Chloroplasts in early eukaryotes to form green plants (Margulis 1970, 1975, 1976, 1981)

Algal symbionts in Protozoa, Ceolenterata, Mollusca, and Platyhelminthes (Taylor 1974, Vandermeulen and Muscatine 1974)

Sacoglossan slugs and algal chloroplasts (Greene 1974)

Symbiosis in some luminous animals (Buchner 1965)

Intraorganismal

Endosymbiosis in insects (Buchner 1965)

Endotrophic mycorrhizae (Harley 1969, Ruehle and Marx 1979, Werner 1992)

Lichens (Ahmadjian and Hale 1973, Lawrey 1984)

Nitrogen-fixing bacteria and algae in higher plants (Mengel and Pilbeam 1992, Werner 1992)

Digestive systems in artiodactyls, termites, roaches, and so on (e.g., Hungate 1966, 1975)

Biological warfare (e.g., viroids and ichneumonids, Edson et al. 1981)

Figs and fig wasps (Wiebes 1979, Janzen 1979)

Symbiosis in some luminous animals (Buchner 1965)

Between organisms

Ectotrophic mycorrhizae and plants (Harley 1969, Ruehle and Marx 1979, Werner 1992)

Insect–fungus mutualisms (Batra 1979)

Crabs and anemones (Ross 1974)

Ants and plants (Janzen 1967b, Bentley 1977, Beattie 1985)

Between populations

Cleaning symbiosis (Limbaugh 1961)

Ants and insects (e.g., aphids and scales) (Wilson 1971, Cushman and Whitham 1989, 1991, Cushman and Beattie 1991, Cushman et al. 1994)

Pollination systems (e.g., Grant and Grant 1968, Richards 1978, Jones and Little 1983)

Intracellular Mutualisms

At the level of intracellular mutualisms, the combination of cells to eventually form a eukaryotic cell with mitochondria (Margulis 1975, 1976, 1981, 1993) was fundamental to all higher life and has resulted in the most extensive adaptive radiations on this earth. In a similar way, the combination of a eukaryotic cell with a photosynthesizing prokaryote, resulting in green plants, provided the basis for another far-reaching adaptive radiation. These mutualisms are so old and so intimate that they seem to function entirely as single cellular units.

Many other intracellular relationships have been observed. Marine algal symbionts occur in Protozoa, Coelenterata, Mollusca, and Platyhelminthes (Taylor 1974), and virtually all reef-dwelling corals have dino-flagellate mutualists (zooxanthellae), which seem to increase the rate of calcification significantly (Vander-meulen and Muscatine 1974). The algal symbionts also pass much of their fixed carbon to the coral, mostly in the form of glycerol (Barnes 1987). Adaptive radiation in corals may well have resulted from the acquisition of algal symbionts. More than 60 genera of stony, or scleractinian, corals contain symbiotic zooxanthellae, and these corals form the largest order of anthozoans (Barnes 1987).

Intraorganismal Mutualisms

Intraorganismal mutualisms have resulted in major adaptive radiations at the level of higher taxonomic units, such as orders. For example, the Artiodactyla, or even-toed ungulates, especially the Cervidae and Bovidae, have radiated extensively. The key to this radiation is digestion of high-cellulose forage by fermenting organisms in the complex alimentary canal. The order Isoptera, or termites, has in a similar way radiated onto food sources that are highly refractive, after acquiring gut symbionts with cellulases that can break down cell walls of plants. So many groups in the order Homoptera (cicadas, plant hoppers, aphids, scale insects, and so on), including very primitive species, have mutualistic organisms in special organs or mycetomes that one must wonder if the whole order was able to radiate onto plants and suck plant juices because such microorganisms provided micronutrients essential to the insects but absent in the plants (Buchner 1965). Many butterflies and moths (Order Lepidoptera) in their larval stages may well depend on microorganisms for digestion of plant materials (see Jones et al. 1981a, 1981b, Jones 1984). The orders Siphonaptera (fleas) and Phthiraptera (lice) depend on mutualists, evidently for micronutrients that enable them to cope with their diets of blood and dry skin.

At the familial level, other radiations have occurred: Orchidaceae and endomycorrhizae, Leguminosae and nitrogen-fixing bacteria (although there is an intracellular phase in this association), lichens (although not regarded as a taxonomic group), and Ichneumonidae and viroids (Edson et al. 1981), in which the success of the parasitic wasps as endoparasites of insects may have developed from the acquisition of virus mutualists that suppress the cellular immune response of the host.

Mutualisms Between Free-Living Organisms

Mutualisms between free-living organisms and populations of species have generally been less extensive than in the more intimate relationships. However, some families and lower taxa have clearly radiated largely as a result of mutualist relationships (no higher taxonomic entities seem to have evolved from this type of mutualism): the family Agaonidae (fig wasps) with the genus *Ficus* (figs), in clearly co-evolved complementary radiations (also involving an intraorganismal stage in the mutualistic relationships, when the wasp larva develops within a gall formed by the fig ovary); the family Lycaenidae (butterflies) with ants; the family Membracidae (tree hoppers) with ants; the genus *Yucca* with yucca moths; and the families Scolytidae and Platypodidae (bark and ambrosia beetles) with ambrosia and other fungi. Smaller taxonomic groupings involving mutualistic relationships include ants and plants (Acacias, Cecropia, and so on); ants and insects (aphids, scale insects, and so on); insects and fungi (ants, termites, primitive gall midges, and so on); crabs and anemonies; cleaning symbiosis between fish species; and probably some pollination systems such as the relationships between legumes and bees, the potato family with moths, and so on. The general prediction seems to be upheld: the more intimate the coevolved system, the greater the adaptive radiation resulting from the initial beneficial interaction between species.

Another pattern emerges from this set of examples: nutrition is frequently involved. As a result, population growth rates may increase frequently as a result of mutualists, while carrying capacity remains unchanged. Therefore, Addicott's models II and III are likely to be closer to reality than model I in many natural systems. Populations are likely to return to equilibrium more rapidly in the presence of a mutualist that improves nutrition than in its absence.

20.5 HOW COMMON IS MUTUALISM?

Although it is difficult to realistically estimate the commonness of mutualistic relationships in nature, the effort seems worthwhile, if just to yield some feeling for the frequency. It is helpful to employ the well-known insect fauna of the British Isles, of which even the smallest and most cryptic species seem to have been searched out and identified. Using the checklist of British insects by Kloet and Hincks (1945), we can estimate the proportion of species that are likely to be involved with mutualists (Table 20-2). Even this very crude estimate involves some fairly certain approximations, such as in the Homoptera, in which practically all species examined have

TABLE 20-2 An estimate of the numbers of insects involved with mutualistic microorganisms in the British Isles

Order	Total Species	Number with Mutualists
Orthoptera	39	8
Phthiraptera	308	308
Thysanoptera	183	183
Hemiptera	411	288
Homoptera	976	891
Lepidoptera	2,233	1,116
Coleoptera	2,844	709
Hymenoptera	6,224	2,874
Diptera	3,190	811
Siphonaptera	47	47

Based largely on information in Buchner 1965.

mycetomes except the Typhlocybidae (Buchner 1965). The estimate also contains what are, frankly, guesses. For example, we have arbitrarily listed half of the Lepidoptera as having mutualists, for there are indications that mutualists may be important in this group (as one would expect for species with a high-cellulose diet), but the relationships may be loose, depending on the bacteria that a particular individual caterpillar happens to ingest (cf. Jones et al. 1981a, 1981b, Jones 1984). A list of species numbers may also underestimate the number of mutualisms, because many species have more than one. Buchner (1965) notes the great diversity of mutualists associated with the Homoptera. Of the 405 species studied, about 5 percent have one mutualist, 55 percent have two, 31 percent have three, 4 percent have four, 2 percent have five, and 0.5 percent have six.

Although personal estimates of the extent of mutualism in this fauna may vary considerably, the conclusion that a significant proportion have mutualistic relationships is inescapable. Of the 20,244 species in the British fauna, we estimate that 7,235, or 36 percent of the species, have mutualists (see also Jones 1984, Price 1984).

Several points are noteworthy:

1. Obligate mutualisms between insects and microorganisms are very common, even in the British Isles with its north temperate climate. If we accept this estimate of 36 percent and add to it the species involved in looser kinds of mutualisms involving pollinators, ants associated with Homoptera and lycaenid butterflies, and so on, the value may well creep close to 50 percent of species (Price 1984).

2. The high proportion of cases of mutualism between insects and microorganisms should focus our attention on the importance of microbial ecology in many natural systems, a subject in which most ecologists have virtually no experience (Price 1988).

3. Whenever a mutualist is involved with a parasite, we would be wise to consider this a three-species system, for host defense may well work against the mutualist and, indirectly, against the parasite. Thus, understanding defense—say, in plants—probably entails much more knowledge than we now have about effects on the two-species system of herbivore plus mutualist.

4. A very high proportion of the insects that probably have mutualistic microorganisms are parasitic on either plants or animals. Major adaptive radiations into such adaptive zones as blood feeding, plant-sap feeding, feeding on skin, and feeding on high-cellulose materials have occurred repeatedly.

20.6 WELL-STUDIED CASES OF COEVOLVED MUTUALISMS

Mutualistic systems provide some of the clearest cases of coevolved systems, in which unusual characteristics devoted to sustaining the mutualism are identifiable in both of the interacting species. In the remainder of this chapter we will deal with four examples, examining the essential features of the coevolution in ants and acacias, *Heliconius* butterflies and the cucurbits they pollinate, figs and fig wasps, and termites and intestinal microorganisms.

Ants and Acacias

More than 90 percent of species in the genus *Acacia* in Central America are protected from herbivores by cyanogenic chemicals in their leaves (Rehr et al. 1973). The remainder of the *Acacia* species are associated with ants in the genus *Pseudomyrmex*, and a well-developed, clearly coevolved mutualism exists between these ants and acacias (Figure 20-4). Each species of swollen-thorn acacia meets all the needs of the ant colony living on the plant: (1) shelter in the form of swollen stipular thorns into which an ant can bore to make its rearing sites for the colony; (2) extrafloral nectaries on petioles, which provide a carbohydrate source; and (3) protein-rich Beltian bodies at the tips of new leaves, which the ants gather and return to the nest sites. The plant is heavily committed to supporting an ant colony, and the ants need never leave the plant.

Why is this costly support of ants adaptive for the plant? The ants act as allelopathic agents, creating

FIGURE 20-4

The characteristics of a bull's horn acacia in relation to its mutualistic association with ants. (**Left**) A young leaf with Beltian bodies at tips of some leaflets, an extrafloral nectary on the petiole, and thorns that look like bull's horns. (**Right**) A shoot with thorns. *Pseudomyrmex* ants have bored into some of the hollow thorns to form entrances to nests. *(From Wheeler 1910)*

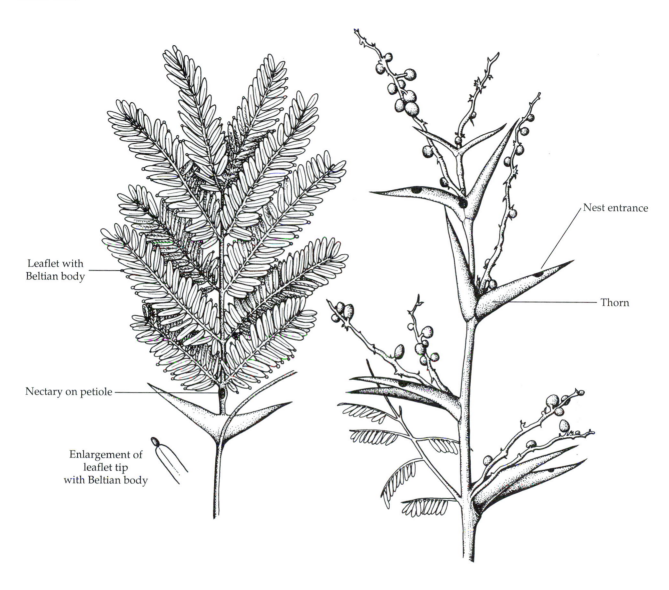

Leaflet with
Beltian body

Nectary on petiole

Enlargement of
leaflet tip
with Beltian body

Nest entrance

Thorn

a very efficient defense against the plant's enemies (Janzen 1966, 1967a, 1967b). The ants play several important roles:

1. They are very aggressive; they constantly patrol the plant and attack any herbivores, from insects to ungulates, driving them away from the plant.
2. They chew at the growing tips of plants that are rising up around the acacia, thereby suppressing potential competitors.
3. The areas of poor vegetative growth around acacias produced by ants isolate the plants from fires

that sweep through this savannah vegetation, greatly increasing the probability of survival during fire.

Janzen's experiments clearly demonstrated the mutualistic relationships between ants and acacias, and the unusual plant structures and unusual behavior of the ants clearly demonstrated the true coevolutionary nature of these relationships.

Similar relationships exist between *Azteca* ants and *Cecropia* plants in Central America (Janzen 1969), where the fine tuning of coevolution is evident in the production of glycogen by extrafloral nectaries—the only known case in the higher plants of synthesis of this "animal" sugar (Rickson 1971).

In East Africa, ants in the genus *Crematogaster* associate with *Acacia drepanolobium* (Hocking 1970, 1975) in a way similar to the mutualism of *Pseudomyrmex* species on acacia, indicating that these complex systems have emerged repeatedly (see also Bentley 1977).

Heliconius Butterflies as Pollinators of Cucurbits

In many pollination systems, it is difficult to assert with confidence that traits improving the mechanisms are mutually induced. For example, the interesting correlations between nutrients in nectar that favor certain pollinators, studied by Baker and Baker (1975), suggest a convergence on the part of the plant species toward the needs of the pollinators without any reciprocal evolutionary change in the pollinators. There is no doubt that the extensive radiations of pollinators such as hummingbirds, bees, and butterflies involved intricate coevolutionary adaptations, with the evolution of sticky pollen, tubular flowers, bifurcated body hairs, and other morphological traits in bees; the long, tubular sucking mouthparts of the higher Lepidoptera; the long, thin beaks of hummingbirds; and many others. When we examine specific relationships today, however, it is difficult to find characteristics in both mutualistic species that are unique or sufficiently rare to prove their own coevolution. So many characteristics, such as lengths of mouthparts, are general adaptations and are inherited from progenitors, which may well have coevolved with their mutualistic partners. Perhaps we need to distinguish between paleocoevolution, in which grand patterns were shaped that now influence many characteristics in pairwise associations between species, and more recent coevolution, in which the currently interacting species have probably produced the selective pressure for evolutionary change.

The pollination of cucurbits in the genera *Anguria* and *Gurania* by *Heliconius* butterflies (Gilbert 1975, 1991) provides examples of unusual traits that suggest tightly coevolved adaptations. The particular traits of the cucurbits include the following.

1. Pollen and nectar production in male flowers is very high (145 g dry sucrose and 20 g pollen per year by a male *Anguria umbosa* in a greenhouse).
2. Pollen grains are unusually large, enabling more efficient collection by *Heliconius* butterflies (*Anguria* pollen is about 80 microns in diameter).
3. Males flower throughout the year even when females are not in flower, providing the butterflies

with an important sustained supply of carbohydrates in nectar and proteins in pollen (the male *Anguria umbosa* just mentioned produced 10,000 flowers in one year).

The ability of *Heliconius* butterflies to exploit these cucurbits depends on a suite of traits.

1. Since butterflies are long-lived (about 6 months), they learn the routes along which the relatively rare cucurbits are sustained and repeatedly follow those routes—a form of traplining.
2. Groups of butterflies roost overnight in long-established locations, enabling young adults to learn trapline routes from experienced butterflies and providing a consistent point of departure for the trapline during the poorly lit early morning hours.
3. A unique ability among *Heliconius* butterflies is their use of pollen as food (Gilbert 1972, 1975, 1991). They collect pollen, moisten it with nectar, and ingest the amino acids released from the grains into solution. Pollen feeding increases egg production to as much as five times that achieved on a pure nectar diet. This improved nutrition contributes significantly to longevity, which in turn contributes to efficient traplining on a relatively sparse cucurbit population.

Gilbert (1977, 1991) emphasized the strong stabilizing influence each mutualist has on the population of the other. Butterflies increase the probability of seed set and reduce variation, while the plants improve *Heliconius* nutrition and constancy of egg production. In neither case does one of the coevolved species increase the carrying capacity of the environment for the other. The mutualism increases not the equilibrium number of individuals in the system but the stability around that equilibrium. Gilbert argued that this type of interaction represents an important and wide-ranging class of pollinator–plant mutualisms.

Figs and Fig Wasps

The more than 900 recorded species of *Ficus* are pollinated by very specific wasps in the family Agaonidae—the fig wasps. Ramirez (1970) listed host relationships for 20 species of New World fig wasps. Each of the 20 species is known to pollinate only one species of fig, and each fig supports only one species of wasp—except one, which supports two species. These fig–wasp relationships are more specific than those of many parasite–host systems (see Price 1980 and Table 19-4). Wiebe's (1966) host catalogue also showed high specificity between figs and their wasps. Many properties are evidently closely coevolved and no doubt have resulted in the extensive adaptive radiation of the genus *Ficus* and the family Agaonidae.

The fig is a false fruit formed by the enlarged receptacle of the inflorescence. The flask-shaped receptacle encloses a large number of flowers, usually of both sexes, as in Figure 20-5. Each fig inflorescence passes through the following stages (Galil and Eisikowitch 1968).

Phase A—prefemale. The young inflorescence before the ostiole opens.

Phase B—female. The ostiolar scales loosen, the female flowers open, and female agaonid wasps penetrate into the inflorescence and oviposit into the ovaries.

Phase C—interfloral. The wasp larvae and unattacked fig embryos develop, and attacked ovaries develop into galls.

Phase D—male. Male flowers mature; wasps reach maturity and emerge from the galls; male wasps inseminate females and bore holes in the receptacle; and females collect pollen, emerge through the holes bored by the males, and fly to other fig inflorescences that are in phase B.

Phase E—postfloral. Seeds ripen, and the receptacle ripens and becomes attractive to fruit-eating animals, which disperse the seed (Figure 20-6).

The life cycle of the fig wasps is also intricate. A very generalized outline follows. Before each female leaves the inflorescence, it loads up with pollen from the now-mature male flowers by packing pollen into special receptacles in the coxae of its front legs and its abdomen. It then exits through one of the holes bored by the males, flies to a fig in phase B on another tree, and enters the ostiole, in the process losing its wings and part of its antennae. The male is wingless and does not leave the fig in which it was reared.

Females then pierce the stigmas and the lengths of the styles of female flowers, ovipositing in the ovaries (Figure 20-7), and scrape pollen out of the receptacles onto the stigmatic surface with their legs, thus fertilizing the flower and others that they do not oviposit in. Eggs hatch, and larvae develop in a gall formed from ovary tissue. Males emerge first, cut holes in the sides of ovaries, and mate with the females inside. Females then leave their gall, collect pollen, and leave the fig, thus completing the cycle.

The more one reads the literature on figs and fig wasps, and the more research accumulates, the more remarkable seem the coevolved adaptations in figs and wasps. Theirs is a unique system in which coevolution is obviously a strong force. As we might expect with high specificity (one pollinator, one host), adaptations uncompromised by the presence of other species, involving coevolved features, have gone so far as to become almost bizarre.

Some coevolved traits of figs include the following.

1. The false fruits of figs are unique and allow only agaonids and a small number of closely related parasitic wasps to enter the inflorescence.
2. Extreme protogyny, with female flowers receptive several weeks before the stamens dehisce, is clearly adapted specifically to the generation time of the fig wasp.
3. The inflorescence contains both stalked and unstalked flowers with short and long styles, respectively, making seeds more and less available to ovipositing wasps. Short-styled flowers usually produce galls and wasps, whereas long-styled flowers usually produce seeds (Figure 20-7).

Coevolved traits in fig wasps should include almost everything about the wasps, for they seem to be totally adapted for pollinating figs and living in figs.

1. These wasps have very specialized morphologies in both the males and females (Figure 20-8). The female body is adapted for squeezing through the

FIGURE 20-5

Diagrammatic cross section of a fig inflorescence, showing the distribution of male and female flowers (drawn as if both were mature synchronously). (After Galil and Eisikowitch 1968)

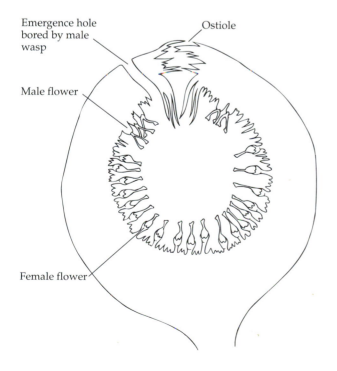

Emergence hole bored by male wasp

Ostiole

Male flower

Female flower

FIGURE 20-6

Phases in fig development. Sycone refers to the false fruit, or syconium. *(Top, from Galil and Eisikowitch 1968; bottom, from Wiebes 1979)*

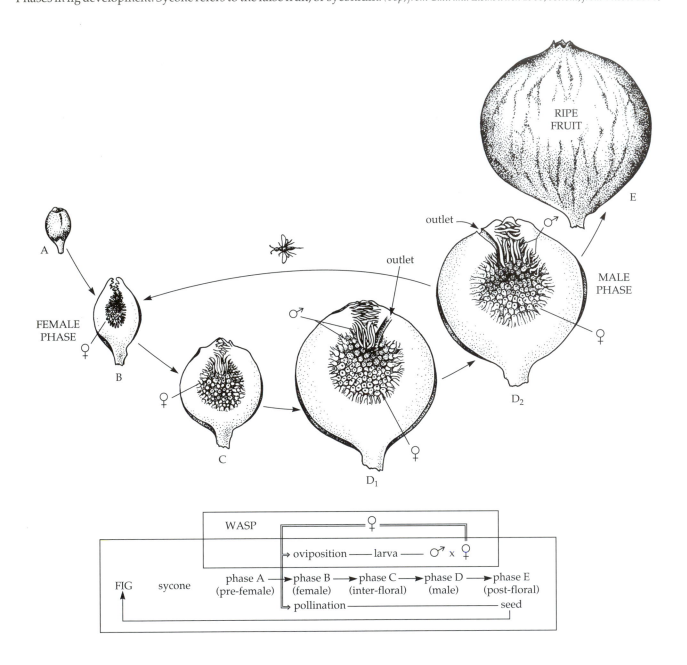

narrow ostiole and has the special pollen receptacles. The male is wingless, with a long abdomen for mating with the female in the gall and an elongated, flattened head and prothorax for rasping an entrance into the gall to reach the female.

2. The female has specialized behavior for loading and releasing pollen.

3. The female produces specialized secretions that initiate gall formation.

The *Ficus*–Agaonidae system is fascinating and, as Janzen (1979) reminded us, is open to study by population biologists, who have hardly scratched the surface of its intricate mutualism.

Termites and Intestinal Microorganisms

Termites are well known to have protozoa in their intestines, with cellulolytic enzymes enabling utilization of wood as food (see Breznak 1975 for details). Cleveland (1924; Cleveland et al. 1934) established the mutualistic nature of this relationship, in which the protozoa digest wood and the termites provide

FIGURE 20-7

A female fig wasp, *Ceratosolen arabicus*, ovipositing in a short-styled flower of *Ficus sycomorus* and extracting pollen from a pouch to fertilize the flower. (After figures in Galil and Eisikowitch 1968 and 1969)

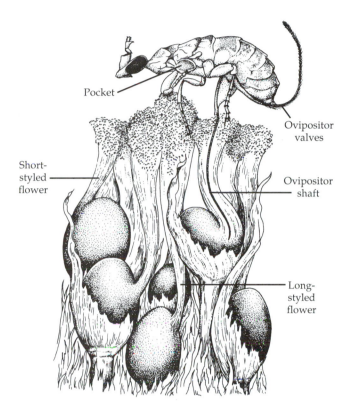

Pocket

Ovipositor valves

Short-styled flower

Ovipositor shaft

Long-styled flower

an anaerobic chamber and a constant supply of wood for the protozoa. But it was not until 1936 and 1938 that Hungate demonstrated that the energy in cellulose is nicely partitioned between mutualists. The protozoa engulf wood particles and ferment the cellulose intracellularly, and the end products are carbon dioxide, hydrogen, and acetic acid. The termite absorbs acetic acid through its hind gut wall and oxidizes it for energy, while the protozoans gain energy from the anaerobic fermentation of cellulose. The cellulolytic protozoa in the gut digest about 70 percent of the cellulose in the diet.

The coevolutionary aspects of this mutualistic relationship are clear.

1. The radiation of some 2,000 species of termites onto a high-cellulose food resource clearly depended on protozoan mutualists with cellulolytic enzymes.
2. The radiations of the various groups of protozoa inhabiting termite guts clearly depended on the termites, for whole genera are unique to termites

and the closely related wood-eating roach *Cryptocercus.*

3. The division of labor in the digestion of cellulose represents a nicely adjusted complementary system.
4. Part of the hind gut in the termite is enlarged into a bulbous "paunch" that houses the protozoa.
5. The specialized behavior of coprophagy ensures infection of protozoa in newly emerged nymphs and reinfection after moulting—a behavior that has been suggested as a driving force in the evolution of eusociality in termites, originally by Cleveland et al. (1934) (see also Wilson 1971). Termites are the only group of insects outside the Hymenoptera to reach this high level of social organization.

This intriguing relationship has additional facets (Breznak 1975). Large populations of bacteria live with the protozoa. Some of the bacteria fix nitrogen and provide it to an animal with a low-nitrogen diet. Nitrogen fixation is most rapid in young termites, which have the highest requirements for protein synthesis, and so some of the bacteria are mutualists. In addition, when the bacteria are killed selectively, the protozoa die off within a few weeks—indicating, for unknown reasons, that the bacteria are important to the protozoa. Thus, the beneficial relationships become more complex.

The whole digestive system is very similar to Hungate's (1972) "cecal model," seen in nonruminant vertebrates. Partial digestion of food and absorption by the host are followed by microbial fermentation in an enlarged caecum, or "paunch," and the animal absorbs and oxidizes the waste products as a source of energy. Through the practice of coprophagy, then, the massive microbial population in the hind gut becomes a source of food, rich in proteins and vitamins.

Yet again, nutrition is central to the mutualism in all three major components of this system. Not only do the termites accumulate mutualists until their nutritional requirements are met—rather as the acacia accumulates adaptations until the nutritional requirements of its ant mutualists are met—but the essential coprophagy and inoculation of the young have selected for close aggregation of individuals and, ultimately, the highest level of social organization seen on this earth. Mutualism is so often creative, leading to novelties that enable the participants to exploit new adaptive zones.

Our intention for this chapter was to convince the reader that mutualism is frequently involved in coevolutionary relationships and is a very common interaction in nature. Without a doubt, mutualism has been as creative a force in evolution as either predation or competition and ranks in importance with the more commonly studied interactions—

FIGURE 20-8

The adaptive morphology of a mutualistic fig wasp, *Ceratosolen arabicus* (Agaonidae), which is the pollinator. *Sycophaga sycomori* (Agaonidae) is a nonpollinating parasite of the fig. *Apocrypta longitarsus* (Torymidae), in which the female oviposits through the receptacle with its 4 millimeter ovipositor, is an inquiline that uses galls formed by *Ceratosolen* and *Sycophaga*. *(After Galil and Eisikowitch 1968)*

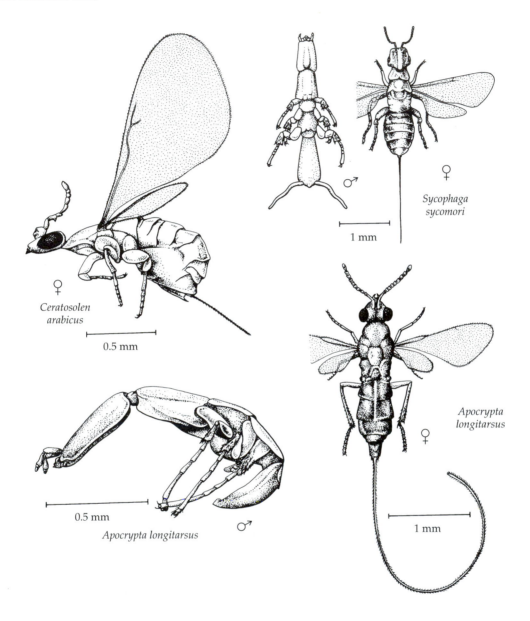

competition, parasitism, and predation (cf. Margulis and Fester 1991, Douglas 1992). Many mutualistic systems may have evolved from originally parasitic relationships, as discussed in Chapter 7. It would be incorrect to regard most close and obligate mutualisms as anything but two-species systems, for if we did, we would probably not understand the ecology of either species adequately.

SUMMARY

The close association between mutualistic organisms disposes them to a coevolutionary relationship that awards increasing benefits to each interacting species.

Although conventional wisdom and some models seem to discount mutualisms as major forces in ecology and evolution, it is readily apparent that mutualisms have provided major breakthroughs to adaptive radiation, and models illustrate the kinds of relationships most likely to provide a stable interaction between participants.

We may classify mutualisms in nature into **intracellular mutualisms,** involving occupation of one mutualist within the cells of its host; **intraorganismal mutualisms,** in which one mutualist lives within the organs of a host; and **mutualism between organisms,** in which essentially free-living individuals of species or populations of species interact. Radiations resulting from these kinds of mutualisms show a pattern, with major evolutionary breakthroughs resulting from intracellular mutualisms and generating new kingdoms of organisms. At the intraorganismal level, radiations of vertebrates and invertebrates have been extensive in higher taxonomic units, such as orders and families. Mutualisms between free-living organisms have generally been less extensive in coevolved systems, but some family-level radiations have resulted.

Many species in nature are involved with mutualisms beyond the early associations that resulted in the evolution of the eukaryotes. The proportion of insects in mutualistic associations is estimated at 36 percent, even for a small island well out of the tropics.

Associations of insects with microorganisms are particularly common.

Some well-researched examples of coevolved mutualisms illustrate the **mutually induced adaptive responses** between interacting species. Ants associated with *Acacia* plants protect the plants from herbivores, weeds, and fire while the plant provides shelter, carbohydrates, and proteins that sustain the ant colony. *Heliconius* butterflies pollinate sparse cucurbit flowers with high pollen production, large pollen grains, and prolonged flowering phenology. The butterflies live unusually long, learn to trapline for pollen, and can digest pollen. In fig–fig wasp systems, unique design of fig inflorescences, extreme protandry, and morphological variation in female flowers demand high specificity of fig species and wasp pollinator species. Fig wasps have very specialized morphology, and females load pollen for transport to new flowers and cause galling of ovaries. Many termites have long been associated with intestinal microorganisms, which enable them to utilize refractive diets rich in cellulose. There are even mutualistic associations between protozoa and bacteria in the paunch of the termite.

Coevolved mutualisms have been very creative as forces in the adaptive radiation of many taxa.

QUESTIONS FOR DISCUSSION

1. On the basis of your own knowledge, can you add examples to the different categories of mutualism discussed in this chapter?

2. Would you argue that mutualism is as important an interaction in nature as are competition, predation, and parasitism?

3. When two species enter into a mutualistic relationship, such that both species remain together indefinitely, as with the origin of the eukaryotic cell, would you accept that these cases should be viewed as examples of saltational evolution, and in some cases saltational speciation?

4. In the evolution of larger organisms, do you think that many biochemical problems have been solved by the chance colonization of a preadapted microorganism, resulting in a mutualistic relationship?

5. Given the clear benefits of mutualisms—such as that between ants and acacias or that between figs and fig wasps—why do you think that there are not many more mutualisms in nature?

6. How would you develop a test of the hypothesis that lineages that acquire mutualists have radiated more extensively than sister groups that have not? Which taxa would you use for the test?

7. In addition to rumination and caecal fermentation in mammals, most arboreal herbivorous mammals, including marsupials, have complex stomachs with similar functions as the rumen. Can you develop a scenario of interacting factors that would result in considerable convergence in the manner in which food is digested in the larger herbivores?

8. To what extent do you think that it is valid to argue that most mutualisms have probably evolved in the tropics, even if subsequent successful colonization of colder latitudes has been common?

9. Does the study of mutualism help with the general conceptual development of evolutionary theory, in your opinion?

10. In your reading of his works, what is your impression of Darwin's position on the role of mutualism in evolution?

REFERENCES

Addicott, J. F. 1981. Stability properties of 2-species models of mutualism: Simulation studies. Oecologia 49:42–49.

Ahmadjian, V., and M. E. Hale (eds.). 1973. The Lichens. Academic Press, New York.

Baker, H. G., and I. Baker. 1975. Studies on nectar-constitution and pollinator–plant coevolution, pp. 100–140. *In* L. E. Gilbert and P. H. Raven (eds.). Coevolution of Animals and Plants. Univ. of Texas Press, Austin.

Barnes, R. D. 1987. Invertebrate Zoology. 5th ed. Saunders College Publishing, Philadelphia.

Batra, L. R. (ed.). 1979. Insect–Fungus Symbiosis: Nutrition, Mutualism, and Commensalism. Allanheld, Osmun, Montclair, N.J.

Bentley, B. L. 1977. Extrafloral nectaries and protection by pugnacious body-guards. Ann. Rev. Ecol. Syst. 8:407–427.

Beattie, A. J. 1985. Evolutionary Ecology of Ant–Plant Mutualisms. Cambridge Univ. Press, Cambridge, England.

Boucher, D. H. (ed.). 1985. The Biology of Mutualism: Ecology and Evolution. Oxford Univ. Press, New York.

Breznak, J. A. 1975. Symbiotic relationships between termites and their intestinal microbiota. Symp. Soc. Exp. Biol. 29:559–580.

Buchner, P. 1965. Endosymbiosis of Animals with Plant Microorganisms. Wiley, New York.

Cleveland, L. R. 1924. The physiological and symbiotic relationships between the intestinal protozoa of termites and their host, with special reference to *Reticulitermes flavipes* Kollar. Biol. Bull. 46:178–227.

Cleveland, L. R., S. R. Hall, E. P. Sanders, and J. Collier. 1934. The wood-feeding roach *Cryptocercus*, its protozoa, and the symbiosis between protozoa and roach. Mem. Amer. Acad. Arts Sci. 17:184–342.

Cushman, J. H., and A. J. Beattie. 1991. Mutualisms: Assessing the benefits to hosts and visitors. Trends Ecol. Evol. 6:193–195.

Cushman, J. H., and T. G. Whitham. 1989. Conditional mutualism in a membracid–ant association: Temporal age-specific, and density-dependent effects. Ecology 70:1040–1047.

———. 1991. Competition mediating the outcome of a mutualism: Protective services of ants as a limiting resource for membracids. Amer. Natur. 138:851–865.

Cushman, J. H., V. K. Rashbrook, and A. J. Beattie. 1994. Assessing benefits to both participants in a lycaenid–ant association. Ecology 75:1031–1041.

Douglas, A. E. 1992. Symbiosis in evolution, pp. 347–382. *In* D. Futuyma and J. Antonovics (eds.). Oxford Surveys in Evolutionary Biology. Vol. 8. Oxford Univ. Press, New York.

Edson, K. M., S. B. Vinson, D. B. Stoltz, and M. D. Summers. 1981. Virus in a parasitoid wasp: Suppression of the cellular immune response in the parasitoid's host. Science 211:582–583.

Futuyma, D. J. 1973. Community structure and stability in constant environments. Amer. Natur. 107:443–446.

Galil, J., and D. Eisikowitch. 1968. On the pollination ecology of *Ficus sycomorus* in East Africa. Ecology 49:259–269.

———. 1969. Further studies on the pollination ecology of *Ficus sycomoros* L. Tijdschr. Ent. 112:1–13.

Gilbert, L. E. 1972. Pollen feeding and reproductive biology of *Heliconius* butterflies. Proc. Nat. Acad. Sci. USA 69:1403–1407.

———. 1975. Ecological consequences of a coevolved mutualism between butterflies and plants, pp. 210–240. *In* L. E. Gilbert and P. H. Raven (eds.). Coevolution of Animals and Plants. Univ. of Texas Press, Austin.

———. 1977. The role of insect–plant coevolution in the organization of ecosystems, pp. 399–413. *In* V. Labyrie (ed.). Comportement des Insectes et Milieu Trophique. C.N.R.S., Paris.

———. 1991. Biodiversity of a Central American *Heliconius* community: Pattern, process and problems, pp. 403–427. *In* P. W. Price, G. W. Fernandes, T. M. Lewinsohn, and W. W. Benson (eds.). Plant–Animal Interactions: Evolutionary Ecology in Tropical and Temperate Regions. Wiley, New York.

Grant, K. A., and V. Grant. 1968. Hummingbirds and Their Flowers. Columbia Univ. Press, New York.

Greene, R. W. 1974. Sacoglossans and their chloroplast endosymbionts, pp. 21–27. *In* W. B. Vernberg (ed.). Symbiosis in the Sea. Univ. of South Carolina Press, Columbia.

Harley, J. L. 1969. The Biology of Mycorrhiza. Leonard Hill, London.

Hocking, B. 1970. Insect associations with the swollen thorn acacias. Trans. Royal Ent. Soc. London 122:211–255.

———. 1975. Ant–plant mutualism: Evolution and energy, pp. 78–90. *In* L. E. Gilbert and P. H. Raven (eds.). Coevolution of Animals and Plants. Univ. of Texas Press, Austin.

Hungate, R. E. 1936. Studies on the nutrition of *Zootermopsis*, [Part] I: The role of bacteria and molds in cellulose decomposition. Zent. Bakt. ParasitKde Abt. II 94:240–249.

———. 1938. Studies on the nutrition of *Zootermopsis*, [Part] II: The relative importance of the termite and the protozoa in wood digestion. Ecology 19:1–25.

———. 1966. The Rumen and Its Microbes. Academic, New York.

———. 1972. Relationships between protozoa and bacteria of the alimentary tract. Amer. J. Clin. Nutr. 25:1480–1484.

———. 1975. The rumen microbial ecosystem. Ann. Rev. Ecol. Syst. 6:39–66.

Janzen, D. H. 1966. Coevolution of mutualism between ants and acacias in Central America. Evolution 20:249–275.

———. 1967a. Fire, vegetation structure, and the ant × acacia interaction in Central America. Ecology 48:26–35.

———. 1967b. Interaction of the bull's horn acacia (*Acacia cornigera* L.) with an ant inhabitant (*Pseudomyrmex ferruginea* F. Smith) in eastern Mexico. Univ. Kansas Sci. Bull. 47:315–558.

————. 1969. Allelopathy by myrmecophytes: The ant *Azteca* as an allelopathic agent of *Cecropia*. Ecology 50:147–153.

————. 1979. How to be a fig. Ann. Rev. Ecol. Syst. 10:13–51.

Jones, C. E., and R. J. Little (eds.). 1983. Handbook of Experimental Pollination Biology. Van Nostrand Reinhold, New York.

Jones, C. G. 1984. Microorganisms as mediators of plant resource exploitation by insect herbivores, pp. 53–99. *In* P. W. Price, C. N. Slobodchikoff, and W. S. Gaud (eds.). A New Ecology: Novel Approaches to Interactive Systems. Wiley, New York.

Jones, C. G., J. R. Aldrich, and M. S. Blum. 1981a. 2-Furaldehyde from bald-cypress: A chemical rationale for the demise of the Georgia silkworm industry. J. Chem. Ecol. 7:89–101.

————. 1981b. Baldcypress allelochemics and the inhibition of silkworm enteric microorganisms: Some ecological considerations. J. Chem. Ecol. 7:103–114.

Kloet, G. S., and W. D. Hincks. 1945. A Check List of British Insects. Kloet and Hincks, Stockport.

Lawrey, J. D. 1984. Biology of Lichenized Fungi. Praeger, New York.

Limbaugh, C. 1961. Cleaning symbiosis. Sci. Amer. 205(2):42–49.

Margulis, L. 1970. Origin of Eukaryotic Cells. Yale Univ. Press, New Haven, Conn.

————. 1975. Symbiotic theory of the origin of eukaryotic organelles: Criteria for proof. Symp. Soc. Exp. Biol. 29:21–38.

————. 1976. Genetic and evolutionary consequences of symbiosis. Exper. Parasit. 39:277–349.

————. 1981. Symbiosis in Cell Evolution. W. H. Freeman, San Francisco.

————. 1993. Symbiosis in Cell Evolution: Microbial Communities in the Archean and Proterozoic Eons. W. H. Freeman, New York.

Margulis, L., and R. Fester (eds.). 1991. Symbiosis as a Source of Evolutionary Innovation: Speciation and Morphogenesis. MIT Press, Cambridge.

May, R. M. 1976. Models for two interacting populations, pp. 49–70. *In* R. M. May (ed.). Theoretical Ecology. Saunders, Philadelphia.

Mengel, K., and D. J. Pilbeam (eds.). 1992. Nitrogen Metabolism of Plants. Clarendon Press, Oxford.

Orians, G., and team. 1974. Tropical population ecology, pp. 5–65. *In* E. G. Farnworth and F. B. Golley (eds.). Fragile Ecosystems. Springer–Verlag, New York.

Price, P. W. 1980. Evolutionary Biology of Parasites. Princeton Univ. Press, Princeton, N.J.

————. 1984. Insect Ecology. 2d ed. Wiley, New York.

————. 1988. An overview of organismal interactions in ecosystems in evolutionary and ecological time, pp. 369–377. *In* C. A. Edwards, B. R. Stinner, D. Stinner, and S. Rabatin (eds.). Biological Interactions in Soil. Elsevier, Amsterdam.

Ramirez, W. 1970. Taxonomic and biological studies on neotropical fig wasps (Hymenoptera: Agaonidae). Univ. Kansas Sci. Bull. 49:1–44.

Rehr, S. S., P. P. Feeny, and D. H. Janzen. 1973. Chemical defense in Central American non-ant acacias. J. Anim. Ecol. 42:405–416.

Richards, A. J. (ed.). 1978. The Pollination of Flowers by Insects. Linnean Soc. Symp. Ser. 6. Academic, London.

Rickson, F. R. 1971. Glycogen plastids in Mullerian body cells of *Cecropia peltata*—a higher green plant. Science 173:344–347.

Ross, D. M. 1974. Evolutionary aspects of associations between crabs and sea anemones, pp. 111–125. *In* W. B. Vernberg (ed.). Symbiosis in the Sea. Univ. of South Carolina Press, Columbia.

Ruehle, J. L., and D. H. Marx. 1979. Fiber, food, fuel, and fungal symbionts. Science 206:419–422.

Taylor, D. L. 1974. Symbiotic marine algae: Taxonomy and biological fitness, pp. 254–262. *In* W. B. Vernberg (ed.). Symbiosis in the Sea. Univ. of South Carolina Press, Columbia.

Vandermeulen, J. H., and L. Muscatine. 1974. Influence of symbiotic algae on calcification in reef corals: Critique and progress report, pp. 1–19. *In* W. B. Vernberg (ed.). Symbiosis in the Sea. Univ. of South Carolina Press, Columbia.

Werner, D. 1992. Symbiosis of Plants and Microbes. Chapman and Hall, London.

Wheeler, W. M. 1910. Ants: Their Structure, Development and Behavior. Columbia Univ. Press, New York.

Wiebes, J. T. 1966. Provisional host catalogue of fig wasps (Hymenoptera: Chalcidoidea). Zool. Verh. Leiden 83:1–44.

————. 1979. Co-evolution of figs and their insect pollinators. Ann. Rev. Ecol. Syst. 10:1–12.

Williamson, M. 1972. The Analysis of Biological Populations. Arnold, London.

Wilson, E. O. 1971. The Insect Societies. Belknap Press of Harvard Univ. Press, Cambridge, Mass.

IV

CONCLUSION

21 IS THE THEORY OF EVOLUTION ADEQUATE?

21

IS THE THEORY OF EVOLUTION ADEQUATE?

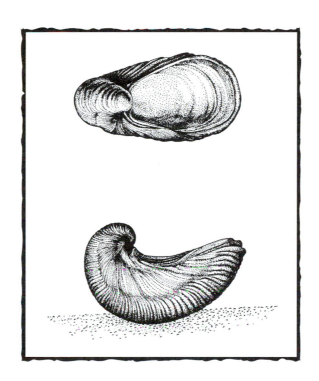

The bivalve mollusc *Gryphaea*, a genus named by Lamarck in 1801. The large, free-lying oyster is shown in top and side views.

21.1 BIOLOGICAL EVOLUTION IN THE SCIENCES

Science is insatiable! Addressing the history of the cosmos—a period of perhaps 15 to 20 billion years—it dictates a human endeavor vastly greater and more challenging than any other. History, literature, religion, politics, civilization, and culture, however captivating and compelling in our daily lives, deal with only a brief episode (in the past and probably in the future) in the life of the cosmos. In the beginning was a pure source of enormous energy. How that source developed into the cosmos, the galaxy, and the planet we live on should be our intellectual concern "forever."

Biological evolution is but a small part of science, and yet it involves some 4 billion years of earth's history and perhaps greater expanses of time on the planets in our galaxy and beyond. By its temporal magnitude, the subject still dwarfs all other human endeavors.

Biological evolution seeks a more temporally encompassing view of the world than even most sciences. For example, computer science addresses a time span that relates to the period covered by evolution as an eye blink relates to a year. The social sciences hardly explore backward beyond the social insects and presocial arthropods, well after life on earth was flourishing. The political sciences are "babes in arms."

The challenges for the science of biological evolution therefore go well beyond those of many other sciences. Even in sciences exploring longer-term phenomena, such as astronomy and geology, exploration occurs largely in the more orderly physical world of law-abiding particles. For living cells and organisms, we hardly know what the laws are.

Given these considerations, how satisfactory should we expect the science of biological evolution to be? Does it have substance and a sound foundation? Is there a conceptual perspective on the development of life on earth consistent with the imperatives of the scientific method? Is the field more metaphysical than scientific? Is it actually the religion of biologists, to be compared with the great religions of the world?

21.2 EVIDENCE FOR EXTENSIVE RELATIONSHIPS AMONG ORGANISMS

The theory of evolution argues for descent with modification, resulting in long-term phylogenetic linkages among organisms. What evidence has this book presented that such tangible relationships are detectable? We can employ this question to review the many cases presented.

"All our information, scanty though it may be, leads us to assume that the same unity of design of which we observe evidence in the modern world extends also across the enormous time gaps of the past." This was Baron Cuvier's statement, translated by Eiseley (1958, 85–86). And by 1837, after a detailed study of the comparative anatomy of modern and prehistoric animals, Cuvier reinforced his own assumption. "At the voice of comparative anatomy every bone and fragment of a bone resumed its place" (Eiseley 1958, 84) (see Chapter 2). Darwin also noted relationships among animals on the Galápagos Archipelago—such as the finches—and relationships at the species and subspecies levels between island dwellers and their mainland counterparts (see Chapter 2).

In Chapter 4, we saw the history of associations unfold in the genus of crickets variously named *Gryllus, Nemobius,* and *Allonemobius.* Eventually, six species with close relationships emerged from what De Geer had thought was one species in 1773. We discussed evolutionary links between circum-arctic gulls in the genus *Larus* and the complexities of reticulate relationships in the plant genus *Clarkia.*

The treatment of speciation in Chapter 5 presented many varied arguments for relationships among organisms: *Dendroica* warblers, *Erythroneura* leafhoppers, *Rhagoletis* fruit flies, morabine grasshoppers in the genus *Vandiemenella,* and plants in the genera *Gilia, Clarkia, Galeopsis, Dahlia,* and *Chrysanthemum.* Theories on the evolution of life (Chapter 6) indicate long-standing unity in the structure of RNA, which enables researchers to trace relationships from early bacteria to fungi, animals, and plants, reinforcing the endosymbiotic theory of eukaryotic cell evolution.

Although many associations remain obscure because of missing links, a remarkable story of relationships unfolded in Chapter 7. Arguments for the early divergences of fungi, animals, and plants are tentative because they occurred so long ago, but RNA analysis suggests extensive relationships among annelids, onychophorans, and arthropods, and embryological development shows ancient associations among protostomes in one large group and among the deuterostomes, including those from echinoderms to vertebrates. It is difficult to follow the evolution of the vertebrates, but scientists have embraced a remarkably coherent explanation of evolution from fishes to amphibians, reptiles, and mammals, with tetrapod limbs having originated in the crossopterygian fishes of 350 million years ago and the coelacanth, *Latimeria,* being something like a "living fossil" of Devonian lobe fin fishes. Scientists have yet to fully unravel the stepwise evolution of terrestrial plants, but an understanding of chronology and phylogenetic relationships is developing rapidly: rhyniophytes to trimerophytes, to progymnosperms, to conifers and seed ferns, to cycads and angiosperms.

Chapter 8, "Adaptive Radiation," illustrated many cases of phylogenetic relationships: ammonites, echinoids, foraminiferans, Hawaiian honeycreepers and silverswords, and the cichlid fishes of African Rift Valley lakes. Chapter 12 explored human origins, using embryology, vestigial organs, skeletal design, deficiencies in design, and molecular comparisons of protein and DNA similarities. The surprisingly close biochemical similarities between apes and humans supported Darwin's original arguments on human descent. While the systematic relationships among organisms leave room for hot debate; as discussed in Chapter 13, "Biological Classification," the debate is as much on methodology as on real phylogeny. Chapter 13 described clear and extensive relationships of *Larus* gulls, mygalomorph spiders, and amniotes.

We addressed cases of relationships that involved chromosomal similarities of parasitic wasps in the genus *Torymus* (Chapter 14), decreasing allozyme similarities in diverging *Drosophila* lineages, and close allozome relationships among species of *Rhagoletis* fruit flies, *Thomomys* pocket gophers, and *Spalax* mole rats (Chapter 15). We learned that scientists have traced very ancient lineages, perhaps to 2 billion years ago, using cytochrome P450 molecules (Chapter 16). Chapter 17 discussed extensive similarities in amino acid sequences in proteins across a wide range of organisms, especially the well-studied globins (including hemoglobins), showing the degrees of relationship between such diverse groups as plants, worms, molluscs, and many vertebrates. Chapter 18 described rapid transitions in design in the Equidae and Artiodactyla, the evolution of the Stylinodontinae, and relationships in the trilobite *Phacops rana* and the oyster *Gryphaea*. Associations of insect herbivore lineages with plant lineages (Chapter 19) and the extensive mutualistic associations of organisms (Chapter 20) provide evidence for tracking among disparate groups through time.

Although such recounting of evidence for objectively detected relationships among organisms can seem tedious, it adds up to a formidable literature supporting descent with modification—and this book describes only a small sampling of the available cases. Scientists may still debate the precise interpretation of results and the precise avenues of evolution, but the big view is clear enough: ample modern evidence exists for extensive phylogenetic relationships among all organisms. Darwin's theory of evolution predicted

this view with precision at a time when phylogenetic relationships had been inadequately explored.

21.3 THE PROCESSES IN EVOLUTION

So many processes result in evolutionary change that inevitably some uncertainty and therefore debate occur about which processes are more or less important in a given case. However, we find remarkable agreement among scientists about the fact that many processes are involved, each with rather predictable results—as we discussed in regard to speciation, genetic systems, change in gene frequencies, molecular evolution, rates of evolution, and coevolution (Chapters 5, 14, 15, and 17–20). In most cases, scientists resolve the issues and divergent views settle into the main fabric of evolutionary theory.

Dichotomous interpretations enrich the eventual picture; they do not detract from it. We have examined a few such debates on evolutionary processes: Darwinian and non-Darwinian evolution, gradualism and punctuated equilibrium, and stabilizing aspects of coevolution (Chapters 17, 18, and 20). After all, the scientific method (Chapter 1) involves the testing of hypotheses and testing among alternative explanations of processes.

21.4 THE ROLE OF DEBATE IN SCIENCE

Debate in science is commonly misinterpreted as a breakdown of conventional wisdom and the decay of a scientific discipline. This is especially true in biological evolution, because some are motivated to discredit the field. Lack of debate, however, results in dogma. The scientific method demands debate on hypotheses, the validity of data and interpretations, and the adequacy of the evidence supporting scientific theory. Debate is essential; it commonly results in major breakthroughs and new avenues for research. Debate can never destroy a science. On the contrary, it keeps the science alive, vibrant, and exciting while it advances knowledge.

This book has not always presented the full arenas of debate on issues, because some shortcuts are necessary, and it is important to grasp the big picture before approaching a volume that attempts to capture the importance of the alternative views. We have, however, discussed some issues and their resolutions, illustrating the important roles of doubt, disagreement, and argument; examples include spontaneous generation (Chapter 1), issues surrounding the transition in thinking from catastrophism to uniformitarianism (Chapter 2), modes of speciation (Chapter 5), the evolution of life (Chapters 6 and 7), processes of extinction (Chapter 9), the neutral theory of evolution (Chapter 17), and hopeful monsters (Chapter 18).

New ideas do not undermine the science but enrich and refine it. The more controversial an issue becomes, the more likely it is to be resolved rapidly, either by exclusion from the main body of theory or by refinement of that theory. For instance, alternative hypotheses refined the neutral theory of evolution and punctuated equilibrium. Scientists aspire toward controversy for their theories, because the resultant attention to the topic promotes rapid resolution of alternative views.

21.5 IS THE THEORY OF EVOLUTION PREDICTIVE?

The argument frequently arises that *real* theory is predictive and, because the theory of evolution lacks predictive power, it is inadequate. However, we have listed in this chapter the many cases cited in this book that support one of Darwin's major predictions:

> There is grandeur in this view of life, with its several powers, having been originally breathed into a few forms or into one; and that, whilst this planet has gone cycling on according to the fixed law of gravity, from so simple a beginning endless forms most beautiful and most wonderful have been, and are being, evolved. (Darwin 1859, 490).

Modern science has repeatedly reinforced Darwin's prediction far beyond what he could have envisioned. Recall that the issue of spontaneous generation was finally resolved after *The Origin of Species* was published (Chapter 1). Note the several figures in this book that show how all life on earth is related. Note also that scientists since Darwin have reinforced the principle of extended relationships, and no evidence suggests an alternative view.

The theory of evolution predicts that species will evolve under the influence of natural selection. We have seen this prediction upheld in cases of which Darwin was unaware: evolution of industrial melanism (Chapter 1), sickle-cell trait (HbS) (Chapter 16), and declines in melanism and HbS when selective pressures changed. The evolution of resistance to pesticides and antibiotics provides additional examples. The theory of evolution also predicts evolutionary change under artificial selection, which is amply documented in many cases (cf. Chapter 1).

Another prediction is that a particular path in evolution depends on chance, involving the random events of mutation and the probability of sampling error or genetic drift. In this sense the theory predicts unpredictability.

The theory of evolution is strongly predictive.

21.6 IS THE THEORY OF EVOLUTION REFUTABLE?

The accumulation of evidence supporting the theory of evolution has been rapid and extensive. Have scientists spent enough time attempting to refute this theory? Surprising new discoveries of life outside the known domains may at first be discredited, but soon the big picture assimilates them. Such a case was the discovery of a "third form of life" (Woese 1981, 114), comprising methanogenic bacteria and their relatives, which was initially placed in its own group, the Archaebacteria. Later, three domains of life were recognized: the Bacteria; the Archaea, including the methanogens; and the Eucarya, including the eukaryotes (Figure 6-9). Other surprises, such as the deep-sea vent communities (for example, along the Galapagos Rift) discovered beginning in 1977, revealed organisms that fitted well into the known systematics. At 2,800 meters below sea level, scientists aboard the deep submersible *Alvin* documented the existence of molluscs, crabs, polychaete worms, acorn worms, and large pogonophoran worms (e.g., Childress et al. 1987). And what about the Va Quang ox and its associated large animals, the giant muntjac and the slow-running deer, recently revealed to science in the dense and wet forests along the Laos–Vietnam border (Dung et al. 1993, Linden 1994)? While they seem to belong to a relictual fauna, and perhaps new genera, they fit right into the current systematics of the mammals. Speaking of giants, Fishelsen et al. (1985) reported a completely enigmatic symbiont in the surgeonfish. It was so large (>600 × 80 μm) that it almost had to be a eukaryotic cell, and yet it had a bacterial arrangement of DNA. In fact, it turned out to be a bacterium, 1 million times larger in volume than a typical bacterium and much larger than any other bacterium. It is, however, clearly genetically related to the gram-positive group of bacteria in the genus *Clostridium* (Angert et al. 1993, Sogin 1993). Related bacteria appear to have become endocellular symbionts in protozoa (Grim 1993). Even the most unexpected finds seem to have their places in the evolutionary system and are related to well-known groups.

We can dream up the kinds of damning evidence that would refute the presently accepted scheme of evolution. A dinosaur from the Devonian would be a shock. Any microbial organism with a totally novel genotype would be cause for concern. But what is the likelihood of finding even several such examples, and how would such evidence weigh against the large body of knowledge supporting the theory of evolution?

Below the scale of the grand scheme of evolution, refutation of hypotheses has been an important element in the advance of evolutionary theory and will no doubt remain so. Various hypotheses discussed in this book have been refuted—for example, those on spontaneous generation (Chapter 1), arboreal evolution of primates (Chapter 12), and the germ-line hypothesis in the generation of immunoglobulin diversity (Chapter 19).

We must make an important distinction between refuting an evolutionary hypothesis with a good, scientifically based alternative supported by strong evidence and concluding that evolutionary hypotheses are inadequate because they cannot explain everything. Nobody claims that evolutionists have all the answers or even that we are close to the correct explanations; the evolution of life (Chapter 6) and the evolution of humans (Chapter 12) are cases in point.

The most important quality of the science of evolution is that all elements of the field are available for examination via the scientific method. Therefore, the available methodologies have the potential to refute hypotheses in all aspects of the science.

21.7 ALTERNATIVE HYPOTHESES TO THE THEORY

Grand Design

No other scientific theory competes with the theory of evolution as an explanation of the unity and diversity among organisms. However, nature in general and the beauty of many plants and animals do inspire a certain wonder, and many have felt that evolution is an inadequate theory. Perhaps there is more to the wonders of life than the passive responses of populations to mutation, natural selection, drift, mass extinction, and so on. Is there a grand design stemming from a grand designer that has shaped the course of evolution in mysterious ways?

Rational people quite commonly search beyond the theory of evolution for a more majestic view of the history of life. But tangible evidence that would force us as scientists to search for something grander than the theory of evolution is elusive. Darwin saw grandeur, beauty, and power in the view of life provided by his theory of evolution: "When I view all beings not as special creations, but as the lineal descendants of some few beings which lived long before the first bed of the Silurian system was deposited, they seem to me to become ennobled" (Darwin 1859, 488–489). He also noted problems with grand design: "We can plainly see why nature is prodigal in variety, though niggard in innovation. But why this should be a law of nature if each species has been independently created, no man can explain" (471).

Progressive Evolution

There remains the beguiling idea that evolution has progressed from simple to complex, from worms to humans, and that humans represent the zenith of the evolutionary process. "And as natural selection works solely by and for the good of each being, all corporeal and mental endowments will tend to progress towards perfection" (Darwin 1859, 489). Darwin, however, rejected the concept of progressive evolution in most of *The Origin*. He emphasized *descent* with modification, not *ascent* to more complex form (i.e., *The Ascent of Man* as envisioned by Bronowski [1973]). Even so, many biologists, including Huxley (1942, 1953), have thought of biological progress as increasing independence from environmental forces, with humans reaching a state that may really and finally connect the physical earth with the metaphysical.

Consideration of the human body, discussed in Chapter 12, should help to undermine any argument for progressive evolution.

> As a piece of machinery we humans are such a hodge-podge and makeshift that the real wonder resides in the fact that we get along as well as we do. Part for part our bodies, particularly our skeletons, show many scars of Nature's operations as she tried to perfect us. (Krogman 1951, 54)

In humans, nature fell far short of achieving perfection. We can confidently reject orthogenesis, or the predetermined path of evolution, as a viable concept.

21.8 BEYOND EVOLUTIONARY THEORY

Writers since Darwin have followed up on his assertion that bodily and mental traits progress toward perfection, by weaving creative arguments about the nature of perfection. While espousing the theory of evolution, they have developed a path from the physical to the metaphysical and from the scientific realm to the spiritual realm (cf. Chapter 1). For example, Teilhard de Chardin (1959, 1964) described an orthogenetic evolutionary progression from prelife to life, to thought, and to superlife, or the omega point—an ethereal unity. Hardy (1965) also bridged the gap between evolutionary theory and what he called natural theology by moving from behavior to psychology, to telepathy, to psychic phenomena, and to religious perceptions.

> I have attempted to demonstrate how, geared into the physical framework of the Darwinian–Mendelian-DNA scheme, the ever inquisitive, exploratory nature of animal behavior, leading to new habits and ways of life, has had an increasingly important role

in this living stream. It is with this, the psychic side, if you like, of animal nature, that I believe Man's religious feelings are related. (Hardy 1965, 262)

More recently, many scientists have alluded to a god, adopting a rather nebulous concept (Lederman 1993; Wright 1989, 1992; Hawking 1988). Hawking, for example, building on advances since Einstein, referred to God to counter Einstein's statement that "God does not play dice with the universe." Hawking responded, "God not only plays dice, but sometimes he throws them where they cannot be seen" (i.e., in a black hole) (Boslough 1985). Charles Darwin was originally a religious man, and explorations beyond science have had their appeal and even their compulsion for other scientists. However, it is always clear when the move beyond science takes place in an argument, just as it is clear that debates on evolution and metaphysics should acknowledge that the two approaches involve different kinds of human endeavor. Lord Morley said, "The next great task of Science is to create a religion for humanity" (Huxley 1923). Any pathway from science to religion must cross an uncertain terrain between the scientific theory of evolution and the metaphysical arena of religious philosophy.

Another extension of evolutionary theory involves the argument that natural selection has operated on a global scale so that the earth is maintained homeostatically in ways advantageous to life. The Gaia hypothesis is named after the earth goddess of the ancient Greeks. Through the ages, equilibrial states have shifted as a result of biotic and abiotic activity, as if the whole earth had a physiology of its own—geophysiology (Lovelock 1979, 1986a, 1986b, 1988, 1991). The concept of Gaia is appealing because of its scope and its heuristic value (Margulis 1993), but it has been criticized for its lack of crisp, and therefore testable, hypotheses (Kirchner 1989) and because of the evidence that homeostasis appears to collapse roughly every 26 million years, resulting in mass extinctions (Stanley 1988) (cf. Chapter 9). Just as human physiology reveals problems with the concept of progressive evolution, so geophysiology indicates an uncertain world with flawed homeostatic mechanisms.

21.9 THE SCIENTIFIC THEORY OF EVOLUTION

This chapter asserts that compelling evidence exists for a mechanistic scientific theory of evolution that accounts for both the unity and diversity of life. Research has reinforced this remarkable accomplishment at increasing rates over the last few decades. New techniques have provided new methods of testing hypotheses, and it is still remarkable that in each case the unity of life appears closer and clearer. In fact, no alternative hypothesis within the grasp of the scientific method can challenge the validity of the theory of evolution. It is very natural, even a basic imperative, for humans to go beyond the scientific method in search of a profounder view of life and a grander vision of this planet. The extent to which we can apply scientific methodology to advance the theory of evolution into realms that now seem metaphysical remains to be explored. However, as Wright (1992, 44) said, "If you admit that we can't peer behind the curtain, how can you be sure there's nothing there?"

SUMMARY

Biological evolution is a relatively small part of the sciences, but it is central to the whole of biology because it alone can account for both the unity among organisms and the great diversity of species that have occupied and do occupy this planet. The time period in which biological evolution has occurred is still vast compared with the periods of many other interests of humans.

Because of the theory of evolution, even in the relatively short time since 1859, a vast and accelerating accumulation of compelling evidence has documented the extensive and long-standing relationships among organisms. Scientists have identified and widely agreed on the processes involved with evolutionary change. Debate on processes enriches the general theory. Darwin made many predictions on the basis of his theory of descent with modification, and modern science has tested many and found them valid. No alternative scientific hypothesis has undermined the theory of evolution.

Humans inevitably search beyond scientific theory and the "cold, hard facts" for deeper meanings in life, frequently entering realms that so far appear impossible to test by the scientific method. The appealing concepts of a grand design underlying evolution, and progressive evolution toward perfection, or a zenith, can be questioned on the basis of obvious inadequacies in design of the human body. Those who see links between the scientific realm and the spiritual or metaphysical realm also strive for a deeper ultimate knowledge, but the extent to which science can engage in this aspect of understanding is limited. The requirements of the scientific method, and not human belief and imagination, set the limits.

The Gaia hypothesis is a reasonable extrapolation of evolutionary theory to encompass a globe covered with integrated biological systems, creating homeostasis and fostering life. Feedback mechanisms among abiotic and biotic systems result in a geophysiology of gigantic proportions, with organism-like properties. Although clearer hypotheses are needed, and problems such as repeated mass extinctions undermine a homeostatic view, the exploration of a grander ultimate view of life will continue.

All things considered, the theory of evolution was a remarkable breakthrough in scientific knowledge. Research continues to reinforce it at an accelerating pace, with no scientific alternatives available.

QUESTIONS FOR DISCUSSION

1. In your opinion, will science expand its present boundaries to use the scientific method on subjects now considered to be metaphysical?

2. Is there any approach you can imagine that would enable a test of the Gaia hypothesis?

3. Given that every intact and viable human society has spiritual beliefs central to its culture, is there an argument to be made that an adaptive evolutionary basis exists for acceptance of a metaphysical world? Would the basis of such acceptance involve organic evolution, cultural evolution, or both?

4. Is there any biological, scientifically sound basis for arguing that the human species is the most advanced organism ever to exist on this earth, and that "the ascent of humans" is a apt phrase?

5. Do you think there are ways in which scientists could have made a better case for both the existence and importance of evolution and its ability to explain the unity and diversity of life?

6. If you were provided with one hour in which to present the theory of evolution, how would you structure your talk, and what would you emphasize?

7. Which would you select as the most compelling example of evolution at each of the scales of microevolution and macroevolution?

8. Would you agree with the argument that acceptance of biological evolution as a science actually involves a system of beliefs that are not documented by the scientific method, just as in other fields of human endeavor like ethics, politics, and religion?

9. Do you think that all courses in the biological sciences would benefit from an evolutionary theme?

10. To what extent do you think that evolutionary theory has strong predictive ability? Which specific examples would you develop for or against the assertion?

REFERENCES

Angert, E. R., K. D. Clements, and N. R. Pace. 1993. The largest bacterium. Nature 362:239–241.

Boslough, J. 1985. Stephen Hawking's Universe. Avon Books, New York.

Bronowski, J. 1973. The Ascent of Man. Little, Brown, Boston.

Childress, J. J., H. Selbeck, and G. Somero. 1987. Symbiosis in the deep sea. Sci. Amer. 256(5):114–120.

Darwin, C. 1859. The Origin of Species by Means of Natural Selection, or The Preservation of Favoured Races in the Struggle for Life. Murray, London.

Dung, V. V., P. M. Qiao, N. N. Chinh, D. Tuoc, P. Arctander, and J. MacKinnon. 1993. A new species of living bovid from Vietnam. Nature 363:443–445.

Eiseley, L. 1958. Darwin's Century. Doubleday, New York.

Fishelsen, L., W. L. Montgomery, and A. A. Myrberg. 1985. A unique symbiosis in the gut of tropical herbivorous surgeonfish (Acanthuridae: Teleostei) from the Red Sea. Science 229:49–51.

Grim, J. N. 1993. Endonuclear symbionts within a symbiont: The surgeonfish intestinal symbiont, *Balantidium jocularum* (Ciliophora), is host to a gram-positive macronuclear inhabiting bacterium. Endocytobiosis Cell. Res. 9:209–214.

Hardy, A. 1965. The Living Stream: Evolution and Man. Harper and Row, London.

Hawking, S. W. 1988. A Brief History of Time: From the Big Bang to Black Holes. Bantam Books, New York.

Huxley, J. S. 1923. Essays of a Biologist. Chatto and Windus, London.

———. 1942. Evolution, the Modern Synthesis. Allen and Unwin, London.

———. 1953. Evolution in Action. Harper, New York.

Kirchner, J. W. 1989. The Gaia hypothesis: Can it be tested? Rev. Geophys. 27:223–235.

Krogman, W. M. 1951. The scars of human evolution. Sci. Amer. 185(6):54–57.

Lederman, L. M. 1993. The God Particle: If the Universe Is the Answer, What Is the Question? Houghton Mifflin, Boston.

Linden, E. 1994. Ancient creatures in a lost world. Time 143(25):52–54.

Lovelock, J. E. 1979. Gaia: A New Look at Life on Earth. Oxford Univ. Press, New York.

————. 1986a. Gaia: The world as a living organism. New Sci. 112:25–28.

————. 1986b. Geophysiology: A new look at Earth science, pp. 11–23. *In* R. E. Dickinson (ed.). The Geophysiology of Amazonia: Vegetation and Climate Interactions. Wiley, New York.

————. 1988. The Ages of Gaia: A Biography of Our Living Earth. Norton, New York.

————. 1991. Healing Gaia: Practical Medicine for the Planet. Harmony Books, New York.

Margulis, L. 1993. Symbiosis in Cell Evolution: Microbial Communities in the Archean and Proterozoic Eons. W. H. Freeman, New York.

Sogin, M. L. 1993. Giants among the prokaryotes. Nature 362:207.

Stanley, S. M. 1988. Government of the planet. Nature 336:270–271.

Teilhard de Chardin, P. 1959. The Phenomenon of Man. Harper, New York.

————. 1964. The Future of Man. Harper, New York.

Wright, R. 1989. Three Scientists and Their Gods: A Search for Meaning in an Age of Information. Harper Collins, New York.

————. 1992. Science, God and man. Time 140(26):38–44.

GLOSSARY

An alphabetical listing of terms, usually starting with nouns or adjectives for rapid reference, in the same word order as in the text. Most words or phrases are explained in the text; a few additional terms are included to help those without a background of elementary biology. Synonyms (*Syn:*) follow some definitions.

acrocentric chromosome A chromosome with its centromere at one end.

adaptionist program(me) A label applied by Gould and Lewontin to the overuse of adaptively oriented explanations of nature that have not undergone critical testing.

adaptive radiation Relatively rapid evolutionary divergence of members of a single phyletic line into a series of rather different niches or adaptive zones.

adaptive zone A set of similar ecological niches.

agamic system A reproductive process without gamete fusion.

agamospermy Seed production in plants without fertilization of the ovule by pollen cells.

age of fishes The Devonian Period, in which fishes radiated extensively and became important components of marine ecosystems.

agglomerative method A clustering method for discovering pattern in a similarity matrix, which groups similarity coefficients into successively fewer groups until a single set that contains all operational taxonomic units is reached.

agrarian revolution A revision in methods of agriculture that coincided with the industrial revolution and included the enclosure of land and selective breeding of stock for increased production to feed an increasingly urbanized society.

allele **1.** A gene variant that segregates in a Mendelian way from its counterpart at the same locus, providing the basis for particulate inheritance and segregation of unit characters. **2.** One of two or more variants of the same gene in a population, such as an allele for white flower color or an allele for purple flower color.

allocyclic forcing mechanism A phenomenon that results in major extinctions in a locality although it is centered in a different region; for example, oceanographic changes that affect terrestrial environments.

allopatric speciation Evolution of new species in different localities, resulting from geographic isolation between populations. *Syn:* geographic speciation.

allopolyploid A polyploid organism that originated in the doubling of the chromosomes of a zygote with two unlike chromosome sets, usually resulting from hybridization of two species. *Syn:* amphiploid.

alternative hypothesis A hypothesis, different from others, that offers an explanation in answer to a question; used especially as an alternative to the null hypothesis, which states that there is no cause-and-effect relationship between two given variables.

altruistic behavior An action by one individual that favors another, while reducing the fitness of the actor.

amictogametic reproduction Reproduction with eggs or spores from mitotic cell division, without the fusion of haploid gametes.

amphiploid A polyploid organism that originated in the doubling of the chromosomes of a zygote with two unlike chromosome sets, usually resulting from hybridization of two species. *Syn:* allopolyploid.

apomictic reproduction Asexual reproduction, often involving development of progeny from unfertilized eggs or from somatic cells of the parent. *Syn:* apomixis.

apomictic thelytoky Production of diploid female progeny without meiosis.

apomixis Asexual reproduction, often involving development of progeny from unfertilized eggs or from somatic cells of the parent. *Syn:* apomictic reproduction.

apomorphic Having a recently homologous basis of similarity.

arboreal theory of primate evolution The argument that the essential ingredient in primate evolution and the emergence of humans was an arboreal lifestyle, which resulted in selection for binocular vision, manual dexterity and sensitivity, increased brain size, and ultimately upright posture.

Archaebacteria One of the two subkingdoms of the Monera, including the methanogens and the halophilic and thermophilic groups.

arrhenotoky Parthenogenetic production of haploid males by a female, from unfertilized eggs.

artificial phylogeny A computer-generated lineage based on randomized origination and extinction events, usually employing conventions that limit lineage diversity at any one time unit.

artificial selection The process of making choices that favor individuals with desired traits in human-controlled breeding programs.

automictic thelytoky   Production of diploid female progeny with the involvement of both meiosis and the fusion of two haploid cells to form a diploid cell.

automixis   Production of diploid female progeny by an unfertilized female with the involvement of both meiosis and the fusion of two haploid cells, derived from an unmated female parent, to form a diploid zygote.

autopolyploidy   A condition resulting from a doubling, tripling, and so on, of the same chromosome set within an individual lineage.

balanced polymorphism   The maintenance of two alleles in a population by a fitness advantage for the heterozygote.

Baldwin effect   The situation in which a genotype comes to code for an adaptive phenotypic change through selection that ensures expression of that phenotype. *Syn:* genetic assimilation.

biological determinism   The false concept that humans are genetically destined to belong in certain segments of society—for example, leaders or servants.

biological species concept   The view, on the basis of arguments by Dobzhansky and Mayr, that a species is a group of actually or potentially interbreeding natural populations that is reproductively isolated from other such groups.

blastopore   The first external opening that develops during cell division, after zygote formation; formed by the pushing in of the developing ball of cells, called the blastula.

blending inheritance   The pre-Mendelian concept that progeny resulted from a blending of the traits of two sexually reproducing parents, with consequent loss of genetic variation on which natural selection could act.

brachydonty   The condition of having low-crowned teeth.

Cambrian explosion   The rapid increase in complex marine metazoans that occurred in Cambrian times; it is represented especially well in the Burgess Shale fauna.

candelabra model   The hypothesis that *Homo erectus* fanned out over Africa, Europe, and Asia, and populations evolved somewhat independently but in parallel to become *Homo sapiens*, giving a candelabra-like appearance to the phylogenetic tree, with *Homo erectus* as the stem. *Syn:* multiregional hypothesis.

carrier of sickle-cell trait   A human individual that is heterozygous for the sickle-cell trait (HbA/HbS).

catastrophic selection   Strong selection at marginal sites that favors those rare genotypes best adapted to marginal conditions, resulting in population crashes during unfavorable episodes and flushes in numbers when selection is weak.

catastrophism   An explanation of the stratigraphic features of the earth that invoked sudden, violent, and drastic changes, interspersed between long periods of stability, through geologic history.

cellular immunity   Recognition of nonself, involving cells that are adapted to attack and destroy foreign bodies in an organism.

centriole   A cytoplasmic organelle in the eukaryotic cell, involved with the organization of spindle fibers during mitosis and meiosis.

character analysis   Comparison of taxonomic characters within a taxon to identify homologous and nonhomologous relationships.

character displacement   Mutually induced selection for reduced similarity in species' characters under the influence of interspecific competition. *Syn:* character divergence.

character divergence   Mutually induced selection for reduced similarity in species' characters under the influence of interspecific competition. *Syn:* character displacement.

chiasmata formation   The development of cross-like attachments between chromatids from different chromosomes as the chromatids separate during meiosis.

chromatid   A replicated chromosome joined to its sister by a centromere.

chromosomal rearrangement   A mutation causing a change in the chromosome, such as a translocation or an inversion.

chromosome   A deeply staining body, visible during mitosis and meiosis, that carries part of the genetic material of a cell.

circular overlap   A phenomenon in which a chain of contiguous and intergrading populations curves back until the terminal links overlap each other and behave as good species.

clade   A group of all the species descended from a single ancestral (stem) species. *Syn:* monophyletic group.

cladistics school of classification   An approach to classification based exclusively on the branching pattern in a phylogenetically related group of organisms. *Syn:* phylogenetics school of classification.

cladogram   A summary of phylogenetic relationships in a group of organisms, derived from a cladistic analysis.

classification   The ordering of organisms into groups or sets on the basis of their relationships.

clone   All individuals derived by uniparental reproduction from a single parental individual and therefore having the same genotype.

coacervate droplet   A nonmiscible aggregate of colloidal protein molecules that forms a ball-like structure in water.

coelacanth effect   The persistence of one survivor of an ancient clade for a long time in geological history, after extinction of most of the members (e.g., the coelacanth and the tuatara).

coevolution   Reciprocally induced evolutionary change between two or more species or populations.

cohesion mechanisms   Processes that keep members of the same species from diverging—including gene flow, stabilizing selection, developmental constraints, and reproductive isolation from other species.

cohesion species concept   The view that the species is the most inclusive population of individuals, having the potential for phenotypic cohesion through intrinsic cohesion mechanisms.

comparative method   A system of inspection, involving several or many cases, to discover similarities or differences (in, for example, design, development, or morphological traits) between groups of organisms.

concept of evolution   The idea that change in species can occur; does not encompass an understanding of the mechanisms involved.

congruence   Similarity between a phylogenetic cladogram and a geographic area cladogram, in a monophyletic group of taxa.

conservative recombinational reproduction   Sexual reproduction in which haploid gametes unite to form a diploid zygote, involving isogamy in haploid protists without formation of polar bodies.

Cope's rule   The general statement that members of a phylogeny tend to become larger through the fossil record.

cospeciation   Pairwise divergence of lineages of, say, hosts and their parasites; the two lineages diversify together without necessarily coevolving. *Syn:* parallel cladogenesis.

crossing over   A process in which chromatids from different chromosomes intersect and exchange corresponding segments.

cyclical parthenogenesis   A complex life cycle involving both sexual and asexual reproduction in different generations. *Syn:* heterogonic life cycle.

Darwin–Muller hypothesis   The proposition that, although much evolution results from natural selection, further evolution can involve mutations that are irrelevant to fitness and natural selection.

Darwin's model of speciation   Darwin's view of origination of new species, involving: a parental species spreading over several different environments with new variation in each environment, differences in natural selection, independent divergence of populations, and ultimately the emergence of a new species in each environment.

Darwin's principle   The proposition that traits are more important in classification if they are uninvolved with special habits or adaptations.

Darwin's species concept   The view that species were simply well-marked varieties, such that species blended into each other; a convenient perspective for a mechanistic understanding of the origin of new species.

deduction   Reasoning from the general to the particular.

deductive logic   A formal system of argument or reasoning involving extrapolation from the general situation to particular cases.

descent with modification   The changing characteristics in a lineage, from parental to descendent forms, under the influence of natural selection.

deuterostome   A large group of organisms, starting chronologically with the echinoderms, in which the blastopore develops into the anus or the anus is closely associated with the blastopore, and the mouth develops secondarily and independently of the blastopore.

deuterotoky   Parthenogenetic production by females of either male or female diploid progeny from unfertilized eggs.

developmental constraints   Limitations on the morphological expression of the genotype, imposed by development processes that keep members of a species from diverging from the normal form.

dialectic   Discussion and reasoning by dialogue as a method of intellectual investigation.

digenomic   Formed by the symbiosis of two prokaryotic species contributing one genome each; applied to a eukaryotic cell.

diplohaplontic life cycle   A life cycle involving alternation of generations, with diploid and haploid phases represented sequentially in different individuals.

diploid   A cell or organism with a chromosome set from each parent or two homologous sets from one parent, usually designated as $2n$ chromosomes.

directional selection   Natural selection that favors one end of a normal distribution of a trait, thereby resulting in a shift of the normal distribution in that direction.

disruptive selection   Natural selection against the most common genotypes in a population, causing divergence in the population into two peak frequencies in a trait.

diversifying coevolution   Mutually induced evolutionary change between interacting species that results in the development of reproductive isolation between species in a lineage.

divisive method   A cluster analysis for discovering pattern in a similarity matrix that groups similarity coefficients into one group and then divides them sequentially into smaller and smaller groups.

ecological refuge   An environment that is particularly favorable for species during major extinction events because of ecological factors such as nutrient supply or habitat heterogeneity.

Ediacaran fauna   A Precambrian fauna of simple soft-bodied organisms, first described at a fossil site at Ediacara in South Australia.

endomembrane systems   Membrane complexes in prokaryotic cells that partition cellular activities and may have given rise to the nuclear membrane of eukaryotes.

endosymbiotic theory   A theory that uses symbiotic association among prokaryotic organisms to account for much of the evolution of eukaryotic cells from prokaryotes. *Syn:* serial endosymbiosis theory.

epigamic selection   Natural selection operating between the sexes of a species, favoring traits that improve attraction and access to mates. *Syn:* intersexual selection.

equilibrium theory of island biology   The concept that the number of species increases with island size, based on different equilibrial numbers set by the balance between colonization and extinction rates. Large islands have higher immigration rates and lower extinction rates than small islands.

escape-and-radiation coevolution   Intermittent, reciprocally induced evolutionary changes in a group of similar interacting species, such that a lineage escapes from selection and radiates, then coevolutionary interactions between lineages are reestablished.

Eubacteria   One of the two subkingdoms in the Monera, including most of the well-known species such as spirochaetes, cyanobacteria, and the nitroxen-fixing bacteria.

eurybiomic   Foraging over two or more biomes with relatively broad feeding and habitat utilization patterns.

evolution   **1.** (Before Darwin's time) A term coined by Albrecht von Haller to mean development from a preformed representation in the embryo to the final organism. **2.** (In Darwin's time) The progressive development of life from the simple to the complex. **3.** (After Darwin) Genetically based, heritable change in one or more characteristics in a population or species through time.

evolutionary school of classification   A school of classification based on observed similarities and differences among groups of organisms, evaluated in the light of inferred evolutionary history. *Syn:* traditional school of classification.

evolutionary species concept   A view of species, applied to the fossil record and uniparental and biparental lineages, that emphasizes the unity of a lineage, its distinctive evolutionary role, and stabilizing selection pressures common to all its members.

extrinsic evolutionary factor   A process that acts on individuals and populations, external to the organisms, such as natural selection.

Fahrenholz's rule   The assertion that, in groups of permanent parasites, the classification of the parasites usually compares directly with the natural relationships of the hosts.

falsification   A necessary component of the scientific method whereby objective tests of a hypothesis can establish that it is invalid.

family   A taxon containing a single genus or a monophyletic group of genera, separated from other families by a decided gap between constellations of characters.

fitness   **1.** The ability of an individual or population to leave viable and reproductively effective progeny, relative to the abilities of other individuals or populations. **2.** The average contribution of an allele or genotype to the next or subsequent generations, compared with those of the other relevant alleles or genotypes.

fixity of species   The concept that species are intact and invariable and do not change in time or space.

flush–crash cycle   A sequence of population events involving rapid increases in density followed by rapid declines, usually to very low numbers.

founder effect   The founding of a new population by one or a few individuals, which therefore carry only a small proportion of the genetic variation of the parental population.

Fuhrmann's rule   The assertion that each order of birds has a particular cestode fauna.

fundamental theorem of natural selection   The theory, formulated by Fisher, that the amount of genetic variation in an evolving population limits the rate of evolution.

gametic disequilibrium   Nonrandom association of alleles at different loci into gametes.

gametophyte generation   The haploid phase in the alternation of generations, which produces haploid gametes. The gametes fuse to form the diploid sporophyte generation.

gene   A genetic factor that determines a character, such as flower color.

gene conversion   Change in genes that results from nonreciprocal recombination following crossing over between homologous chromosomes during meiosis.

gene duplication   The doubling of a gene in a genotype so that the cell carries two sets of DNA for the same haploid genotype.

gene flow   Movement of genes between populations, resulting from dispersal of individuals.

gene-for-gene coevolution   Reciprocally induced evolutionary change between interacting species in which hosts and parasites have complementary gene loci for resistance and virulence, susceptibility and avirulence, with each gene in the host matched by a gene in the parasite.

gene-for-gene hypothesis   The concept that, during the evolution of a host and its parasite, the two develop complementary genic systems, with each host gene that affects defense matched by a parasite gene that affects attack.

gene mutation   A change in a gene, resulting from an error in nucleotide sequence replication.

genet   The genotype of all individuals in a clone.

genetic assimilation   The situation in which a genotype comes to code for an adaptive phenotypic change through selection that ensures expression of that phenotype. *Syn:* Baldwin effect.

genetic drift   Change in gene frequency between generations, due to sampling error that becomes stronger as populations decline in size.

genetic hitchhiking   Travel of a more-or-less neutral allele through the generations because of its close association, on a chromosome, with an allele at another locus that has an adaptive advantage.

genetic identity   Similarity in genes between two populations or two species.

genetic recombination   An exchange of genetic material between homologous chromosomes, involving synapsis and crossing over.

genetic system   All the factors that affect the hereditary behavior of a species and its evolutionary potential.

genotype   The genetic constitution of an individual, either in its entirety or at a particular locus or loci.

genus   A taxon containing a single species or a monophyletic group of species, separated from other genera by a decided gap between constellations of characters.

geographic cladogram   A cladistic representation of the global distribution of a monophyletic taxon, using the known tectonic movements of the plates of the earth's crust since the breakup of Pangaea about 200 million years ago.

geographical barrier   A large geographic feature, such as a mountain range or an ocean, that strongly inhibits the free movement of individuals, making dispersal rare or absent.

geographical isolation   The effective separation of two populations, resulting from a geographic obstacle to free migration such as a mountain range, an ocean, or some other inhospitable feature.

geographical refuge   A geographical region that serves as a haven for certain taxa during major extinction events.

geographic speciation   Evolution of new species in different localities, resulting from geographic isolation between populations. *Syn:* allopatric speciation.

germ cell   A cell of the body involved with the production of gametes.

germ-line hypothesis   The argument that all genes coding for the variable-region sequences of amino acids in the immunoglobulin molecules arose during evolution through gene duplication, and all are present in the germ-cell DNA.

giga annum   A thousand million, or a billion, years.

gradual speciation   Slow development of new species under the influence of natural selection. *Syn:* phyletic gradualism.

Greek species concept   The view that a species was an unstable grouping based on the outward appearance of its member organisms, which could readily change (with age or environment, for example).

grouping   Aggregation of lower taxa, such as species and genera, into higher taxa, such as families and orders.

guild coevolution   Reciprocally induced evolutionary change in groups of similar interacting species.

haplodiploidy   Parthenogenetic production by a female of haploid males from unfertilized eggs, coupled with the production (involving male sperm) of diploid females from fertilized eggs.

haploid   A cell or organism with a single chromosome set derived from one parent, not two; usually denoted as having n chromosomes.

Hardy–Weinberg equilibrium or law   The principle that, in the absence of evolution, the genotype frequencies in generation $t + 1$ are the squares of the gene frequencies among gametes produced in generation t.

hermaphrodite   An organism with both male and female functional reproductive organs.

heterochrony   A shift in the timing of developmental events relative to those in the parental population.

heterogonic life cycle   A complex life cycle involving both sexual and asexual reproduction in different generations. *Syn:* cyclical parthenogenesis.

heterospory   The existence of two spore types: a small spore with a male function and a large spore with a female function.

heterothallic   Exhibiting isogamy, but with fusion to form a zygote taking place only between gametes from different thallus types (as in many fungi).

heterotrophic nutrition   Feeding on organic material originally synthesized by plants, in the absence of synthesis of food from inorganic chemicals.

heterozygote advantage   A condition in which the heterozygote has higher fitness than the two alternative homozygotes in a population.

heterozygous   An individual with different alleles at one locus or at all loci under consideration.

holophyly   A cladist's meaning for monophyly—all the species descended from a single ancestral (stem) species—as distinguished from the meaning used by the evolutionary school of classification.

homologous chromosome   A chromosome that matches another, usually derived from the other parent.

homology   A similarity in a trait or traits of two organisms, resulting from common origin.

homoplasy   A similarity in a character or characters of two organisms, resulting from convergent evolution or parallel evolution without a common ancestor.

homozygon   An individual organism with identical alleles at a particular locus.

homozygous   An individual with identical alleles at the same locus or at all loci under consideration.

hopeful monster   A new individual in a lineage, bearing a mutation or mutations that cause a radical shift in phenotype from the parental type, with the implication that the new form is poorly adapted for most available environments.

humoral immunity   Recognition of nonself, involving protein molecules that are adapted to attack and destroy foreign bodies in an organism.

hybrid   An individual that results from gametes produced by different species.

hybrid complex   A group in which discrete species are difficult to discern; the product of natural hybridization among species that results in a loss of morphological discontinuities between the originally divergent ancestral forms.

hybridization   Union of the genomes of two different species, resulting in a living organism.

hybrid speciation   The origination of a new species directly from a natural hybrid.

hybrid swarm   A complex mixture of two or more parental species, F_1 hybrids, backcross types, and segregation products.

hypermutation   Prolific generation of somatic mutants in the genes coding for immunoglobulin molecules.

hypothesis   A tentative explanation of a question related to science, stated in such a way that testing via the scientific method is possible, and such tests can either yield answers consistent with the hypothesis or falsify the hypothesis.

hypsodonty   The condition of having high-crowned teeth.

immigration curve   The trajectory followed on an empty island as species colonize, starting at a relatively high rate and declining to zero when all species from the source of immigrants have arrived.

inclusive fitness   Relative success at leaving genes in the next or subsequent generations, based on an individual's own reproductive success plus the success of relatives that carry the same genes.

independent assortment   Separation of characters on different chromosomes into different gametes, independently of each other, during meiosis.

individual fitness   Relative success at leaving genes in the next or subsequent generations, based only on an individual's reproductive success.

inducer   A compound that causes a reaction in the body, resulting in increased production of enzymes that break down that compound.

induction   Reasoning from a part to the whole or from particulars to generalities.

inductive logic   A formal system of argument or reasoning involving the development of generalities from particular facts.

industrial melanism   Evolution of dark-colored forms of species under natural selection for improved camouflage against substrates darkened by industrial contamination.

industrial revolution   The radical shift that took place in European society from 1760 to 1840, based on the invention of the steam engine, the consequent mechanization of manufacturing and transport, and massive urbanization centered around factories.

instantaneous speciation   Rapid or abrupt origination of new species.

intersexual selection   Natural selection operating between the sexes of a species, favoring traits that improve attraction and access to mates. *Syn:* epigamic selection.

intracellular mutualism   A mutually beneficial relationship between two species in which one lives within the cells of the other; endosymbiosis involving mutualists.

intraorganismal mutualism   A mutually beneficial relationship between two species in which one lives within the body of the other but not in the cells.

intrasexual selection   Natural selection among members of the same sex that improves access to mates.

intrinsic evolutionary factor   A process within the individuals of a population that results in evolution, such as a change in the timing or rate of developmental events, or heterochrony.

introgression   Incorporation of genes from one species into the gene pool of another.

inversion   A reversal in the order of genes in a chromosome.

isogamy   A condition in which gametes of both sexes in a species are alike in appearance and size.

isolation species concept   The name Paterson gave to the biological species concept, to contrast it with the recognition species concept.

isomorphic   Having the same shape.

karyotype   A frequently unique set of chromosomal morphologies characteristic of a species.

kinetosome   A self-replicating body at the base of a flagellum or cilium.

kingdom   Often considered to be the highest rank in the taxonomic hierarchy of classification. Scientists generally recognize five kingdoms at this time: Monera, Protista, Fungi, Plantae, and Animalia.

kin selection   Natural selection on characteristics that favor genetically related individuals, as in a family.

Latin binomial   The two-word latinized name for a species.

law of the unspecialized   The generalization proposed by Cope on the basis of the fossil record, that unspecialized groups persist and advance through time but specialized groups reach evolutionary dead ends.

Lazarus effect   A phenomenon in which a species is unobserved in the fossil record for a long period of geological time, then reappears as either a fossil or an extant species (such as the coelacanth).

Lazarus species   A species that is unobserved in the fossil record for a long period of geological time and that appears again in higher strata, as if risen from the dead.

locus   The position of a gene in a chromosome.

Ludwig theorem   The concept that a genotype utilizing a novel subniche can be added to a population even if it would be of inferior viability in the normal niche of the species.

macroevolution   Evolution above the species level, involving new species and higher taxa.

macromutation   A large genetic alteration in a lineage, resulting in a major evolutionary change. *Syn:* systemic mutation.

marginal population allopatric speciation model   Mayr's alternative to sympatric speciation, to account for many closely related herbivorous insects within the same genus. In populations that are marginal relative to the parental range, host shifts and speciation occur in allopatry, with subsequent dispersal back into the ambit of the parental stock.

mega annum   One million years.

megaton   A unit used to denote the power of nuclear warheads; the explosive equivalent of a million tons of trinitrotoluene (TNT), commonly used as an explosive in conventional bombs.

meiosis   Reduction division of diploid cells during gametogenesis, resulting in four daughter cells with a haploid number of chromosomes.

meristic character   A countable trait such as number of hairs per sclerite on an insect.

metacentric chromosome   A chromosome with the centromere at its center.

metric character   A measurable trait such as the length of a wing or leg.

microbial mat   A layering of inorganic particles held together with microbes, composed of intertwined microbial filaments and sedimentary particles.

microevolution   Evolution below the species level.

microspecies   **1.** A uniform population that is slightly different phenetically from related uniform populations. **2.** A group of individuals that is distinctive on the basis of morphological characteristics with uniparental reproduction so that each lineage diverges independently.

microtubule-organizing centers   Structures or loci in the eukaryotic cell, including centrioles and kinetosomes, that give rise to microtubular arrays such as mitotic spindle fibers.

migration   The movement of individuals in a population to and fro between localities.

mitosis   A process that results in the equal division of the genetic material of a cell into two daughter cells during cell division in a diploid eukaryotic organism, usually producing genetic copies of the parental cell.

modern synthesis   A theoretical synthesis in the field of biology, with the theory of evolution as its central theme and encompassing the fields of comparative morphology, cytology, genetics, animal and plant systematics, and paleontology; achieved during the 1940s.

molecular clock   The more-or-less constant rate of mutational change of molecules over long periods of time, simulating a timepiece.

monophyletic group   A group of all the species descended from a single ancestral (stem) species. *Syn:* clade.

monopodial branching   A growth pattern in plants in which one main stem is the axis.

Mosaic geology   The biblical account of the Creation in the book of Genesis, traditionally attributed to the prophet Moses.

Muller's ratchet   The hypothesis that sexual reproduction is an object of selection because it purges deleterious mutations from some lines of progeny. Recombination produces some progeny that have reduced loads or are load-free.

multidimensional species concept   The view that a species is a group of populations that occur in separate localities (allopatric) and/or at different times (allochronic).

multiregional hypothesis   The view that *Homo sapiens* evolved in many regions of Africa, Europe, and Asia, from *Homo erectus* populations that had spread from the point of origin. *Syn:* candelabra model.

mutation   Any novel genetic change in the gene complement or genotype relative to the parental genotypes, beyond that achieved by genetic recombination during meiosis.

mutational load   the sum of all deleterious mutations in a lineage.

mutational variation   A term used by DeVries to denote changes in a lineage, beyond variation within a species, that separated the carrier of the change from its parents as a completely new species.

natural selection   The process in nature that causes evolution through differential reproductive success among members of a population; that success depends on genetically based and heritable variation in characteristics that confer relative advantage or disadvantage to the bearer.

neutralist   One who espouses evolutionary change through random mutational change in molecules that are neutral under the influence of natural selection.

neutral theory of molecular evolution   The theory that molecules evolve under the influence of random mutation rather than natural selection. *Syn:* non-Darwinian evolution.

Noah's Ark model   The view that modern humans evolved in one locality, spread to other parts of the Old World, and replaced all other extant populations of *Homo sapiens*, as if they had walked off Noah's Ark. *Syn:* replacement hypothesis.

non-Darwinian evolution   The theory that molecules evolve under the influence of random mutation rather than natural selection. *Syn:* neutral theory of molecular evolution.

nondimensional species concept   The view that a species is restricted to one place and time and is distinct from others in the same locality.

nonrecombinational reproduction   Reproduction without meiosis, with progeny resulting from mitotic cell division, yielding mostly genetic copies of the female parent.

null hypothesis   A tentative suggestion that there is no statistically significant relationship between two given variables.

numerical phenetics   Use of quantified phenetic traits and standard computer-based analysis in classification, employed by the phenetics school of classification. *Syn:* numerical taxonomy.

numerical taxonomy   Use of quantified phenetic traits and standard computer-based analyses in classification, employed by the phenetics school of classification. *Syn:* numerical phenetics.

obstetrical dilemma   An evolutionary phenomenon that occurred during the development of the human lineage, in which brain size at birth increased while the size of the birth canal decreased as the pelvis became more robust. Selection then occurred that mitigated the "dilemma."

Oparin–Haldane hypothesis   A conjecture, formulated independently by Oparin in Russia and Haldane in England, to account for the spontaneous generation of living organisms on earth through gradual prebiotic evolution of molecules and aggregates.

operational taxonomic unit (OTU)   A term applied to a group of organisms to be classified, implying no preconceived ideas about the group's status as a population, species, genus, or other taxon.

ordinary variation   Normal variation within a species.

orthologous gene   A gene that can be tracked to a common ancestor after passing down a lineage over many generations as that lineage diversified.

outcrossing   Union of gametes derived from different individuals.

outgroup   A taxon that is outside the group under close scrutiny, having diverged before the studied group diversified, and that provides a reference point for recognition of basic ancestral characteristics.

paleopolyploid   A species whose lineage has a polyploidization event in its phylogenetic background.

pangenesis   A hypothesis advanced by Darwin to explain the hereditary process, involving the passage into the gametes of a gemmule from each part of the body, conveying information about the exact nature of that body part.

Panglossian paradigm   A label applied by Gould and Lewontin to the overuse of adaptively oriented explanations of nature as if all organisms were perfectly adapted in all ways, in the sense of Dr. Pangloss's philosophy that all is for the best in this best of possible worlds.

parallel cladogenesis Pairwise divergence of lineages of, say, hosts and their parasites; the two lineages diversify together without necessarily coevolving. *Syn:* cospeciation.

paralogous gene A gene resulting from a duplicated ancestral gene, such that an orthologous gene can produce multiple copies that evolve to perform different functions.

parapatric speciation Speciation at the edge of the parental species' range, resulting from a chromosomal mutation that causes reproductive isolation and permits the carrier to invade new adjacent territory in the absence of a geographic barrier. *Syn:* stasipatric speciation.

paraphyletic group A group of species that includes a common ancestor and some, but not all, of its descendants.

parthenogenesis Unisexual reproduction by a female, without male gametes, from an egg.

particulate theory of inheritance The theory, derived from Mendel's research, that characters are inherited as if they were particles, separating independently into gametes so that each variant of a character is maintained in a population in the absence of evolutionary change.

pathogen ratchet The hypothesis that sexual reproduction is an object of selection because it reduces parent–offspring transmission of parasites. The rare progeny most different from the parental genotypes are least likely to be infected by the parent's parasites. *Syn.* Rice's ratchet.

phenetic bottleneck A frequently necessary shortcut in testing the biological species concept; it employs phenotypic characters alone and does not examine, for example, effective reproduction among allopatric populations.

phenetics school of classification A school of classification that employs many numerical, exclusively phenetic characters, and standard numerical procedures for sorting species into clusters.

phenogram A diagram of a phenetic classification, showing the relationships among the organisms under study.

phenotype The actual appearance of an individual, resulting from an interaction of the genotype and the environment in which it develops.

phyletic gradualism Slow development of new species under the influence of natural selection. *Syn:* gradual speciation.

phylogenetic constraint The evolutionary history of a species' lineage, which limits the evolutionary path that species can follow.

phylogenetic species concept An irreducible cluster of organisms, diagnosably distinct from other such clusters and containing a parental pattern of ancestry and descent.

phylogenetics school of classification An approach to classification based exclusively on the branching pattern in a phylogenetically related group of organisms. *Syn:* cladistics school of classification.

plesiomorphic Having an anciently homologous basis of similarity.

polyphyletic group A group of species united on the basis of convergent similarity; the common ancestor is not included in the group.

polyploid complex An array of polyploid types within a group of species that includes a large variety of karyotypes.

polyploidy A condition in which the number of chromosome sets in the nucleus is a multiple greater than 2 of the haploid number.

postmating mechanisms of reproductive isolation Biological characteristics of species that prevent effective reproduction after transfer of gametes, during either copulation in animals or pollination in plants.

preadaptation An organismal characteristic that happens to become adaptive under new conditions.

premating mechanisms of reproductive isolation Biological characteristics of species that prevent the transfer of gametes between species.

primary speciation Origination of new species via divergence from one common ancestral species.

primordial recombinational reproduction Reproduction in which the maternal genotype is not preserved because mechanisms of genotypic maintenance are poorly developed.

progressionism The concept that life has developed from simple forms to more complex organisms through the geological past. After each cataclysmic event on the earth, life was created again but in more advanced forms; ultimately, a progression of such creations resulted in the human and a unity of nature, developed from a supernatural plan.

progressive occupation Gradual speciation and divergence of a single lineage into different adaptive zones through geological time.

proteinoid microsphere A spherical, nonmiscible aggregate of protein-like molecules in water.

protostome An organism in which the mouth is derived from or is near the blastopore during embryogenesis.

pseudoarthrosis A false joint.

pseudogamy Production of progeny in an apomictic species that nevertheless requires the arrival of a male gamete at the surface of the egg.

pseudomonopodial branching A growth pattern in plants in which dichotomous branching occurs, with one branch becoming much stronger and central—developing into a main axis—and the other becoming lateral.

punctuated equilibria Long periods without evolution, interspersed with brief episodes of rapid evolutionary divergence in a lineage.

Punnett square A matrix, devised by Reginald Punnett, with all possible gamete genotypes from the male parent on one axis and all those from the female on the other, so that the frequency of possible combinations among the progeny can be identified when gametes combine at random.

quadrigenomic Having four genomes from four different prokaryotic species within a eukaryotic cell.

qualitative character   A trait such as color or pattern that has no obviously numerical basis for classification.

qualitative multistate character   A trait that cannot be ordered numerically but can exist in several to many different forms, such as sculpting on an arthropod exoskeleton.

quantitative multistate character   A single numerical value, or a coded value, within a range of possible values in the organisms under study, such as the number of segments in the antenna of an insect.

quantum evolution   Rapid transit from one adaptive peak to another in a diverging lineage.

quantum speciation   The budding off of a new and very different daughter species from a semi-isolated peripheral population of the ancestral species, in a cross-fertilizing organism. *Syn:* speciation by saltation.

ramet   One individual member in a cloning organism.

ranking   The placing of higher taxa into their correct positions in the taxonomic hierarchy.

reaction norm   The usual range of phenotypes seen in a population under normal conditions.

rearrangement of genes   Spontaneous variation through a chromosomal mutation—such as an inversion, deletion, or translocation—which repositions genes relative to each other.

recognition species concept   A view of species that emphasizes the importance of specific-mate recognition systems involving the complexities of sexual biparental eukaryotic reproduction.

recombination   A process, involving synapsis, crossing over, and chiasmata formation, that results in exchange of segments between homologous chromosomes during meiosis.

recombinational reproduction   Reproduction involving meiosis, crossing over, and recombination.

Red Queen's hypothesis   The argument that sexual reproduction is an object of selection because it results in rapid evolution when biotic forces such as predators, parasites, competition from other species, prey, or food plants exert strong selective pressures.

replacement hypothesis   The view that modern humans evolved in one locality, spread to other parts of the Old World, and replaced all other extant populations of *Homo sapiens*. *Syn:* Noah's Ark model.

reproductive isolating mechanism   Any biological attribute of a species that prevents effective reproduction with members of another species.

reproductive isolation   Prevention, by a biological attribute of a species, of effective reproduction between its members and the members of another species.

resemblance matrix   A two-dimensional display of similarity coefficients between all operational taxonomic units (OTUs) under study. *Syn:* similarity matrix.

reticulate phylogeny   A phylogenetic web of related species, derived from primary speciation, which hybridize in parts of the species group; the phylogeny therefore shows both divergence and convergence of lineages in the group.

Rice's ratchet   The hypothesis that sexual reproduction is an object of selection because it reduces parent–offspring transmission of parasites. The rare progeny most different from the parental genotypes are least likely to be infected by the parent's parasites. *Syn:* pathogen ratchet.

saltational speciation   Origination of a new species involving a rapid jump in characteristics from one state to another in a lineage.

science   A field of study concerned with observation and classification of facts and especially with the establishment of verifiable general laws and scientific theories.

scientific law   A statement of a relationship of phenomena that, so far as is known, is invariable under given conditions.

scientific method   A method of studying questions that draws on accessible knowledge through observations and experiments and that results in the erection of hypotheses and, ultimately, scientific theory.

scientific theory   A conception, proposition, or formula relating to the nature, action, cause, or origin of a phenomenon or group of phenomena, formed by deduction and generalization from facts.

secondary contact   Renewed coexistence of populations derived from a common parental population, after the splitting of the original species into two or more geographically isolated populations over a period of time and a subsequent change in conditions (because of dispersal or a shift in climate, for example).

secondary speciation   Origination of new species, often rapidly, without divergence from one common ancestral species and involving hybridization, polyploidy (or both), or a shift in breeding from outcrossing to an amictic system of reproduction.

segregation of unit characters   Mendel's discovery that, during gametogenesis, variants of a character at the same location on a chromosome separate out to different gametes.

selectionist   One who advocates evolutionary change through natural selection.

selfing   Fertilization of a gamete by other haploid cells from the same individual.

semispecies   **1.** Populations of the same species with partial reproductive isolation. **2.** Groups of populations intermediate between good races and good species, connected by a reduced amount of interbreeding and gene flow.

serial endosymbiosis theory   A theory that uses symbiotic association among prokaryotic organisms to account for much of the evolution of eukaryotic cells from prokaryotes. *Syn:* endosymbiotic theory.

sexual selection   Natural selection on traits related to obtaining mates for sexual reproduction, involving both the ability to compete with members of the same sex for a mate and attraction between the sexes.

shift in sexuality   A change in reproductive mode, usually from outcrossing to parthenogenesis.

sickle-cell anemia   A disease in humans caused by a single-gene mutation for hemoglobin molecules that distorts red blood cells, reduces blood flow, and causes symptoms such as tiredness, headache, muscle cramps, and so on. The sickle-cell gene is homozygous (HbS/HbS).

sickle-cell trait The gene in humans for β-hemoglobin, resulting from a mutation at the sixth codon from CTC to CAC, which produces sickle-cell anemia when homozygous.

Siluro-Devonian explosion A period from about 439 million to 362 million years ago, in which the initial radiation of terrestrial plants occurred.

similarity matrix A two-dimensional display of similarity coefficients between all operational taxonomic units (OTUs) under study. *Syn:* resemblance matrix.

sister groups Groups of species descended from a common ancestral group.

social Darwinism The misguided application of Darwinian natural selection to ranking in society, which argued that the upper classes in a social hierarchy had biological attributes conferring superiority in the struggle for existence.

somatic The body or parts of it; usually used in reference to cells of the body, as opposed to germ cells.

somatic mutation Genetic mutation in a somatic cell.

somatic mutation hypothesis The proposition that a small pool of genes passes along the germ line, and the numerous variants in immunoglobulin molecules arise by a process of random somatic mutation during ontogeny of the individual.

speciation by macrogenesis The sudden origination of new species, involving a rapid shift in characters from one state to another in a lineage.

speciation by saltation The budding off of a new and very different daughter species from a semi-isolated peripheral population of the ancestral species, in a cross-fertilizing organism. *Syn:* quantum speciation.

species A group of actually or potentially interbreeding populations that are reproductively isolated from other such groups.

specific coevolution Reciprocally induced coadaptation of two interacting species, without a gene-for-gene relationship.

specific-mate recognition systems Fertilization mechanisms involving courtship, timing, habitat selection, coloration, the endocrine system, copulatory organs, and other factors related to the recognition of compatible mates within a species.

spontaneous generation The appearance of a living organism from nowhere and out of nothing.

sporophyte generation A diploid that embodies part of a life cycle and is formed by the fusion of two haploid gametes; it produces haploid spores through meiosis and is interspersed with a haploid gametophyte generation.

stabilizing selection Natural selection that keeps the mode of the normal distribution of a trait more or less constant. *Syn:* normalizing selection.

stasipatric speciation Speciation at the edge of the parental species' range, resulting from a chromosomal mutation

that causes reproductive isolation and permits the carrier to invade new adjacent territory in the absence of a geographic barrier. *Syn:* parapatric speciation.

stasis A stable period without evolutionary change in a lineage.

stenobiomic Having a relatively narrow habitat range and food utilization pattern.

stromatolite A many-layered set of microbial mats making up a stable composition of sedimentary material; in highly saline aquatic environments it forms boulder-like structures, which may ultimately be preserved in a fossilized form that is as stable as most other sedimentary rocks.

struggle for existence Darwin's term, derived from Malthus's essay on population, implying a fight for life among competing members in and among populations and species in an overcrowded environment.

subspecies Populations of the same species in different geographic locations, with one or more distinguishing traits.

Superkingdom Eukaryota One of two taxa above the kingdom level, including all eukaryotic species.

Superkingdom Prokaryota: One of two taxa above the kingdom level, including all prokaryotes.

survival of the fittest Herbert Spencer's term for natural selection.

survivorship curve A record of the decline in numbers of a cohort as time passes, from 100 percent surviving at the beginning of the time scale to the death of the last member.

sympatric speciation model The establishment of new populations of a species in different ecological niches within the normal cruising range of the individuals of the parental population, accompanied by reproductive isolation from the parental species.

symplesiomorphy A shared ancestral character.

synapomorphy A shared derived character.

synapsis Alignment of a pair of sister chromatids with their homologous pair.

syngameon A condition in which phenetically different species interbreed extensively so that the most inclusive interbreeding group is larger than the species.

synthesis of biology A blending in the 1940s of formerly disparate aspects of biology—including the fields of comparative morphology, cytology, genetics, animal and plant systematics, and paleontology—into the unifying concepts of the theory of evolution.

systematics Scientific study of the kinds and diversity of organisms and of any and all relationships among them.

systemic mutation A large genetic alteration in a lineage, resulting in a major evolutionary change. *Syn:* macromutation.

taxonomy The theoretical study of classification, including its bases, principles, procedures, and rules.

telocentric chromosome A chromosome with the centromere off center.

thelytoky Parthenogenetic production of diploid female progeny by a female, from unfertilized eggs.

theory of evolution   A factually based mechanistic explanation for the unity and diversity of life on earth.

Tommotian fauna   An early Cambrian fauna comprising small-shelled organisms with no apparent relationship to extant phyla, first described from Tommot in Siberia.

track synthesis   The search for replicated patterns of geographic distribution among different monophyletic taxa.

traditional school of classification   A school of classification based on observed similarities and differences among groups of organisms, evaluated in the light of inferred evolutionary history. *Syn:* evolutionary school of classification.

translocation   Movement of a chromosome segment from one location to another in the karyotype.

transmutation of species   Darwin's term, now replaced by the word "evolution," meaning the change from one species to another—literally, the process of altering form, appearance, or nature.

trigenomic   Having three genomes from three different prokaryotic species; applied to a eukaryotic cell.

two-state character   One of a pair of unit characters that are "all or none," such as winged and wingless.

type specimen   A single specimen used to represent and describe a species and preserved (as far as possible) in a museum.

typological species concept   The idea that species are fixed in time and do not evolve, so that one specimen of each species can be an adequate representative.

undulipodium   A eukaryotic flagellum or cilium.

uniformitarianism   A scientific school of thought that argued that geological phenomena were the products of natural forces operating over very long periods of time, with considerable consistency in the factors involved.

uniparental reproduction   Propagation involving one parent, by vegetative methods or by a degenerate form of sexual reproduction, such as agamospermy.

unit character   As used in numerical taxonomy, an attribute of an organism about which one statement can be made, yielding a single piece of information.

urkaryotes   The presumed ancestral line of bacteria that became eukaryotic partially through endosymbiosis.

vegetative reproduction   Development of new individuals from the somatic cells of the parent.

vicariance biogeography   The study of cladistic classification using geographic separation of a continuous biota into two or more geographic subunits, such as the separation of continents during plate-tectonic activity.

visual predation hypothesis   The proposition that the distinctive features of primate evolution include a predatory habit in an arboreal environment, with the use of coordinated vision and prehensile appendages to capture prey.

Wallace effect   Natural selection resulting in increased reproductive isolation between newly sympatric species.

weighting of characters   A method for defining the relative importance of the phyletic information content of a character.

xenobiotic   A foreign chemical introduced into the body of an organism.

SUBJECT INDEX

TAXONOMIC INDEX

AUTHOR INDEX

Boldface numerals indicate the pages on which complete references are given.